Peterson Reference Guide to

SPARROWS
of North America

Area Covered by This Book

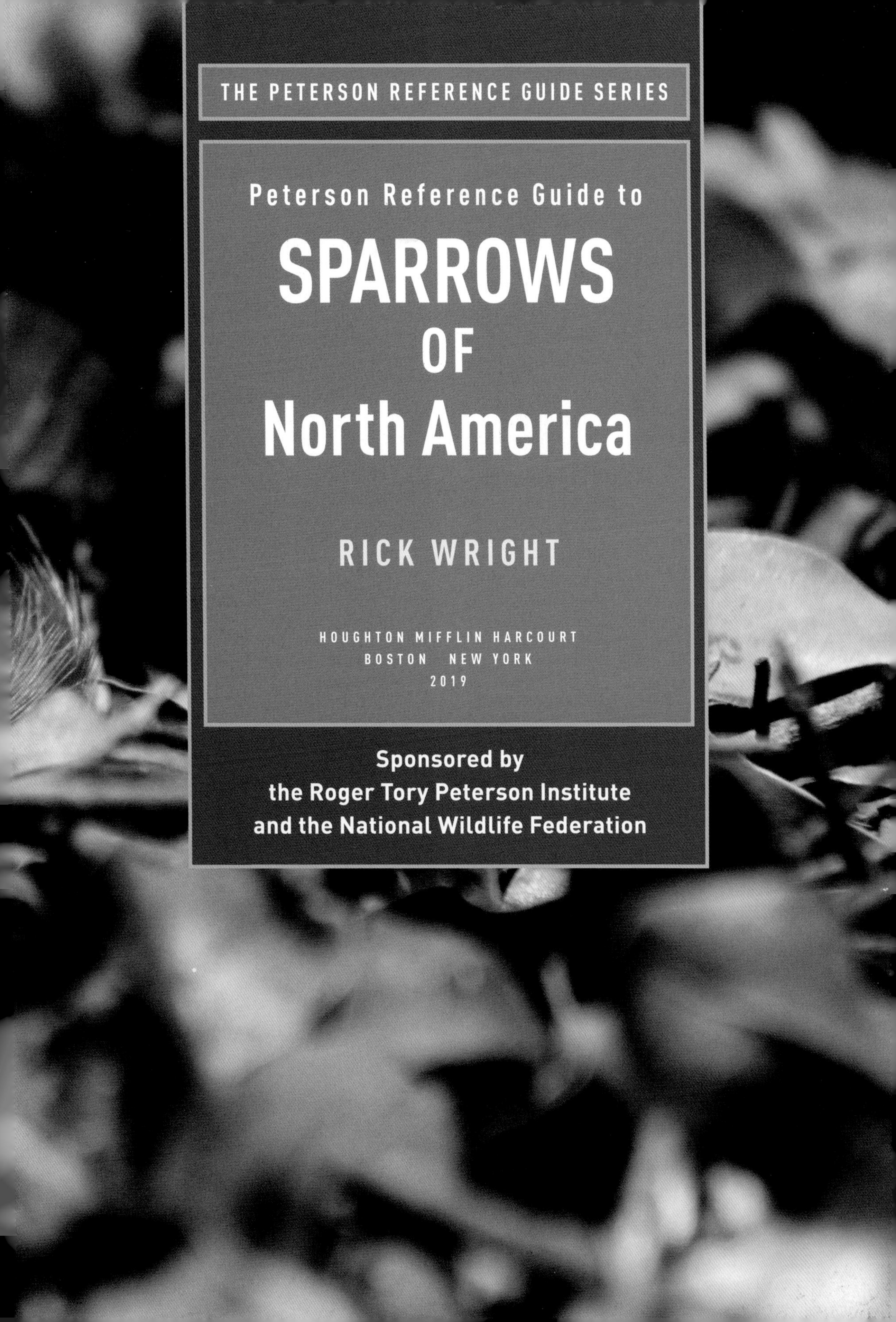

THE PETERSON REFERENCE GUIDE SERIES

Peterson Reference Guide to

SPARROWS
OF
North America

RICK WRIGHT

HOUGHTON MIFFLIN HARCOURT
BOSTON   NEW YORK
2019

Sponsored by
the Roger Tory Peterson Institute
and the National Wildlife Federation

Library of Congress Cataloging-in-Publication Data is available.
ISBN 978-0-547-97316-6

Book design by Eugenie S. Delaney

Printed in China

SCP 10 9 8 7 6 5 4 3 2 1

ROGER TORY PETERSON INSTITUTE
OF NATURAL HISTORY

Continuing the work of Roger Tory Peterson through Art, Education, and Conservation

In 1984, the Roger Tory Peterson Institute of Natural History (RTPI) was founded in Peterson's hometown of Jamestown, New York, as an educational institution charged by Peterson with preserving his lifetime body of work and making it available to the world for educational purposes.

RTPI is the only official institutional steward of Roger Tory Peterson's body of work and his enduring legacy. It is our mission to foster understanding, appreciation, and protection of the natural world. By providing people with opportunities to engage in nature-focused art, education, and conservation projects, we promote the study of natural history and its connections to human health and economic prosperity.

Art—Using Art to Inspire Appreciation of Nature

The RTPI Archives contains the largest collection of Peterson's art in the world—iconic images that continue to inspire an awareness of and appreciation for nature.

Education—Explaining the Importance of Studying Natural History

We need to study, firsthand, the workings of the natural world and its importance to human life. Local surroundings can provide an engaging context for the study of natural history and its relationship to other disciplines such as math, science, and language. Environmental literacy is everybody's responsibility—not just experts and special interests.

Conservation—Sustaining and Restoring the Natural World

RTPI works to inspire people to choose action over inaction, and engages in meaningful conservation research and actions that transcend political and other boundaries. Our goal is to increase awareness and understanding of the natural connections between species, habitats, and people—connections that are critical to effective conservation.

For more information, and to support RTPI, please visit rtpi.org.

To Alison

Peterson Reference Guide to

SPARROWS
of North America

CONTENTS

INTRODUCTION 1

A Note on Notes 1

What Is a Sparrow? 1

Taxonomy and Classification 3

A Note on English Names 3

Historical Approaches to Sparrow Identification 4

Ruling Out the Non-Sparrow 9

The Genera and Species of North American Sparrows 14

SPECIES ACCOUNTS 29

Striped Sparrow 30

Song Sparrow 32

Swamp Sparrow 46

Lincoln Sparrow 52

Sierra Madre Sparrow 57

Baird Sparrow 61

Henslow Sparrow 65

Large-billed Sparrow 70

Belding Sparrow 73

San Benito Sparrow 78

Savannah Sparrow 80

Ipswich Sparrow 87

Vesper Sparrow 91

LeConte Sparrow 96

Seaside Sparrow 100

Nelson Sparrow 108

Saltmarsh Sparrow 115

Bell Sparrow 119

Sagebrush Sparrow 125

Canyon Towhee 130

California Towhee 135

Abert Towhee 140

White-throated Towhee 143

Rusty-crowned Ground Sparrow 145

Rusty Sparrow 149

Rufous-crowned Sparrow 153

Rufous-capped Brush Finch 157

White-naped Brush Finch 159

Eastern Towhee 161

Spotted Towhee 167

Guadalupe Towhee 176

Socorro Towhee 178

Collared Towhee 180

Green-tailed Towhee 184

American Tree Sparrow 188

Red Fox Sparrow 193

Sooty Fox Sparrow 198

Slate-colored Fox Sparrow 204

Thick-billed Fox Sparrow 208

Guadalupe Junco 212

Baird Junco 214

Yellow-eyed Junco 217

Red-backed Junco 221

Gray-headed Junco 225

Pink-sided Junco 230

Oregon Junco 235

Cassiar Junco 241

Slate-colored Junco 244

White-winged Junco 249

Golden-crowned Sparrow 254

White-crowned Sparrow 261

Harris Sparrow 269

White-throated Sparrow 275

Rufous-collared Sparrow 280

Chestnut-capped Brush Finch 283

Green-striped Brush Finch 285

Black-throated Sparrow 287

Five-striped Sparrow 291

Lark Sparrow 294

Lark Bunting 298

Chipping Sparrow 303

Clay-colored Sparrow 308

Black-chinned Sparrow 313

Worthen Sparrow 318

Timberline Sparrow 321

Brewer Sparrow 324

Field Sparrow 329

Grasshopper Sparrow 333

Olive Sparrow 339

Bridled Sparrow 342

Black-chested Sparrow 344

Stripe-headed Sparrow 346

Bachman Sparrow 349

Cassin Sparrow 355

Botteri Sparrow 360

Rufous-winged Sparrow 365

Acknowledgments 373

Notes 375

Indexes 427

INTRODUCTION

Most bird books treat their subject as one entirely separate from the cultural world that humans inhabit, focusing exclusively on what for the past 2,500 years we have called "natural history": identification, behavior, and ecological and evolutionary relationships. But birds have a human history, too, beginning with their significance to Native cultures and continuing through their discovery by European and American science, their taxonomic fortunes and misfortunes, and their prospects for survival in a world with ever less space for wild creatures.

That human history is made up not just of facts and measurements but of stories. Some of those stories are amusing, some sobering, but all should remind the reader of one important truth: everything we think we know, someone else had to learn. A fuller awareness of the slow evolution of ornithological knowledge over the centuries can inspire modern birders both to greater ambition and to greater patience with their own development. If scientific ornithology is still debating the status, indeed the very existence, of, for example, the Cassiar Junco a century after its discovery, we field observers can be more comfortable in our own uncertainties.

Much of a bird's human history is revealed in its changing taxonomy—the names (scientific and vernacular) assigned a species over time. Tracing the nomenclatural career of a bird over the decades and centuries is one of the best routes to track the ways that ornithological and popular observers alike have tried to come to terms with a new, odd, or particularly interesting bird. Read in this way, even the driest of synonymies becomes a trove of bird and birding lore that can only deepen our appreciation of our forebears' efforts to untangle some of the knottiest problems in American natural history.

It is impossible, especially in a volume as reluctantly slender as this one, to tell all of the stories associated with all of the names assigned to all of the American sparrows over all time. Instead, for each species, subspecies, or "flavor" of sparrow, we have selected one or two anecdotes that illuminate the confrontation between humans and birds over the years. Virtually every sparrow taxon's human history could fill a book of its own, and it is hoped that readers will find their interest piqued sufficiently to dig deeper into the great store of birding and ornithological knowledge of these only apparently bland little birds.

This apparent Cassiar Junco gives every indication of confidence in its own existence. *Cathy Sheeter*

A NOTE ON NOTES

The end notes, beginning on page 375, provide bibliographical citations to the sources of quotations, paraphrases, and other explicit references, with the exception of eBird data and identification information drawing on Peter Pyle's *Identification Guide to North American Birds,* Vol. 1 (Salinas: 1998), both of which are cited parenthetically in the text.

WHAT IS A SPARROW?

Some English names for birds are unequivocal. All birders (and many non-birders) have essentially the same mental image of a pelican, a duck, or a flamingo, and a guide dedicated to waxwings or kingfishers would need nothing more than a sketch and a single sentence to satisfactorily identify its subject. In all those and many other cases, a clearly defined term is precisely applied to a neatly delimited avian group.

Other bird names are more difficult to pin down. Notoriously, the same English word can denote birds that are entirely different—and often not even closely related. "Sparrow," unfortunately, is a particularly complicated example of the ways in which the conventional scientific use of a term can diverge significantly and confusingly from its natural usage in everyday language.

The original "sparrows" were the chunky brown

396 BULLETIN 50, UNITED STATES NATIONAL MUSEUM.

Rocky Mountain district of United States and British Columbia, breeding from the more eastern ranges in Colorado, etc., west to the White Mountains in southeastern California, mountains of northeastern California (Lassen and Modoc counties), eastern Oregon (near Camp Harney), etc.; north to interior of British Columbia (Nelson, etc.); during migration south to New Mexico and Arizona, west to Los Angeles County, California (casual); and western slopes of Sierra Nevada; east to western Kansas, etc.

Passerella schistacea BAIRD, Rep. Pacific R. R. Surv., ix, 1858, 490, part ("type from head waters of the Platte," Colorado; U. S. Nat. Mus.), 929, 927 (Fort Bridger, Wyoming); ed. 1860 ("Birds N. Am."), atlas, pl. 69, fig. 3; Cat. X. Am. Birds, 1859, no. 376.—COOPER, Orn. Cal., 1870, 223 (figs. of head, bill, and feet).—SNOW, Birds Kansas, 1873, 7 (1 spec.).—BAIRD, BREWER, and RIDGWAY, Hist. N. Am. Birds, ii, 1874, pl. 28, fig. 9.—RIDGWAY, Orn. 40th Parallel, 1877, 486 (Carson City, Nevada, Mar.; Wahsatch Mts., Utah, breeding).

Passerella iliaca, var. *schistacea* ALLEN, Bull. Mus. Comp. Zool., iii, July, 1872, 168 (Ogden, Utah).—RIDGWAY, Bull. Essex Inst., v, 1873, 183 (Colorado); vii, 1875, 37 (Nevada).—HENSHAW, Ann. Rep. Wheeler's Surv., 1877, 1318.
var. *schistacea* RIDGWAY, Bull. Essex Inst., v, Nov., 1873, 191.
Passerella iliaca var. schistacea COUES, Key N. Am. Birds, 1872, 147, part.
[*Passerella iliaca*] *schistacea* RIDGWAY, Bull. Essex Inst., vi, Oct., 1874, 174; Man. P.[*asserella*] N. Am. Birds, 1887, 434.
Passerella iliaca schistacea RIDGWAY, Bull. Essex Inst., vii, Jan., 1875, 22 (op. Humboldt Valley, Nevada, Sept.); Proc. U. S. Nat. Mus., iii, 1880, 181; Nom. N. Am. Birds, 1881, no. 235c.—HENSHAW, Bull. Nutt. Orn. Club, iii, 1878, 7 (crit.).—COUES, Check List, 2d ed., 1882, no. 284.—AMERICAN ORNITHOLOGISTS' UNION, Check List, 1886, no. 585c.—TOWNSEND, Proc. U. S. Nat. Mus., x, 1887, 220 (e. base Mount Lassen, n. California, breeding).—BENDIRE, North Am. Fauna, no. 3, 1890, 97 (foot of San Francisco Mt., Arizona, Sept. 29).—MERRIAM, v, 1889, 113 (breeding range, habits; descr. nest and eggs).—FISHER, Goss, Birds Kansas, 1891, 478 (w. Kansas, rare winter visitant).—WHITE N. Am. Fauna, no. 7, 1893, 102 (Panamint Mts., California, Mar.; Mts., California, July).—RIDGWAY, Proc. Ac. Nat. Sci. Phila., 1893, 51, 64 (Nelson, int. British Columbia).—COOKE, Birds Colorado, 1897, 107 (rare summer resid.); Bull. Col. Agri. Coll., 1898, 187 (Florissant, Colorado, July; near Glenwood Spring, Grand River, June).—MERRILL, Auk, xv, 1898, 17 (Fort Sherman, n. w. Idaho, May).—GRINNELL, Publ. ii, Pasadena Acad. Sci., 1898, 40 (Los Angeles, California, 1 spec. Dec. 14, 1898).
Passerella iliaca, γ, *schistacea* RIDGWAY, Proc. U. S. Nat. Mus., i, March 21, 1879, 418 (Murphys, Calaveras Co., California, Jan. 4).
P.[*asserella*] *i.*[*liaca*] *schistacea* COUES, Key N. Am. Birds, 2d ed., 1884, 386.
Passerella townsendii, var. *schistacea* BAIRD, BREWER, and RIDGWAY, Hist. N. Am. Birds, ii, 1874, 56, pl. 28, fig. 9.—BENDIRE, Proc. Bost. Soc. N. H., xix, 1877, 120 (Camp Harney, e. Oregon, breeding; descr. nest and eggs).
Passerella townsendii . . . var. *schistacea* COUES, Check List, 1873, no. 189c, part.—HENSHAW, Rep. Orn. Spec. Wheeler's Surv., 1874, 118 (s. of Apache, Arizona, Sept.); Annot. List Birds Utah, 1874, 6; Zool. Exp. W. 100th Merid., 1875, 295 (Provo, Utah, July; s. of Apache, Arizona).
Passerella townsendii . . . . var. *schistacea* HENSHAW, Rep. Orn. Spec. Wheeler's Surv., 1872 (1874), 15 (Provo, Utah, July).
[*Passerella townsendii* var. *schistacea*] b. *schistacea* COUES, Birds N. W., 1874, 162, part (synonymy).

If any science out-dismals economics, it is bibliography. But following up on the fine print can lead us deep into the human history of natural history. *Robert Ridgway*, The Birds of North and Middle America. *Image from the Biodiversity Heritage Library. Digitized by Cornell University Library.*

---

birds still known today as the House and Eurasian Tree Sparrows, cheerful and familiar inhabitants of British towns and farms. Over the ages, though, almost any small, stout-billed, brown or gray bird could be and has been called a "sparrow"; one relic of that catholic usage is the still current alternative name "Hedge Sparrow" for the Dunnock, a Eurasian accentor only distantly related to any sparrows in a strict, scientific sense.

British explorers and colonists took their bird names with them, and when they were confronted in other lands on other continents with unfamiliar drab songbirds, they often called them "sparrows," too, whatever the birds' true affinities. In the Old World, many of the birds given the name were in fact, coincidentally, close relatives of the House Sparrow, but that family, known as the Passeridae or (rather unhelpfully) the Old World sparrows, is not represented among the native birds of the Americas.

The New World is home to plenty of other small brown birds, though, and the British colonists in

North America, not knowing or (more likely) not caring that they were in most cases not the same as the dooryard birds of home, simply called the smallest and the brownest of the lot "sparrows." For larger, more colorful, or more distinctively patterned birds, such equally vague traditional names as "bunting," "finch," and "junco" were also available. The New World orioles and vultures and, of course, the American Robin—none of them especially closely related to their Eurasian eponyms—got their misleading secondhand names in the same way.

Scientific terminology, by abandoning English (or any other vernacular) names in favor of the conventionalized use of latinized terms, is intended to clear up the confusion created by the use of one common name for many birds—a situation known as polysemy—or of many common names for one bird—polylexy. The goal of scientific nomenclature, since its modern beginnings in the 1840s, has been to assign one unique and invariable name to each organism; those names are also intended to indicate relationship, rather than mere superficial similarity.

Unfortunately, schemes of scientific classification, like all human endeavors, are subject to revision: new knowledge, new technology, and even fashion can change the way we group birds. Thus, the New World sparrows have over the years been variously treated as allied to the "true" or "winter" finches of the family Fringillidae, joined with the classic colorful Eurasian buntings of the family Emberizidae, and tossed together with a wide range

---

Superficially similar to New World sparrows, the House Sparrow is a bulky, coarsely marked bird with a wide tail and thick bill. *Texas, April. Brian E. Small*

of otherwise dissimilar songbirds into a sprawling assemblage of nine-primaried oscine passerines. The answer to the question "What is a sparrow?" depends on when and where it is asked.

Currently, ornithologists classify the New World and the Old World sparrows in two separate families, thought to be only somewhat distantly related. The New World birds—which go by a muddling variety of English names, including "sparrow," "towhee," "bunting," and "junco"—are now assigned to a well-delimited family Passerellidae, its closest affinities probably with the Old World buntings of the genus *Emberiza*.

For the purposes of this book, the Gordian knot is sliced by considering "sparrows" to be only the members of the family Passerellidae as understood by Klicka et al. in their 2014 biochemical assessment of evolutionary relationships in the songbirds. Their classification, which has met with widespread acceptance among ornithologists and birders, is almost certain to be adopted eventually by the American Ornithological Society's (formerly the American Ornithologists' Union's) Committee on Taxonomy and Nomenclature, which publishes the "official" checklist of North American birds. Thus, though there are many other species in the world whose English name contains the word "sparrow," they are not here considered "real" sparrows unless they belong to the family Passerellidae as defined by Klicka and his coauthors; conversely, birds whose English name does not include "sparrow"—such as the juncos, the towhees, and the tropical brush finches—are nevertheless accounted sparrows so long as they are considered members of that family. The sole exception for present purposes is the "bush tanagers" of the genus *Chlorospingus*, some species of which are included among the passerellids by Klicka et al. but all of which are omitted here as radically different in appearance and behavior from the other sparrows.

TAXONOMY AND CLASSIFICATION

In restricting "sparrows" to the narrowly defined family Passerellidae, this guide follows the progressive phylogeny set forth by Klicka et al. in 2014. In answering the vexed question of which "kinds" of sparrows should be given separate treatment here, however, we take a decidedly eclectic approach. Not all of the taxa considered here are accorded species status in the *Check-list* published (and regularly updated) by the American Ornithologists' Union (AOU), now the American Ornithological Society (AOS), but all are of sufficient historical note, and most are sufficiently distinctive in appearance, to be of interest to birders in the field.

The scientific determination of species status is now made largely on the basis of biochemical stud-

The very distinctive Large-billed Sparrow of the American Southwest is currently treated by the American Ornithological Society as merely a subspecies group of the Savannah Sparrow. *California, January. Brian E. Small*

ies that claim to measure the divergence between the genetic material of two or more populations. If the difference is great enough, the populations are treated as separate species; if the difference is too little, they are treated as conspecific. Although such studies may be able to quantify those differences, they cannot answer the critical question about their significance: How much difference, over how much time, is required for us to recognize two populations as distinct at the level of species? Reasonable scientists can offer different and equally reasonable answers to this question, with the result that there is not complete agreement about the species status of some of the sparrows included in this guide. Some of the populations currently lumped by the AOS under the name "Savannah Sparrow," for example, or some of those at present considered to make up the AOS's "Fox Sparrow" are treated by other authorities as fully distinct species, and are treated as such here.

A NOTE ON ENGLISH NAMES

The past two decades or so have seen an increasing effort to infuse English names with taxonomic force, creating complicated and ever-changing systems bristling with unnatural hyphens and peculiar

John James Audubon named this species for his young colleague Spencer F. Baird. After considerable taxonomic wandering, the Baird Sparrow once again occupies the genus Baird himself created for it, *Centronyx*. *North Dakota, June. Brian E. Small*

spellings to indicate genetic relationships. But such indications are rightly the burden borne by scientific binomials, not by vernacular names, and in this guide we avoid such awkward and abstruse neologisms as "brushfinch" and "fox-sparrow" in favor of forms more at home in English prose.

We do, however, follow recent precedent in capitalizing the English names of avian species—not because those names are somehow "proper nouns," but to make the text easier for the impatient eye to scan.

A large number of sparrows have been assigned patronyms over the years, names commemorating an explorer, scientist, or companion somehow associated with the bird. In scientific names, these patronyms are usually in the genitive case, and it has become the custom to form English names on the same pattern, putting the human honoree's name in the possessive. That practice misunderstands the true signification of the Latin genitive in such situations: it serves as a marker not of possession but of a particular kind of attribution. There is con-

siderable reason to abandon the false possessive in English names, which otherwise forces us into such barbarous constructions as "the Baird's Sparrow" and potentially confuses the species name with circumstances of actual ownership, such that it can be impossible to know whether a given specimen is an example of *Centronyx bairdii* or a sparrow that happens at one time to have belonged to Spencer Baird. This guide returns to the tradition of presenting English patronyms without the possessive "s."

HISTORICAL APPROACHES TO SPARROW IDENTIFICATION

Almost a century ago, Neltje Blanchan noted in her tendentiously titled *Birds Worth Knowing* that "in this 'much be-sparrowed country' of ours, familiarity is apt to breed contempt for any bird that looks sparrowy." Blanchan was writing in a very different context about a very different bird, but that contempt—dread, even, at times—persists in the minds of many North American birders who grow anxious even today at any encounter with a brown bird. There is nothing natural or inevitable about that anxiety, however. It is instead the historically contingent result of an identification method that is

The common and familiar Song Sparrow may have been "discovered" for European science nearly two centuries ago by the English naturalist Mark Catesby. Or perhaps not. *Maine, June. Brian E. Small*

The handsome White-crowned Sparrow was one of 13 species known to Alexander Wilson, the "Father of American Ornithology." *Texas, November. Brian E. Small*

as traditional as it is outmoded—and the origins of which can be traced ultimately less to ornithological reflection than to economic necessity.

The earliest efforts to illustrate all of North America's sparrows were not intended to ease their identification in the field. Mark Catesby, traveling through the American Southeast in the early eighteenth century, and both Alexander Wilson and John James Audubon, a century or more later, were, understandably, more intent on documenting the existence of the birds than in facilitating their identification; consequently, in none of those early illustrated works were the sparrows (or any other taxonomic complex) presented systematically—or even grouped together—as a family. Wilson, writing in the first dozen years of the nineteenth century, found such a systematic presentation "altogether impracticable," arguing that whatever taxonomic system he might adopt for the volumes of his *American Ornithology* would inevitably be disrupted as "numerous species, at present entirely unknown, would come into our possession long after that part of the work appropriated for the particular genera to which they belonged had been finished." In the better-known and extreme case of Audubon's *Birds of America*, the illustrations and

texts were even printed separately in installments issued over a significant span of years; the plates were issued not in any "natural" sequence but in commercially convenient groupings determined by the size of the birds depicted.

By the late 1820s, even as the Audubon juggernaut steamed on, much of the preliminary inventorying of the North American avifauna was complete. Some discoveries awaited, of course, especially in the West and Southwest, but the birds of the eastern United States and Canada were largely "known." Furthermore, thanks to the work, much of it carried out in North America, of such dedicated taxonomists as Louis Pierre Vieillot and Charles Lucien Bonaparte, some of the wild diversity of nomenclatural opinion that had obtained before this period—"a source of great perplexity to the student," in Wilson's words—had been settled for the time being, and for the first time, most authorities could agree in most cases on at least the higher-level affinities of North American birds.

With the first, centuries-long phase of ornithological prospecting nearly complete and at least a temporary and partial truce declared in the matter of avian classification, the time was ripe for the first serious attempts at systematically organized, illustrated handbooks of bird identification for amateurs.

The earliest of the influential works on that model were published, predictably, in Philadelphia and Boston, further solidifying those cities' shared claim to the title of cradle of American ornithology. In 1828, fifteen years after the author's untimely death, the Philadelphia publisher Harrison Hall reissued Alexander Wilson's *American Ornithology* in three volumes, incorporating and reorganizing all of the textual and graphic material from Wilson's original and the ninth, posthumous volume prepared by George Ord. Hall adopted the taxonomic system used in John Latham's *General Synopsis* for his reprint, "not because"—as Hall writes 40 years after the original publication of the *Synopsis*—"he considers it the best" but rather because Wilson had mentioned it approvingly in the introduction to the *Ornithology*.

While Hall's systematic approach resulted in

The Brewer Sparrow, a classic species of sagebrush flats in the American West, is named for the wealthy Boston collector and naturalist Thomas M. Brewer, coauthor with Baird and Ridgway of one of the nineteenth century's most important bird books. *California, June. Brian E. Small*

grouping into two genera the textual accounts of the 13 sparrow species Wilson had described, the reuse of the plates from the original *American Ornithology* meant that the illustrations of those same species were still as widely scattered as they had been a decade and a half before. The Field Sparrow, for example, occupies Plate 16, while the reader must leaf through to Plate 31 before encountering an image of the White-crowned Sparrow. Furthermore, the fact that, as Hall recounts with justifiable pride, all of the reprinted plates were, like the originals published during Wilson's lifetime, hand-colored to ensure their "permanency, brilliancy and accuracy" added significantly to the cost of the work at the same time as its usefulness to identification was diminished.

More useful to the sparrow watcher—and more practical in its portable octavo format—was Thomas Nuttall's *Manual of the Ornithology of the United States and of Canada*, first published in 1832. Leaving aside the embarrassing Ambiguous Sparrow (in fact a Brown-headed Cowbird), 14 species (again, split over the two ancient genera *Fringilla* and *Emberiza*) were included, each introduced by a list of "specific characters," most of which are visible in the living and unrestrained bird—one of the very first sets of field marks ever made available to the field birder.

Unfortunately, in neither the first nor the second, 1840 edition of his admirable work did Nuttall illustrate most of his birds. Not until 1874, a lifetime after Wilson's *American Ornithology* and a long generation after Nuttall, would a simply written, systematically organized, and extensively illustrated guide to North America's birds be made available "for the masses."

A History of North American Birds by Spencer Fullerton Baird, Thomas Brewer, and Robert Ridgway—each of those coauthors, incidentally, already or soon to be the eponym of at least one sparrow taxon—shared with the almost contemporary first edition of Elliott Coues's *Key* a highly technical introduction to birds as a scientific class, but its treatment of individual species, though still focused in the first instance on the identification of specimens in the laboratory or the cabinet, was significantly more user-friendly than anything that had come before.

The most significant innovation in the *History* was the supplementing of the text with colored illustrations, the first time such a feature was offered in any comprehensive and (relatively) inexpensive identification guide to the birds of North America. While most of the great handbooks of the past had relied on engraving and hand-coloring for their illustrations, the 64 colored plates ornamenting the *History* were produced with the recently introduced technology of chromolithography. Although the process resulted in attractive and colorful images that added far less to the cost of the book than the older, more labor-intensive procedures, certain economies still had to be observed. As Baird had explained a few years before, the publication of accurately illustrated natural histories posed

> a difficult problem, namely, that of furnishing the means of identifying the species, without making the work very bulky and expensive. The plan here adopted of giving as far as possible life-size figures of the heads of each species . . . will, we trust, enable even the tyro to refer correctly to genus and species such specimens as may be collected, since the most characteristic parts will be found figured with scrupulous accuracy.

Full figures and details showing the structure of bills and feet are scattered through the text of the *History* in the form of cheap woodcuts, but the colored pictures, from originals by Ridgway and by Henry W. Elliott, were gathered onto plates—plates that depict only the heads of the species illustrated.

It was the financial exigencies of color publishing that required the depiction of only the heads of the *History*'s sparrows, but the practical effect on the development of bird identification in North America can hardly be overstated. For the nearly century and a half that followed the appearance of the *History*, birders have been directed to begin their identification efforts at the head of the bird and work back in search of the field mark or combination of characters that lead to the correct diagnosis. Sixty years after the publication of the *History*, Roger Tory Peterson's first *Field Guide* pointed—literally, with those trademark field-guide arrows—to

The handsome sparrow plates in the 1874 *History* depicted only the heads of adult birds. *Spencer Baird et al., A History of North American Birds. Image from the Biodiversity Heritage Library. Digitized by Smithsonian Libraries.*

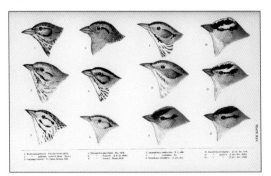

The strong, sharp head markings of the Clay-colored Sparrow can draw an observer's attention away from the species' important structural characteristics. *Minnesota, June. Brian E. Small*

the head as the site of the distinguishing marks for sparrows, and in 1966, the *Golden Guide* invoked even more clearly the head-first heritage by introducing its accounts of the passerellids with two full pages of sparrow "busts." The traditional injunction to start at the head remains a standard piece of advice in field guides and on websites today—all because of the cost associated with the reproduction of color illustrations in the early 1870s.

The fine details of head markings can be useful, and in a very few cases important, in pinning down the identification of some species of American sparrows. But *starting* with the examination of features that are often subtle and sometimes difficult to see can mislead inexperienced observers (and, often enough, others) to confuse birds that are otherwise dissimilar. To take one extreme example of many, even such distinctively plumaged and distantly related birds as the Clay-colored and Lark Sparrows are, surprisingly enough, regularly mistaken one for the other simply because their face patterns are superficially similar.

By beginning with the fine marks of the head, birders have been taught to invert, intellectually and physically, the process of identification. It is easier and more efficient, given the choice not always afforded by small, active, sometimes furtive creatures, to focus at first on the anatomical structures responsible for the one distinctive character shared by all the living birds of North America: flight.

Every bit as distinctive as the Lark Sparrow's strikingly marked head is the bird's bulky, long-tailed structure. *Arizona, April. Brian E. Small*

"How providential—from the bird student's point of view—that birds have tails!" exclaimed Blanchan in her *Birds Worth Knowing*. Tails and wings are the most crucially functional parts of any flighted bird's body, and in many cases they differ noticeably in length, shape, and pattern between species that might otherwise strike the observer as remarkably similar. Returning to the example cited just above, although the Clay-colored Sparrow and the Lark Sparrow have a similar distribution of streaks and stripes and lines on the head, potentially leading to confusion, they differ dramatically at the rear end of their bodies: the long, narrow, noticeably notched tail of the Clay-colored Sparrow could never be mistaken for the broad rounded fan of the Lark Sparrow, even when (as is the usual case in the perched bird) the "lacy" white corners of the Lark Sparrow's tail are not visible.

Most sparrows' wings, though equally providential for the birds that use them, are less immediately distinctive from the observer's point of view. Because most of our sparrows are relatively short-distance migrants—flying hundreds rather than thousands of miles with the seasons—or, as in many of the more southerly species, entirely sedentary, their wings tend to be short and rounded. This often makes their shape unhelpful in differentiating similar species in the field (though very useful, as discussed below, in distinguishing sparrows from potentially similar non-sparrows). Even here, how-ever, there are exceptions in which wing structure can immediately identify an otherwise problematic bird.

For example, Lark Buntings, in the brown plumage worn by females and, for much of the year, males alike, can vaguely recall any number of other streaked sparrows, from which they are quickly and reliably told by their distinctive wing structure. Like several other ground-dwelling birds of open, sandy, brightly lit spaces—among them various pipits and

The Lark Bunting's long, broad tertials almost entirely cover the primaries when the wing is folded. *Colorado, June. Brian E. Small*

larks—the Lark Bunting has evolved notably long, broad tertial feathers to protect the primaries from mechanical abrasion and the effects of strong light. As a result, the outermost flight feathers can at times be almost completely concealed on the folded wing, giving the bird an unmistakably "skirted" appearance entirely unlike that of any other brown, streaked sparrow.

Even when the shape and pattern of wing and tail are less distinctive at the species level, these are the structures that experienced observers, consciously or not, use first to narrow their consideration from 10 or 20 or 40 possible identifications to the three or four most similar birds. In practice, in the case of the passerellid sparrows, this usually leads to a set of identification contenders belonging to just one or two genera.

This procedure, now second nature to many birders, of eliminating possibilities based on shape and structure was first codified in print in 1990, when Kenn Kaufman in his *Advanced Birding* laid out what he called "the generic approach"—more idiomatically styled the "genus approach." Those few pages of brief prose and rough sketches, by making explicit the departure from outmoded strategies concentrating on "field marks," set a turning point not just in sparrow identification but in American birding. In a recent cogent and concise formulation, that of Marshall Iliff, the new technique is this: "To identify a sparrow, consider first its shape, then its habitat and habits, and finally its [plumage] field marks."

Taking this more thoughtful approach, most sparrows will be readily identifiable by careful observers using any of the standard guides, whether digital or print. Very few species can be said to offer any identification problems at all, and true subtlety and any sort of deeper knowledge are called for at the species level in only a very small number of especially thorny cases; not coincidentally, most of those "look-alike" challenges have their origin in recent taxonomic innovations, forcing birders to look more closely to distinguish birds that we have been used to thinking of as identical—or not thinking of at all.

RULING OUT THE NON-SPARROW

A number of birds from other families can be mistaken at first glance, and sometimes even at second, for passerellid sparrows. It is essential that those potential confusion species be ruled out before proceeding to identify an unfamiliar sparrow.

The Old World sparrows of the family Passeridae have a legitimate claim to being the original "sparrows," but these birds—two species of which have been successfully introduced to North America—are now thought to be only distantly related to the

The House Sparrow may resemble some New World sparrows at first glance, but its coarse markings, wide tail, and thick bill can help identify it in any plumage. *Texas, November. Brian E. Small*

(New World) family Passerellidae (or, for that matter, to the Eurasian buntings of the family Emberizidae). Both the widely established House Sparrow and the more locally distributed Eurasian Tree Sparrow are classic small brown birds, and both are superficially very similar to many birders' mental image of a typical passerellid sparrow. The female House Sparrow in particular, lacking the more distinctive plumage characters of the male, is a continual source of confusion, especially for new birders and for observers still relying on a traditional "field marks" approach.

Both sexes and all ages of the House Sparrow are immediately distinguished from the passerellid sparrows by their relatively short and very broad tails, heavy bodies, and very thick bills. The most important and the most obvious plumage character is—all earlier field guides to the contrary—the back pattern. Where superficially similar brown American sparrows are neatly and finely streaked black and brown above, House Sparrows have broad orange stripes running parallel down the back; if their upperparts were maps, American sparrows would chart only rural highways, and House Sparrows show the New Jersey Turnpike.

House Sparrows are also quickly identified by their loud, ringing chirps, which are quite unlike the clearer musical notes or high, lisping *tseet*s given by most passerellid sparrows. The American birds are generally silent unless they are alarmed or going to roost, but House Sparrows may call continuously, even while feeding. That apparent need

The Eurasian Tree Sparrow resembles the more familiar House Sparrow, but has a brown crown and small black cheek patch. *France, April. Rick Wright*

to maintain contact extends to the species' flocking behavior. When flushed, a flock of passerellid sparrows typically flies into cover—trees, bushes, or brush, depending on the species—where individuals space themselves evenly and discreetly. House Sparrows, in contrast, lift off in a befuddled panic, each member of the flock intent on reaching the center of a low bush or tree, where they gather in irregular clumps of three or four birds, sometimes so close as to be nearly touching one another. This behavior is easily learned by even the beginning birder, and makes it possible to identify House Sparrows with considerable confidence even from a fast-moving vehicle.

Introduced from Europe more than a century and a half ago, House Sparrows are now common over nearly all of North America. The Eurasian Tree Sparrow, originally released in smaller numbers and at fewer places, has spread much more slowly, and remains restricted to a small range within about 300 miles of St. Louis, Missouri. Odd individuals occasionally seen elsewhere in the United States and Canada may be pioneering strays from the established population, recently released or escaped captives, or even, in the case of birds seen in the Pacific Northwest, conceivably wild vagrants from the Old World.

Though they are slightly smaller and slimmer than their House Sparrow cousins, Eurasian Tree Sparrows differ from North America's passerellid sparrows in the same ways, namely, in shape and structure, behavior, and back pattern. Unlike the House Sparrow, both sexes and all age classes of this species display the distinctive black and chestnut head pattern—paler and more subdued in

juveniles—that is different from anything seen in our native sparrows.

Not long ago, the sparrows now classified in the family Passerellidae were assigned to a very broadly defined family Fringillidae, which also included the cardinals and grosbeaks and the true, or "winter," finches. The family Fringillidae now comprises only the finches in the narrow sense, with the cardinals and grosbeaks occupying their own family, Cardinalidae. Influenced by their taxonomic history, though, some observers still confuse certain members of these groups with sparrows.

The cardinals and male Rose-breasted and Black-headed Grosbeaks are distinctive enough that such confusion is rare, but females and young birds of those two grosbeak species are brown and streaked; they can be immediately distinguished from all passerellid sparrows by their consistently arboreal habits, large size, fistlike shape, and enormous conical bills.

The smaller, smaller-billed members of the family Cardinalidae—known in English, confusingly enough, as buntings—pose greater identification

Female and juvenile Rose-breasted Grosbeaks differ from sparrows in the huge bill; short, notched tail; and intricately patterned wings. Black-headed Grosbeaks are similar but more colorful, with darker bills. *Texas, April. Brian E. Small*

Quite sparrowlike on first acquaintance, female and juvenile Lazuli Buntings can be identified by their overall plainness. Indigo Buntings are even duller, usually without the wing bars but with faint streaking below. *California, May. Brian E. Small*

Like many birds, the Dickcissel is almost unmistakable—once an observer has had a chance to get to know it well. Females and winter birds are very plain, with large bills and usually a yellow supercilium. *Texas, April. Brian E. Small*

challenges. Even the dullest Painted Bunting is decidedly greenish, but female and young male Indigo, Varied, and Lazuli Buntings are largely unmarked brown and can strike the inexperienced observer as remarkably "sparrowlike," especially when migrants gather to feed on the ground. That very lack of conspicuous markings, relieved only by dull, uneven breast streaking in Indigo Buntings and by variably contrasting whitish wing bars in Lazulis, is a good clue, as are the rather thick bill, continually twitching tail, and rich metallic calls typical of these species; many individuals are also likely to show blue in the wing, tail, or rump.

The Dickcissel has routinely been deemed aberrant in any of the families to which it has been assigned over the years. Currently occupying an uncertain taxonomic seat among the Cardinalidae, this is a very colorful, distinctive bird in some plumages, but more puzzling in others. Young birds and many individuals in winter lack the conspicuous yellow and chestnut tones that so readily identify most Dickcissels in alternate plumage, leaving them a nondescript tan-brown. These dull birds can be distinguished from passerellid sparrows by their short, wide tails; rather long wingtips; coarsely marked, broadly striped backs; and heavy, pale bills. As that brief list of characteristics suggests, basic-plumaged Dickcissels more closely resemble House Sparrows than they do any of our native American sparrows, from which they can be distinguished in the same ways.

The "new" and more narrowly defined Fringillidae also includes, alongside such unmistakable birds as the Pine and Evening Grosbeaks, several species that might be mistaken for passerellid sparrows. The yellow wing and tail patches of Pine Siskins are not always readily visible in the field, but the very small size, deeply notched tail, long wings, peculiarly sharp bill, and acrobatic feeding behavior identify these finches quickly.

Even the dullest Pine Siskin shows distinctively long wings and a thin, sharply pointed bill. *California, January. Brian E. Small*

The brown House Finch can cause confusion, but the combination of long, notched tail; blurry streaked underparts; and thick, curved bill matches none of the American sparrows. *New Mexico, January. Brian E. Small*

The red finches—House, Cassin, and Purple—can be more confusing. All three are roughly sparrow-sized, and though adult males of these species are red, females and young birds of all three can be brown and streaked. The Cassin and Purple

Finches, with their strong face patterns, heavy bills, long wings, and short, deeply notched tails, are distinctive if seen well, but House Finches are more anonymous, and more superficially sparrowlike, in both shape and plumage.

Female and juvenile male House Finches are dull brown with broad, blurry streaks and bland faces surrounding beady black eyes. The medium-long tail is broad and squared at the tip, and the wing is short and rounded. The very round head ends in a swollen bill with a decidedly curved culmen, quite unlike the bill of any passerellid sparrow. The loud, chirping calls and hoarse but musical warbled song are further distinctions from sparrows, as is their habit of traveling and feeding in noisy, tightly clumped flocks; unlike most sparrows, House Finches also readily perch on rooftops and roost and even nest under the eaves of houses or in hanging flowerpots.

Three or four members of the family Icteridae—the blackbirds, orioles, cowbirds, and grackles—are also commonly confused with passerellid sparrows. As familiar and nearly unmistakable as the adult male Red-winged Blackbird is, females and young birds of that species are continually mistaken for large sparrows; many a beginning birder has gone home with her confidence in the leader of a field trip shaken by the insistence that that shy, streaky bird in the cattails was "just" a Red-winged Blackbird. In truth, no sparrow is as bulky and as large

The juvenile Brown-headed Cowbird is heavily scaled above and streaked below, recalling a brown House Finch or a sparrow. *Oregon, August. Brian E. Small*

as a Red-winged Blackbird, and none is as dark and as heavily and thoroughly striped above and below. Even if the long, sharp bill and pink or yellow throat are invisible, the habit of nervously twitching and spreading the long, broad, very black tail with each low-pitched *chuck* note gives this species away. On the Pacific Coast, the increasingly scarce Tricolored Blackbird poses the same risk of misidentification and can be distinguished in the same ways.

The Bronzed, Shiny, and Brown-headed Cowbirds are smaller and structurally more sparrowlike at first glance than are the other blackbirds. Adult males of all three species are unlikely to be misidentified, but the females and, especially, the streaked and scaled juveniles are often taken for large members of the family Passerellidae. Not only new birders make the error. In 1832, no less a naturalist than Thomas Nuttall described a "new" species that he named *Fringilla ambigua*, the Ambiguous Sparrow; it took several years for that bird to be definitively and correctly reidentified as a fledgling cowbird.

Though superficially recalling those of some sparrows, the conical bills of cowbirds are actually significantly larger and longer, and—just as in the case of the Red-winged Blackbird—their tails are longer, broader, and decidedly blackish. A very good identification clue is the long, stout, coarsely scaled black tarsi of the cowbirds; the feet of sparrows are finer and the scales covering them so much

This bird's streaked underparts and heavily marked head might send a hasty birder to the sparrow section of the field guide—but this is a Red-winged Blackbird. *California, January. Brian E. Small*

Very unlike adult males in their natty breeding plumage, brown Bobolinks are often mistaken for New World sparrows, but differ in structure, behavior, and plumage. *Florida, April. Brian E. Small*

more delicate as to be hardly noticeable in the field. Cowbirds in their briefly held juvenile plumage are also often still dependent on their foster parents, whom they follow around, beaks agape, with long, powerful striding steps (and not elastic bounding hops, as in the case of most sparrows).

The most sparrowlike of the blackbirds, the Bobolink, is also in some ways the least blackbird-like. Adult males in their backward-tuxedo alternate plumage are like nothing else. In late summer, however, they molt into a yellowish brown streaked plumage—the "traveling suit"—like that of the females and young. Viewed quickly and in isolation—in a photograph, for example—such birds, with heavily patterned heads, neatly striped backs, and streaked underparts, can mislead even experienced observers. In the field, however, brown Bobolinks are readily distinguished from even the most similarly patterned sparrows by their larger size, pointed tail feathers, and longer bills, creating a "snouty" impression of the face. The loud, nasal calls are also distinctive, as is the habit of Bobolinks to feed in large, nervous flocks; large numbers of migrants are often seen flying high overhead in the early morning, a habit entirely unlike the more discreet, more solitary, and more nocturnal movements of most passerellid sparrows.

Until very recently, the longspurs and "white"

buntings were classified with the sparrows in an extended family Emberizidae. Biochemical analyses conducted in the early twenty-first century, however, have revealed that the four longspur species and the Snow and McKay Buntings are in fact not especially close relatives of the sparrows, but rather

The Chestnut-collared Longspur shares with the other members of its family a long, pointed wing and short tail; longspurs are both more gregarious and more secretive than most New World sparrows. *Colorado, June. Brian E. Small*

"form a well-supported clade that diverged early in the radiation of the New World nine-primaried oscines." Consequently, these six (or five, depending on one's view of the precise relationship of the two "white" buntings) species have been removed from the sparrows and buntings and been assigned their own, new family, Calcariidae.

Even the darkest Snow and McKay Buntings show distinctively extensive white in the wing and tail, but the longspurs over much of the year are patterned in brown and black, and even breeding-plumaged males are sometimes mistaken for sparrows. Structurally, longspurs are large, short-tailed, long-winged, pudgy-bellied, large-headed birds with thick (or, in the case of the McCown Longspur, very thick) conical bills. Chestnut-collared and McCown Longspurs have tails that are largely white, while the Smith and Lapland Longspurs show white outer tail feathers that can recall those of a Vesper Sparrow or a junco.

The identification of female and winter longspurs is notoriously subtle, but they are in general readily distinguished from passerellid sparrows by their larger size and almost invariable preference for windswept, open habitats, where they swirl in large, loosely organized flocks from field to barren field, uttering loud rattling and whistling calls. On the ground, too, longspurs move differently from sparrows, crouching and shuffling mouselike through sparse vegetation, heads held low and legs bent. The equally inconspicuous grassland sparrows are more actively furtive: they move by hopping quietly

Classified in half a dozen genera over the past two centuries, the Grasshopper Sparrow's taxonomic instability is similar to that undergone by many other sparrow species over the years. *Brian E. Small*

through dense grass and forbs, and when alarmed fly singly low over the ground before seeking shelter by diving into the thickest available vegetation, where they disappear until flushed again. Longspurs and Snow Buntings are more likely to seek the open safety of the next county.

A number of tropical species once included in the Passerellidae have now been reassigned to the "true" tanagers, the Thraupidae. The seedeaters and grassquits of the genera *Volatinia*, *Sporophila*, and *Tiaris* are tiny, stub-billed birds of open country with distinctive and sometimes bright male plumages; females and young birds of most species are plain dull brown, less heavily marked than any sparrow.

THE GENERA AND SPECIES OF NORTH AMERICAN SPARROWS

This guide treats the passerellid sparrows that are known to breed in zoogeographic North America, or the Nearctic region, from arctic America south to the volcanic belt that crosses Mexico from Jalisco in the west to Veracruz in the east; species found only on Caribbean islands are omitted. It is anticipated that the sequence followed here, based on the phylogeny produced by Klicka et al., will be adopted eventually for use in the AOS (formerly AOU) *Check-list*.

Widespread over much of the Northern Hemisphere, the Lapland Longspur is a hefty, short-tailed, very long-winged bird of vast open flats, a habitat regularly used by very few New World sparrows. *Colorado, October. Brian E. Small*

Phylogenies are usually laid out in print as branching trees, a graphic form that is able to indicate relationship with great precision. Obviously and inevitably, the linear presentation required by a conventional list is less eloquent, and can even suggest false affinities between only distantly related species. The Bell and Sagebrush Sparrows, for example, are clearly depicted in the two-dimensional tree produced by Klicka et al. as several branches away from the large brown towhees—but the "sage" sparrows and those towhees necessarily take adjacent places in the one-dimensional space occupied by a list. To avoid misleading implications of that sort as much as possible, the list below, which names all of the species included in this guide, inserts an ornament to indicate each "jump" from one major branch, or clade, of the family tree to the next.

Just as the sequence here differs from that in the current edition of the AOU (now AOS) *Check-list*, this guide also carries out almost all the suggestions made by Klicka et al. for the generic reassignment of a few species. The result is the resurrection of two genus names for certain of the grassland sparrows at present classified by the AOS in *Ammodramus*; experienced observers will recognize the names of these "new" genera, one of which was widely in use until the 1970s.

The three species in the genus *Melospiza* share chunky bodies and long tails; these "shade sparrows" are most frequent in brushy, often damp habitats. *Swamp Sparrow, Maine, June. Brian E. Small*

STRIPED SPARROW, *Oriturus superciliosus*
This bird of high-elevation Mexican grassland and brush is a rather large, chunky sparrow with a thick bill and short wings. Most closely related to the

The Striped Sparrow is the only member of the genus *Oriturus. Texas, January. Loch Kilpatrick*

"shade" sparrows in *Melospiza* and certain of the "grass" sparrows, it is assigned to its own genus, of which it is the only species. Astonishingly, this species has appeared in the United States, a single individual discovered in central Texas in early 2015.

SONG SPARROW, *Melospiza melodia*
SWAMP SPARROW, *Melospiza georgiana*
LINCOLN SPARROW, *Melospiza lincolnii*
The genus *Melospiza* includes one of the most familiar and most widespread passerellids, the Song Sparrow; with nearly 30 generally accepted subspecies, this is the most geographically variable of all North American songbirds. That species and its close relatives, the Lincoln and Swamp Sparrows, share fairly long, rounded tails and deep bellies and a preference for nesting in thick cover, which males leave to sing their loud, cheerful songs. All three are common and confiding visitors to bird feeders outside of the breeding season, the Song Sparrow virtually continent-wide and the Swamp and Lincoln Sparrows in the East and West, respectively.

SIERRA MADRE SPARROW, *Xenospiza baileyi*
The sole species in the aptly named genus *Xenospiza*, this sparrow has puzzled ornithologists and field observers since its discovery. Not formally described until the mid-twentieth century, visually

The unusual and very range-restricted Sierra Madre Sparrow is the sole representative of the genus *Xenospiza*, meaning "odd sparrow." *Manuel Grosselet*

Both species in the small genus *Centronyx* are named for friends of John James Audubon. This is a Henslow Sparrow. *Ohio, May. Brian E. Small*

this species combines certain plumage features of the "grass" sparrows with those of some of the *Melospiza* sparrows. This pot-bellied, fairly long-tailed sparrow inhabits high-elevation grasslands in its very restricted range.

BAIRD SPARROW, *Centronyx bairdii*
HENSLOW SPARROW, *Centronyx henslowii*

Sharp-tailed, short-winged, and big-billed, these secretive sparrows of dense grassland are remarkably similar in plumage, with richly colored wings, backs, and heads; their ranges are normally entirely non-overlapping.

Spencer Baird, in treating "his" sparrow in 1858, determined that it merited its own genus, *Centronyx*, the "spurred claw." He distinguished that genus from *Passerculus*—the Savannah Sparrows—not only by the elongated, deeply curved nail of the hind toe but by its longer, more slender bill and "quite unusually long" wings. Anticipating Elliott Coues's later surmise, Baird described the plumage as essentially that of a female Smith Longspur.

On the basis of his experience with the species in the field, Coues in 1873 pronounced the Bairdian genus "scarcely tenable," and returned the sparrow to *Passerculus*, as "the species is so much like a savanna[h] sparrow."

The first to merge Baird's *Centronyx* with *Ammodramus* (spelled, in a simple slip of the pen, *Ammodromus*) appears to have been Christoph Gottfried Giebel, a German ornithologist and bibliographer who almost certainly never had the opportunity to examine a specimen of the sparrow he was synonymizing. Coues, no doubt already planning his own great bibliography, immediately denounced Giebel's work as "slovenly" and "peculiarly exasper-

ating," but his was the name adopted by the AOU in the first edition of the *Check-list*, in which *Centronyx* was demoted to the rank of subgenus. Thus, with the exception of the period between the publication of the Twelfth (1903) and the Fifteenth Supplements (1909), when *Coturniculus* was briefly restored at the level of an independent genus, the AOU has consistently called the species *Ammodramus bairdi(i)*, the ending varying from edition to edition.

Today, the genus *Ammodramus* as used in the AOS *Check-lists* is considered polyphyletic, grouping together species that are not in every case one another's closest relatives. The question thus arises as to the names properly to be applied to the reconstituted groups of species.

Harry C. Oberholser's review of the synonymies showed that *Ammodramus* had first been used to name one of the populations of the Grasshopper Sparrow, and thus must remain with that species and its closest relatives. The Baird Sparrow and the Henslow Sparrow, each other's closest relative and only distantly related to the true *Ammodramus* in this strict sense, are therefore in need of a new generic name expressing their relationships. *Coturniculus*, coined by Bonaparte in 1838, is the earliest contender—but that name, too, was first applied to the Grasshopper Sparrow and so must be "retired" as merely a younger synonym of *Ammodramus*. Another genus name, *Passerherbulus* ("little grass sparrow"), coined by C. J. Maynard in 1895, has as its type species, designated by Maynard himself, the LeConte Sparrow; that name can be used for the Baird Sparrow—the recommendation offered by Klicka et al.—only if that and the LeConte are deemed congeneric, a relationship contradicted by the genetic assessments carried out by those same authors.

What remains, then, is *Centronyx*, coined in 1858 for the Baird Sparrow and now, by priority, the correct genus name to be used for that bird and for its closest relative, the Henslow Sparrow; this solution was adopted, properly, in the latest edition of Howard and Moore's *Complete Checklist* and is followed here. More than 150 years on, the Baird Sparrow thus now bears Baird's scientific name for it, *Centronyx*.

LARGE-BILLED SPARROW, *Passerculus rostratus*
BELDING SPARROW, *Passerculus guttatus*
SAN BENITO SPARROW, *Passerculus sanctorum*
SAVANNAH SPARROW, *Passerculus sandwichensis*

Generally similar in plumage to some of the smaller and more secretive birds in the genera *Centronyx*, *Ammospiza*, and *Ammodramus*, these short-tailed sparrows range from small and dark to large and strikingly pale; some have slender, pointed beaks, others heavy bills swollen at the base. Most are fairly easy to see in their preferred habitats, which range from salt marsh and coastal dunes to native grasslands and farm fields. Though the Savannah Sparrow is currently the only species included by the AOS in the genus *Passerculus*, several of the more distinctive birds in this widespread genus—which

Small, brown, and terrestrial, the Savannah Sparrow is perhaps the most "sparrowy" of all. Its congeners in the small genus *Passerculus* ("little sparrow") are somewhat more distinctive. *California, May. Brian E. Small*

The Vesper Sparrow's classic white outer tail feathers are only infrequently visible on perched birds. This open-country sparrow is the only member of the genus *Pooecetes. Maine, June. Brian E. Small*

are treated separately in this guide—are likely to be once again elevated to the full species status they have been accorded in the past.

VESPER SPARROW, *Pooecetes gramineus*

A medium-sized, rather plain sparrow, the Vesper Sparrow is the only bird in its genus. In overall plumage pattern, this species may recall a Song Sparrow at first glance; in its open-country habits, it more closely resembles the Savannah Sparrow. Not particularly shy, Vesper Sparrows are highly terrestrial and inconspicuous when they are not singing on the breeding grounds; flocks of hundreds gather in September on the western Great Plains for the southbound migration.

LECONTE SPARROW, *Ammospiza leconteii*
SEASIDE SPARROW, *Ammospiza maritima*
NELSON SPARROW, *Ammospiza nelsoni*
SALTMARSH SPARROW, *Ammospiza caudacuta*

Small to medium-sized, large-headed, and short-tailed, the members of the genus *Ammospiza* are some of the most colorful sparrows—and at the same time some of the most elusive, preferring dense grassland and marsh habitats, where they can be much easier to hear than to see. Even when these birds are seen well, they can sometimes be difficult to identify. In the case of the Seaside Sparrow, the identification challenge is complicated by extensive plumage variation across a wide geographic range; some populations may represent distinct species. All of the members of this genus face a variety of conservation challenges, from rising sea levels to habitat destruction caused by agricultural activities and oil and natural gas extraction.

The beautiful LeConte Sparrow is the smallest and perhaps the most colorful member of this genus of sharp-tailed sparrows. *Minnesota, June. Brian E. Small*

William Swainson transferred what Wilson and Audubon had called *Fringilla maritima* into *Ammodramus* in 1837; the Seaside Sparrow shared that genus, which Swainson had named himself ten years earlier, with the Saltmarsh Sparrow and the Swamp Sparrow, an association probably suggested more by habitat preference than any of the physical characters Swainson adduces. While the Swamp Sparrow would eventually find a more congenial home alongside the Song and Lincoln Sparrows, the Seaside, LeConte, Nelson, and Saltmarsh Sparrows continued to be classified as members of *Ammodramus* through the rest of the nineteenth century and into the first two editions of the AOU *Check-list*.

In 1909, *Ammodramus* was restricted so as to apply only to the Baird and Grasshopper Sparrows, and *Passerherbulus*, up to then recognized only as a subgenus, was raised to full generic rank. In the third edition of the *Check-list,* published the following year, the Seaside, Henslow, LeConte, Saltmarsh, and Nelson Sparrows were moved into *Passerherbulus.* By 1931, when the fourth *Check-list* appeared, that group, too, was deemed artificial and "composite," and several of the erstwhile *Passerherbulus* species were assigned to their own genus, *Ammospiza*, a name coined in 1905 by Harry C. Oberholser; they were joined there in 1973 by the LeConte Sparrow. The circle of genera through which all these birds moved would close—temporarily—in 1982,

The more somberly colored Seaside Sparrows have sometimes been assigned to their own genus, but recent molecular study has confirmed their close evolutionary relationship to the other *Ammospiza* sparrows. *Texas, February. Brian E. Small*

when the AOU returned them to a once-again broadened *Ammodramus*, comprising all of the species listed here and the Baird Sparrow.

The impression that *Ammodramus* in its new, and its old, broad sense represents a miscellany rather than a "natural" group has been confirmed by recent biochemical studies. Klicka et al. discovered that the

Grasshopper Sparrow—Swainson's original *Ammodramus*—was only relatively distantly related to the others, and they proposed that Oberholser's *Ammospiza* be revived for the Seaside, LeConte, Saltmarsh, and Nelson Sparrows. That recommendation has been followed in a recent world checklist, and is almost certain to be widely adopted. Should even further refinement ever be needed, the name *Thryospiza*, coined by Oberholser in 1917, applies only to the Seaside Sparrows.

BELL SPARROW, *Artemisiospiza belli*
SAGEBRUSH SPARROW, *Artemisiospiza nevadensis*

Only recently returned to distinct species status after decades of being lumped under the name Sage Sparrow, these two species prefer running to flying, darting from bush to bush in the sparsely vegetated habitats they favor. When they do flush to a perch, they share with the superficially similar Black-throated Sparrow the habit of nervously flicking the tail upward. The differences between the two newly recognized species are subtle, and there may in fact be a third "Sage" Sparrow species in the mix, as the identification notes in the species accounts explain.

The wonderfully named genus *Artemisiospiza* ("sage sparrow") may comprise three, two, or one species; that taxonomic uncertainty is mirrored in the challenge of field identification. *Sagebrush Sparrow, New Mexico, December. Brian E. Small*

CANYON TOWHEE, *Melozone fusca*
CALIFORNIA TOWHEE, *Melozone crissalis*
ABERT TOWHEE, *Melozone aberti*
WHITE-THROATED TOWHEE, *Melozone albicollis*
RUSTY-CROWNED GROUND SPARROW, *Melozone kieneri*

These drab, long-tailed, short-winged birds are by far our largest sparrows, easily mistaken in a brief view for a thrasher or a large ground-feeding animal. All *Melozone* sparrows are relatively reclusive,

The most colorful *Melozone* in our area, the Rusty-crowned Ground Sparrow is a reclusive forest dweller, most often seen only when it sings from an exposed perch. *Jalisco, June. Chuck Slusarzcyk, Jr.*

but Canyon and California Towhees readily become accustomed to human interruption, and all are able to overcome their reticence when offered seed or suet. As expected of birds that frequent dark places with poor visibility, their voices are loud, pairs often communicating with raucous chatters and rattles.

RUSTY SPARROW, *Aimophila rufescens*
RUFOUS-CROWNED SPARROW, *Aimophila ruficeps*

Long a "grab-bag" genus comprising many species not particularly closely related to one another, *Aimophila* is now limited to just three species of fairly plain sparrows with rusty caps and heavy bills; one of the three, the Oaxaca Sparrow, occurs only south of the area treated here. Our two more northerly *Aimophila* are handsomely somber birds of dry slopes and mountains; their terrestrial habits can make them inconspicuous, but they tend to be quite vocal and often permit close approach by the quiet, careful observer.

Heavy-bellied and thick-billed, the relatively monotone Rufous-crowned Sparrow shares the genus *Aimophila* with two Middle American species. *Texas, March. Brian E. Small*

Only two of the colorful tropical sparrows of the genus *Atlapetes* enter our area. This is the handsome Rufous-capped Brush Finch. *Sinaloa, January. Amy E. McAndrews*

WHITE-NAPED BRUSH FINCH, *Atlapetes albinucha*
RUFOUS-CAPPED BRUSH FINCH, *Atlapetes pileatus*

Our *Atlapetes* brush finches, representatives of a large and widespread tropical genus, are big, fairly slender, short-winged sparrows with long tails and small bills. Their dull gray upperparts and muted head patterns are brightened by yellow underparts.

EASTERN TOWHEE, *Pipilo erythrophthalmus*
SPOTTED TOWHEE, *Pipilo maculatus*
COLLARED TOWHEE, *Pipilo ocai*
GREEN-TAILED TOWHEE, *Pipilo chlorurus*

This small genus combines two small, green-backed, quite brush finch–like southern sparrows with the large, strikingly colored, and familiar "rufous-sided" towhees. Though all are characteris-

---

This Eastern Towhee (of a southern, white-eyed population) is collecting seeds from the messy forest floor, typical behavior for all the species in the genus *Pipilo*. *Florida, February. Brian E. Small*

---

tic birds of deep brush, where they feed on shady ground beneath the vegetation, none is especially shy.

AMERICAN TREE SPARROW, *Spizelloides arborea*

This rather large, long-tailed, colorful species has a genus entirely to itself; it is an arctic breeder, descending to lower latitudes across much of the United States and Canada in the winter. Its plumage combines features typical of certain *Peucaea* sparrows with characters of some of the much smaller *Spizella* species. Gregarious, vocal, and tame, this is one of the most common winter birds over much of the American Midwest.

Like so many other passerellid sparrows discovered in the early days of American ornithology, this species was first assigned to the great Linnaean genus *Fringilla*. In 1817, the French ornithologist Louis Pierre Vieillot, who had come to know the species well during his residence in New York and New Jersey, moved it into the genus *Passerina,* which he had erected the year before to encompass the Indigo Bunting, the Bobolink, and the Snow Bunting.

Vieillot's classification was not followed by British or American ornithologists, who preferred a more conservative scheme retaining the New World sparrows in *Fringilla*. In the 1830s, however, both William Swainson and John James Audubon shifted this and a number of other brown species into *Emberiza*. Even that newly revised genus was too miscellaneous for Charles Bonaparte, who created a new genus *Spizella* to include the Field Sparrow and, soon thereafter, the Clay-colored, Chipping, and American Tree Sparrows. Jean Cabanis objected to

The bright, cheerful-voiced American Tree Sparrow has been placed in several genera over the past two centuries, but is now treated as the sole occupant of *Spizelloides. Alaska, June. Brian E. Small*

The portly, broad-tailed sparrows of the genus *Passerella* owe their English name to the fox-red tail and upperparts of the Red Fox Sparrow. *Alaska, June. Brian E. Small*

the unclassical sound of Bonaparte's genus name, which joined a Greek root with a Latin suffix, and created his own better-formed replacement, *Spinites* ("related to a finch-like bird"). That name, of course, had no standing, however linguistically barbarous Bonaparte's coinage, and the American Tree Sparrow shared the genus *Spizella* with six other rusty-capped or narrow-tailed passerellids until 2014, when Slager and Klicka determined that the American Tree Sparrow was not closely related to the others.

Many birders and ornithologists have remarked on the divergences in size, bill shape, and voice between the American Tree Sparrow on the one hand and the remaining *Spizella* sparrows on the other. In this century, biochemical studies have produced phylogenies that consistently confirm the suspicion that the American Tree Sparrow is evolutionarily only distantly connected to its putative congeners. To reflect that distance, Slager and Klicka coined the genus *Spizelloides*, containing the single species *Spizelloides arborea*.

RED FOX SPARROW, *Passerella iliaca*
SOOTY FOX SPARROW, *Passerella unalaschensis*
SLATE-COLORED FOX SPARROW, *Passerella schistacea*
THICK-BILLED SPARROW, *Passerella megarhyncha*

The broad-tailed, deep-bellied, thick-billed fox sparrows are the largest brown sparrows over most of their range, dwarfing all but the towhees and the Harris Sparrow. The songs and most calls are equally imposing, loud and complex; the song of

most individuals has a remarkably wide range in pitch for a sparrow. These are some of the most variable birds in North America, and even experienced observers can be momentarily stymied when confronted with a fox sparrow of a population different from the ones they are used to. It is likely that the Fox Sparrow as currently constituted by the AOS in fact comprises three or four species, but it remains unclear in some cases just which of the currently recognized subspecies should be assigned to which of the "new" species. The birds are treated here as four identifiable forms, each of which is probably a distinct species.

GUADALUPE JUNCO, *Junco insularis*
BAIRD JUNCO, *Junco bairdi*
YELLOW-EYED JUNCO, *Junco phaeonotus*
DARK-EYED JUNCO, *Junco hyemalis*

Over much of Canada and the United States, juncos are among the most abundant of wintering passerines, well known to birders and non-birders alike—often, even today, as "snowbirds," one of the very few genuine folk names for birds to survive in English-speaking America. These confiding, chubby, medium-sized sparrows are easily identified at the species level as currently understood, and, accordingly, they are often ignored after the first welcome sighting each autumn of the bird Howard Elmore Parkhurst described as "reflecting the leaden skies above and the snow below." But for observers willing to go beyond nineteenth-century literary kitsch and the mere determination of eye color, the juncos pose some of the most subtle of identification

The juncos are among the most familiar feeder birds over much of their vast range, which reaches from the Arctic to southern Middle America. This Yellow-eyed Junco shares its white outer tail feathers with all but the Volcano Junco of Panama and Costa Rica. *Arizona, May. Brian E. Small*

The large, boldly patterned *Zonotrichia* sparrows rank high among many observers' favorites. The Harris Sparrow, named for one of Audubon's generous friends, has one of the most geographically restricted breeding ranges of any North American bird. *Manitoba, June. Brian E. Small*

challenges. In addition, these tame birds are well suited to scientific study, and it is to them that we owe some of the most exciting recent advances in our knowledge of passerine adaptation and evolution.

GOLDEN-CROWNED SPARROW, *Zonotrichia atricapilla*
WHITE-CROWNED SPARROW, *Zonotrichia leucophrys*
HARRIS SPARROW, *Zonotrichia querula*
WHITE-THROATED SPARROW, *Zonotrichia albicollis*
RUFOUS-COLLARED SPARROW, *Zonotrichia capensis*

Big, bold, and gregarious, the "crowned" sparrows share long, straight tails, deep bellies, and round, heavily marked heads atop short, thick necks. They are strongly terrestrial feeders, preferring dark, heavily vegetated edges, but some species have the obliging habit of flying to an exposed perch when agitated. All five species sing beautiful but simple whistling or buzzy songs; the White-crowned Sparrow, thanks to the vocal variation it exhibits across its wide range, may be the most studied singer of any bird in the world.

CHESTNUT-CAPPED BRUSH FINCH, *Arremon brunneinucha*
GREEN-STRIPED BRUSH FINCH, *Arremon virenticeps*

The *Arremon* brush finches are a large group of big, chunky, short-winged sparrows with long, rounded tails and thin but long bills. Our species, not especially closely related to the other, more southerly members of the genus, are matte green above, with well-marked heads. They tend to be shy, but can sometimes be lured out of the tropical forest edge with a handful of millet or other small grain.

BLACK-THROATED SPARROW, *Amphispiza bilineata*
FIVE-STRIPED SPARROW, *Amphispiza quinquestriata*

These strikingly marked birds of the American Southwest are medium-sized, rather squat sparrows with simple, bold head patterns. The Black-throated Sparrow is not particularly shy, but Five-striped Sparrows are strongly terrestrial for much of the year, making them hard to glimpse except when the males are singing after spring or summer rains.

Taxonomically, the Five-striped Sparrow has been notable for its extreme generic mobility. Sclater and Salvin named the species expressly for "the five white lines which originate from the bill." They described their new *quinquestriata* as a member of the genus *Zonotrichia*, not out of conviction but because they "knew not where else to place it." On examining the type specimen, Robert Ridgway urged that the Five-striped Sparrow be removed from the

Most *Arremon* sparrows are South American, but two species, including the splendidly patterned Green-striped Brush Finch, occur north of Mexico's volcanic belt. *Manuel Grosselet*

"very well circumscribed group of purely Nearctic species" that made up the true *Zonotrichia* and that "this very peculiarly marked and exceedingly distinct Sparrow" be assigned instead to *Amphispiza* with the similarly marked Black-throated Sparrow. Twenty years later, however, Ridgway admitted to "considerable doubt" about the correct placement of the Five-striped. Observing that "its style of coloration so strongly resembles that of [the Black-throated Sparrow] that it seems almost unreasonable to place it in a different genus," he nevertheless concluded that it was "wholly out of place" in *Amphispiza* and that on the basis of the rounded wing it belonged in fact in *Aimophila*, "a generic, or supposed generic, group . . . among which there are very great differences of coloration and considerable differences of form," principally in wing structure.

The allocation to *Aimophila* was generally accepted through most of the twentieth century, in spite of the similarity of the Five-striped's plumage to that of the Black-throated Sparrow. By the time the bird entered the AOU *Check-list* in 1983, though, it was once again classified as an *Amphispiza*, along with the Black-throated and Sage Sparrows, a determination apparently based on the conclusions in an unpublished thesis and one rejected as premature by a subsequent study. Accordingly, in the 1998 edition of the *Check-list*, the AOU returned the Five-striped Sparrow to *Aimophila*, at least pending more rigorous investigation.

Found in only a small area of northwestern Mexico and a tiny corner of the southwestern United States, the Five-striped Sparrow shares the genus *Amphispiza* with the more widespread and equally handsome Black-throated Sparrow. *Arizona, July. Brian E. Small*

That more rigorous investigation was published in 2009. Using mitochondrial DNA sequencing, Jeffrey M. DaCosta and his coauthors discovered that the Five-striped Sparrow was not particularly close evolutionarily to the other species in the old genus *Aimophila*, but was in fact the nearest relative of the Black-throated Sparrow. A year later, the AOU finally reunited the two in the genus *Amphispiza*, an action the wisdom of which was confirmed by a subsequent molecular study that sampled more taxa than had the earlier, 2009 investigation.

LARK SPARROW, *Chondestes grammacus*

The genus *Chondestes* includes only this gregarious species, a large, broad-tailed, colorful sparrow of open areas with sandy substrates. Still abundant over most of its range, it has greatly decreased in the East with the reforestation that followed the decline of farming. This sparrow often flies relatively high during the day, a behavior unusual in many other species; the bright underparts and very high-pitched flight calls are conspicuous when birds pass overhead.

LARK BUNTING, *Calamospiza melanocorys*

The only member of the genus *Calamospiza* is a large, heavy, short-tailed sparrow with a large, almost grosbeak-like bill. The pattern of white in the wing coverts is unique and distinctive, but it can be overlooked on the folded wing of some females and young birds. Once the observer is moderately

The male Lark Bunting in alternate plumage is unlike any other sparrow, its spectacularly contrasting plumage well justifying the old English name "white-winged prairie blackbird." *Colorado, June. Brian E. Small*

The "mascara sparrow" is the only member of its genus; its closest relative appears to be the Lark Bunting, another sparrow of open grassy country. *California, October. Brian E. Small*

The *Spizella* sparrows, such as this alternate-plumaged Chipping Sparrow, are tiny, slender, long-billed sparrows with trilling songs. *Michigan, May. Brian E. Small*

The Grasshopper Sparrow is best appreciated at close range, where its colorful, intricately patterned plumage is obvious. *Ohio, May. Brian E. Small*

familiar with the species, the manner of flight—fast, erratic, and slightly twinkling—identifies flocks even from distances at which plumage characters are useless.

CHIPPING SPARROW, *Spizella passerina*
CLAY-COLORED SPARROW, *Spizella pallida*
BLACK-CHINNED SPARROW, *Spizella atrogularis*
WORTHEN SPARROW, *Spizella wortheni*
TIMBERLINE SPARROW, *Spizella taverneri*
BREWER SPARROW, *Spizella breweri*
FIELD SPARROW, *Spizella pusilla*

With the American Tree Sparrow removed to its own genus, the *Spizella* sparrows as now understood are tiny and slender, with long tails and fine, short bills; their backs are finely streaked black, the ground color richer chestnut in some species and grayer in others. The odd Black-chinned Sparrow is structurally like its congeners, but differs in its uniformly gray head and throat; the other *Spizella* resemble one another rather closely in plumage, generally differing more in the intensity than in the precise distribution of their markings.

GRASSHOPPER SPARROW, *Ammodramus savannarum*

Heavy-bodied and heavy-billed, this species and its South American congeners spend their entire lives in grassland, making them extremely vulnerable to habitat loss; indeed, the Florida subspecies of the Grasshopper Sparrow may be the most endangered passerine in the United States.

As in the case of the *Centronyx* and *Ammospiza*

sparrows, this species' generic allocation has been less than stable over the centuries. First named by Gmelin and Wilson as a *Fringilla*, the bird was subsequently placed in *Passerina* by Vieillot and in *Emberiza* by William Jardine and by Audubon. The path to the taxonomic future was laid out by William Swainson, who in 1827 described a new sparrow from southern Mexico as *Ammodramus bimaculatus*. Spencer Baird lumped Swainson's *bimaculatus* with the other subspecies of the Grasshopper Sparrow in 1858—and transferred them all to the genus *Coturniculus*, which Charles Bonaparte had erected 20 years earlier, with this as the genus's type species.

The species entered the AOU *Check-list* as an *Ammodramus,* but in 1903, following the lead of Robert Ridgway, the checklist committee restored *Coturniculus* to full generic status, assigning to it the Baird, Grasshopper, LeConte, Saltmarsh, and Nelson Sparrows. Six years later, the AOU synonymized *Coturniculus* with *Ammodramus*, retaining the Baird and Grasshopper Sparrows in the "new" genus but recognizing *Passerherbulus* for the Henslow, LeConte, Saltmarsh, Nelson, and Seaside Sparrows; because these decisions were made between the publication of the second edition of the *Check-list* in 1895 and of the third in 1910, the generic back-and-forth to which the Grasshopper Sparrow was subjected remained limited to the pages of *The Auk,* where all supplements were (and still are) published. The *Check-list* itself has always listed the species under the genus *Ammodramus*, the treatment adhered to today in other major lists as well. The most recent biochemical analyses have confirmed that *Ammodramus* is not closely related to the Baird, Henslow, LeConte, Seaside, Saltmarsh, or Nelson Sparrows, all of which have at one time or another been assigned to that genus and all of which cluster in most birders' minds as similar to the Grasshopper Sparrow. Instead, the nearest relatives of the Grasshopper Sparrow and its congeners, the tropical Yellow-browed and Grassland Sparrows, are the visually quite different *Arremonops* sparrows of tropical forest.

OLIVE SPARROW, *Arremonops rufivirgatus*

A member of a tropical genus of mid-sized, dull greenish sparrows, the Olive Sparrow inhabits dense thorn scrub, where its rattling song is more conspicuous than its bland olive upperparts and subtly marked head. As its genus name suggests, this species resembles a small, dull brush finch, in both appearance and retiring habits.

BRIDLED SPARROW, *Peucaea mystacalis*
BLACK-CHESTED SPARROW, *Peucaea humeralis*
STRIPE-HEADED SPARROW, *Peucaea ruficauda*
BACHMAN SPARROW, *Peucaea aestivalis*
CASSIN SPARROW, *Peucaea cassinii*
BOTTERI SPARROW, *Peucaea botterii*
RUFOUS-WINGED SPARROW, *Peucaea carpalis*

Until recently lumped with the *Aimophila* sparrows of the southern and southwestern portions of North America, the genus *Peucaea* is likely to be further subdivided as more genetic work is carried out on

The modestly plumed Olive Sparrow is the only member of its genus north of southern Mexico. *Texas, February. Brian E. Small*

these birds. Robert Ridgway pointed out well more than a century ago that *Aimophila* as it had come to be construed was an inconveniently miscellaneous assemblage of "something more than a dozen species (not counting subspecies), among which there are very great differences of coloration and considerable differences of form."

Over much of the twentieth century, ornithology wrestled with the *Aimophila* problem, attempting to

The Black-chested Sparrow is one of the most ornate species in the varied genus *Peucaea. Manuel Grosselet*

sort the species into groups of evolutionarily allied taxa and to create a scheme that would reflect the relationships among those groups. In 1955, Robert W. Storer identified two clusters of species, one inhabiting arid tropical scrub and the other inhabiting temperate grassland and savanna. Unfortunately, Storer found that four of the species traditionally assigned to *Aimophila* did not fit comfortably into either group, and he thus declined to recommend that the genus be split, preferring the status quo to any equally unsatisfactory replacement.

An ambitious study by Larry L. Wolf, published in 1977, identified three species groups (plus the Five-striped Sparrow, which was then still considered an *Aimophila*), "which had separate evolutionary histories and probably are not as closely related to each other as some earlier authors thought." Like Storer, Wolf concluded that radiation within each group had proceeded in a different ecological setting: the thorn forests of Middle America, pine-oak woodland in Middle America, and "weedy, open country of Middle America and [the] United States." He too stopped short of proposing a new generic classification, suspecting that one or the other of these groups might prove more closely related to other passerellid genera than to the remaining members of *Aimophila* and therefore unwilling to impose a new affiliation until those suspicions could be further tested.

More substantial progress came as morphological analysis yielded to molecular study in taxonomy. In 2003, Rebecca J. Carson and Greg S. Spicer analyzed three mitochondrial genes across 34 taxa of passerellid sparrows, producing several alternative phylogenetic trees that all placed the Rufous-crowned Sparrow nearer the brown towhees and the Cassin and Bachman Sparrows nearer the Grasshopper Sparrow, confirming the miscellaneous nature of the traditional *Aimophila* at least for those three species.

Six years later, another study, this one treating all 13 of the *Aimophila* species then recognized, affirmed that the Rufous-crowned, Rusty, and Oaxaca Sparrows were only distantly related to the other members of the genus as traditionally understood, and that South America's Tumbes and Stripe-capped Sparrows likewise stood apart. The remaining *Aimophila*—the Bachman, Cassin, Botteri, Black-chested, Bridled, Stripe-headed, Rufous-winged, and Cinnamon-tailed Sparrows—were found to clus-

With a resident range similar to that of the Five-striped Sparrow, the modest but charming Rufous-winged Sparrow is a sought-after species of the American Southwest. *Arizona, April. Brian E. Small*

ter neatly with the Grasshopper Sparrow and its South American relatives. DaCosta and his coauthors suggested that the name *Aimophila*, founded originally on the Rusty Sparrow, should apply only to the first three species, while *Rhynchospiza* should be revived for the Tumbes and Stripe-capped Sparrows. The clade gathering together the rest of the species would require a name of its own: *Peucaea*, erected by Audubon in 1839 for the Bachman Sparrow.

The American Ornithologists' Union endorsed the change in its 2010 Supplement to the *Check-list*, removing the Bachman Sparrow and its clade mates from the old, broad *Aimophila* and formally placing them in the resurrected *Peucaea*. As currently constituted, the genus *Peucaea* comprises birds that share medium size, round tails, and large bills, but otherwise range in plumage from the demure to the dazzling. All prefer dry habitats, whether open pine woods, oak scrub, or high desert; the breeding cycles of western species appear to be synchronized with precipitation patterns, such that at least the males of these otherwise furtive sparrows are most easily seen in high and late summer, when monsoon rains trigger territorial behavior and breeding.

SPECIES ACCOUNTS

A surprising vagrant away from its Middle American range, this Striped Sparrow remains the only member of its species ever detected north of Mexico. *Texas, January. Loch Kilpatrick*

STRIPED SPARROW
Oriturus superciliosus

This still relatively little-known sparrow enjoys the distinction of having been first described in one of the most startlingly titled works in ornithological history. Published in 1838 in London, William Swainson's *Animals in Menageries* introduced the reading public to some 150 of the mammals and birds judged best suited to being held in European zoos, from the Whistling Marmot to the now extinct Pink-headed Duck. At the end of the volume, Swainson appended scientific descriptions and drawings of 225 more birds, "either new, or hitherto imperfectly described." Among them was a species he named *Aimophila superciliosa*, the specific epithet calling attention to the conspicuous stripe above the eye, "very broad and cream coloured, beginning at the nostrils and passing beyond the ears," as depicted in Swainson's none too successful sketch. Swainson had visited the American tropics some years earlier, but he, and European science, knew this particular species only from a specimen in his own extensive collection.

The Striped Sparrow remained a scarce curiosity in European cabinets for much of the rest of the nineteenth century. Twenty-five years after Swainson first described it, Philip Lutley Sclater went so far as to deny that this sparrow existed as a separate species at all: in the printed catalog of his vast collection of ornithological specimens, Sclater lumped this with another large Mexican passerellid, the Rusty Sparrow—a reflection, perhaps, of the quality and state of preservation of the skins he was working with.

The turning point came in the late 1880s, when Frederick DuCane Godman and a team of collectors made a tour of southern Mexico and the Yucatán Peninsula. They found the Striped Sparrow "to be an exceedingly common species" from Mexico City south, where they collected both adult and juvenile examples. Walking in Godman's footsteps a decade later, Frank Michler Chapman observed that this sparrow was "a bird of much character in pose, and when excited mounts to the top of a bush, partly erects its tail and chirps vigorously," a behavior praised by Witmer Stone as making it "very easy to shoot." By the turn of the twentieth century, skins, skeletons, nests, and eggs were fairly well represented in American and European museums and private collections.

Any sparrow originally assigned to Swainson's ragbag genus *Aimophila* can be expected to have a complicated naming history. For the first 60 years from its description, though, the Striped Sparrow enjoyed relative nomenclatural stability—apart from the tendency of European authorities to spell the genus name *Haemophila*, a hypercorrection apparently inspired by the bizarre belief, still encountered today in the popular literature of bird names, that this group of sparrows was so called after a supposed fondness for blood (Greek *haima*) rather than for dense shrubbery (*aimos*). In 1870, George Robert Gray subsumed *Aimophila* under an even larger, even more miscellaneous genus, *Embernagra*, a lump that happily found no wider favor.

Not until 1898 did Robert Ridgway establish the need to remove the Striped Sparrow from its traditional genus. Ridgway admitted that this bird resembled other *Aimophila* sparrows, but its proportionally shorter tail and more rounded wing convinced him that it should be separated; he named the new genus *Plagiospiza*, the "unstraightforward sparrow."

There was general agreement even then that the genus *Aimophila* was excessively disparate, gathering together birds not all closely related to one another. Eventually, however, a more controversial question arose: had Ridgway needed to coin his new genus name, or was there an older name already available to replace the ill-suited *Aimophila*?

On the eve of World War II, A. J. van Rossem, as part of his ongoing effort to unravel some of the taxonomic mysteries of American ornithology, visited collections in France, Belgium, the Netherlands, Germany, and Britain. In Leiden, van Rossem was able to examine the specimen on which Charles Lucien Bonaparte in 1850 had based his genus *Oriturus*—and was able to confirm that that mounted bird was, as Bonaparte himself later recognized, in fact an adult Striped Sparrow. Thus, if the Striped Sparrow was not properly to be considered an *Aimophila*, its new genus name had necessarily to be Bonaparte's *Oriturus*, a name published nearly half a century before Ridgway's coining of *Plagiospiza*.

FIELD IDENTIFICATION

Round-bodied, long-tailed, and quite short-winged, this stout sparrow resembles closely no other species in its range; at a great distance and in a fleeting view, however, it has been mistaken for the much smaller, much more slender Chipping Sparrow, which shares a much weaker version of the Striped Sparrow's bold head pattern. At close range, the coarsely barred central tail feathers, rusty secondary panel, and rich back pattern of black stripes and pale scaling are visible. The adult's strong head pattern combines a rusty crown with a finely streaked median stripe and a broad whitish supercilium; the ear coverts are blackish, bordered above by a black eye line that is continuous with a black lore. These markings are made the more striking by the black bill and the absence of a conspicuous lateral throat stripe, creating a first impression of front-heaviness and incompleteness.

Juveniles are similar, but with less pronounced head markings and paler, less rusty upperparts and secondaries. The dull gray underparts may have a

This large sparrow's bold markings are distinctive in its range. *David Krueper*

slight yellowish wash, and are narrowly and irregularly streaked dusky across the breast.

This is usually the only sparrow found in its high-elevation habitat of bunchgrass meadows and Sierra Madrean forest openings between 6,000 and 14,000 feet. It is often encountered in flocks of five to ten birds. When disturbed, Striped Sparrows often fly up to perch on fences or dirt piles, where they call loudly and repeatedly.

The calls are bright and sharp, with a strong attack and rapid decay, *TEEek*, often given in a long series that may accelerate into a buzzing stutter. The brief but loud song begins with similar ticking notes, followed by one to four abrupt piping whistles and a loose, rattling trill, *teekteek breeb breeb drdrdrdr*.

RANGE AND GEOGRAPHIC VARIATION

The Striped Sparrow breeds only in Mexico, from extreme eastern Sonora and the Chihuahua mountains south to Veracruz and Oaxaca. Although the species is presumably sedentary or nearly so in its range, an unexpected stray appeared in January 2015 in south-central Texas. This modestly plumed sparrow seems an unlikely cage bird, and the Texas bird may have been a genuine vagrant from the south.

The southern subspecies *superciliosus* is darker overall, with a less pronounced reddish tone to the upperparts than northern *palliatus*. The central tail feathers of *palliatus* are gray, while in *superciliosus* the outer vanes of those feathers are olive-brown. Striped Sparrows from Zacatecas, San Luis Potosí, Jalisco, Nayarit, and Durango are intermediate in plumage.

SONG SPARROW
Melospiza melodia

This Song Sparrow, of the subspecies *merrilli*, is typical of birds breeding in the northern inland portions of the species' range. It is more reddish than the grayer subspecies of the East. *British Columbia, June. Brian E. Small*

Like so many other common and widespread American birds, this familiar species appears to have been discovered, depicted, and named over and over in the early days of natural history in the New World. Today, the earliest scientific description is credited to Alexander Wilson, the "father of American ornithology," who in 1810 published a good engraving—the specimen he illustrated most likely still exists—and two pages of prose treating this species, "the most numerous, the most diffused over the United States, and by far the earliest, sweetest, and most lasting songster" of all the sparrows known to him. The following year, Wilson demonstrated quantitatively the abundance of this species when on "little more than eight acres" belonging to his patron and friend William Bartram he tallied no fewer than five nesting pairs of Song Sparrows.

Living and working in Philadelphia, one of the cradles of American science, Wilson had excellent libraries at his disposal, but even so, and even after having given the sparrow a name, he warned the reader that

> So nearly do many species of our Sparrows approximate to each other in plumage, and so imperfectly have they been taken notice of, that it is absolutely impossible to say, with certainty, whether the present species has ever been described or not.

Wilson was right to worry. Almost a hundred years earlier, Mark Catesby had painted a bird that he called the Little Sparrow. Catesby's illustration is hardly more eloquent than the name, showing a small, indistinctly streaked and spotted sparrow with a long, broad tail and dark bill and feet. The brief accompanying text, though, offers several suggestive details: these birds are "usually seen single, hopping under Bushes," are "most common near Houses," and "breed and abide" all year in Virginia and the Carolinas. It seems more likely than not that Catesby was already describing the bird that Wilson would eventually name.

Two years before Wilson's birth, in 1764, the very young William Bartram had sent a sparrow skin to the British ornithologist and engraver George Edwards. Edwards, a friend and sometime disciple of Catesby, prepared a thorough description of what he, too, called the Little Sparrow, a bird clearly recognizable in his excellent engraving as a Song Sparrow. Bartram can thus be accounted the "second discoverer" of the species, following Catesby; but oddly, he seems never to have encountered another—or, more likely, not to have recognized it as different from any other brown bird—at his Philadelphia home or on his extensive journeys through the American Southeast in the 1770s.

Edwards's illustrated works were an important source for the next generation of ornithologists, and his Little Sparrow was duly recorded in both John Latham's *General Synopsis of Birds* in 1783 and Thomas Pennant's *Arctic Zoology* the year following. In fact, Edwards's sparrow appears twice in both those works, under two different names, as the Ferruginous Finch and the Fasciated Finch. Both authors had access to new specimens held in the collections of Anna Blackburne, sent to her by her brother, Ashton Blackburne, from his travels in New Jersey, New York, and Connecticut. Pennant's description of the Ferruginous Finch includes the first published transcription of any of the Song Sparrow's vocalizations—not, however, of the song, but of the familiar husky chirp, *shep, shep,* whence the bird is "called in New York, the Shepherd."

Catesby, Latham, and Pennant all contented themselves with assigning the "new" Blackburnian sparrows an English name, leaving it to Johann Friedrich Gmelin to translate their names into scientific binomials as *Fringilla fasciata* and *Fringilla ferruginea*. Alexander Wilson knew Gmelin's list in its English translation of 1802, but in identifying his new sparrow, Wilson adduced only Pennant's Fasciated Finch, which he marked with a query; no doubt encouraged, too, by the omission of any mention of a similar bird in Bartram's 1791 *Travels*, he felt himself free to give the species a label of his own selection. Wilson named the sparrow, in English and in scientific Latin, for its song, "short but very sweet, resembling the beginning of the Canary's song, and frequently repeated, generally from the branches of a bush or small tree, where it sits chanting for an hour together." Such careful attention to a bird's vocalizations was still relatively rare in the ornithology of Wilson's day, making the names he assigned the species that much more notable: the Song Sparrow, *Fringilla melodia*, the "melodious finch."

Thanks to its adoption by Charles Lucien Bonaparte and by John James Audubon in the 1820s and 1830s, Wilson's scientific name was the standard for much of the rest of the century. As early as 1840, however, Thomas Nuttall, in the second edition of his influential *Manual*, determined that Wilson's Song Sparrow and the Fasciated and Ferruginous Finches were all one and the same bird, and while

In eastern birds, the gray head and back are heavily marked with gray-brown to dull reddish. *Ohio, June. Brian E. Small*

Along with the long tail and plump body, the very large, wedge-shaped lateral throat stripe is an important field character. This individual is of the small subspecies *cooperi. California, April. Brian E. Small*

he retained Wilson's felicitous English name, he changed the scientific name—back—to Gmelin's *Fringilla fasciata*. Nuttall offered no explanation, but in 1876, the Illinois ornithologist D. W. Scott set forth a vigorous argument that Gmelin's *fasciata* referred to the same species as the Wilsonian *melodia*, and that as *fasciata* antedated Wilson's by 22 years, it was, as Scott pointed out in his essay's tendentious title, "the proper specific name of the Song Sparrow." When both Robert Ridgway and Elliott Coues simultaneously pronounced their agreement with Scott, the matter was settled: *fasciata*, a reference to the fault bars shown by the tail feathers of poorly nourished birds, was to be the official name.

Accordingly, and with the removal of the species from *Fringilla*, the first two editions of the AOU *Check-list* used the name *Melospiza fasciata*. Shortly after publication of the second, however, Harry C. Oberholser determined that the name *fasciata* had been coined for another bird, the Pine Siskin, in 1776, a dozen years before Gmelin applied it to the sparrow, making it "necessary to revert to the long-familiar name of Wilson as the tenable specific designation of the Song Sparrow."

Two years after this restoration of *melodia*, the "specific designation" was changed again, but for reasons of a different kind. On the very same page on which he named *fasciata*, Gmelin in 1788 had described another bird, *Fringilla cinerea*, on the basis of specimens taken on Unalaska Island and shipped to Sir Joseph Banks, president of the Royal Society of London. In 1903, the AOU concluded that Gmelin's Aleutian sparrow was of the same species as the more familiar Song Sparrows to the east and south. Alexander Wilson's name of 1810 fell once more to the force of priority, banished to the ever-longer list of synonyms applied to the brown bird with the cheerful song.

The AOU's churning bibliographic enterprise continued unabated in the years leading to the appearance of the third edition of the *Check-list*, and updates continued to appear nearly up to the day of its publication. The Supplement of 1908 once again dealt with the scientific name of the Song Sparrow, announcing tersely that

> the Song Sparrows again become . . . *Melospiza melodia*, by reason of the preoccupation of *Fringilla cinerea* Gmelin.

Just as *fasciata* was already spoken for, so too had *cinerea* already been applied to a different species, in this case an Abyssinian weaver. The new *Check-list*, issued in 1910, left unmentioned any of the alternative names and their tangled relationships, declaring Wilson's *melodia* the only correct choice.

Nearly 120 years later, it appears that that name

Except in northwestern and Pacific island populations, the dark bill is relatively small; the base can appear swollen at some angles. *Maine, June. Brian E. Small*

has at last vanquished its rivals. There is one remaining contender, however. Gmelin's sources for his *Fringilla ferruginea*, Latham and Pennant, clearly—as clearly as possible in a century when so many of America's birds were still new—describe their Ferruginous Finches as Song Sparrows. Pennant's transcription of the call, quoted above, is consistent with only the hoarse, low-pitched chirps of a Song Sparrow. Latham's brief but more detailed account provides a description of the bird's size, its bill color, and its plumage details, all of which fit the Song Sparrow; he also refers explicitly to Edwards's earlier engraving of the Little Sparrow, an image in turn assignable only to the Song Sparrow.

This historical evidence notwithstanding, *ferruginea* is now treated as an invalid name for the Red Fox Sparrow, an error ultimately traceable, ironically or not, to none other than Alexander Wilson, who misapplied Gmelin's name to that species for the first time in 1811. But if the Ferruginous Finch was in fact a Song Sparrow, and if the weight of Wilsonian tradition can be thrown off, then *melodia* might once again be destined for the list of merely alternative scientific names, trumped one more time by the principle of priority.

Happily, the generic allocation of this species has been considerably more stable. For the first two and a half centuries of modern scientific ornithology, virtually any small bird with a thick-based bill was considered a "finch" and assigned to the enormous, and enormously varied, genus *Fringilla*. Even as late as Audubon's *Synopsis* of 1839, all of the passerellid sparrows known to him, with the exception of the

This Song Sparrow, of the large, sooty subspecies *kenaiensis*, shows the uniform gray ground color and heavy, blurry streaking characteristic of northwestern and island types. The bill is very long and sharply pointed, almost blackbird-like. *Alaska, March. Brian E. Small*

colorful towhees and the Lark Bunting, were still classified as *Fringilla*. It was about that time that the unpacking of the old catch-all genera finally began in earnest, as the full diversity of the American avifauna was gradually unrolled.

Charles Lucien Bonaparte assigned the Song Sparrow to *Zonotrichia* in 1838, a practice followed until the publication of Spencer Baird's monumental *Birds* 20 years later. Baird separated the Song, Swamp, and Lincoln Sparrows and placed them in their own genus, differing from the *Zonotrichia* sparrows in tail and wing shape and in the plumage patterns of the underparts and crown. He named the new genus *Melospiza*, a straightforward translation of Wilson's English name.

This removal of the Song Sparrow from *Zonotrichia* faced its most significant challenge in 1964, when Raymond A. Paynter argued forcefully that Baird's criteria and the others adduced in the intervening hundred years in support of the distinction "all fail to separate the genera"; tail and wing shape are "useless," underparts streaking variable, head patterns difficult to assess, and the condition of the squamosal region of the skull "probably . . . of no taxonomic value at the generic level." Accordingly, Paynter merged *Melospiza* back into *Zonotrichia*, a genus concept followed, naturally, in the volumes he prepared for the world checklist begun by James L. Peters. Paynter's view was not widely accepted in the ornithological and birding literature, which would retain the distinct genus *Melospiza*, but it was adopted for use in many, even most, museum

collections, such that even today specimens of the Song Sparrow are catalogued and stored under the enlarged genus *Zonotrichia*.

Whatever the "correct" name of the Song Sparrow, the nomenclatural travails of this common bird over the centuries remind birders once again that we stand in a tradition as rich as it is sometimes complicated.

FIELD IDENTIFICATION

Common and familiar over most of its wide range, this species nevertheless poses an identification challenge for many observers. The vast span of geographic variation is an important element to the challenge, as is the age- and season-related plumage variation shown by a single individual. Furthermore, several of the classic "confusion species" are locally uncommon or shy, sometimes making it difficult to gain the familiarity necessary to confidently rule out those similar species.

The traveling birder is often startled by how unSong-Sparrow-like the Song Sparrows of a new region can seem. The most important way to forestall confusion is simply to remember that such plumage variation exists and to resist the temptation to assign a confusing sparrow to a different—any different—species simply because it does not look like the Song Sparrows "back home."

That temptation can be particularly grave for observers who rely on such hoary field characters as color, underpart streaking, or central breast spots to identify sparrows. Instead, like other birds, Song

Birds from farther south in the Pacific Northwest are very dark and rusty, with coarse brown streaking below. This winter bird may be a member of the subspecies *morphna*, the most widespread of the rusty Song Sparrows. *Washington, January. Brian E. Small*

Sparrows, in whatever part of their range, belonging to whatever subspecies, whatever their size and color tones, should be initially identified using a combination of structural, behavioral, and habitat clues, with plumage color and markings serving only as confirmation.

All Song Sparrows are fairly long-tailed; the folded tail is neither strikingly narrow nor noticeably broad, and usually appears evenly square-tipped or slightly notched on perched birds. Feeding birds may hold the tail still and just above the horizontal, or they may raise it high above the back; that behavior appears for whatever reason to be more frequent in western and southwestern populations. Active or nervous birds flip the tail at irregular intervals, usually with a decided sideways movement, but the tail is rarely flirted or fanned when the bird is on the ground.

The wing is short and rounded, the primaries of the folded wing extending only slightly beyond the tertials; the primary extension is significantly shorter than the longest tertial. In spite of the fairly long tail, most Song Sparrows make a stocky impression, perched and in flight; the largest races may appear somewhat lankier. Because of the short wing and full breast and belly, the body may appear almost spherical on relaxed birds with feathers fluffed. The head is large, with a high crown and dis-

tinct forehead; a short crest is occasionally raised at the front of the crown.

The seasonal variation in the plumage aspect of many Song Sparrows is underappreciated. This species has only one molt a year, taking place in late summer or early autumn once breeding activities have concluded. As a result, many individuals, especially first-cycle birds still wearing their juvenile wing and tail feathers, are badly worn by July and August, with many feathers abraded or even visibly broken.

Feathers are abraded by contact with branches, grass, soil, and other elements of the physical environment; their quality is also degraded by exposure to sunlight. Dark feathers and parts of feathers are more resistant to such wear than are pale feathers or parts of feathers: melanin, the pigment that provides the color of dark structures, also physically strengthens the feather. In those Song Sparrow subspecies with broad paler edgings to the feathers of the back, this differential wear reduces the edgings, leaving the brown, chestnut, or black markings intact and making them more prominent. This effect can be especially noticeable in several of the most widespread Song Sparrow subspecies, including *melodia* and *atlantica* of eastern North America, with their buff and pale gray upperpart edgings, respectively, and *montana* and *merrilli* of the West, with their gray feather edges. In worn plumage, birds of these subspecies are significantly and noticeably darker above than in fresh feather, when the brown and black of the upperparts are relieved by brighter highlights.

Observers can also be misled by the rather different plumages of juvenile Song Sparrows. The juvenile plumage—the feather dress worn from the time of fledging to the time of the preformative molt—is much less neatly and regularly marked than that of adults. This is in part due to the lax texture of juvenile body feathers, but results mostly from the fact that most birds do not "invest" the energy required to produce large amounts of melanin in regular patterns in feathers that will be held for just weeks; thus, the dark markings of juvenile Song Sparrows are paler and more diffuse than those of adults, combining with the feathers' loose structure to create a scruffy, messy appearance quite unlike the textbook neat streaks and blotchy breast spot observers might be expecting. In family groups in summer, adults and young can often be immediately distinguished by the darkness or paleness of the overall plumage. Juveniles unaccompanied by a parent can be more confusing; again, structure, size, and behavior offer the best guidance.

Among the species commonly confused with Song Sparrows, the Savannah Sparrow is the most abundant and the most widespread. It is also one of the most geographically and individually variable of sparrows, such that even experienced observers can find themselves faced with a bird that does not match their mental image of "the" Savannah. Newer sparrow watchers often focus on traditional field marks such as crown pattern or supercilium color that are difficult to see and difficult to assess without a fairly complete knowledge of the full range of variation in both species.

Fortunately, structure and behavior separate Song and Savannah Sparrows almost immediately. There is some overlap in size, with large Savannah Sparrows noticeably larger than small Song Sparrows. Starting at the rear of an unidentified sparrow, however, the observer notices immediately that where the Song Sparrow has a long, more or less even-tipped tail, all Savannah Sparrows have a noticeably short tail, deeply notched and with obviously attenuated feather tips. Dragged through the imaginary dust, a Savannah Sparrow would trace two sharp lines; a Song Sparrow would scoop a shallow channel. The most widespread Savannah Sparrow populations have distinct dull whitish edges to the tail feathers as well, visible in flight and on the ground, another feature distinguishing them from Song Sparrows.

The extent and brightness of the feather edges on the back and lower back of Song Sparrows vary

The small, almost completely dark bill and uniform upperparts, without strong contrast between the back and head or the back and tail, distinguish the Song Sparrow of rusty western populations from the fox sparrows. *Washington, January. Brian E. Small*

with subspecies and season, but in no population are those edgings white. In the Savannah Sparrows found over most of the continent, the back feathers are conspicuously edged a very pale grayish white, and those edges often line up to form neat white tracks down the upper back, especially conspicuous in fresh autumn birds.

The ground color of the underparts can also be helpful in distinguishing these two species. Even the creamiest-bellied of Song Sparrows does not appear as harshly bright white below as most Savannah Sparrows in the field. The breast streaking is highly variable in both species, and in both it may or may not coalesce into a central breast spot. The streaks are usually finer and neater in the Savannah Sparrow, with the exception of the otherwise very distinctive subspecies known as the Ipswich Sparrow.

The most telling characteristic of the Song Sparrow's underparts pattern is the lateral throat stripe: large, rich brown-black, and wedge-shaped on Song Sparrows, and paler, narrower, and less strikingly broadened at the base in most Savannah Sparrows.

In ideal conditions, the fine details of head pattern can also be used to distinguish these two species, but the size and color of the bill are easier to see and to assess. Savannah Sparrows have small, dull straw-colored bills, with a straight culmen and neat sharp tip; Song Sparrows of whatever race have larger, blackish bills, with a slight but noticeable curve to the culmen and a blunt bill tip. The frequent impression in the field is that an aggressive

The juvenile Song Sparrow is buffier and more coarsely marked than the adult. This Colorado bird is of the subspecies *montana*, a widespread brown Song Sparrow of the interior West. *Colorado, June. Cathy Sheeter*

Savannah Sparrow would inflict a puncture wound, while a Song Sparrow would chew.

Experienced observers usually distinguish these two species by behavior. Song Sparrows are not as obsessively gregarious as Savannah Sparrows, which can occur in loose flocks of several dozen outside of the breeding season; even when large numbers of Song Sparrows are present at a single site, as on a day of heavy southbound migration, individuals are obviously scattered through the vegetation. When flushed from the ground, Song Sparrows retreat into the nearest dense vegetation, while Savannah Sparrows fly off in low swooping flight to a distant fence line or clump of grass. Savannah Sparrows often call when flushed, a distinctively high, thin, short lisp; Song Sparrows tend to flush silently—but are on average more vocal when perched.

Another streaked brown sparrow of open country, the romantically named Vesper Sparrow is often confused with the Song Sparrow, especially in the easternmost portions of its range, where the species is generally uncommon and often unfamiliar.

Potential confusion can be resolved immediately by a good view of the tail. While Song Sparrows have an even-width tail, that of the Vesper Sparrow often gives the impression in the field of being slightly compressed at the base and broader toward the tip; that tip is usually clearly notched, whereas that of a Song Sparrow is variably round, square, or quite shallowly notched. The patterns of the tail differ as well: the Song Sparrow's rectrices are uni-

formly brown, but the outermost tail feather of the Vesper Sparrow is almost entirely white on the outer web and largely white on the inner, which is marked with a variably broad blackish stripe. These white areas, appearing as a narrow, well-defined white edge to the tail, are visible from above only when the tail is spread in flight or in preening; from below, the Vesper Sparrow's white is visible even on the closely folded tail, but it is important to be aware of the effects of light and angle: if the remainder of the tail feathers show through the white, even a Vesper Sparrow can look dark-tailed, and reflections from snow or water can give the undertail of a Song Sparrow a misleadingly silvery sheen.

The Vesper Sparrow's subtle wing pattern is equally distinctive. In fresh plumage, the bland gray-brown greater coverts have dull, diffuse buffy tips, creating a fairly broad lower wing bar; the median coverts are black with white tips, creating a neat upperwing bar. Both bars are subject to heavy wear, by late summer often leaving a relatively plain wing relieved only by the black bar of the median coverts. Song Sparrows lack such conspicuous wing bars at any season; in most subspecies, the remiges are darker rust than the grayer flight feathers of Vesper Sparrows, and the greater coverts are marked with larger black "teardrops."

The underparts of most Vesper Sparrows are streaked with dull brown on the breast and flanks; the streaking usually appears fine and somewhat irregular, and occasionally seems to coalesce into a poorly defined blotch at the center of the breast. The ground color of the underparts is often decidedly creamy, resulting in relatively low color contrast between the upperparts and underparts of many individuals. Some juvenile Song Sparrows are superficially similar, but most are whiter beneath, with heavier, more regular streaking that extends onto the belly, a feather region usually only slightly marked, if at all, on Vesper Sparrows.

Vesper Sparrows usually avoid the damp, dense thickets preferred by Song Sparrows in favor of open grasslands, deserts, beaches, and agricultural fields. Birds feeding on the ground can be approached very closely; once flushed, they may fly into an isolated bush or low tree or land on a fence, but unlike Song Sparrows, only exceptionally will they seek refuge in heavy vegetation. The flight call, less frequently given by Vesper than by Savannah Sparrows, is low-pitched but short.

As might be expected, the two species with which the Song Sparrow shares the genus *Melospiza* are also frequently confused with their congener. All three—the Song, Swamp, and Lincoln Sparrows—share the preference for deep, shady habitats and a heavy build, long tail, and uniform wing coloration, but adults are readily distinguished. The chestnut-red wings, dark gray face, and bright white throat of the Swamp Sparrow identifies even those birds—most of them probably in their first plumage cycle—with heavy streaking below; post-juvenile Lincoln Sparrows are equally distinctive, with very fine shaft streaks on the feathers above and below and an odd and beautiful yellowish wash on the breast and jaw stripe. Both species are shorter-winged than the Song Sparrow, a difference sometimes perceptible in the field, and both have longer, more slender bills. In mixed autumn flocks, the distinctive buzzing *tzeer* of both Swamp and Lincoln Sparrows, often reminiscent of the flight call of a *Passerina* bunting, is a good way to pick these slightly less common birds out from what are usually the hordes of Song Sparrows.

Juvenile *Melospiza,* on the other hand, pose some of the most badly underappreciated identification challenges of any passerellid sparrows. Fortunately for the human observer, that plumage is only briefly held; both Lincoln and Swamp Sparrows complete their preformative molt in the breeding range, before the southbound migration, as do most Song Sparrows. Thus, most juvenile *Melospiza* sparrows are likely to be seen in the company of adults of the same species, most often in typical nesting habitat: Lincoln Sparrows nest in willow-choked ponds and bogs, Swamp Sparrows in extensive cattail marshes; Song Sparrows may occur in both those habitat types, but except for California's marsh-dwelling subspecies, tend to prefer slightly drier substrates.

Structurally, juvenile Song Sparrows with fully grown tails have longer tails and shorter, more rounded wings than either Lincoln or Swamp Sparrows; bill length and shape are also suggestive, though it must be recalled that the colorful, fleshy "flanges" at the gape of juvenile birds' bills can make those characters more difficult to assess. The most useful plumage marks include the color and pattern of the crown, the throat markings, and the width of the lateral throat stripe. Juvenile Song Sparrows have a brown, faintly streaked crown with or without a gray median stripe, the throat is barely or not at all streaked, and the lateral throat stripe is broad and contrasting. Lincoln Sparrow juveniles are very similar but have clear dark streaking on the crown and throat; the lateral throat stripe in both that species and the Swamp Sparrow is thin and poorly defined. The juvenile Swamp Sparrow's crown is blackish, without a median stripe, and its throat is unstreaked; in both of those characters, juvenile Swamp Sparrows are more similar to their parents than is the case for either of the other *Melospiza* species.

The fox sparrows are nearly as variable in appearance as the Song Sparrow. The Red Fox Sparrow, the widespread northern and eastern representative of the complex, is a large, fat, heavy-billed bird with distinctively rusty red tail, rump, and streaking above and below. Though it does not actually resemble any of the Song Sparrows, the Red Fox Sparrow seems to be the desperate default identification for inexperienced observers faced with a Song Sparrow different from "their" local birds. The large size, yellow lower mandible, and pale gray, rusty-marked head should separate this fox sparrow from all Song Sparrows.

A greater challenge is posed by the large, dark

Juveniles of the dark, reddish northwestern subspecies can be very dark rusty overall. *British Columbia, June. Cathy Sheeter*

Song Sparrows of Alaska and the Pacific Northwest, which overlap with several populations of Sooty Fox Sparrows. Those birds differ from even the biggest and darkest of sympatric Song Sparrows in their still larger size, nearly or completely unstreaked backs, and solidly dark head and neck with no or only obscure markings; they also lack the thick, dark lateral throat stripe of the Song Sparrows. The underparts of Sooty Fox Sparrows often show a "vested" effect, with the streaking heaviest across the breast, in some forming a nearly solid blackish patch.

One easily overlooked source of potential confusion is the endangered and very locally distributed Sierra Madre Sparrow, aptly said by its original describer to resemble both a *Melospiza* and a *Passerherbulus* sparrow. The Sierra Madre Sparrow is shorter-tailed and shorter-winged than Song and Lincoln Sparrows, and differs further from both in its small, slender bill and colorful upperparts, the chestnut feathers of back, rump, and scapulars with large black spots and conspicuous buffy edges. The Sierra Madre Sparrow also has a yellow wing flash, lacking in the *Melospiza* sparrows.

RANGE AND GEOGRAPHIC VARIATION

One of the most widespread birds in the Americas, the Song Sparrow breeds from the westernmost Aleutian Islands across the boreal forest region to Newfoundland and Labrador, south to the mid-Atlantic Coast and the central Great Plains in the East and to Sonora and southern California in the West; there are disjunct resident populations in western and central Mexico. Northern mainland breeders move south to winter across northern Mexico and nearly all of the United States, with the exception of southernmost Texas. Within this vast range, only the driest, most sparsely vegetated deserts and alpine areas, the deepest forests, and the most urban of human landscapes are without Song Sparrows at one time of the year or the other.

With their short, relatively rounded wings, those Song Sparrow populations that are not entirely sedentary are only short- to mid-distance migrants. Nevertheless, individuals have overshot their usual wintering destinations to appear in the Bahamas and on Hispaniola, and have even crossed the North Atlantic, apparently under their own power, reaching Norway, Great Britain, and the Low Countries. The North Sea records are from late spring and early summer, an unexpected season for transatlantic vagrants. Presumably at least some of those strays wintered quietly after arriving in the preceding autumn, and remained undetected until the urges of spring made themselves known. Europe's first, discovered and netted on Fair Isle, Scotland, at the end of April 1957, mounted a low perch and sang for a full week after it was banded.

Just as surprisingly, individuals of at least one of the large Song Sparrow races of the Aleutians, *maxima, insignis,* and *sanaka*, have flown west to land in the Russian Far East and Japan. Those island races are normally entirely sedentary.

Perhaps the most unusual extralimital record of any Song Sparrow is that of a single *rufina* from southeastern Arizona in January 1963. The distance involved in wandering from the moist coastal Pacific northwest to the arid Sonoran desert is noteworthy in itself, but what is most remarkable about this individual is the circumstance of its eventual capture, a story tersely summed up by Allan Phillips in the memorable phrase "Marshall, sling-shot, ornithology class, and faithful swimming dog."

One of the best pieces of advice ever given to the new birder is to learn the local birds first and to learn the local birds well. Helpful as it is in most cases, that directive can be of little use when applied to many observers' experience of this species: no matter how conscientiously they have studied the Song Sparrows of their backyards and parks, they can still be stymied by the encounter with a member of the same species from a different part of its range.

Over the years, this familiar bird has been split into more than 50 subspecies, several originally described as species themselves, and some two dozen of which are still recognized by most authorities. Ranging from small to extraordinarily large, from obscurely blotched to delicately streaked, from dark to pale, and from rusty to chocolate to slaty, this is one of the most geographically diverse birds in the world. No less systematic a cataloguer than the systematist Robert Ridgway admitted to "great difficulty . . . in the attempt to characterize satisfactorily the different subspecies" that have arisen from this sparrow's "plasticity of organization." If Ridgway found diagnosis challenging in the museum drawers, how much more difficult it is in the field, above all in the nonbreeding period, when birds from two or more discrete breeding populations may be seen in the same area.

The 1957 edition of the AOU *Check-list*, the last to provide complete entries for subspecies, tallied 31 subspecies of the Song Sparrow north of Mexico. Peter Pyle's *Identification Guide*, the best modern guide to the sexing and aging of United States and Canadian birds in the hand, cites 29, with 10 others in Mexico. Most recently, the authors of the species account in *Birds of North America* list 24 over the species' entire range, as does the authoritative and regularly updated world checklist curated by the International Ornithological Congress.

In 2009, Michael A. Patten and Christine L. Pruett undertook a thorough revision of the subspecific taxonomy of this species. Their examination of more than 5,000 specimens from throughout the Song Sparrow's entire vast range produced a list of 25 diagnosable subspecies, differing from the *Birds of North America* and IOC lists by the retention of the race *zacapu,* resident in an extremely limited range in Michocoán. This guide follows the taxonomy set forth by Patten and Pruett, and the subspecies descriptions below depend heavily on their diagnoses and key in combination with the descriptions in Pyle, Ridgway, and *Birds of North America.*

Given the migratory habit of most Song Sparrow populations, field identifications to subspecies outside the breeding season can be tentative at best. Furthermore, all but the island subspecies must be expected to intergrade with adjacent populations, producing intermediate forms unassignable to a single subspecies even on the nesting grounds and even in the hand. Only breeding individuals encountered in the heart of a subspecies' known summer range, or distinctively plumaged insular birds known to be resident, can be more or less safely assigned to race.

The following overview is arranged in subspecies groups, gathering different subspecies on the bases of visual similarity and geographic proximity. Group names should be used whenever an individual bird cannot be identified to subspecies with complete certainty—in other words, nearly all the time. The scheme here recognizes the subspecies recognized by Patten and Pruett. It is almost certain that a system based on "true" genetic relatedness rather than on human-perceived affinity would differ from the arrangement that follows.

The English group names for the various Song Sparrows set forth in Pyle, Patten and Pruett, *Birds of North America*, and elsewhere are based on geography. As useful as such a nomenclature is for some people, those labels are neither memorable nor always convenient for the field observer, and they are replaced here with more visually descriptive English group names, some available from their use in older literature and most of them already current among North American birders.

"CHOCOLATE" SONG SPARROWS

The nominate subspecies, *Melospiza melodia melodia,* has a spotted throat, distinct black-brown streaks on fairly clear white underparts, and reddish brown back feathers with broad black shaft streaks; the lower back is clearly streaked black. The supercilium is white, the lateral throat stripe deep rusty brown. The bill is stout but not obviously "outsized."

The race breeding on the mid-Atlantic Coast, *atlantica,* is slightly larger-billed and slightly less distinctly marked above and below. The ground color of the back is grayer, and the dark shaft streaks of the back feathers are edged with gray rather than brown.

The third subspecies in this group, *montana,* is similarly proportioned to *melodia,* but the wing is longer and the bill slightly more slender. It is paler and grayer above, with gray edges to the back feathers and less crisply black, browner-toned streaking below.

Three Middle American subspecies can also be grouped among these dark brown birds. A permanent resident just to the north of Mexico's Transvolcanic Belt, *mexicana* is a slender-billed, brown-backed bird with attenuated extremities: the wing, the tarsus, and the bill are all proportionally longer than in the similarly colored northern races.

Apparently restricted to a small part of western Durango, *goldmani* is larger than all the subspecies in this group but the next. It is dark above, with reduced and diffuse blackish streaking on a dark brown ground color. The diffuse reddish streaking of the underparts shows low contrast with the dull buffy-white ground color; Patten and Pruett note its overall similarity to *morphna.*

The largest and darkest of the southern chocolate Song Sparrows is *villai,* described in 1957 from the state of Mexico. The upperparts are sooty brown, with little contrast between the ground color and the streaking; the flanks show a gray tinge.

The chocolate Song Sparrows, richly colored, neatly marked, and "normal" in size, are the familiar Song Sparrows over most of the continent. Taken together, they breed from the mountains of eastern Oregon to Newfoundland, south to northeastern Arizona, North Carolina, and the southern edge of the Nearctic region. In winter, *atlantica* moves south along the coast as far as Georgia, while *montana* and *melodia* spread into the southern tier of the United States in the West and the East, respectively. The Mexican subspecies are resident within their ranges.

"SOOTY" SONG SPARROWS

This group includes large to very large Song Sparrows, with notably dark plumage above and relatively poorly defined streaking below; the overall color impressions of the various subspecies included here range from dull gray-brown to gray to somber dark gray. The bill size ranges from large to extraordinarily large and thick in some island populations.

The two largest subspecies of all, *sanaka* and *maxima,* have wings and tails up to 30 mm longer than those of the smallest "olive" Song Sparrows (described below). Their bills are large and long; the bill of *sanaka* is proportionally more slender but still stouter than the noticeably longer, thinner bill of the slightly smaller *insignis.*

All three are gray above with streaks and dull grayish below. *Melospiza melodia maxima* and *sanaka* have sparse sooty streaking on the back, darker on the lower back; *sanaka* is generally grayer. *M. m. insignis* is on average darker above and less streaked. The blurry streaking of the underparts is reddish in *maxima* and *sanaka*, colder brown in *insignis*.

The remaining three races in this group are smaller, though still large compared to other Song Sparrows. The darkest, *kenaiensis*, is dark gray above with darker gray streaks, browner on the lower back; the breast streaking is dull dark gray-brown against a gray ground color.

The upperparts of the longer- and thinner-billed *caurina* are somewhat paler, with warmer brown streaking, and the breast streaks are more reddish than the colder brown of *kenaiensis*. The supercilium and jaw stripe are gray.

The smallest-billed member of this group is *rufina*, dark gray above with deep brown streaks and off-gray underparts with low-contrast sooty brown markings. This subspecies is smaller and more reddish than the otherwise similar *caurina*.

The sooty Song Sparrows breed in Alaska and the Pacific Northwest, south to British Columbia's Haida Gwaii. The Aleutian Island populations *maxima* and *sanaka* are resident in their breeding ranges, making those the easiest of the Song Sparrow subspecies to identify even apart from their extreme bulk and heavy bills. Some members of the race *insignis* leave their breeding areas in the Kodiaks to winter farther south along the Alaskan coast.

The remaining three, *kenaiensis*, *caurina*, and *rufina*, breed, respectively, on Alaska's central coast, Alaska's southern coast, and the islands of southeastern Alaska and northern British Columbia. All three migrate to winter as far south as the Washington coast, with the nonbreeding range of *caurina* and possibly of *rufina* extending into northern coastal California.

The overall darkness of plumage shown by the subspecies in this group appears to be due to a combination of adaptive characters. Birds inhabiting humid environments are often more darkly colored than their dryland counterparts, a tendency codified as Gloger's rule; comparative study of two Song Sparrow populations, one from the northern Pacific Coast and the other from the southwestern deserts, suggests that the correlation between humidity and feather darkness has evolved in response to the much greater abundance of feather-degrading bacteria in humid habitats. Increased production of melanin, the pigment that makes dark feathers dark, is also associated with increased territoriality and aggressive behavior in birds—such as the Song Sparrows of the Aleutians—breeding on islands, where competition for resources is intense.

"RUSTY" SONG SPARROWS

These six medium-sized to fairly large subspecies, *morphna*, *merrilli*, *adusta*, *zacapu*, *rivularis*, and *fallax*, share the overall reddish shade of their plumage; they differ from one another chiefly in the darkness and intensity of those tones. This western and southwestern group, too, like the sooty Song Sparrows, provides a neat illustration of Gloger's rule, with the palest subspecies occupying the most arid environments.

The longest-billed of the six, *morphna*, is also the darkest, deep reddish brown above with coarse dark streaking; the gray underparts are heavily marked with broad, blurry reddish streaks. Of all the non-insular Song Sparrows, *morphna* is the race most likely to give the first impression that it is uniformly colored and marked above and below. It is redder than either *rufina* or *caurina*, and shorter-billed than the latter.

The more inland race *merrilli* has a slightly longer tail, less strikingly slender bill, and paler, colder brown-gray upperparts; its streaking, above and below, is clearer and darker than that of *morphna* or the next subspecies. This is the commonest of the migrant subspecies over much of California.

The pale, sparsely marked *fallax* is a resident of desert wetlands. *Arizona. Caroline Lambert*

The breeding range of this race is adjacent to that of the rather similar *montana*; in fact, Patten and Pruett explicitly label *merrilli* intermediate in plumage between *montana* and *morphna*. The lower back is more clearly marked in *montana* than in *merrilli*, and *merrilli* is more noticeably reddish than the deep brown *montana*, but the two meet and presumably intergrade in eastern Oregon and Washington and central Montana, and like all non-insular, non-sedentary Song Sparrows, neither should be identified with certainty in the field away from the core of the breeding areas.

The representative of this subspecies group in the dry Southwest of the United States and northernmost Mexico is one of the most strikingly plumed of all Song Sparrows. In 1854, Spencer Baird named a new passerellid from Arizona *Zonotrichia fallax*, the "deceptive sparrow." Baird's newly named species revealed "a very close resemblance to *Z. melodia*," the Song Sparrow, but with a smaller bill, longer tail, and very different plumage. The back of this desert dweller is fairly pale gray with broad reddish streaks; the feathers of the upper back lack visible black shaft streaks, giving the bird's upperparts an appearance of softness. The creamy white underparts are sparsely marked with reddish streaks, on some birds nearly restricted to the sides of the breast. Birds of this subspecies also differ in behavior from most other Song Sparrows, much more frequently cocking the tail while on the ground and readily walking as often as hopping as they feed on the shady edges of desert wetlands.

Because of these differences, *fallax* was treated, on Baird's authority, as a distinct species for nearly 20 years after its first description. Elliott Coues lumped this subspecies with the chocolate Song Sparrows in 1872, but, in uncharacteristic confusion, used the name *fallax* to denote not just this population but all of the Song Sparrow races of mainland western North America. Not until Robert Ridgway included the bird in his *Manual* was *fallax* correctly treated as a single subspecies of the Song Sparrow, restricted in its range to the low deserts of California, Arizona, Nevada, and Utah. Individuals of this pale reddish subspecies still live up to their name today, deceiving observers familiar with only the more widespread and more "normally" plumaged Song Sparrows into misidentifying these delicately marked birds as sparrows of another species entirely.

The somewhat larger Baja California subspecies *rivularis* resembles *fallax*, from which its breeding range is separated by the Sea of Cortez. Even paler and less contrasting in plumage pattern than its mainland counterpart, *rivularis* is larger with a longer, thinner bill and narrower streaking below.

The southernmost member of this visual group

Photographed in southwestern Alberta, this rusty Song Sparrow is probably a representative of the widespread northwestern subspecies *merrilli*. *Alberta. Caroline Lambert*

is *adusta*, a fairly large resident of the west-central states of Guanajuato and Michoacán. The upperparts are rusty brown with relatively broad dusky streaks. Beneath, the blackish streaking takes the form of blotchy spots, which join to create a necklaced appearance across the upper breast.

The extremely local race *zacapu* is similar but slightly shorter-tailed; it is darker above and lacks the pale edges of the inner secondaries, a feature that cannot be expected to be seen in the field. Both of these races share with *villai* and *mexicana* the tendency of the underpart markings toward blackish spotting rather than streaking, neat and highly contrasting with the white ground color of breast and belly.

Rusty Song Sparrows breed from southern British Columbia to Alberta and western Montana, south to the desert marshes and streams of northern Sonora and southern Baja California; *adusta* and *zacapu* occupy very small, isolated ranges in west-central Mexico. The dark *morphna* occupies the coasts of southern British Columbia, Washington, and Oregon, moving south in winter to coastal California. Breeding away from the coast and south and east to northwestern Montana, *merrilli* spreads in winter as far as southern California, Arizona, and New Mexico, thus overlapping with several other Song Sparrow subspecies. The desert Southwest's *fallax* is, as expected for a bird of cool shady marshes and riverbanks, very locally distributed in southern California, Nevada, and southwest Utah south to the Sonoran desert of Arizona and Sonora. Baja California's *rivularis* is found in mountain canyons of the southern portion of the peninsula.

The west Mexican subspecies *zacapu* and *adusta*, along with *goldmani* (here treated as a "chocolate" race, above), have very limited ranges. *Melospiza melodia goldmani* appears to be restricted to the immediate vicinity of El Salto, Durango, where the type specimen was taken in July 1898. The same expedition collected the first specimen of *adusta*; it is found only "along the Río Lerma drainage in Michoacán from Pátzcuaro upstream to Lago Yuriria, Guanajoto." Only some 20 miles separates the range of *adusta* from that of *zacapu*, resident at the eponymous locality. *M. m. zacapu* is darker and redder than the paler and browner *adusta*. Patten and Pruett note that the Song Sparrows resident at nearby Lake Chapala in Jalisco represent either an isolated population of *zacapu* or another, still undescribed subspecies.

While the eastern half of North America must content itself with just two currently recognized Song Sparrow races, California and its islands are occupied by no fewer than seven extremely local resident subspecies, a diversity resulting from the state's famously wide range of habitat and climate types. Several of these subspecies occupy dangerously small ranges; indeed, the resident populations of the Channel Islands race *Melospiza melodia graminea* once found on San Clemente and Santa Barbara Islands are now extinct.

These California specialties are the smallest of the Song Sparrows, generally faintly olive-tinged above and with distinct, often blackish streaks below. Because all of these sparrows are permanently resident within their circumscribed ranges, they are by definition identifiable to subspecies when breeding; in migration and winter, however, they may share the marshes and brushy areas with seasonal immigrants of the subspecies *caurina*, *morphna*, *merrilli*, and *montana*, all of which are larger and darker than the local breeders.

Resident on the Pacific Coast from southern Oregon to northern California, *Melospiza melodia cleonensis* is a small-billed Song Sparrow with

The greenish gray face and collar contrasting with the rustier wing suggest the subspecies *pusillula*, the Alameda Song Sparrow. *California, April. Caroline Lambert*

The distinct yellowish tones on the face of this San Francisco Bay Song Sparrow are typical of the smallest subspecies, *pusillula*, the Alameda Song Sparrow. *California, April. Caroline Lambert*

California's largest and heaviest-billed Song Sparrow is the fittingly named *maxillaris*, known in English as the Suisun Song Sparrow. *California. Caroline Lambert*

a distinctly streaked lower back and rust-streaked grayish white underparts. In plumage, bill size, and range, this subspecies can be considered the "link" between the rusty Song Sparrows and the present group.

Other than in the San Francisco Bay area, central coastal California is inhabited by *gouldii*, which is proportionally longer-tailed than *cleonensis* and differs further in the crisper, blacker streaking of its white underparts. The upperparts are paler than in *cleonensis* and the reddish back shows a saturated olive cast, the feathers of the upper back without gray edges.

The immediate area of San Francisco Bay has three distinctive subspecies to itself. The Song Sparrow of the bay's northern salt marshes, *samuelis*, has a chestnut-streaked olive-tinged back and a dull buffy white breast with high-contrast, neat streaking. Its southern counterpart, *pusillula*, is the smallest of all Song Sparrows; it is distinctive in having yellowish, crisply streaked underparts and a yellow-olive tone to the gray upperparts. The supercilium is also tinged yellow.

The brackish marshes of Suisun Bay, upstream from San Francisco Bay proper, are inhabited by the largest-billed of California's specialty Song Sparrows, the aptly named *maxillaris*. Less olive than the others in this group, *maxillaris* is heavily and distinctly streaked black above and below.

Central California from Merced to Kern Counties is the home of *heermanni*, fairly large-billed but without the strikingly swollen bill base of *maxilla-*

The quite long-billed subspecies *heermanni* is streaked deep reddish above. *California. Caroline Lambert*

ris. Though the ground color of the upperparts is also olive-tinged, it is grayer than in *maxillaris* and the streaking ruddier and not as heavy. The underparts are streaked with strongly contrasting reddish brown.

The insular race *graminea* is now resident from San Miguel to San Nicolas and Santa Catalina Islands; it no longer occurs on Santa Barbara or San Clemente. The pale olive-gray upperparts are streaked with brown, the white underparts clearly marked black. The breast streaking is sparse.

Shaped very like a Song Sparrow but thinner-billed, this species is distinctly gray-faced and bright-winged. *Maine, June. Brian E. Small*

SWAMP SPARROW
Melospiza georgiana

"The history of this obscure and humble species is short and uninteresting." Those are startling adjectives, but Alexander Wilson, describing the Swamp Sparrow in 1811 for what he thought was the first time, was making a point less about the bird than about his Old World colleagues.

Wilson makes no explicit claim to have discovered the Swamp Sparrow—an English name of his coining. But he does point out that his account in the *American Ornithology* "now for the first time introduces" a bird recorded nearly 40 years earlier by his American mentor and patron William Bartram, who on his southern journeys in the 1770s had encountered a bird he named *Passer palustris*, the Reed Sparrow. Wilson snidely remarks that the species is "very numerous" but has somehow gone "unknown or overlooked by the naturalists of Europe."

In his usual eagerness to trumpet the superiority of American science in America, Wilson overlooked or did not know that the English ornithologist and collector John Latham had indeed published a description of the Swamp Sparrow in 1790, the year before the appearance in print of Bartram's *Travels*

and two decades before Wilson brought the bird, as he believed, "to the notice of the world." Latham's Latin analysis, which he later translated into English in his *Supplement II to the General Synopsis of Birds* (1802) and *General History of Birds* (1823), is immediately recognizable as pertaining to this species, and indeed provides a rather better description than Wilson's. It seems that Latham knew the bird from specimens viewed "at Mr. Humphries's," probably George Humphrey, the famous London dealer in naturalia; the specimens had been shipped from "the internal parts of Georgia, in America," and so Latham, perhaps with an eye to the favor of the reigning monarch, named it *Fringilla Georgiana*, the Georgian Finch.

The two names—Latham's *georgiana* and Bartram's (and Wilson's) *palustris*—led parallel existences for the next 20 years. Naturalists writing in Europe generally preferred Latham's name, while those working in America, whether born in the New World or temporary expatriates, usually adopted or at least acknowledged Bartram's. In 1815, writing in London, James Francis Stephens simply copied out Latham's account, not even mentioning Wilson; in Germany, eight years later, Hinrich Lichtenstein, though he knew Wilson's *American Ornithology* well, cited only Latham as his authority for this species. In contrast, Charles Bonaparte, ornithologizing in Philadelphia and New Jersey in the early 1820s, credited Wilson exclusively, with no reference to Latham. Thomas Nuttall, the British author of an authoritative handbook to American birds, knew both names, but gave priority to Latham's in both editions of his *Manual*.

Nuttall was the last ornithologist for almost half a century, on either side of the Atlantic, to use his compatriot's name for this species. In part, it was the weight of nomenclatural authority brought to bear by Bonaparte, Audubon, and others that pushed Latham's *georgiana* aside in favor of the more descriptive Wilsonian name; in part, though, it was also a punctilious concern about the adequacy of Latham's original description. Spencer Baird, with access at the Smithsonian to the largest collection of sparrows in the New World, expressed his doubts most clearly when he moved the Swamp Sparrow into his new genus *Melospiza*: "In the uncertainty whether the *Fringilla georgiana* of Latham be not rather the [Bachman Sparrow] than the Swamp Sparrow, I think it best to retain Wilson's name."

The nomenclatural committee of the new American Ornithologists' Union debated the case in preparing the first edition of the *Check-list*, and in 1885 announced its decision: *georgiana* was correctly applied to the Swamp Sparrow and antedated Wilson's *palustris*.

Notably, though, one of the members of that committee, the great and irascible Elliott Coues, had long been an advocate of William Bartram and the names published a century before in Bartram's *Travels*. Even while urging on others strict adherence to the AOU's names, Coues himself retained the invalidated *palustris*—Bartram's name for the Swamp Sparrow—in the third (1887) and in all three printings of the fourth edition of his *Key to North American Birds* (1890, 1894, 1896). In his typically orotund style, and referring to himself in the third person, Coues justified his flouting of the decisions reached by a majority of the committee members:

> The naming of our birds . . . has lately been pitched in a key so high that the familiar notes of the [names Coues used in earlier editions of his *Key*] might jangle out of tune . . . . During the confusion unavoidably incident to such sweeping changes in nomenclature as we have recently made . . . the present edition . . . leaves the names . . . untouched in the body of the text . . . . The author's . . . reserving to himself, as he certainly does, the right of individual judgment in every question of ornithological science . . . the occasion for individual dissent on the part of

any member of [the AOU committee] . . . arises when in his private capacity as an author he has, as it were, to pass upon and approve or disapprove any results of the labors of others.

Coues's disapproval explicitly extended to the use of the name *georgiana* for the Swamp Sparrow. Only at the very end of his short life did he finally acknowledge that "there is no doubt that this is *Fringilla georgiana* Latham," an admission that would no doubt have dismayed Alexander Wilson 90 years before.

Unlike the scientific name coined by Wilson, his English name was promptly adopted and has been authoritative ever since. In the late nineteenth century, apparently on the instigation of Elliott Coues, this species was occasionally known as the "Swamp Song Sparrow" in an attempt to make its relationships clear; that name never became especially widespread, and its last use appears to have been in *Birdcraft* by Coues's sometime collaborator Mabel Osgood Wright, first published in 1895.

Before Spencer Baird moved this species into the newly erected *Melospiza* in 1858, the Swamp Sparrow endured several different generic assignments. Like the Song Sparrow, it was included in an enlarged *Zonotrichia* in the mid-nineteenth century

Even adults can show streaking beneath and a central breast spot, but the white throat is bordered by only a fine lateral throat stripe. *Maine, June. Brian E. Small*

(and like the Song Sparrow, is retained in that catch-all genus in Paynter's relevant volume of the Peters *Check-List*). Somewhat ironically, William Swainson, the coiner of the genus *Zonotrichia*, placed this sparrow in his genus *Ammodramus*, apparently misled by its marshy habitat and dark plumage into grouping it with the Saltmarsh and Seaside Sparrows. A scant year later, Charles Bonaparte assigned the Swamp Sparrow to the genus *Passerculus* alongside the Savannah Sparrow, but Swainson's classification would prevail for the next 20 years, until the creation of *Melospiza* to comprise the Song, Lincoln, and Swamp Sparrows.

Adult Swamp Sparrows are easily recognized, but the misidentification of juveniles and formative-plumaged birds has a long tradition in American birding—and not just in the field. At the end of the nineteenth century, the conservationist and early Audubonite T. Gilbert Pearson had occasion to reexamine the first specimen record for North Carolina of the Clay-colored Sparrow. He immediately

> became convinced that an error had been committed in the identification, and at once sent it to the Smithsonian Institution. Prof. [Charles W.] Richmond identified it as being simply *Melospiza georgiana*,

a species that resembles the much smaller, sleeker, long-tailed and tiny-billed Clay-colored Sparrow only by a very great stretch of the imagination.

A generation earlier, three of the best-known ornithologists of the nineteenth century had committed an even grosser blunder. Sometime before 1874, Dr. Samuel Cabot collected an immature sparrow in Essex County, Massachusetts; part of the specimen—the skin of the head, nape, and upper back—was sent to the Smithsonian in Washington, where it was identified as a new species and named, in honor of its discoverer, *Passerculus caboti*. The bird's head was painted by Robert Ridgway, if none too well, and the Cabot Sparrow was included in the 1874 *History of North American Birds* he prepared with Spencer Baird and Thomas Brewer—all three of whom, incidentally, served as the eponym of a sparrow.

There was no mention of the mysterious bird again for nearly a decade. Seven years after painting the fragmentary sparrow, however, Ridgway pronounced the species "untenable." In revising his *Key* for a second edition, Elliott Coues had occasion to inspect the remnants of the Cabot Sparrow, and discovered immediately that the bird was a Swamp Sparrow, "in a plumage hitherto unrecognized, in which there is a decided yellow loral spot, and a vague yellowish suffusion of the cheeks and throat." In fact, the olive-yellow overlay of the cheek, lore,

and supercilium in some formative-plumaged Swamp Sparrows continues to mislead observers today who look first to the visual "field marks" of the head. Such birds are sufficiently frequent that they can be thought of as constituting a "yellow morph" of the species at this age.

FIELD IDENTIFICATION

As the scientific name of one of the widespread subspecies, *ericrypta*—"well hidden"—suggests, the greatest identification challenge posed by many Swamp Sparrows is their reclusive habit. Breeding birds and winterers alike are partial to dense, dark vegetation, and they can be difficult to see well at any season in thick cattails and overgrown brushy fields.

That very furtiveness can serve as a first clue. Seen even fleetingly, this oval-bodied, medium-large, long-tailed, short-winged sparrow makes a distinctively dark impression in the field. The folded tail, proportionally slightly shorter than in Song Sparrows, usually appears square-tipped or very slightly notched. Swamp Sparrows on the ground usually hold the tail still and at or just below the horizontal, but it may be twitched upward by nervous birds; the tail is not often fanned by perched individuals. Flying birds hold the tail steady or, especially toward the end of the flight, flip it like a Song Sparrow's.

The wing is shorter and more blunt-tipped than that of the Song Sparrow; the primaries extend only slightly beyond the tertials of the folded wing, their extension significantly shorter than the longest tertial. The inland subspecies show considerable deep bright rufous in both the wing coverts and the secondaries.

The striping of the back is bold and bright, mixing black and chestnut with white. *Maine, June. Brian E. Small*

At any distance, the Swamp Sparrow is a chunky, dark, bright-winged bird with a bright white throat. *Maine, June. Brian E. Small*

Though both wing and tail are shorter than in the Song Sparrow, Swamp Sparrows as a rule appear less globular than that species, perched and in flight, probably as a result of the slightly smaller, somewhat flatter head and, especially, the longer, more slender bill. This species raises its short crest much less frequently than either the Song or the Lincoln Sparrow.

Unlike the Song Sparrow, adult Swamp Sparrows undergo two molts a year: a complete prebasic molt in late summer or early fall, on the breeding grounds, and a prealternate molt in late winter, on the wintering grounds. The prealternate molt is limited to the feathers of the head and nape, and it is this molt that produces the "classic" rusty cap shown in field guides. On most alternate-plumaged males, the crown is nearly entirely rufous red, with a small black patch (larger in *nigrescens*) on the forehead split by an incipient white median crown stripe. The crown of most alternate-plumaged females is considerably duller brown, with dark brown or black streaking on each side of a plainer gray-brown median crown stripe. In either plumage, adult Swamp Sparrows are distinctively gray elsewhere on the head, including the broad supercilium and sides of the neck, more clearly isolating the dark brown or black line behind the eye than in the Song Sparrow.

In comparison to Song Sparrows, adult Swamp Sparrows of the subspecies *georgiana* and *ericrypta* appear more neatly black-streaked on the back; in fresh plumage, the pale edges of the mantle feathers may align to form whitish stripes, creating a very colorful back pattern that contrasts with the plain dull rusty rump. The fringes of the back feathers are less conspicuous in *nigrescens*, making the upperparts appear darker and without white striping, and thus slightly more Song-Sparrow-like.

Many illustrations present adult Swamp Sparrows as immaculate below, but in fact many or even most show fine, scattered streaks on the flank; formative birds, and perhaps some adults, are more heavily streaked on the belly and breast, and often show a central breast spot. Those streaks are never as broad or as concentrated as in the Song Sparrow, however. Swamp Sparrows share with the Lincoln Sparrow finely streaked buff-gray undertail coverts and vent.

The throat and jaw stripe, conspicuously white in adults and dull buffy white in formative birds, are separated by a fine black lateral throat stripe, evenly narrow throughout its length. At any distance, the bright white throat contrasting with the somber gray of the head and the dark rust of some or all of the wing coverts provides the most conspicuous and distinctive visual character of this handsome bird.

Basic-plumaged adult Swamp Sparrows and birds in their formative plumage are buffier and less rusty overall, usually with little or no rufous on the crown, which has a broad and conspicuous gray median stripe. They also average more heavily streaked below. The major confusion species in this plumage is the Lincoln Sparrow, which differs most

Juvenile *Melospiza* sparrows are challenging indeed. On this very young Swamp Sparrow, note the richly colored wing, contrasting back pattern, and extensively buffy flank. *Alberta, June. Caroline Lambert*

notably in its grayer wing and upperparts, finer back pattern, yellow-washed breast and jaw stripe, and marginally more slender bill. Lincoln Sparrows of those age classes are more neatly and regularly streaked on the breast and flanks than are Swamp Sparrows. These differences are usually clear-cut to an observer with experience with both species, but they can be difficult to learn, as Swamp Sparrows are scarce in much of the West and Lincoln Sparrows scarce in much of the East.

The potential for misidentification is more severe in juveniles. Thirty years ago, Roger Tory Peterson reminisced about "an identification game" presided over by Ludlow Griscom:

> Twenty skins of problem birds—their labels hidden—were laid out on a tray for us to identify . . . . I won hands down with a perfect score. The reason? I already knew the clichés! A supposed "Lincoln's" Sparrow was certain to be a juvenile Swamp Sparrow.

The salient facts for the field birder are that these two species overlap in the breeding season only in Canada and northernmost New England—and that both species molt out of juvenile plumage into the preformative plumage before they begin their first southward migration. Thus, a juvenile "Swincoln" sparrow anywhere in the eastern United States south of the Adirondacks is presumptively a Swamp Sparrow. In northern New York, Vermont, Maine, and much of the Swamp Sparrow's extensive Canadian range, however, both species are present in the breeding season, and unattended juveniles—birds wearing their first plumage out of the nest—can be very difficult to distinguish.

Compared with juvenile Lincoln Sparrows, the juvenile Swamp Sparrow is minutely shorter-winged and longer-tailed, differences not usually discernible in the field. The Swamp Sparrow's crown is black, often with a gray median stripe (contra Pyle), and variably streaked with dark olive; the paler, browner crown of the Lincoln Sparrow is distinctly black-streaked. Perhaps the best distinction is the throat pattern: streaked in Lincoln and unstreaked in Swamp Sparrows, a difference that persists in adult birds as well.

RANGE AND GEOGRAPHIC VARIATION

The species as a whole breeds from the southern Northwest Territories and eastern British Columbia east to Quebec and Newfoundland, south in the eastern half of North America to northern

Missouri and West Virginia and in brackish tidal marshes from northern New Jersey to Chesapeake Bay. The wedgelike extension of the breeding range into western Canada east of the Rockies is a typical distribution pattern for many songbirds, an echo of the longer persistence of glacial ice on the southern edges of their breeding range in the West.

The core winter range is in the southeastern United States, west to central Texas and north to Iowa and Virginia; the southernmost winterers reach the very edge of the area covered by this guide, in Veracruz, Guerrero, and Jalisco. Very small numbers winter regularly on the coasts of Washington, Oregon, and California; on the lower Colorado and upper Rio Grande Rivers; and increasingly in urban wetlands and natural marshes in southeastern Arizona. A classic "half-hardy" species, many stay north of the wintering area until November, with the odd individual overwintering on the breeding grounds in milder years. The Coastal Plain Swamp Sparrow winters from Chesapeake Bay south to South Carolina. Though Swamp Sparrows are very rare in peninsular Florida, strays wander as far as Bermuda, where as many as four have been recorded in a single winter.

This species presents one of the numerous cases in which seemingly arcane taxonomic decisions can affect conservation efforts. While the IOC and the Peters *Check-List* tally only two Swamp Sparrow subspecies—the nominate *georgiana* and *ericrypta*—other authorities (among them Pyle and the accounts in *Birds of North America* and *Handbook of the Birds of the World*) follow the fifth edition of the AOU *Check-list* in recognizing a third, *nigrescens*, breeding in brackish tidal marshes and bogs on the mid-Atlantic coastal plain, from northern New Jersey to the upper Chesapeake Bay. For that last subspecies, with a population of probably less than 50,000 individuals, official recognition or nonrecognition can be the difference between survival and extinction:

> Although the mid-Atlantic Coast of North America has some of the most extensive tidal marshes in the world, the habitat is nonetheless limited and is subject to intensive management, such as burning, ditching, and open pond creation. This habitat is downstream from agricultural, urban, and industrial runoff, and it faces direct threats from invasive species (for example, *Phragmites*), rising sea levels, and climate change.

If *nigrescens* is not deemed a "good" subspecies, interest in the bird is lessened—and with it the motivation to protect it and its imperiled habitats.

Fortunately, from the point of view of conservation, the Coastal Plain Swamp Sparrow is visually distinctive in the field—far more so, in fact, than the other two "very weakly defined" races that make up the species.

The two inland-breeding subspecies are both small-billed and brown-flanked. The two are best considered indistinguishable in the field away from the breeding grounds. In alternate plumage, the crown of the more northerly and westerly breeder, *ericrypta*, is paler reddish; that of the more southerly and easterly *georgiana* is darker rufous. In both subspecies, the rusty crown of alternate-plumaged birds shows little or, in many males, no black streaking. The warm-colored flanks are reddish-tinged brown in *ericrypta*, darker and more rufous in *georgiana*. The nesting ranges of the two subspecies approach each other in eastern Canada; in winter, *georgiana* is essentially restricted to the southeastern United States, while *ericrypta* can be found at that season from California to the Atlantic Coast. Because Latham's type was a nonbreeding bird from Georgia, where both of these subspecies can be expected, it is impossible to know which of the two his original description applies to.

The Coastal Plain Swamp Sparrow, *nigrescens*, is more distinctive. Above, the ground color of back and rump is paler—the name *nigrescens* notwithstanding—and grayer than the rust of the inland races, but the black streaks of the mantle feathers are on average wider. The flanks are olive-gray, with significantly less reddish overlay than in the other two subspecies. In both alternate and basic plumages, the crown is more heavily streaked and flecked with black than in other Swamp Sparrows; breeding birds, especially males, have a more extensively black forehead and more black on the dark gray nape. Perhaps the best feature, one sometimes visible in the field, is the color of the eye line: mostly black in *nigrescens*, largely rusty brown in the others. The flight feathers of the wing are duller in *nigrescens*, making the reddish wing coverts more conspicuously contrasting.

The longer, deeper, and broader bill of *nigrescens* is often readily noticeable in the field, particularly in males; its length at times creates a flat-foreheaded, "spike-billed" impression recalling that of the Saltmarsh or Seaside Sparrow. The slightly larger overall body size of this subspecies is unlikely to be discernible in life even if it is seen together, in migration or winter from Delaware to the Carolinas, with individuals representing *ericrypta* or *georgiana*. The most important characters, in which the Coastal Plain Swamp Sparrow is "virtually nonoverlapping" with the other two subspecies, are the larger bill, grayer flanks, and greater amount of black on the head and nape.

This slender-billed, modestly elegant bird wins many birders' votes for favorite sparrow. *Texas, April. Brian E. Small*

LINCOLN SPARROW
Melospiza lincolnii

On July 4, 1833, one of the young hunters on John James Audubon's Labrador expedition wrote in his journal that

> Mr. A. finished a drawing of a new finch which I shot at Esquimaux Islands, there are several rare and beautiful plants peculiar to that country represented upon it.

Thomas Lincoln was 21, barely six months older than John Woodhouse Audubon, who was also a member of the company. The two were inseparable in the field and on the boat:

> These two always go together, being the strongest and most active, as well as the most experienced shots . . . . Now we are sailing in full sight of the northwestern coast of Newfoundland . . . John and Lincoln are playing airs on the violin and flute.

The elder Audubon shared his son's affection for Lincoln. Indeed, it is hard not to read his account of the young men's adventures—and their hardships—as a piece of nostalgia for his own early explorations. Audubon spent most of his Labrador days drawing, painting, and writing on board the boat, and a hint of wistful envy of the more active life led by his collectors shines perversely through even in passages such as these:

> The Caribou flies have driven the hunters on board; Tom Lincoln, who is especially attacked by them, was actually covered with blood, and looked as if he had had a gouging fight with some rough Kentuckians . . . . Our young men returned from Port Eau fatigued, and, as usual, hungry . . . . The young men went off with the Indians this morning, but returned this evening driven back by flies and mosquitoes. Lincoln is really in great pain.

Audubon rewarded Lincoln's efforts by naming for him that new "finch," first collected on June 27, 1833, near the mouth of Quebec's Natashquan River. Lincoln is often given credit for the discovery of the species, but Audubon's published account makes it pointedly clear that he himself was the discoverer—and Thomas Lincoln only incidentally the collector. Audubon first heard the song of the bird in a lush valley, a series of "sweet notes"

> surpassing in vigor those of any American Finch with which I am acquainted, and forming a song which seemed a compound of those of the Canary and Woodlark of Europe. I immediately shouted to my companions . . . . Chance placed my young companion, Thomas Lincoln, in

a situation where he saw it alight within shot, and with his usual unerring aim, he cut short its career . . . . I named it *Tom's Finch*, in honour of our friend Lincoln, who was a great favourite among us.

In the official description, Audubon was more formal, assigning the new species the name Lincoln's Finch, *Fringilla Lincolnii.*

Audubon's party collected at least ten specimens during their time in Labrador, finding the birds "more abundant and less shy the farther north we proceeded." In spite of its abundance on the boreal breeding grounds, the migratory path of the new sparrow and its winter range remained unknown; even five years on, Audubon could add only that others had meanwhile been secured in the vicinity of New York City.

It is a fair measure of how uncommon and how secretive this bird is in the great population centers of the eastern United States that Audubon did not encounter another specimen until fully a decade later, on the Missouri River near Fort Leavenworth, a "strange place for it, when it breeds so very far north as Labrador." Apart from those records, the Lincoln Sparrow remained in ornithological imagination a mysterious creature of the far north until the very end of Audubon's lifetime. In the 1850s, however, the bird began to be recorded virtually everywhere in North America, from Guatemala to Pennsylvania, from Oaxaca to Utah. The next decades filled in the gaps, both in this retiring sparrow's breeding range and in the vast sweep of its migrations.

Surprisingly enough in the case of a brown bird with few conspicuously distinctive markings, the Lincoln Sparrow, once in hand, seems almost invariably to have been correctly identified even in those early days. A notable exception was the explorer Heinrich von Kittlitz, who collected three specimens in what is now southeastern Alaska and named them *Emberiza gracilis*, describing his new species as characterized by "the small, slender structure and cinereous crown marked with several black lines." About 1835, W. G. Pape drew this bird for Johann Friedrich von Brandt's fragmentary ornithology of Russian America; probably as the result of a printing error, Brandt's caption inscrutably labels the bird "*Emberiza spinoletta*, Kittl." It took almost thirty years for Otto Finsch to clear up the confusion, when he pronounced the figure in Brandt's work "readily recognizable" as a Lincoln Sparrow.

Kittlitz's view was partly vindicated, though, in 1906, when Harry C. Oberholser determined that the Lincoln Sparrow subspecies described by William Brewster in 1889 was in fact none other than the bird Kittlitz had collected in the 1830s—making his the scientific name properly to be applied to the

The black markings above and below are finer than in the Song or Swamp Sparrows, and the buffy flank, breast, and jaw stripe neatly set off the white of the throat and belly. *Texas, April. Brian E. Small*

The contrast between rusty wing and tail and the colder, grayer back is typically greater than in Song or Swamp Sparrows. *Maine, June. Brian E. Small*

Lincoln Sparrows of the Pacific Northwest. Thus, the discovery of one race of this species named by the Franco-American Audubon for the Mainer Lincoln is now credited to a Silesian naturalist: *Melospiza lincolnii gracilis* (Kittlitz).

FIELD IDENTIFICATION

The Lincoln Sparrow is slightly smaller, more slender, and shorter-tailed than Swamp or (most) Song Sparrows; the rounded, square-tipped, or faintly notched tail can sometimes give an impression of narrowness. Feeding birds may hold the tail still and at the horizontal, or they may cock it slightly. Active

The chestnut crown is finely streaked black, with a broad gray central stripe. *Texas, April. Brian E. Small*

or nervous birds flip the tail, usually with a decided sideways movement, but the tail is rarely flirted or fanned when the bird is on the ground.

The rounded wing is slightly, but sometimes noticeably, longer and more pointed than that of the Song and Swamp Sparrows. This longer wing appears to cancel out the effects of the slightly shorter tail, and Lincoln's Sparrows usually strike the observer as smaller and leaner than similar species; the long, fine bill can make the bird seem quite small-headed in the field. A fairly long, ragged crest is often raised.

With the exception of juveniles, the plumage of the Lincoln Sparrow is distinctive when seen well—Spencer Baird went so far as to call the species "easily known among the American sparrows." The overall somber appearance, relatively fine streaking above and below, yellowish breast and jaw stripe, and neat, crisply drawn head markings make the Lincoln Sparrow look like no other sparrow—once, that is to say, the observer has had some experience with the species. For many observers in the East, however, such experience can be elusive.

Few nonjuvenile Lincoln Sparrows probably go misidentified, but the species' name is not infrequently misattached to Song and Swamp Sparrows. Peterson, no doubt remembering Ludlow Griscom's museum quizzes, warned that "the immature Swamp Sparrow in spring migration is continually misidentified as" a Lincoln Sparrow. The salient differences from both Swamp and Song Sparrows, in juvenile and adult plumages, are described in the accounts for those species. The most significant source of confusion in most cases is an inappropriate emphasis on the details of the head pattern: while the Lincoln Sparrow's fine markings are distinctive, some hopeful observers may have no difficulty discovering white eye-rings and gray median crown stripes even in species where those characters are inconspicuous or even absent.

While Song and Swamp Sparrows account for almost all over-reporting of this species, the focus on head patterns has led to even Vesper Sparrows—with their prominent eye-ring and striped crown—being misidentified as Lincolns.

Though the head pattern can serve as confirmation, Lincoln Sparrows are most easily identified by their dark, slightly reddish tail and wing, the neatly pencil-streaked rump and undertail coverts, the narrow streaking of the back, and the medium-gray head. The pattern of the breast and throat in some formative-plumaged (and perhaps some adult) Swamp Sparrows can suggest that of the Lincoln Sparrow, but in the Swamp Sparrow, the yellowish wash is more suffused and often patchy, the streaks thicker and less regularly arranged across the breast and down the flanks. A genuine Lincoln Sparrow has a clear buff-yellow jaw stripe, breast band, and flanks, the breast and flanks more or less densely marked with regularly spaced black streaks of the same width as those marking the rump and vent region.

Fluffier and buffier than its parents, the juvenile Lincoln Sparrow shares the adult's fine streaking above and below, narrow eye-ring, and buffy or yellow jaw stripe. *Colorado, July. Debra Mootz*

The eye-ring varies from white to yellow, but is always narrow and usually conspicuous on the grayer face. *Texas, April. Brian E. Small.*

There is often a larger spot, as in Song and some Swamp Sparrows, at the bottom of the Lincoln Sparrow's yellowish breast band. *California, October. Brian E. Small*

RANGE AND GEOGRAPHIC VARIATION

Widespread across North America's boreal zone, the Lincoln Sparrow breeds from west-central Alaska east to northern Quebec, Labrador, and Newfoundland. In the East, nesting birds occur south to northern Minnesota, Wisconsin, and Michigan, and in the mountains from northern New York to Maine and Nova Scotia.

In common with many other boreal breeders, the Lincoln Sparrow follows the western mountains much farther south. The breeding range extends through the Cascades to south-central and southwestern California, and in the Rocky Mountains

from British Columbia, Alberta, Idaho, and Montana south through central New Mexico; nesting populations are also present in southern Utah and the mountains of central Arizona.

Migrants occur across the continent south of the breeding range, with peak movements in April and September. This species' relative scarcity in the eastern portions of its breeding range and its retiring habits on migration—Roger Tory Peterson memorably called it "a skulker, afraid of its own shadow"— make of the Lincoln Sparrow a usually uncommon sight on the East Coast, but west of the Mississippi this is a common, sometimes abundant migrant, soon familiar to new birders. In fact, much of the prevailing confusion about this species' identification is a cultural and historical artifact: the notion that this is always a difficult task would never have arisen had the authors of certain early field guides been based not in Boston or New York but in the American West. Vagrants have strayed as far as Greenland, the Bahamas, Cuba, Puerto Rico, and other islands in the West Indies; there is one winter record from Iceland.

Lincoln Sparrows winter well into Central America, regularly in Guatemala, El Salvador, and Honduras, and with small numbers infrequently penetrating as far south as Costa Rica and Panama. In the north, the species winters on the Pacific Coast from southernmost British Columbia south. Apparently more sensitive to cold and snow cover than some other sparrows, the Lincoln Sparrow occurs at this season across the southern tier of states from California and southern Nevada to southern Missouri and South Carolina; it is especially common in the warm Southwest. The northern limits of the winter range appear to be moving steadily northward in the Midwest and on the Great Plains.

The nominate race *lincolnii* is the most widespread subspecies, breeding across the boreal forests of North America from Newfoundland to western Alaska. Birds from the eastern part of this range have a ruddy-tinged mantle with narrow black streaks; the ground color of the back generally becomes colder, grayer-brown and the streaking coarser to the west.

The northern Pacific Coast's *gracilis* is smaller, with broad black streaking above on a yellowish brown ground color. Lighter edgings on the back feathers produce a strongly contrasting pattern. The black lateral crown stripes "crowd together . . . nearly obliterat[ing] the median gray crown region."

The coastal mountains of the West and the southern Rockies are occupied by breeding *alticola*. This is a large Lincoln Sparrow with gray-brown upperparts with narrow black streaks.

Birds of intermediate appearance are to be expected where subspecies meet on the breeding grounds; even at the core of each subspecies' breeding range, visual identification is complicated by considerable individual variation and the occurrence of browner and grayer birds within a single population (Pyle).

The relatively weak variation across this species' wide range contrasts notably with the extreme differentiation of the Song Sparrows. The much longer distance traveled each year by migrating Lincoln Sparrows presumably allows for more and more regular mixing of populations on the wintering grounds.

The fine streaks of the underparts continue onto the whitish undertail coverts. *California, April. Brian E. Small*

A rare bird with a very restricted range, this handsome sparrow recalls a short-tailed, bright-winged Song or Lincoln Sparrow. *Manuel Grosselet*

SIERRA MADRE SPARROW
Xenospiza baileyi

In the spring of 1931, Alfred M. Bailey, H. B. Conover, and W. F. Ardis briefly visited the west Mexican state of Durango. The party "had not planned on collecting birds but . . . took a few for identification." Among the specimens secured was "a small dark Bunting," which Bailey sent to Outram Bangs in Cambridge. To his "joy and surprise," Bangs recognized the bird as identical to one collected by William B. Richardson in Jalisco more than 40 years before. That earlier specimen, and seven others Richardson had supplied to the British Museum, were widely believed to be hybrids, but Bangs, on the strength of the new skin sent to him by Bailey, determined that they in fact represented an undescribed sparrow species and, indeed, an undescribed genus. Bangs named the species for Bailey, and he christened the genus *Xenospiza*, the "peculiar sparrow."

The fact that this species languished unrecognized in the museum drawers for more than 40 years points clearly to the challenges involved in its field identification. Not a few cautionary examples have been set over the years by eager searchers who triumphantly announced that they had found this rare bird, only to acknowledge later that they had been deceived by its superficial resemblance to a Song or Lincoln Sparrow.

Those who are confused by this peculiar sparrow are in good company. What would become the type specimen had been examined by any number of famous museum scientists before its formal description by Bangs.

> Ridgway wrote saying that he had never seen anything like it, and ventured no guess as to its parentage, simply calling it a hybrid. [Edward William] Nelson did likewise. Oberholser thought it sprang from a union of *Coturniculus henslowi* [a former name for the Henslow Sparrow] and *Passerculus savanna* [a former name for the Savannah Sparrow]. This was also [William] Brewster's opinion. A.K. Fisher suggested *Passerculus savanna* and *Melospiza georgiana* as possible parents.

The strong white face pattern and bright underparts with a fine breast band are characteristic. *Manuel Grosselet*

As that list suggests, the most "peculiar" thing about Bangs's *Xenospiza* is its puzzling close resemblance to several other sparrow genera—and its failure to fit comfortably into any of them. Frank Pitelka's examination of the slender specimen material then available led him "to doubt that this sparrow should be segregated in a monotypic genus." He found "superficial" similarities between the Sierra Madre Sparrow and the grass sparrows of the genera *Passerherbulus*, *Ammospiza*, *Ammodramus*, and *Passerculus*; his "total impression," however, was that the Mexican sparrow was "a species which is closer to *Melospiza* than to any other genus." Pitelka stopped short of moving the Sierra Madre Sparrow into *Melospiza* pending the availability of "additional specimens, particularly skeletons, and data on habitat, song, and behavior."

Beginning in 1954, in the course of a study of the Sierra Madre Sparrow's habitat, song, and behavior, Robert Dickerman, Allan R. Phillips, and Dwain W. Warner collected more than 50 new specimens from the La Cima population discovered by Helmuth Wagner. Their review of the greatly expanded specimen material led them to conclude that *Xenospiza* was "close" to the same grass sparrows listed by Pitelka but "not close" to *Melospiza*. They, too, declined to lump the genus with any other, suggesting that it be retained as monotypic and listed next to *Ammodramus*.

Raymond Paynter, in the sparrow volume of the Peters *Check-List*, went one step further and reassigned the bird outright to a greatly enlarged *Ammodramus*. Though it was followed in a small number of more popular works, most ornithologists rejected that generic concept in favor of a monotypic *Xenospiza*; the chief exception was Steve N. G. Howell, who found that the species was "better placed in *Ammodramus*." Recent molecular study of the Sierra Madre Sparrow and its relatives has supported either its traditional treatment as the sole species in its genus—or the more radical merger of *Xenospiza* and several other currently recognized genera into a vastly expanded *Passerculus*. At present, all world checklists list *Xenospiza* as a distinct genus.

FIELD IDENTIFICATION

The "curious little" Sierra Madre Sparrow resembles each of the species adduced over the years in the arguments about its taxonomic position. Medium-sized, with a very short wing and moderately long, rounded or faintly notched tail, this bird recalls a *Melospiza* sparrow in its overall structure; the white underparts, with a hint of a buff breast band and distinct blackish streaking, and broad grayish supercilium are also suggestive of that genus. The Song Sparrows of central Mexico are larger and

The chestnut wing coverts can be especially striking in flight. *Manuel Grosselet*

longer-tailed than this species; their underparts are dingier and their flanks tinged gray-brown, unlike the relatively bright white breast and belly of this species. The breast streaks, often crowding together to form a spot in the center of the breast, are wider and blacker than in the Lincoln Sparrow, finer and less extensive than in most Song Sparrows; at close range, the markings of the feathers at the side of the breast are clearly triangular. The yellow wing flash, created by the marginal coverts of the underwing, is a further important distinction, but not always visible in the field.

The wings and back of this species are deeper, brighter rust-red than those of a Mexican Song Sparrow or a Lincoln Sparrow, making an impression superficially more like the rich upperpart colors of a Swamp Sparrow. In spite of its nomination as a "possible parent," that species, which, like the Lincoln, winters in the range of the Sierra Madre Sparrow, is easily distinguished. No Swamp Sparrow is as crisply and contrastingly marked below as the Sierra Madre Sparrow, instead showing a more somber blend of gray, buff, and white on the underparts; any streaking is irregular and diffuse. The Sierra Madre Sparrow lacks the yellowish tinge shown by many winter Swamp Sparrows on the breast and jaw stripe.

The precise pattern of the back distinguishes the Sierra Madre Sparrow from any of the three *Melospiza* species; observers faced with a silent bird should make every effort to confirm the distribution of colors on the mantle feathers. Especially in fresh plumage, those feathers are generously edged and tipped with dull buffy gray, creating a decidedly scaled effect on the upperparts, quite different from *Melospiza* sparrows but similar to the "shingled" appearance of such grassland species as the LeConte Sparrow—which does not occur in central Mexico. The colorful and complex back pattern of the Grasshopper Sparrow can also be a source of confusion; the Sierra Madre Sparrow's tail is longer and rounder, and its bill much smaller. Both share the yellow wing flash. The Savannah Sparrow is also shorter- and sharper-tailed, and noticeably paler above.

The throat of the Sierra Madre Sparrow appears to be always immaculate white, without the streaking and spotting typical of a Lincoln Sparrow. A black lateral stripe separates the throat from a clear white jaw stripe, which curls back beneath the whisker and ear coverts to join a narrow, finely streaked grayish nape; the collared appearance is thus much more conspicuous than on a Song Sparrow.

The remainder of the head markings are remarkably—deceptively—similar to those of many Song Sparrows. The grayish ear coverts are bordered above by a black eye line; above that, the wide supercilium is unstreaked gray. The crown has broad blackish brown lateral stripes and an often inconspicuous gray median stripe. The lores are dark smudgy brown, setting off the much paler anterior portion of the supercilium. The bill is fairly small and slender, more closely recalling that of a Lincoln than of a Song Sparrow; both the upper and lower portions of the bill are dark dull gray, unlike the yellowish lower mandible seen on Swamp Sparrows.

Newly hatched Sierra Madre Sparrows fledge between mid-June and mid-August, and apparently wear their juvenile plumage into August or September, in some cases into October, when more northerly sparrow species are beginning to arrive. In this plumage, they are generally duller than adults. The feathers of the back have broad black shaft streaks and olive-brown tips and edges, giving a neatly scalloped appearance to the upperparts; the tertials are edged with buffy brown. Below, juveniles are extensively soft warm buff, with poorly defined, irregular streaks and spots on the breast. The crown is indistinctly streaked, and the supercilium is rich brown-gray, tending to yellow above the lore. The bill is paler than in adults, washed with dull yellowish pink. Juveniles are darker above than either Lincoln or Savannah Sparrows of the same age, and differ from both in the pattern of the upperparts.

RANGE AND GEOGRAPHIC VARIATION

The individuals Richardson shot in August 1889 may have been the last ever seen in Jalisco. Fourteen years after the species' (re)discovery in Durango by Bailey, Conover, and Ardis, however, in April 1945, the German artist, zookeeper, piano tuner, and ornithologist Helmuth Wagner collected a new specimen of the Sierra Madre Sparrow in La Cima, some 350 miles from the type locality, in Mexico's Distrito Federal. A second Durango site was discovered by John Davis in June 1951, and in July 2004 a third. As both the Bailey locality and the Davis locality have been almost entirely degraded by crop farming and livestock grazing, the third Durango site, a marshy bunchgrass meadow between the cities of Durango and Mazatlán, is now most likely the only remaining nesting area in the northern part of the species' presumed historic range; in 2004, this location was occupied by all of three breeding pairs. The species was still present here as late as 2013, with as many as six individuals reported.

In the decades since Wagner took his La Cima bird, scattered populations of this species have continued to be discovered in the southern Valle de México, on the remnant high-elevation grasslands of the Distrito Federal and the state of Morelos. Each of the surviving colonies is small, and the total global population of this species has been estimated at well under 7,000 individuals. The Sierra Madre Sparrow is a bird to see now.

Frank Pitelka described a new subspecies on the basis of Wagner's first specimen from La Cima, taken in 1945. That male skin differed consistently from each of the three specimens he examined from Jalisco, with darker, more crowded breast markings and clearer black and gray patterning on the head. Pitelka named this southeastern subspecies *sierrae*, for "the type locality, in high mountains near Mexico City in the Distrito Federal."

In comparing birds they collected with seven of the old Jalisco specimens, including the type, Dickerman et al. "found complete overlap in all characters cited by Pitelka," and concluded that in spite of the distance separating the two populations, Sierra Madre Sparrows from the southeastern and from the northwestern portions of the species' presumed range could not be distinguished as subspecies. This view has prevailed, and the species is now considered monotypic.

BAIRD SPARROW
Centronyx bairdii

In late July of 1843, John James Audubon and his traveling companions Edward Harris and John Bell were hunting buffalo above the banks of the Missouri River near Fort Union, in what would much later become Montana. Harris—who had already had a sparrow named for him by Audubon—and Bell—later to be commemorated in the name of a vireo—heard an unfamiliar sound. The party at first thought that it was the song of a Marsh Wren, but Bell determined that the vocalizations were "softer and more prolonged," and the two young men crashed off through the tall grass in pursuit.

> They had much difficulty in raising them from the close and rather long grass, to which this species appears to confine itself; several times Mr. Bell nearly trod on some of them, before the birds would take to wing, and they almost instantaneously re-alighted within a few steps, and then ran like mice through the grass.

At length, they succeeded in shooting three birds, and Bell's suspicion was borne out: the mysterious song was that of an unknown sparrow. This would be the final new bird Audubon described and painted, and in the very last account of the last volume of the octavo edition of the *Birds of America*, he named it, both in the English name and in the scientific epithet, for his "young friend Spencer F. Baird, of Carlisle, Pennsylvania." Baird was 20 years old at the time, but already a fast-rising star in American ornithology; as Elliott Coues would put it long after Audubon's death, "the glorious Audubonian sun had set . . . but the scepter was handed to one who was to wield it with a force that no other ornithologist of America has ever exercised."

Over the next decades, not even the sceptered Baird could force the reclusive new sparrow to give up any more of its secrets. Doubts were even raised as to the existence of the species, some ornithologists for a time believing that Audubon's type, a "faded specimen in worn plumage, preserved in the Smithsonian," was in fact nothing more than a young longspur in "some obscure plumage." It took nearly 30 years for the Baird Sparrow to be seen again—and even then it was not recognized as such.

In April 1873, writing in the *American Naturalist,* Robert Ridgway identified a skin taken by Charles E. Aiken in El Paso County, Colorado, as a new species "evidently closely related to *C. Bairdii*," but differing "specifically in quite different proportions, and also apparently, in different coloration." Ridgway named it *ochrocephalus,* "ochre-headed," but the bird had no time to acquire an English name.

The short tail, complex back pattern, and heavy-based bill help distinguish this rare prairie breeder from similar species. *North Dakota, June. Brian E. Small*

In September of that same year, David Scott rather vehemently informed the readers of that same journal that "this nominal species . . . is neither entitled to specific rank, nor even to a name as a well marked variety or race." Baird himself quickly confirmed Scott's view, writing to Coues that *ochrocephalus* was simply the fresh, autumn plumage of *bairdii*, and by the end of the year, Ridgway, too, had issued a typically gracious retraction:

> *Centronyx bairdii* (Aud.) = *C. ochrocephalus* Aiken. Mr. Aiken has collected a second specimen of this bird at the same locality where the first one was procured . . . . This one, collected May 6, 1873, being in spring plumage is decidedly intermediate between Audubon's original type of *C. Bairdii* (in worn, faded midsummer dress) and the autumnal specimen which Mr. Aiken characterized as *C. ochrocephalus* . . . there is every probability of all three specimens being the same species in different seasonal stages. Mr. Aiken is not to blame for describing his first specimen as a new species, for he . . . trusted the identification of the specimen to me, and at my suggestion described it as new.

At the base of all this doubt and confusion, even 40 years after Bell shot the first birds on the prairies of Montana, was the extreme sparseness of the specimen material available to the ornithologists of the East. That circumstance changed abruptly and dramatically in 1873, just as Aiken was discovering that the species migrated through central Colorado. That summer, Elliott Coues undertook a collecting trip to the grasslands of Dakota Territory; he returned with some 75 skins, representing all age and sex classes, and the surprising affirmation that the Baird Sparrow was the most abundant bird in northernmost Dakota, "in some places outnumbering all the other birds together." The first nest and eggs were found in July by Joel A. Allen. That autumn, the Smithsonian's Henry W. Henshaw discovered Baird Sparrows "in immense numbers" on the high grassland of southeastern Arizona and southwestern New Mexico; he collected at least 18 specimens, some of which he suspected of having bred in the region.

Within little more than a decade, the Baird Sparrow's true breeding range would be more or less outlined, its migration periods defined, and its range found to extend into northwestern Mexico. In 1885, Ernest Thompson Seton was able to offer the first life history notes on this species, which he found "exceedingly abundant" in western Manitoba. Coues posed the obvious question: why, after four

decades, had the floodgates opened to let so many Baird Sparrows pour forth?

> Has the bird really been, all this time, so common and widely dispersed, or has it only recently become so?

In response, he suggested that the bird's populations were cyclical, subject to "a kind of rotation in [the species'] abundance." If such was the case, the number of Baird Sparrows present—and shot—on the prairies of the northern Great Plains in the 1870s and 1880s represented a point in the cycle higher than has ever been reached since by what is now a scarce and only locally distributed bird.

FIELD IDENTIFICATION

This species is now only rarely confused with the Marsh Wren, but opportunities for misidentification abound in the case of a bird so small, so brown, so furtive, and so little-known as the Baird Sparrow. Though Spencer Baird, in his office at the Smithsonian, was able to discern a number of features that would "distinguish this species very readily," others, with actual field experience of the bird, found it more challenging. Elliott Coues thought the species "so much like a savanna[h] sparrow that it was some days before [he] learned to tell the two apart, at gunshot range, often shooting one by mistake for the other." At least one pair of Kansas Henslow Sparrows gave their lives when they were mistaken for Baird Sparrows, and Alexander Wetmore's later search for the species in the eastern part of that state "entailed a considerable mortality among obscurely marked individuals of Le Conte's Sparrow." The AOU *Check-list* almost always limits itself to matters of taxonomy and distribution, but the editors of the most recent, seventh edition issue a most unusual and most clear warning in the account for the Baird Sparrow: "Many sight records refer to misidentified Savannah Sparrows."

That abundant grassland sparrow, which breeds and winters in enormous numbers at the same sites as its much scarcer cousin, remains the major source of identification errors on the part of eager searchers for Baird Sparrows. The abundance of the Savannah Sparrow over so much of North America means that many observers have never looked closely at the bird—not, that is, until they are in search of a rarity. At that point, in the high grasslands of southeastern Arizona or on the open prairies of eastern Montana, some birders are confronted for the first time with the wide range of variation shown by Savannah Sparrows, and some birders decide that a bird that looks somehow "different enough" must be the sought-after Baird. As the AOU's admonition suggests, there is no easier and no surer way to find a "Baird" Sparrow than not to know the Savannah.

The Baird Sparrow is a rather small, chunky sparrow with a short but fairly broad tail, short wings, and a large head; the large bill and flat, uncrested crown give it a forward-leaning look. In the field, this species often gives the impression of a contrastingly dark-streaked back and mostly white or off-white underparts, sometimes with visibly buffier flanks. Where both species occur, Savannah Sparrows in the areas of range overlap seem longer and slightly longer-, narrower-tailed, with a smaller, often short-crested head and bill. Their back usually strikes the observer as paler and less heavily patterned; the underparts are clear white.

In good views—more easily obtained in the nesting range than in winter—adult Baird Sparrows show blackish back feathers, scapulars, and wing coverts with fine deep chestnut edges and broad yellowish cream borders; juveniles are similar, but the edges and tips of the feathers are even more conspicuous, creating a distinctive scaled or scalloped appearance. The tail feathers are edged with light cream, with the result that Baird Sparrows in flight away from the observer, as they are most often seen, appear to be pale-tailed.

Savannah Sparrows within the range of the Baird Sparrow are much plainer and more uniformly colored above, the dull gray-brown back streaked with black and white stripes. The tail pattern is variable, but even in those individuals that show whitish edgings, the overall impression made in flight is of a dark tail.

Harder to see and, at least for observers without experience, harder to assess is the head pattern, traditionally touted as the best way to identify the Baird Sparrow. Unlike the more crisply marked Savannah Sparrow, this species has an "open" face, tinged with pale buff and contrasting strongly with the blackish lateral crown stripes and the very narrow but well-defined whisker streak; the rear portion of the pale buffy ear coverts often shows one or two black spots. A small but significant number of Savannah Sparrows also have colorful faces, but with a decidedly greenish yellow cast rather than the soft peach of the Baird; even those bright-headed Savannahs show darker, more heavily streaked ear coverts that are almost completely enclosed by a black whisker, rear border, and eye line. The supercilium of the Savannah Sparrow is thus better defined and more clearly separated from the cheek than is that of the Baird Sparrow, which merges into the rest of the buffy face.

The Baird Sparrow's ochre median crown stripe can be bright and conspicuous, but it can also be difficult to glimpse, especially in winter, when these birds are very shy. A more useful mark, first made

The yellowish tan of the head plumage varies from deep and extensive, as here, to sometimes barely noticeable. The central crown stripe can be dull orange or pale buffy white. *North Dakota, June. Brian E. Small*

The fine black lateral throat stripe and whisker border a whitish or buffy jaw stripe; the lore is pale. *North Dakota, June. Brian E. Small*

explicit by Kevin Zimmer 30 years ago but still underappreciated by most birders, is the color and pattern of the nape. In Baird Sparrows, the nape is rich buffy yellow, with scattered black streaking, contrasting with both the back and the thick lateral crown stripes. The ground color of the nape of most Savannah Sparrows is dull grayish, and the black streaking is finer, denser, and more regular, such that there is little contrast with the remainder of the upperparts.

On the wintertime grasslands of the desert Southwest, behavior reliably serves to indicate the presence of a Baird on fields covered by Savannah Sparrows. Savannah Sparrows flush easily, and fly low over the ground with extravagant, almost larklike swooping; they frequently give their distinctively high, thin call in flight. They land at a distance, but usually in sight, perched atop a grass stem, on a fence wire, or even on bare ground. Baird Sparrows are more reluctant to fly, and seek safety in direct, usually silent flight close to the ground, virtually never landing in the open or taking a conspicuous perch; instead, they drop to the base of a grass clump and run, at times far enough never to be located again.

Baird and Henslow Sparrows barely overlap in distribution, even in migration, but the two can be difficult to distinguish, and particular care should be taken in identifying a bird encountered outside the expected range of either species. In addition to a pronounced fondness for native grassland and a secretive nature, these closely related sparrows also share a large, square head and fairly large bill; the "open" face pattern, in which the ear coverts are only indistinctly set off from the supercilium, is remarkably similar in both. In the bright, harsh light typical of their favored habitat, the difference between a golden buff and an olive-green face can be difficult to see. The best and often most easily visible character separating the two is the wing pattern. While the Baird Sparrow has dull, medium-brown primaries, secondaries, and greater coverts—the last with small black spots—and gray-edged black tertials, the Henslow is far more richly colored, with blackish brown primaries and extensive deep rusty brown on the secondaries, greater coverts, and tertials. In flight, a Henslow Sparrow is dark ruddy, while a Baird is paler and more obviously variegated above.

Juveniles of both species can be seen away from the nesting areas in late summer, in the case of Baird Sparrows into late September. The juvenile Baird Sparrow is darker-winged than the adult, but lacks the deep chestnut of the Henslow; juvenile Henslow Sparrows are also distinctively greenish yellow above, unlike the variegated, scalloped back pattern of Baird Sparrows of the same age.

The flight calls of the Baird Sparrow are very high, short, and thin *tseep*s, with almost no audible decay. Perched birds give a variety of lower-pitched chips, sometimes dry like a Lincoln Sparrow's, sometimes fuller and more musical. The simple, beautiful song begins with several soft notes resembling the musical chip, followed by a light, loose trill; there may be a shift to a different pitch in the middle of the trill. At a distance, poorly heard Savannah Sparrows can sound similar.

RANGE AND GEOGRAPHIC VARIATION

The Baird Sparrow breeds only on the northern Great Plains, from south-central Alberta, southern Saskatchewan, and southeastern Manitoba south to central Montana, northern South Dakota, and, irregularly in very small numbers, northwestern Minnesota. Stray males occasionally establish territories south and east of the breeding range in Wyoming, Colorado, and Nebraska, with reports east to Wisconsin and Ontario. Three of the four records for British Columbia have been in early June, suggesting a similar phenomenon of "pioneering" males; there are accepted nesting records for none of those states and provinces.

The specimen record shows that in the 1870s this species was a fairly early autumn migrant, with several August birds from the wintering grounds in southeastern Arizona; the numerous early reports of suspected breeding from that region are all based on such early migrants. Today, this species is only very infrequently encountered south of the breeding range before October, with most winterers seeming to arrive in Arizona at the end of that month. This apparent change may be in part a matter of method: it seems likely that more early-arriving Baird Sparrows elude identification today than they did 140 years ago, when the shotgun provided greater certainty.

The known winter range has shrunk considerably since this species' apparent heyday in the mid-1870s. Once fairly widespread over much of eastern Arizona, since the 1880s it has been reliably found only in the extreme southeast of the state. Numbers are very small in the "heel" of New Mexico; the species is more often reported in extreme western Texas. The winter range in Mexico is not well known. Recent surveys by the Rocky Mountain Bird Observatory have confirmed wintering populations in northeastern Sonora and northwestern Chihuahua; the species penetrates south through western Coahuila as far as northern Durango and Zacatecas.

The usual route of migration in this species presumably takes it across the western Great Plains of New Mexico, Colorado, and Wyoming; there are remarkably few records for Oklahoma, Kansas, and Nebraska, though Baird Sparrows breed directly north and winter directly south of the western portions of these states. A relatively robust specimen record from Missouri suggests that migrants regularly occurred at least formerly in that state and Iowa.

Vagrants have occurred west to southeastern British Columbia, Washington, and southern California and east to New York and Maryland. Most such strays are detected in October and November.

No subspecies are recognized.

Big-headed and big-billed, this species is uncommon to scarce over most of its range. *Ohio, May. Brian E. Small*

HENSLOW SPARROW
Centronyx henslowii

The severe tightening of credit in the western United States in 1818 led to the still young nation's first peacetime financial crisis, the Panic of 1819. Among the victims was the struggling young Franco-American merchant John James Audubon, whose declaration of bankruptcy in July 1819 opened the bars of Louisville's debtor's prison, but could not free him from the continuing persecutions of his creditors. The harried entrepreneur turned to teaching French and painting portraits to support his family, until in the late winter of 1820 he found employment at the fledgling Western Museum in Cincinnati. During the half year he spent at the museum, Audubon and the curator, Robert Best, made occasional collecting forays across the Ohio River into Kentucky, where one day the companions encountered a sparrow "on the ground, amongst tall grass . . . . Perceiving it to be different from any which [he] had seen, [Audubon] immediately shot it."

When a decade on it came time to publish a formal account of the new sparrow in the *Ornithological Biography*, Audubon would remember the "many kind attentions" bestowed on him in Eng-land by Thomas Henslow, then professor of botany in Cambridge, where among his most promising students he counted Charles Darwin.

Writing in 1831, Audubon knew nothing of the life history or distribution of the Henslow Sparrow he had collected more than a decade before. In the years following, though, he and John Bachman, his coauthor in the project that would lead to the *Viviparous Quadrupeds*, "procured a great number" of specimens in South Carolina, and "found it in large numbers in all the pine barrens of the Floridas" in winter. James Trudeau and Edward Harris reported to Audubon that this sparrow was numerous in the agricultural fields of New Jersey, where it bred. Audubon went on to describe the range of the "abundant" Henslow Sparrow as the mid-Atlantic Coast from Maryland to New York in the breeding season, moving to the Carolinas, Georgia, Florida, Alabama, and Louisiana for the winter; he thought it merely "accidental" west to Ohio.

It is difficult to imagine anything less like the current status and distribution of the Henslow Sparrow. It is now, and has long been, virtually extinct as a breeding bird in any of the coastal states where Audubon and his contemporaries found it so abundant. To the extent that it remains common anywhere, it is on the grasslands of the old Midwest and the southeastern Great Plains, where Audubon never saw it; indeed, this species was not recorded west of its Kentucky type locality until June 1857, when Ferdinand V. Hayden collected one in what is now central Nebraska.

While the Henslow Sparrows of the midwestern prairies simply went undetected until the dawn of intensive ornithological activity there, the fate of this species in the easternmost portions of its breeding range parallels closely the history of the region's land use. Before European settlement, the Henslow Sparrow was doubtless very rare in most places east of the prairie. The clearing of the Carolinian forests for agricultural exploitation and grazing probably benefited this bird by creating large-scale openings in what had been a densely wooded landscape; it is no coincidence that the reports of "abundant" Henslow Sparrows came at the same time as extensive agriculture was at its zenith in the American Northeast.

The species' dramatic decline in those same areas can be laid to the end of the farming tradition in the East and the subsequent reforestation and urbanization of agrarian landscapes that once hosted such grassland birds as Dickcissels and Lark Sparrows. The encouraging return of the Henslow Sparrow in the old Midwest of Ohio and Pennsylvania has coincided with new management practices that encourage the growth and maintenance of grasslands; the region's history of strip mining has proved a blessing in disguise to these and other open-country birds as reclamation efforts produce new habitats similar to the extensively managed fields and pastures so familiar to Audubon and his contemporaries more than 150 years ago.

FIELD IDENTIFICATION

To observers familiar with the species, adults of this chubby, large-headed, huge-billed sparrow are virtually unmistakable. The experience required to gain such familiarity is not easily had; most birders, especially those living outside of the core midwestern breeding range, rarely see Henslow Sparrows, and very rarely see them well, with the result that most birders' mental image of this bird comes not from experience but from a book.

This is a very old problem, one that plagued even nineteenth-century collectors with the bird in the hand. On receiving two specimens from John Krider in the early 1880s, John W. Detwiller labeled them both as Henslow Sparrows, an identification they carried for a quarter century—until Witmer Stone examined the birds and discovered that they were in fact Baird Sparrows, probably collected by Krider on his Dakota expedition of 1881.

Fortunately for observers relying on their eyes rather than mustard seed shot, those two species only rarely overlap outside of the specimen drawer.

The upperparts of the adult are quite blackish, while the folded wing shows considerable chestnut. The bill is very heavy, with virtually no interruption between the upper mandible and the forehead. *Ohio, May. Brian E. Small*

At all ages, the tail shape differs markedly between the Henslow and the Baird Sparrow: while the latter has a nearly square, somewhat notched tail tip, the central rectrices of the Henslow Sparrow are noticeably longer than the outermost tail feathers, creating a graduated, even cuneate outline that can be visible in the field to the patient sparrow watcher, especially when the bird is landing on a perch or preening.

Details of the wing pattern are also helpful. While the Baird Sparrow has dull, medium-brown primaries, secondaries, and greater coverts—the last with small black spots—and gray-edged black tertials, the Henslow is far more richly colored, with blackish brown primaries and extensive deep rusty brown on the secondaries, greater coverts, and tertials. In flight, even a Henslow Sparrow of the somewhat duller, western subspecies is ruddy, while a Baird is paler and more obviously variegated above.

The head patterns of the two species are quite similar, with a fine black whisker defining the lower ear coverts but the rear and upper edge of those coverts less clearly set off from the rest of the face; as in the Baird, the rear border of the Henslow Sparrow's ear coverts is marked with one or two smudgy black spots, the remnants, as it were, of a poorly defined black eye line. The pattern of the crown is also similar in both species, a median stripe separated from the face by a black lateral crown stripe.

The colors of the head, however, differ between these two grassland sparrows. The head of a Baird Sparrow is overall white, gray, and dull orange; the adult Henslow Sparrow's head is dark dull olive on the jaw stripe, ear coverts, supercilium, central crown stripe, and nape, a somberness that combines at any distance with the deep colors of the back and wing to create a dark impression unlike the often notable paleness of the Baird.

Juveniles of these two sparrows are also extremely similar. In addition to its graduated tail and even bulkier bill, the juvenile Henslow Sparrow differs from the Baird in its much less conspicuously streaked breast and more finely marked reddish upperparts, without the heavy, regular scaling of the Baird. The face is dull olive rather than buffy, as in the juvenile Baird; the dull whitish throat is not separated from the jaw stripe by a well-defined black lateral streak, thus appearing noticeably wider in the field. The broad central crown stripe in both species can have a yellowish cast, obscuring the distinction between the Baird Sparrow's ochre and the Henslow Sparrow's green.

Seen even reasonably well, the pale buffy, orange-lored adult Grasshopper Sparrow is unlikely to be mistaken for a Henslow Sparrow, though both species share the large head and bill and often striking white eye-ring. It is the juvenile Grasshopper Sparrow that occasions more confusion. Plainer above than their parents, juvenile Grasshopper Sparrows with their hint of a lateral throat stripe and neatly streaked necklace across a buffy breast band can recall adult Henslow Sparrows. The narrow white central crown stripe, pale jaw stripe and supercilium, mottled brown back, and plain gray-brown wings with narrow white wing bars of the Grasshopper Sparrow distinguish it neatly from the darker-headed, rustier-backed and rustier-winged Henslow.

If the juvenile Grasshopper Sparrow has been taken for the adult Henslow, the juvenile Henslow has occasionally been misidentified as an adult Grasshopper. As the young Roger Tory Peterson joked, "Something surely went amiss here." Almost unstreaked on the breast, with a pronounced yellowish buffy wash to the underparts, the juvenile Henslow Sparrow is nevertheless easily distinguished from the adult Grasshopper by its dark crown and reddish secondaries.

The winter ranges of the Henslow and the LeConte Sparrows overlap almost entirely. Both adults and juveniles of the LeConte are colorful and pale, with wide straw-colored stripes running down the back and a more or less striking orange outline to the grayer ear coverts; the bill is quite small and slender. These characters can be surprisingly easy to see even in flight, and immediately distinguish this species from the much darker, larger-headed and larger-billed Henslow Sparrow. The Nelson and Saltmarsh Sparrows are less sandy above than the LeConte, but they, too, are conspicuously orange on the face, and their bills are longer, thinner, and more

The short, "sharp" tail is often badly worn by contact with grasses and the ground. *Ohio, May. Brian E. Small*

Fine black streaks form a vague necklace across the upper breast of the adult Henslow Sparrow. The juvenile is nearly unstreaked beneath; its duller upperparts and plainer wings may recall an adult Grasshopper Sparrow. *Ohio, May. Brian E. Small*

pointed than the enormous bill of the Henslow Sparrow.

Like Nelson and Saltmarsh Sparrows, the Seaside Sparrow is decidedly a bird of the marshes, favoring wetter habitats at all seasons than the Henslow. In winter, along the coasts of upper Texas, Louisiana, and western Mississippi, Henslow Sparrows could conceivably occur within distant sight of tidal wetlands inhabited by the dark rust and olive Fisher Seaside Sparrow; the dark reddish wings, buffy breast, and somber grayish olive head of that bird could recall a Henslow Sparrow, but the Seaside is significantly larger, with a long, almost blackbird-like, pointed bill and a well-defined gray cheek patch bordered with dull orange-buff.

All Henslow Sparrows appear stocky and fairly short-tailed, with a flat back, full breast, and remarkably large, square head with an outsized bill; the crown is flat, with almost no visible "stop" at the base of the bill. The usual first impression is of a dark sparrow with a faintly streaked breast, the streaks underlain by a soft tinge of buff separating the white belly from the white throat. Though the back feathers are edged in crisp white when they are fresh, in worn plumage many birds simply look blackish brown above; the wing coverts and secondaries are bright rust, even deeper reddish than the same feathers on most Swamp Sparrows.

In some individuals, the head is truly green; in others, especially juveniles, it appears in the field to be an odd yellowish brown, entirely unlike the yellows and oranges shown by similar sparrows. The central crown stripe can be difficult to distinguish from the broad black lateral stripes, but the adult's neat paired whisker and lateral throat stripe—the so-called "double whisker"—is usually easily seen, particularly on singing birds. Juveniles usually lack the lateral throat stripe, a further distinction from Baird Sparrows.

RANGE AND GEOGRAPHIC VARIATION

This species is now essentially extirpated as a breeding bird in New England and the mid-Atlantic coastal region south to New Jersey. Occasional singing males can still appear at scattered locations even here, but only very exceptionally do those wanderers find mates and produce young. The maps created by eBird are especially eloquent: although not all records of this species are submitted to eBird, of course, it is nevertheless unlikely that over the past ten years in the coastal strip from Maine to Virginia there were significantly more nesting-season Henslow Sparrows than the nine birds reported there.

Breeding Henslow Sparrows are vastly more frequent and more common to the west of the former range of the eastern subspecies *susurrans*. The summer range of this species—now presumably everywhere represented by the nominate race *henslowii*—is expanding in many areas in response to the restoration of grasslands in agricultural areas, disused strip mines, and other formerly inhospitable habits. The Henslow Sparrow now breeds very locally but more or less regularly from southern Minnesota through Iowa, Missouri, and southeastern Nebraska to eastern Kansas, thence east through Illinois, Indiana, Michigan, Ohio, and southern Ontario to western New York, Pennsylvania, Maryland, and West Virginia; smaller numbers are found in at least some summers in the Dakotas, Virginia, Maryland, and North Carolina. One singing male was discovered in July 2009 on James Bay in Ontario, the northernmost record for the species; there was no evidence of breeding. Suitable habitat within this large range is scattered and on the whole scarce, such that Henslow Sparrows may be common on one particularly suitable tract of grassland and entirely nonexistent elsewhere in the area. Further, not all presumably acceptable habitat is used each year; like many other birds of tall grass, this species appears to be erratic in its long-term distribution, an impression heightened by its relative scarcity over so much of its range.

Because of their usually silent and furtive behavior, migrant Henslow Sparrows are rarely detected on the passage between the breeding and wintering grounds. Most southbound birds move in late September and October, with a few lingering in nesting areas into November. Birds in the western portion of the breeding area return to nesting sites from mid-April to May, a date range consistent with the limited phenological data available from the old eastern breeding range.

Wintering Henslow Sparrows are found across the southeastern United States, from eastern Texas across southern Arkansas, Louisiana, Mississippi, and Alabama to the coastal Carolinas, southern Georgia, and most of Florida. Exceptionally, individuals may attempt to overwinter in the breeding range. This species is difficult to detect in winter, leaving the precise outline of its normal nonbreeding range uncertain, especially toward the north. Apparent vagrants have been reported from the Texas Panhandle, Colorado, New Mexico, Nova Scotia, New Brunswick, and the Bahamas; most such strays occur in the fall migration period.

Given the historical pattern of European settlement in North America, the nominate subspecies of most wide-ranging birds is that found breeding on or near the mid-Atlantic Coast. In the case of the Henslow Sparrow, however, the nominate subspecies, *Centronyx henslowii henslowii*, is the western race, while the eastern race was not recognized and described until nearly a century after Audubon collected his type on what was then the western frontier of the United States.

In 1918, William Brewster, one of the great gentleman-amateurs of the American ornithological tradition, determined that the Henslow Sparrows breeding east of the Alleghany Mountains were "easily distinguishable from the Ohio Valley" birds. The new subspecies, which he named *susurrans*, "the whisperer," Brewster described as larger than "the true *henslowi*" of the West, with a larger, thicker bill and significantly more reddish upperparts, the "bright chestnut . . . sometimes so widespread that the dull black central areas of the feathers are . . . narrowed and otherwise obscured." Pyle notes that there is some overlap in bill size between the two recognized subspecies, and that the difference in the color of the upperparts holds true only on the average.

While the breeding ranges of the two subspecies are, by definition, distinct, Brewster in late autumn 1874 had shot "an ultra-typical example of the Ohio Valley form . . . true *henslowi*" in Massachusetts. With the wholesale extirpation of this species from nearly all of its eastern breeding range, there is now some unclarity as to whether its infrequent and irregular reappearances there in recent years represent relic individuals of *susurrans* or wandering birds, like Brewster's, from the western population. It is possible that the nominate *henslowii* has "swamped" and supplanted *susurrans* over its former eastern range. "Indeed, whether the eastern race, *A. h. susurrans*, still exists as such is uncertain."

LARGE-BILLED SPARROW
Passerculus rostratus

By the turn of the twentieth century, the breeding ranges of nearly every bird species of the United States and Canada had been identified. A small number of nests and eggs still stubbornly resisted collection—the Harris Sparrow, for example, guarded its secrets until as late as 1931—but even in those few cases, ornithologists and explorers knew more or less where to look.

In the late summer of 1852, Adolphus L. Heermann returned to Philadelphia with 1,200 bird skins packed in his trunk, the booty from three years of collecting in California. Among them were several specimens of an undescribed finch Heermann had collected the winter before in San Diego. John Cassin at the Academy of Natural Sciences found this "plain-plumaged bird . . . unlike any other finch that I have ever seen," and named it *Emberiza rostrata* for its long, stout bill.

Though Heermann reported that it was common on the wintertime beaches of southern California, the sparrow's summer haunts were still a mystery more than 50 years later, when Joseph Grinnell reported that

> strange as it may seem, there is a land bird of California which abounds at times in suitable places but whose nesting grounds appear to be entirely unknown. This species, our only land bird yet remaining thus distinguished, is the Large-billed Sparrow . . . . The interval between its departure in the spring and arrival in the autumn amounts to a period of four months, during which we know nothing of its whereabouts.

Grinnell's review of half a century's winter records of the Large-billed Sparrow led him to a daring hypothesis:

> That the Large-billed Sparrow breeds somewhere to the *south* of its winter home, and *migrates north* in the fall, returning southwards each spring! Such a suggestion may seem absurd, but nevertheless fits best the data . . . . what a remarkable exception [that] would be to the rule of southward migration in the northern hemisphere!

The riddle was finally answered by Luther Goldman, who in May 1915, more than 60 years after Heermann shot his first in San Diego, discovered the breeding grounds at the mouth of the Colorado River. As Harry Oberholser observed, Goldman's find

> also solves the migration mystery of this species, fully as interesting a result . . . . The great majority begin, by the middle of August, to leave their breeding ground, whence they move in various directions . . . . The winter range extends southward, westward, and northwestward like a very short widespread fan, with its axis at the mouth of the Colorado River. The extreme length of the winter range from northwest to southeast is approximately 1150 miles, while the known breeding area is only about 30 miles long.

With that, the breeding range of the last of North America's terrestrial birds was defined, and ornithology could move on to the next, even more controversial question of just what a Large-billed Sparrow is.

John Cassin's faint bewilderment at the sight of Heermann's San Diego sparrows was apparent in his assignment of those birds to the catch-all genus *Emberiza*—and in his initial comparisons of the specimens with the Bachman and the Baird Sparrows, neither of them a species that springs to the modern birder's mind when encountering a sparrow on the rocky beaches and salicornia flats of the upper Sea of Cortez. By 1855, Cassin had rethought the generic affinities of what he was now calling the Long-billed Marsh Sparrow, determining on the basis of its long bill and partiality for coastal locations that it should be considered the Pacific representative of the group including the Seaside and Saltmarsh Sparrows of the East Coast, with which it would share the genus *Ammodramus*.

"With some hesitation," Spencer Baird took a different approach. Apart from its "enormously large" bill, Baird found that the California bird resembled the Savannah Sparrow more than any of the other streak-breasted sparrows, and in 1858, he placed Heermann and Cassin's sparrow in the genus *Passerculus*, which Charles Bonaparte had founded for the Savannah Sparrow. For the next 30 years, there was near unanimity in following Baird's classification; even Heermann himself adopted *Passerculus* over Cassin's *Ammodramus*.

Of far greater consequence than this more or less expected ebb and flow of generic identities was the AOU's decision in 1944 to henceforth treat three of the four *Passerculus* sparrows recognized in the fourth edition of the *Check-list* as a single species, the Savannah Sparrow, with 16 subspecies north of

The broad, blurry streaks below and relatively plain upperparts and head are nearly as distinctive as this southwestern sparrow's very large bill. *California, January. Brian E. Small*

Mexico; only the Ipswich Sparrow was (temporarily) maintained as a full species, a status it would enjoy until 1973.

For the past three-quarters of a century, the AOU has continued to treat the Large-billed Sparrow as nothing more than a subspecies group of the Savannah Sparrow; in the years since the publication of the 1944 Supplement, the committee on taxonomy and nomenclature has twice rejected proposals to resplit that bird into two, three, or four species. Genetic studies, however, have suggested strongly that the Savannah Sparrow in its broad, conservative sense comprises three distinct clades—a conclusion that aligns neatly with both the experiences of field observers and statistical analyses of plumage pattern and color and of size and structure.

As in so many cases, among the American sparrows and other groups, too, the question is reduced to a simple one: What kinds of difference, and how much difference, constitute significant—species-level—distinctions? Scientific ornithology, with its ever greater emphasis on biochemical methods, can be expected to reach different conclusions from those that satisfy field observers, whose responsibility and whose pleasure consist in recording differences discernible to the eye and ear but perhaps masked in the laboratory. As Joseph Grinnell remarked in his criticism of an earlier revision of these sparrows, "consistency of treatment is impossible." Neither is it always helpful. Most birders will have little difficulty identifying most Savannah-like sparrows to "kind," whether each kind is better considered a species or a subspecies group; ignoring those differences, whatever their ultimate scientific significance, flattens nuance and discourages the collection of knowledge.

FIELD IDENTIFICATION

Medium-large, short-tailed, and extremely heavy-headed, Large-billed Sparrows of both subspecies, reddish or gray, heavily streaked or obscurely marked, are very distinctive in the field. Though *atratus* is rather darker overall than the nominate race, both give an immediate visual impression of uniformity, with relatively little contrast between the gray or gray-brown back and dirty whitish underparts with poorly defined scattered streaks. The plain brown, medium-short tail matches the equally undistinguished color of the very short wing; two thin white wing bars are visible in fresh plumage. At any distance, the most striking plumage feature is the white, faintly speckled throat, which stands out well from the otherwise subdued pattern of the head; the jaw stripe is also white, but the broad supercilium is often only slightly paler yellowish brown than the ear coverts and the solid or nearly solid brown crown. Juveniles are similar, but even less regularly marked below.

The most striking character of this species is the bill, nearly as long as the head is wide and noticeably swollen at the base. The lower mandible and most of the upper are grayish pink. At close range, the long, curved culmen is obviously dark, and there may be a blackish tip to the lower mandible. The sturdy feet and toes average duller pink than in the other *Passerculus* species.

RANGE AND GEOGRAPHIC VARIATION

The Large-billed Sparrow comprises two subspecies: nominate *rostratus* and *atratus*. The southern *atratus* is slightly larger and larger-billed than the decidedly dispersive *rostratus*. As its name suggests, *atratus* ("blackened") is much darker and grayer than the pale, rather washed-out *rostratus*, and shows broader, blacker streaking above and below.

This sparrow breeds only in Mexico, its preference for coastal salt marsh leading to discontinuous nesting range. Southern *atratus* appears to show only limited seasonal movement out of its saltmarsh breeding range in southern Sonora and Sinaloa, though some appear in winter in southern Baja California Sur. The nominate subspecies breeds from the Colorado River delta of northern Baja California Norte south to central Sonora; nesting takes place in early spring, and migrants disperse coastally to southern and central California, northern Baja California, and southern Sonora and Sinaloa starting as early as June. Away from the coast, *rostratus* is an uncommon to scarce July to March visitor to California's Salton Basin. There are also one August specimen and at least two winter sight records from the lower Colorado River in Arizona.

Heermann's first specimens of the "Large-billed Marsh Sparrow," one of which is depicted here in a lithograph by William E. Hitchcock, mystified even so experienced an ornithologist as John Cassin. *Image from the Biodiversity Heritage Library. Digitized by the University of Pittsburgh Library System. | www.biodiversity library.org*

Somewhat larger-billed than the widespread Savannah, the Belding Sparrow is also strikingly dark and crisp in its streaking above and below. *California, March. Brian E. Small*

BELDING SPARROW
Passerculus guttatus

Richard C. McGregor, posted as ornithologist to the Philippine Bureau of Science, in September 1907 wrote home to his California colleagues, "I don't enjoy these birds near as much as I would a good bunch of sparrows." McGregor was no doubt recalling one of his earliest collecting trips, an expedition led by A. W. Anthony in the spring of 1897. The 26-year-old McGregor, on leave from his undergraduate studies at Stanford, had come home from Baja California with "a good bunch of sparrows" indeed, including juvenile specimens of the Socorro Spotted Towhee and the San Benito Sparrow, along with 16 skins, three eggs, and the nest of a bird McGregor would describe as a new species, *Ammodramus halophilus*, the San Ignacio Lagoon Sparrow.

Though McGregor had consulted with Robert Ridgway in preparing his species description, the older scientist reconsidered shortly thereafter. In 1901, Ridgway recognized the Lagoon Sparrow—

meanwhile renamed the Abreojos Sparrow—as a distinct subspecies of the Large-billed Sparrow. A dissenting voice, that of William Brewster, was heard almost immediately: while Brewster agreed that the "new" sparrow was a kind of Large-billed, he suspected that McGregor's discovery was the unrecognized solution to a different, much older puzzle.

In December 1859, the colorful and charismatic Hungarian-American naturalist John Xantus was in San José del Cabo. Employed by the United States Coast Survey, Xantus collected hundreds of specimens for the Smithsonian, including a heavily marked sparrow that would lie on its back in a drawer for nearly eight years, until George N. Lawrence described it as new to science.

What Lawrence named *Passerculus guttatus*, the St. Lucas Sparrow, proved remarkably elusive. Apart from the single specimen taken by Xantus, only three more were killed before 1897—all in southernmost Baja California and all in fall and winter, leaving the breeding grounds and summer plumage of the bird a mystery. On comparing nine of McGregor's sparrows from San Ignacio Lagoon with three skins of the St. Lucas Sparrow, Brewster in

Many Belding Sparrows show only a little yellow on the face; this bird is notably colorful. *California, March. Brian E. Small*

1902 realized what even the excellent eye and keen calipers of Robert Ridgway had missed: that the two "species" were in fact probably identical, and that McGregor had discovered not a new species but the long-sought nesting grounds of Xantus and Lawrence's St. Lucas Sparrow. Two years later, in its 1904 Supplement, the AOU agreed with Brewster, eliminating McGregor's Lagoon Sparrow (Ridgway's Abreojos Sparrow) "from the *Check-list* as equivalent to *P[asserculus] rostratus guttatus,* in summer plumage." (Bizarrely, *halophilus* would reappear without explanation in the third edition, only to be synonymized once again in 1949.)

Meanwhile, 30 years earlier, Robert Ridgway had recognized that another mysterious streaked sparrow from the California coast, named by Spencer Baird *Passerculus anthinus,* was actually "two quite different birds, one of which is a very dark-colored form of" the Savannah Sparrow, the other "so very different in its appearance as to convey at once, in the case of spring and summer birds, the impression of a decidedly distinct species." Ridgway went on to describe that "decidedly distinct" sparrow as *Passerculus beldingi,* the Belding's Marsh Sparrow, and the species entered the AOU *Check-list* in its first edition. Not until 1944 was the Belding Sparrow—along with many other sparrow populations, including *guttatus*—lumped as a mere subspecies of the Savannah Sparrow, the status it has retained in the AOU lists ever since.

Recent study, however, has suggested that *beldingi* and *guttatus* are more closely related to each other than is either to any other *Passerculus* sparrow.

The underparts streaking is made up of broad, usually wedge-shaped blackish feather centers. *California, March. Brian E. Small*

Many authorities now treat *beldingi* and *guttatus* as making up a distinct species, the Belding Sparrow. That English name honors Lyman Belding, "a naturalist of the old school" and "last of the Pioneer ornithologists of California." Belding had come to the study of birds late, and not knowing that there were ornithologists in his home state, he began to send his specimens to the Smithsonian "for Mr. Ridgway's opinion."

> Mr. Ridgway was very patient and prompt in writing, long interesting letters concerning the specimens I had sent . . . . I do not think this kind encouragement was exceptional, for I think Profs. Baird and Ridgway were always glad to assist the student of natural history,

especially, perhaps, a student who would spend the winters of 1881–1882 and 1882–1883 collecting around the Sea of Cortez. It was at San Diego, in the March following his return from Baja California, that Belding shot the two sparrows his mentor Ridgway would eventually name for him.

FIELD IDENTIFICATION

This sparrow is easily identified in spring on its scattered breeding grounds, where it is usually the only sparrow present in its favored coastal salt marshes. In their California range, breeding Belding Sparrows can be seen near Song Sparrows, which differ in their much longer tail, plumper form, browner overall plumage without white on the back, and noticeably browner streaking below; the Song Sparrow's bill is shorter, thicker, and less spikelike, and the head lacks the conspicuous yellow supercilium and the hint of a white central crown stripe shown by some Belding Sparrows.

The breeding range of the northern race *beldingi* is nearly adjacent to that of the Savannah Sparrow subspecies *alaudinus*, resident just to the north on the California coast. The northern edge of the range of *beldingi* appears to be uncertain, but the two may breed within 20 miles of each other in the vicinity of Morro Bay; even there, however, there is not known to be any locus of contact (or, if the Belding Sparrow is considered conspecific with the Savannah, any site of transition). That race of the Savannah Sparrow is quite similar to *beldingi*, but paler and grayer, with lighter, less extensive dark spotting below and a finer, shorter bill; *alaudinus* may show less, or less bright, yellow on the face. The upperparts of *beldingi* are more clearly tinged with olive and less coarsely streaked. Ridgway also notes that the Belding Sparrow's legs and feet are darker gray-brown than those of the Savannah. Nevertheless, if breeding distribution is not taken into account, this is one

The long, sturdy tarsus and alert posture are characters shared with several other marsh-dwelling sparrow species. *California, March. Brian E. Small*

of the most subtle identifications in North American birding, and the visual similarity between these two birds is one important reason that the Belding Sparrow is often considered to represent a subspecies group within the Savannah Sparrows rather than a distinct species.

While Savannah Sparrows of the subspecies *alaudinus* appear to be sedentary, individuals from other populations can occur in winter in the range of the Belding Sparrow. All are paler, shorter-billed, and less heavily streaked, with more obvious white edgings to the back feathers and usually a more conspicuous white median crown stripe.

In the southern portion of the species' range in Baja California, migratory Belding Sparrows can be seen with wintering Large-billed Sparrows of either subspecies. In addition to the evident difference in bill size, Large-billed Sparrows are heavier, paler, and less conspicuously marked above and

The back stripes of the Belding Sparrow are coarse black against a brown background, typically with little white edging. *California, March. Brian E. Small*

below than Belding Sparrows; the Belding Sparrow typically shows brighter yellow on the supercilium and face than the blander, more uniformly colored Large-billed.

Belding Sparrows are medium-small, dark ground-feeders with short tails and wings and long, spikelike bills. The tail is noticeably notched, the feathers edged bright white in fresh plumage. With their bold striping above and below and usually bright yellow head markings, perched birds can momentarily recall a tiny Red-winged Blackbird. In some individuals, the largely black, white-fringed tertials contrast greatly with a reddish wing panel created by the rusty outer webs of the secondaries, making the Belding Sparrow a strikingly colorful bird.

RANGE AND GEOGRAPHIC VARIATION

Much of the uncertainty about the precise taxonomic status of the *Passerculus* sparrows of western Mexico arises from the fact that their breeding ranges are so strikingly disjunct; birds from the different named populations encounter one another only in winter, if even then. These are the circumstances that press the biological species concept virtually to its breaking point: Do birds nesting far apart, with hundreds of miles of unsuitable habitat separating them, retain the potential for interbreeding, or has their isolation resulted in so great a divergence that they would continue to mate with their own "kind" even should they somehow come into secondary contact?

The northernmost Belding Sparrows, of the subspecies *beldingi*, are resident from Point Conception in California south along the coast to El Rosario in Baja California Norte. This is also the breeding *Passerculus* of the Islas Todos Santos, where the birds use dry scrub for nesting in the absence of the otherwise strictly preferred salt marshes.

The next link in the chain of Belding Sparrow being is the population of *anulus*, resident in marshes on the east shores of Sebastián Vizcaíno Bay and surrounding Lago Ojo de Liebre, the type locality.

The breeding range of *guttatus*, the nominate subspecies, surrounds the Laguna San Ignacio in Baja California Sur. Unlike its conspecifics to the north, this subspecies is migratory, at least some individuals leaving the breeding grounds after nesting to move south as far as the southernmost part of the peninsula. The southern race *magdalenae*, breeding in the marshes around Magdalena Bay, shares the dispersive habit of *guttatus*, though in this case, too, only some members of the population appear to move out of the breeding area in winter, others lingering north. Thus, in winter, the southern portion of the Baja California peninsula can host Belding Sparrows representing *guttatus* and *magdalenae*, alongside Large-billed Sparrows of both races.

The smallest and, on average, darkest of the races of the Belding Sparrow is the northern *beldingi*. The olive-washed back is moderately streaked black; the underparts are densely streaked with black.

The southern *guttatus* is slightly larger and heavier-billed. The ground color of the back is duller, tending rather to gray than to brown, and the black streaking is diffuse. The streaking of the underparts is dark brown rather than black. Both subspecies show deep, bright yellow in the supercilium.

In addition to the nominate *guttatus* and *beldingi*, two other subspecies have been described. In 1926, Laurence M. Huey named the rather small, black-streaked birds from Lago Ojo de Liebre, on the west coast of Baja California Sur, *anulus*. Huey found that the gray back of *anulus* was washed with lighter olive and more narrowly marked with black than in the other races then recognized. Van Rossem considered that distinction "minor," but emphasized that

anulus, true to its name ("link of a chain"), connects *beldingi* and *guttatus* in the intermediate size of its bill.

Van Rossem himself described the fourth subspecies, *magdalenae*, which breeds in the marshes and estuaries of Magdalena Bay, on the outer coast of Baja California Sur. Similar in size to *guttatus,* these birds are lighter and more greenish above, with better-defined streaking and more prominent pale edgings on the back. The streaking of the underparts is brown. The bill is shorter and thicker than that of *guttatus*, with a more distinctly curved culmen. Both *magdalenae* and *anulus* show yellow in the supercilium.

While most authorities today continue to recognize all four subspecies, the *Handbook of the Birds of the World* synonymizes *anulus* with *beldingi* and *magdalenae* with *guttatus*.

Characteristically unkempt, juvenile Belding Sparrows are darker and more heavily marked than Savannah Sparrows of the same age. *California, March. Aaron Budgor*

Heavy-billed and dark, this is the only *Passerculus* sparrow resident on the San Benito Islands in the Pacific off Baja California. *Baja California. Tim Stenton*

SAN BENITO SPARROW
Passerculus sanctorum

On the label of the first known specimen of this island specialty are two notes in two different hands: "*P. guttatus* Lawr.! (R.R.)," and underneath that "Scarcely! stet *sanctorum*—C." It is not known when this penciled exchange took place, or how much time intervened between the first shot and the return volley; but even in their terseness, these comments memorialize not just the relationship between two of America's greatest descriptive ornithologists but the decades-long confusion about the status and identity of two small sparrows shot by Thomas H. Streets on a tiny island off Baja California.

Streets, naturalist on the United States Navy's 1873–1875 survey of the North Pacific islands, collected the birds on San Benito Island, the largest of the three rocky islands of the San Benito Archipelago. An oddly jumbled description of the sparrows was published in 1877, in which they were assigned to George N. Lawrence's still mysterious *Passerculus guttatus*, then known only from the type specimen. Though the published report appeared under Streets's name, he expressly credited the identification of the birds to Elliott Coues—the "C." of the specimen tags.

Coues would later complain that the skins brought by Streets "were not in good order, and did not furnish entirely satisfactory indications"; R. C. McGregor, less tactful, affirmed in the same year that the birds had been less skinned than "mummified." In spite of the poor state of preservation of the material before him, Coues included Streets's sparrows in the second edition of his *Key*, noting that they were "like" *guttatus* but larger, with bills as large as that of the Large-billed Sparrow of the Mexican mainland. He speculated that they might in fact represent a new species, which he named, rather tentatively, *Passerculus sanctorum*. (He later explained the unusual name: "There are so many places named in Lower California for [saints] that I concluded to dedicate this Sparrow impartially to the whole calendar of them.")

Robert Ridgway—"R. R." in the conversation on the labels—was of a different opinion. He found the controverted specimens "essentially identical" in color to the type of *guttatus*; the birds' measurements, however, identified them firmly as Large-billed Sparrows, even though, as Ridgway admitted, they were "quite appreciably different in coloration, and also in the form of the bill, from ordinary" members of that species. It is small wonder that exclamation marks erupted on the specimen tags.

Coues complained that his description and his name were ignored by the AOU, but he was vindicated when an expedition organized by A. W. Anthony returned from the San Benitos with "a fine series" of skins, "showing the assigned specific characters to be valid." The AOU's committee on taxonomy and nomenclature "promptly accepted" the species in the Eighth Supplement of 1897, altering the English name from the Couesian "All Saints' Sparrow" to the simpler "San Benito Sparrow."

Coues died in 1899. In the Eleventh Supplement, published two and a half years later, the AOU committee reversed itself to adopt, without comment, Ridgway's assessment of *sanctorum* as "merely an insular form of *P. rostratus*"; the name was accordingly changed to *Ammodramus rostratus sanctorum*. More interestingly, the committee simultaneously changed the author designation for the name *sanctorum*. In 1897, the name had been credited to Coues; now, in 1902, Ridgway was cited as the author, as he would be in the third, fourth, and fifth editions of the *Check-list*. Only recently, with the publication of the *Handbook of the Birds of the World*, has Coues's role in the story once again been acknowledged and his name once again attached to that of the San Benito Sparrow.

FIELD IDENTIFICATION

At all seasons, this is believed to be the only *Passerculus* sparrow on the San Benito Islands. A sturdy, notably short-tailed bird with a large, thick-based bill, the San Benito Sparrow most closely resembles the Large-billed Sparrows of the Gulf of California, which disperse in winter to the Cape region of Baja California and could conceivably stray to the San Benitos, which are 50 miles from the mainland. The best distinctions from that species are the subtly different bill shape, with a straight rather than curved culmen, and the better-defined, darker streaking of the underparts. The absence of yellow on the head and the dark brown rather than black underpart streaking distinguish the San Benito Sparrow from the Belding Sparrow, which is resident on the adjacent mainland.

RANGE AND GEOGRAPHIC VARIATION

This species is restricted to the three islands of the San Benito Archipelago, where it is said to be the most abundant and conspicuous land bird present. The highest densities are on San Benito del Este.

Though resident on its breeding islands, the San Benito Sparrow may engage in some limited seasonal movement. Old records from the southernmost mainland of Baja California rest on uncertain identifications.

No subspecies are recognized.

Robert Ridgway's 1874 illustrations for the *History of North American Birds* represent an early stage in the attempt to unravel the mysteries of the Savannah-like Sparrows. *Image from the Biodiversity Heritage Library. Digitized by the Smithsonian Libraries. | www.biodiversitylibrary.org*

SAVANNAH SPARROW
Passerculus sandwichensis

The New World sparrows seem to have tried Alexander Wilson's considerable patience more than once. In 1811, the Father of American Ornithology published the account of a new sparrow. At the conclusion of his description, he felt obliged to apologize to the reader for its wordiness:

> The very slight distinctions of colour which nature has drawn between many distinct species of this family of Finches, render these minute and tedious descriptions absolutely necessary.

Wilson named his new Savannah Sparrow for the city where he first discovered it, though by the time he came to write the *American Ornithology*, he knew that it was "generally resident" along the Atlantic Coast north to New York; he found the species especially common in the marshy fields of southern New Jersey, where he collected and painted two birds later donated to the "noble collection" of Charles Willson Peale in Philadelphia.

Though he had access to all the relevant handbooks, in preparing his "minute and tedious" account of the Savannah Sparrow Wilson somehow failed to notice the similarity of his New Jersey bird to one described a quarter of a century earlier by John Latham and Thomas Pennant. That bird, which Latham called the Sandwich Bunting, had been "met with at Aoonalashke, and Sandwich Sound, by our late voyagers"—namely, the crew and scientific staff of James Cook's third voyage, aboard the *Resolution*, in search of a Northwest Passage. The specimen, like many others from the ill-fated voyage, had passed into the collections of Joseph Banks. Gmelin named it *Emberiza sandwichensis* for the Alaskan type locality, now known as Prince William Sound, in the thirteenth edition of Linnaeus's *Systema naturae*; Wilson used the English translation of that work, by his contemporary William Turton.

Certain ornithologists took umbrage at the names "*sandwichensis*" and "Sandwich." Otto Finsch, for example, adopted the old name *arctica*, protesting that the bird had "in no way any connection to the Sandwich Islands," meaning Hawaii, and that there was "no rational way" in which the name *sandwichensis* should be applied to it. Others, though, such as Spencer Baird, resigned themselves to abiding by the principle of priority: since "*sandwichensis*, *arctica*, and *chrysops* [coined by Pallas in 1826] all seem to apply equally well, it will be best to take the oldest name as a provisional one at least."

Baird was also the first to point out in print that the Sandwich Bunting was "almost exactly like" Wilson's sparrow, differing only in its slightly larger size and, obviously, far-distant western range. It was up to Joel A. Allen, later the first president of the American Ornithologists' Union, to argue that "the *Emberiza sandwichensis* of Gmelin unmistakably refers to" the Wilsonian Savannah Sparrow. With a large series of Massachusetts specimens at hand, Allen demonstrated that all of the purported differences in size, bill shape, and color distinguishing *sandwichensis* from Wilson's *savanna* recurred within the New England population, and that they should be put down rather to individual and sexual variation than to any true distinction at the level of species. In this, Allen echoed his friend and colleague Elliott Coues, who himself had recently observed that among large numbers of specimens he had "shot about Washington [D.C.], I have found fully as great differences as I have ever detected in comparing the eastern with western forms."

Coues, Allen, and the other members of the committee responsible for the first, 1886 edition of the AOU *Check-list* stood by their taxonomic convictions, lumping *savanna* and two additional West Coast forms as mere subspecies of *sandwichensis*. Simultaneously, the committee moved the genus *Passerculus* into *Ammodramus*, a taxonomic reorganization that gave Coues pause: since the Grasshopper Sparrow was already *Ammodramus savannarum*, then the name *Ammodramus sandwichensis savanna* for the eastern Savannah Sparrow could be considered to be preoccupied, "a particularly awkward and unlucky matter" if so that would cost that latter species "its most distinctive designation—the very one, too, that gives it its common English name." Coues's proposed replacement for *savanna* was *wilsonianus*, a respectful attempt to commemorate the role of the Father of American Ornithology and his "minute and tedious" descriptions in the discovery of the Savannah Sparrow—even if it was for the second time.

FIELD IDENTIFICATION

Abundant nearly everywhere in its range and generally—at least in migration and winter—confiding and tame, the Savannah Sparrow is worth getting to know well. This species fits many birders' notions of the "generic" sparrow—small, brown, and streaked. But its great range of individual and geographic variability and its superficial resemblance to several other species makes a thoughtful encounter with the Savannah Sparrow one of the best opportunities for learning how to look at passerellids.

The Savannah Sparrow's combination of a rather long, pointed wing and short, slightly notched tail serves to eliminate many possibilities, both in flight

Individual variation essentially obliterates geographic variation in this very widespread species. This bird from southeasternmost British Columbia shows the neat fine streaking on white underparts and the thin, pale bill typical of Savannah Sparrows. *British Columbia, June. Brian E. Small.*

and on the perched bird. Over much of its range, too, particularly in the East, the small size and small-headed, slender-billed impression are distinctive from the considerably larger, heavier, more obviously block-headed Vesper and Song Sparrows.

Savannah Sparrows are traditionally distinguished from Vesper Sparrows by the absence of conspicuous white in the tail. In fresh fall plumage, however, many Savannahs have very noticeable pale or whitish edges to the tail feathers, especially evident in flight directly away from the observer. On perched birds in any plumage viewed from beneath, those edges and the relatively pale ground color of the translucent feathers can make the undersurface of the tail look entirely white.

Birders with less experience—or birders in areas where Vesper Sparrows are uncommon and eagerly sought—may be misled by the apparent color and pattern of a Savannah Sparrow's tail. If there is uncertainty, the length and shape of the tail can be used to distinguish these two species: while the Vesper Sparrow has a moderately long, broad tail, Savannah Sparrows always have short, narrow tails that can look like an evolutionary afterthought. The individual feathers are also more conspicuously

pointed than in the Vesper Sparrow, though juvenile Vespers also have narrowly tapered (though significantly longer) outer tail feathers.

The back pattern, too, helps distinguish Savannah from other, superficially similar sparrows. Whatever the ground color of the back, Savannah Sparrows have neater, finer, more "penciled" black streaks than Song or Vesper Sparrows, and the pale to bright white edges of the mantle feathers line up to form two parallel stripes, especially obvious on darker individuals. Both Song and Vesper Sparrows are more coarsely marked above, with much more diffuse patterning, a useful character when birds are actively feeding in grass, heads and tails invisible. On the Savannah Sparrow's folded wing, the more richly colored secondaries and greater coverts from a bright brown panel contrasting with the colder brown or grayish back are most noticeable in paler, grayer individuals; Vesper and Song Sparrows are more nearly uniform on the wing.

Beneath, most Savannah Sparrows give the impression of being very white, with at most a tinge of dull buff restricted to the rear flanks. The streaks are fine and very dark, ranging from deep brown to sooty to blackish, and range from distinctly

The long white back streaks distinguish this species from the Song Sparrow, in all populations of which the upperparts are brown, gray, and black. *British Columbia, June. Brian E. Small*

each species makes its own, recognizably different impression. Song Sparrows are the most uniformly colored and least patterned of the three, creating a "helmeted" appearance that contrasts with the white throat and dark, conspicuous, wedge-shaped or triangular lateral throat stripes; this is least apparent in the pale, southwestern subspecies, but those are at the same time the subspecies least likely to be confused with a Savannah (or a Vesper) Sparrow. The Vesper Sparrow's more complex head pattern combines dark ear coverts with a wide white jaw stripe, which wraps around the coverts to surround them on three sides; if the Song Sparrow is wearing an old-fashioned leather football helmet, the Vesper Sparrow has donned earmuffs.

The Savannah Sparrow is heavily, neatly marked on the head, resulting in a pattern of alternating light and dark stripes that is recognizable even when the individual marks responsible for it are not. The bright white throat (spotted or unspotted) is bordered by a fine, narrow lateral throat stripe, which is bordered by a white jaw stripe, which is bordered by a fine black whisker, which is bordered by paler ear coverts, which are bordered by a fine black eye line, which is bordered by a whitish to yellow supercilium, which is bordered by a finely streaked dark lateral crown stripe, which gives way finally to a pale, often very narrow median crown stripe. This overall dark-light-dark pattern is unlike that of any other sparrow, and is particularly striking in pale individuals viewed head-on, when it is every bit as dramatic as that of a Black-and-white Warbler. Savannah Sparrows have rather pale, pinkish horn–colored bills, adding one more contrasting element to the pattern.

The size and shape of the bill also differ from the bills of Song and Vesper Sparrows. In all Savannah Sparrows, even the smallest-billed birds, the bill is shallow at the base and acutely pointed, usually—unlike the Belding and Large-billed Sparrows—giving the impression of straightness and sharpness. Song Sparrows have darker, bulkier, more obviously curved bills, while the dull pink bill of a Vesper Sparrow is notably swollen at the base, tapering quickly to a blunt tip.

Eager searchers occasionally misidentify Savannah Sparrows as the much rarer Baird or Henslow Sparrow. The plumage differences are briefly discussed in the accounts for those species; even more helpful, though, is the Savannah Sparrow's quite different behavior. In migration and winter, Savannah Sparrows are most often encountered in small to medium-sized groups of up to 15 or 20 individuals; the flocks are not especially cohesive, and may form adventitiously when birds aggregate at an attractive food source. The *Centronyx* sparrows do not flock, and even when they share the same grassy fields

disorganized to long stripes. The extent and density of streaking on the underparts are variable, generally greater in the West than in the East, but in all birds, the streaks cover the entire breast and continue without obvious interruption onto the flanks; the flank streaking is of the same color and width as that of the breast. In adults, the streaking is densest at the center of the breast, commonly forming a spot.

Vesper and juvenile Song Sparrows are far less starkly dark and white below, with browner, broader streaking against a duller, buffier ground. In Vesper Sparrows, the streaks are often organized into a clear band across the upper breast, continuing more sparsely and narrowly down the sides and flanks; there is usually a subtle contrast between the yellowish tinge of the breast and flanks and the purer, unmarked white of the throat and belly. Most juvenile and many adult Song Sparrows are more thoroughly buffy over the entire underparts.

The precise head patterns of these three species can be difficult to see in the field, but each is distinctive; and even when the finer details are not visible,

with Savannah Sparrows, they are usually deep in the dense vegetation, not on the open ground of paths and edges.

RANGE AND GEOGRAPHIC VARIATION

The Savannah Sparrow occupies a vast breeding range, stretching from the Russian Far East across the northern Canadian provinces and territories to Newfoundland and Labrador, south in the East to South Dakota, Iowa, northern Illinois and Indiana, Ohio, West Virginia, Pennsylvania, and New Jersey. In the West, Savannah Sparrows breed to central California, Nevada, Arizona, New Mexico, and Colorado. The disjunct Middle American population is resident from Chihuahua to the southern Mexican Plateau, with rare or former breeders as far south as the highlands of southwestern Guatemala.

Coastal populations from southwestern British Columbia south appear to be resident, though their permanent range is invaded in migration and winter by birds from farther north. Migrants are otherwise found from September to November and again in April and early May in appropriate habitat virtually everywhere in North America, south to western Nicaragua and Honduras; there are two early October records from the Shetland Islands. Savannah Sparrows are considered accidental in Korea, but are regular and rare in Japan in the winter.

The winter range may be pushing slowly north, but most winter in a broad band across the southern tier of the United States, from California to the Carolinas. The species is variably uncommon to scarce in the northern Caribbean.

In a rare moment of bemused modesty, Elliott Coues once wrote, "I must candidly confess . . . that I cannot distinguish P. [*sandwichensis*] *alaudinus* from the common Eastern P. [*sandwichensis*] *savanna*." Joel A. Allen expressed his frustration with the division of skins of *Passerculus* sparrows into geographic races in far stronger terms:

> These specimens are separable to some extent into several series, which may be based either upon difference in general size, the character of the bill, or upon coloration; but these several kinds of variation fail to corroborate each other. If separated upon differences in size, the two or more series thus separated embrace every combination of the other differences; and similar incongruities result when the separation is made upon differences in coloration or other characters. Yet the Massachusetts specimens present among themselves differences as well marked and of the same character as is assumed to distinguish several of the so-

called species [later to be considered subspecies] from the Pacific coast, that have been proposed and adopted by different authors.

Allen determined that whatever true geographic variation there might be in the Savannah Sparrows was entirely masked by individual variation and variation due to age and sex, a conclusion too rarely kept in mind by observers eagerly "sorting through" migrant Savannah Sparrows, light and dark, large and small, snouty and petite, each fall.

Nevertheless, the naming of Savannah Sparrow races remained a thriving enterprise in the first half of the twentieth century, coming eventually to a total of 28 described races. The descriptive tradition reached what was presumably intended to be its acme in the revision carried out by James L. Peters and Ludlow Griscom in 1938. As an exercise in typological connoisseurship conducted by two pairs of keen and experienced eyes, the paper they

The very short, notched tail can show considerable white. The most bruited field mark, the yellow forepart of the supercilium, is quite variable in extent and saturation, and may be entirely invisible in the field. *California, May. Brian E. Small*

produced is a monument to taxonomic methods that would soon be rendered obsolete by advances in genetics, evolutionary theory, and molecular biology. Indeed, there was contemporary—almost immediate—criticism of Peters and Griscom's methodology as too heavily reliant on "changeable personal hunch," an excessive concentration on the appearance of the individual over the characters of its population, and a confusion of the biological with the nomenclatural connections between the forms subjected to analysis.

Just how thoroughly subjective the division of the Savannah Sparrows into races could be was highlighted, perhaps inadvertently, by Peters and Griscom themselves in the introductory comments of their paper: while the senior author convinced himself of the validity of the subspecies *labradorius*, Griscom simultaneously grew equally certain that some of the specimens Peters had assigned to that race "were no such thing," surely a first sign that reasonable men—or even these coauthors— might differ. Nevertheless, Peters and Griscom ultimately determined that 18 subspecies should be recognized, including two new races described in the paper: *oblitus*, breeding from Hudson Bay and western Quebec south to Minnesota; and *crassus*, breeding on the islands of Alaska's Alexander Archipelago. (The numbers given here and in the immediately following paragraph include various populations that are treated in this guide as belonging to the Belding, Large-billed, and Ipswich Sparrows.)

The next edition of the AOU *Check-list* followed almost 20 years later, in 1957. It listed 16 races, deleting *crassus* and the Mexican subspecies *brunnescens* and maintaining separate species status for the Ipswich Sparrow; the *Check-list* adds *Passerculus sandwichensis rufofuscus*, the breeding Savannah Sparrow of Chihuahua, Arizona, and New Mexico, described two years after Peters and Griscom published their revision. Pyle numbers 14 diagnosable races north of Mexico, while the latest edition of the Howard and Moore *Complete Checklist, Birds of North America,* and the list of the International Ornithological Congress tally 17 throughout the range of the species, as understood there to include all members of the genus *Passerculus*.

Allen's venerable warning about variation and variability within a single population suggests that the lines between subspecies could be drawn arbitrarily almost anywhere—and still never produce a consistent system of differences and similarities. On the other hand, the near consensus of the eminently authoritative modern sources just mentioned is encouraging. The subspecies descriptions below are arranged by geography in the sequence set forth by Pyle, whose diagnoses they rely on heavily.

The subspecies *crassus*, named by Peters and Griscom for its stout bill, breeds on the coast of southeastern Alaska, moving south in the winter along the coast to California. The underparts of this medium-large Savannah Sparrow are fairly heavily streaked dark brown; above, it is dark brown with blackish streaking. The supercilium is deep yellow, the ear coverts tinged with the same color.

One of the smallest and most attractive of the Savannah Sparrows, *brooksi* is a familiar breeding bird on the West Coast from British Columbia to northernmost California, with some dispersal in winter to the southern California coast. Small-billed and pale, with a golden cast to the upperparts, this subspecies shows a subdued yellowish supercilium and heavy blackish streaking above and below. Known appropriately enough as the Dwarf Savannah Sparrow, these birds were named in 1915 for their first collector, the British Canadian painter Allan Brooks, "ornithologist, artist, and brave defender of his country."

The coastal salt marshes from Humboldt County south to San Luis Obispo County, California, make up the year-round range of the resident *alaudinus*, a small, long-billed, and dark Savannah Sparrow. This bird and its name were the subject of considerable controversy 80 years ago: Peters and Griscom stripped the name, coined by Bonaparte in 1853, from the interior birds to which it had been traditionally applied and reassigned it to the coastal birds once named *bryanti*, largely on the basis of a reexamination of the type specimen in Paris, "a mounted bird, worn and faded, furthermore 'partially albinistic.'" With heavy dusky streaking above and below and a deep yellow supercilium, *alaudinus* in its post-1938 understanding has been described as the "transition" between the coastal Savannah Sparrows and the Belding Sparrow of the California coast, with which it may interbreed in the vicinity of Point Conception.

The "original" Savannah Sparrow, of the nominate race *sandwichensis*, breeds from the eastern Aleutians east to mainland western Alaska. It is large and large-billed, with brown, black-streaked upperparts and sparsely streaked white underparts. The supercilium is dark yellow. This subspecies winters coastally from British Columbia to central California.

Another originally Bonapartean name, *anthinus* was reapplied by Peters and Griscom to the northwestern population breeding inland from western Alaska to northwestern Manitoba, in winter extending its range south to southern California, Arizona, New Mexico, and Texas. Slightly smaller than the nominate *sandwichensis* and rather small-billed, this subspecies is heavily streaked dark brown above and below; the supercilium is fairly dark yellow.

Breeding in the interior from British Columbia and western Manitoba south across the western Great Plains and Great Basin to eastern California and Arizona, *nevadensis* is the familiar wintering Savannah Sparrow over much of the Southwest. The "standard" Savannah Sparrow for many observers, this subspecies is aptly described by Pyle as "medium" in size, bill size, upperparts color, and underparts streaking. In the field, the supercilium usually appears to be only faintly yellowish.

The race *rufofuscus* breeds in a small range encompassing the mountains of central Arizona and New Mexico, perhaps ranging into Chihuahua; it is lumped by *Birds of North America* with the distantly disjunct Mexican race *brunnescens*, from which the original describer of *rufofuscus* was able to distinguish it by the brighter brown tones of the upperparts and the heavier black streaking above and below. Medium-sized, this subspecies has a deeper yellow supercilium than does *nevadensis*, the more northerly subspecies whose breeding range most nearly approaches that of this race.

Peters and Griscom's *oblitus* is also lumped by *Birds of North America*, with the more easterly breeding *labradorius*. As diagnosed by its describers and as understood by Pyle, *oblitus* is a medium-sized, long-tailed, small-billed Savannah Sparrow with dark, black-streaked upperparts and sparse crisp black streaking below. It breeds from eastern Manitoba east to the western Great Lakes, wintering commonly across the southeastern quarter of the United States, with the exception of Florida. The eastern counterpart, *labradorius*, breeds in Quebec and Newfoundland, but its winter range apparently overlaps largely with that of *oblitus*, a fact that has been adduced, rather irrelevantly, to "argue against the latter's diagnosability." The eastern *labradorius* is very slightly shorter-tailed and more heavily streaked below than *oblitus*.

The equivalent observation—that wintering birds occur in nearly the same numbers in the same area—led the authors of the *Birds of North America* account to lump *mediogriseus* with Wilson's *savanna*. When *mediogriseus* is admitted as a distinct subspecies, it is distinguished from *savanna* by its darker, more grayish upperparts and coarser, more blackish streaking below, and from *oblitus*—when that subspecies is recognized as distinct—by its paler, duller upperparts and finer streaking below. This subspecies breeds to the east of *oblitus* and to the west of *labradorius*, from Ontario and western Quebec south to Illinois, Ohio, Pennsylvania, and New Jersey; it winters, like *oblitus* and *labradorius*, in the southeastern United States, west to the edge of the Great Plains in Kansas, Oklahoma, and Texas.

When *oblitus, mediogriseus*, and *labradorius* are

Many show a richer brown panel in the folded wing, while others, like this California bird, are more uniformly gray-brown. *California, May. Brian E. Small*

accounted "good" subspecies, *savanna*, the race that bears Alexander Wilson's name, is restricted as a breeding bird to Nova Scotia and Prince Edward Island; in the nonbreeding season, this remains a coastal bird, wintering from eastern Massachusetts south to the central and even southern Florida peninsula. This is a medium-sized Savannah Sparrow with a small bill, medium-brown upperparts, fairly sparsely streaked underparts, and a pale to moderately dark yellow supercilium.

The recognition of so many Savannah Sparrow subspecies breeding in the East and wintering, more or less all together, in the Southeast raises the difficult question of just which Savannah Sparrow Alexander Wilson shot, painted, and named *savanna* (the question is rendered happily moot if *oblitus, mediogriseus*, and *labradorius* are considered synonyms). If the female specimen from which he described the species was taken in New Jersey in the breeding season, then Wilson's name would properly be applied to the local breeding race *mediogriseus*; if it was collected in winter or spring migration, the type could represent any one of the four eastern-wintering subspecies. Faced with irresoluble uncertainty—

Wilson failed to provide the date on which he visited Great Egg Harbor—Parkes and Panza affirmed the correctness, if the arbitrariness, of letting *savanna* in the strict sense apply to the birds of Nova Scotia, thus preserving that name independent of the taxonomic validity of the other names assigned eastern populations.

In December 1879, the Indiana naturalist Amos W. Butler collected half a dozen Savannah Sparrows in south-central Mexico, which revealed "certain peculiarities" that led him to describe them as representatives of a new subspecies, *brunnescens*. The birds somewhat approach in darkness the race today known as *alaudinus* (then known as *bryanti*), but are larger and more brightly colored brown, with large bills. The throat, breast, and breast sides are heavily streaked brown or black, and the ground color is buffy rather than bright, pure white. This race is now known to be resident from Chihuahua south to Michoacán and Puebla. John P. Hubbard considered *rufofuscus* inseparable from *brunnescens*, in which case the range of the subspecies extends north to Arizona, New Mexico, and Colorado, where it "intergrades toward" *nevadensis*.

In the proper season and habitat, Savannah Sparrows of the migratory races are often among the most abundant and characteristic birds to be seen. One generally recognized resident subspecies, however, *Passerculus sandwichensis wetmorei*, has been driven nearly to extinction, probably by the agricultural development of its only known locality in extreme southwestern Guatemala. The bird was rediscovered and its song and breeding habits described for the first time in the summer of 2016. Earlier records of the species in Guatemala, from 2002, 2004, and 2006, may have been of *brunnescens* or even, conceivably, of overshooting migrants from even more northerly populations.

The subspecies *wetmorei* is known from eight specimens taken by W. B. Richardson in the high mountains at Hacienda Chancol in a single June week in 1897. Van Rossem described the new form as much darker and browner above than *brunnescens*, with a yellow supercilium becoming white at the rear; the bill is slightly larger. He compared it in general coloration to "the darkest and brownest" individuals of *alaudinus*, but much less heavily streaked below. While all of the specimens collected by Richardson were, naturally, in worn plumage, Van Rossem suspected that fresh autumn and winter birds "must be very richly coloured indeed."

As has been observed from nearly the beginning, the variation among Savannah Sparrows is complex and confusing. Individual variation within geographic populations is great, and for the most part those geographic populations differ from one another not abruptly but clinally, one kind grading imperceptibly into another in color, size, and structure. The more or less consistent differences in color discernible between some populations can be cogently explained as adaptations to habitat conditions, especially in the interest of cryptic coloration. In these circumstances, there appears to be not

> much value in delimiting subspecies on the basis of clinal variation, unless there are well-defined steps in the clines. Chopping clinal variation into subspecies results in more or arbitrarily delimiting overlapping groups on a phenetic continuum. [There is] no virtue in naming subspecies when the only way they can be reliably separated is on the basis of locality.

The response, carried out in practice in the account for this species in the *Handbook of the Birds of the World*, is to treat all of the Savannah Sparrows on that continuum as indistinguishable at the level of subspecies, thus making the species monotypic. (The authors of the *Handbook* account do maintain subspecies status for the Ipswich Sparrow, which is treated in this guide as a full species.)

This solution may appear to be a mild form of taxonomic nihilism, a throwing up of the hands on confronting a taxonomic past and a phenotypic present so badly tangled as to resist any attempt at a sensible unraveling. In fact, however, while it does not prevent the identification of populations by breeding locality, it does sweep away two centuries of nomenclatural nightmares, sparing readers, writers, and observers confusion of the sort warned against by Grinnell and, regrettably, created by Peters and Griscom.

The treatment of the Savannah Sparrow as monotypic (or as comprising only two, very well-marked subspecies) has the further advantage of highlighting the very different taxonomic structures of the coastal Belding and Large-billed Sparrows, which, in comparison to the Savannah Sparrows, show abrupt and relatively clearly diagnosable geographic variation.

What is most significant to the conscientious observer is that the treatment espoused in the *Handbook* removes all temptation to attach a name to the different Savannah Sparrows encountered in the field. Instead, birders can concentrate on discerning the subtle differences between large birds and small birds, dark birds and pale birds, heavily streaked birds and lightly spotted birds, yellow-browed birds and cream-browed individuals, and so on; fixing an arbitrary and poorly understood subspecific name to a bird is often the surest way to preempt close examination.

Large, frosty, and thick-billed, this Atlantic coastal specialty is most often seen on its wintering grounds. *New Jersey, March. Brian E. Small*

IPSWICH SPARROW
Passerculus sandwichensis princeps

"**While the Ipswich Sparrow** has been maintained as a distinct species ever since it was first described, there are no reasons (except possible sentimental ones) that would warrant the continuation of such a course." Whatever the "true" taxonomic status of this highly distinctive *Passerculus* sparrow, it merits separate treatment and full consideration by birders—not merely on "sentimental" grounds, as Peters and Griscom had it, but because the human story behind its discovery and identification says so much about the ornithological enterprise in nineteenth-century America.

Known to the old-time inhabitants of Nova Scotia's Sable Island simply and aptly as "the gray bird," this globally rare passerellid was formally described to science in 1872. C. J. Maynard was a famous taxidermist and collector, and the author of the first guide dedicated to the sparrows and finches of eastern North America. Maynard had actually collected the first specimen of his "Large Barren Ground Sparrow" in the early winter of 1868, in coastal northeastern Massachusetts, but the publication of his new species was delayed—not by printers and editors, but by a case of mistaken identity.

Maynard first encountered this puzzling sparrow in the sand dunes of Ipswich, in Essex County, Massachusetts. Rather than assume from the start that the bird was fully unknown and new to science, Maynard reasonably sought an identification among the known birds of North America. Stymied, he sent one of the specimens to Spencer Baird at the Smithsonian. Maynard reported that Baird identified it as the first record for Massachusetts of the sparrow Audubon had named for him 30 years earlier:

> Previous to the capture of this there was but one specimen extant, which was one of the original birds captured by Audubon upon the banks of the Yellowstone River, July 26, 1843. My specimen, through the kindness of Professor S.F. Baird, has been compared with the original, which is in his possession, and pronounced identical . . . .

Baird had in fact been somewhat more circumspect in his response to Maynard's inquiry, writing that "in all essential points it seems to be the same bird," while noting at the same time that the Massachusetts skin was paler and less clearly marked than the Audubonian specimen, and differed further in the color of the median crown stripe, "paler, not as fulvous as in the type." Those differences, though, Baird suggested might be due to the fact that while Audubon's breeding sparrows were worn and ragged, Maynard's, taken in November and December, still wore the fresh plumage of "clear autumnal birds."

Even in the 1860s, recording a new species for the heavily birded state of Massachusetts was a great accomplishment. Maynard commissioned a woodcut portrait of the Bay State's first Baird Sparrow from his frequent collaborator E. L. Weeks to serve as the frontispiece of his *Naturalist's Guide*, and in the accompanying bird list, he provided a full description of his specimen and a comparative table of the measurements of that bird and the North Dakota type. Maynard shot two more in mid-October of 1871, leading William Brewster to consider Baird Sparrows "regular winter visitants from the North."

In the spring of 1872, Maynard had occasion to visit the Smithsonian collections himself. He brought his Massachusetts sparrows with him, and discovered on comparing them to the type of the Baird Sparrow that they were not identical after all.

Elliott Coues, who also examined the skins, hinted coyly that "a full investigation will reveal something not now anticipated," a suspicion confirmed when Maynard published a formal description of the new sparrow later that year.

In the *Naturalist's Guide* of 1870, when he still believed that the Massachusetts dune skins were Baird Sparrows, Maynard was at pains to point out the distinctions between his birds and the Savannah Sparrow, asserting that the two could not "justly be compared" given the differences they exhibited in overall size and bill size. In 1872, though, satisfied that he had in fact discovered a new species, Maynard could afford to admit the similarity between what he now called the Barren Ground Sparrow and the Savannah Sparrow, which he now went so far as to consider "closely allied"; thus, he assigned the new bird to the genus *Passerculus*, giving it the specific epithet *princeps*, "the first," probably in reference to its superior size, and possibly in allusion to this appealing species' visual primacy among the sparrows.

Baird, Thomas Brewer, and Robert Ridgway ratified Maynard's diagnosis in their 1874 *History*, confirming that even "making all possible allowance for seasonal differences in coloration," it was "impossible to reconcile" the birds from Ipswich with the Baird Sparrow, which was then still known from only the single "rather defective and worn" speci-

The fine streaks of the breast and flanks are cinnamon or black with cinnamon edges. *New Jersey, March.* Brian E. Small

men in Baird's possession. The *History* was the first substantial handbook to use Maynard's *princeps* for this bird, though it changed the English name to the more manageable Ipswich Sparrow, the label under which the bird has been known since.

Maynard published a new edition of the *Naturalist's Guide* in 1877, now calling the bird he had discovered the Pallid Sparrow. A certain resentment against Baird and his role in the earlier misidentification shines through Maynard's revised text: in 1870, he explains, he had indeed listed the Baird Sparrow for Massachusetts,

> but in justice to myself, will say through no fault of mine, I being misled by others, not having an opportunity for comparing the specimens taken with the [type of] *Centronyx*.

The inclusion of the Ipswich Sparrow in the Baird *History* made other ornithologists and collectors aware of the existence of this new and rare bird, and soon "the once prized Ipswich Sparrow" had been recorded as a migrant or winterer from the sandy coasts of Nova Scotia and New Brunswick to Virginia and Georgia.

The breeding grounds, though, remained a mystery until 1884. In that year, Robert Ridgway published a notice of "a considerable series of eggs" labeled as belonging to Savannah Sparrows—but much too large to have been laid by that species. The eggs had been collected in 1862 on Sable Island, Nova Scotia; Ridgway suggested "that they may be in reality those of the Ipswich Sparrow," and requested the help of his colleagues in determining whether that was the case. C. Hart Merriam wrote to a missionary in residence on the island, who promptly sent a specimen of the Sable Island "gray bird"—which turned out "to be an unquestionable Ipswich Sparrow." The first nests and definitively identified egg clutches were collected there by Jonathan Dwight in the summer of 1894.

The first published illustration identified as of an Ipswich Sparrow was the woodcut Weeks prepared in 1870; for the new edition of the *Naturalist's Guide*, Maynard had the cut printed in color (Dwight, with considerable justification, would later call the result "wretched"). But another painting, published six decades before Maynard announced his Massachusetts discovery, suggests strongly that if he was the first to recognize the Ipswich Sparrow as something new, he was not the first ornithologist to shoot one.

In January 1893, Isaac Norris De Haven informally addressed a meeting of the Delaware Valley Ornithological Club, pointing out that Alexander Wilson—the club's adopted patron saint—had pre-

Rock jetties and sparsely vegetated dunes are the best places to look for this bird. *New Jersey, March. Brian E. Small*

pared not one but two paintings of the Savannah Sparrow, a bird Wilson collected, painted, and described as new to science. The type specimen of the new *Fringilla savanna* was a female, a choice Wilson obviously regretted:

> The drawing of this bird was in the hands of the engraver before I was aware that the male was so much its superior in beauty of markings and in general colours.

That male—"a very beautiful" specimen, "delicately marked"—had to wait until the next volume of the *American Ornithology* for its turn at description and depiction. And that male was an Ipswich Sparrow, unmistakable in both text and image, frosty gray and white with "small pointed spots of brown" below. If this individual had not coincidentally been a male, or the other not coincidentally a female, Wilson might have considered the striking differences between them more than sexual—and he might have been more than the merely unwitting discoverer of another new sparrow.

FIELD IDENTIFICATION

Most birders encounter this sparrow on its Atlantic Coast wintering grounds, when the plumage is still fresh after the prebasic molt undergone on Sable Island. In winter, the large size and frosty overall

The upperparts are dominated by frosty gray, with little of the contrast shown by other Savannah Sparrows. *New Jersey, March. Brian E. Small*

aspect are distinctive; even at a distance, a winter Ipswich Sparrow is decidedly gray-toned, without the bright black, white, and brown patterning on the back of mainland Savannah Sparrows. The richly colored "rear quarter" formed by the chestnut secondaries and greater coverts of Savannah Sparrows is lacking in Ipswich Sparrows, which are noticeably more uniform above.

Below, Ipswich Sparrows may be less brilliantly white than are many Savannah Sparrows, with a more extensive wash of buffy brown across the breast and flanks; that color sometimes invades the broad pale jaw stripe as well. The breast streaks are slightly finer than in Savannah Sparrows, especially on the flanks. In many birds, the streaks are light cinnamon-brown, while in others they are subtly two-toned, with a very narrow blackish shaft streak fading to dull rust toward the feather edge.

Rather than the neat black head markings of the Savannah Sparrow, Ipswich Sparrows are marked with more muted, blurrier brown, making the head pattern less dramatic and more obviously dominated by a thin white eye-ring and a broad gray supercilium. In late winter, not long before the birds move north in late March, the prealternate molt results in a clear yellow tinge to the supercilium, usually paler and more diffuse than the neatly defined yellow of a Savannah Sparrow.

As in most sparrows and passerines in general, the prealternate molt involves almost none of the body feathers, replacing only head feathers and, in some individuals, some of the tertials and inner tail feathers. As a result, adult Ipswich Sparrows are at their most ragged and worn in July and August, having endured the travails of two migrations, a winter, and nesting. In the weeks preceding the prealternate molt, adults can be very white below, the streaking blacker and sparser, and the patterns of upperparts and head severely abraded. With the commencement of molt of the forehead and lore, the bill can appear unusually long. Seen only on Sable Island, Ipswich Sparrows in this state can startle observers more used to seeing the species in fresh plumage on cold Atlantic beaches.

Wintering Ipswich Sparrows also share those beaches with small numbers of Vesper Sparrows. The Ipswich Sparrow's relatively plain wing, frequently buffy breast band, blurry streaking, and muted head pattern with an often conspicuous pale eye-ring may recall the similar plumage of the Vesper, but the two can always be distinguished by tail length—short in the Ipswich, medium-long in the Vesper—and bill shape—swollen at the base in the Vesper, more slender in the Ipswich Sparrow.

The short, high-pitched flight call given by flushed Ipswich Sparrows in winter is very similar to that of eastern Savannah Sparrows, though perhaps not as thin on average and more broadly modulated, with a more buzzing attack: *dzeet*. The song is like that of a mainland Savannah Sparrow, with a more obviously descending trill at the conclusion.

RANGE AND GEOGRAPHIC VARIATION

The Ipswich Sparrow breeds on Sable Island, "the ribbon-like crest of a submerged bank" 100 miles off the coast of Nova Scotia. There are infrequent cases of mixed pairings with Savannah Sparrows on mainland beaches in the province.

Though large numbers sometimes linger on Sable Island, nearly all Ipswich Sparrows eventually move south to winter in the outer dunes of the Atlantic Coast, from Nova Scotia to South Carolina, and in small numbers to northern Florida. The highest abundance in winter is usually found on the ocean beaches of Delaware, Maryland, and Virginia, where Christmas Bird Counts regularly report more than 100 individuals.

Dwight, almost certainly correctly, rejected inland reports from New Hampshire and Texas, but there are documented records of this species from early winter in Quebec City and on the Ontario shore of Lake Erie. Northerly strays have occurred on Newfoundland and Saint-Pierre, in the latter case adding a species to the French national list. An April bird in southwest England—at first thought to be perhaps a Little Bunting, an aberrant Meadow Pipit, or a female Yellowhammer—remained for nearly a week, singing from the rocks under which it roosted at night.

VESPER SPARROW
Pooecetes gramineus

In the "Advertisement" with which he introduced his *Arctic Zoology*, Thomas Pennant singled out for particular gratitude his countrywoman Anna Blackburne, to whose

> rich museum of American birds . . . I am indebted for the opportunity of describing almost every one known in the provinces of Jersey, New York, and Connecticut. They were sent over to that Lady by her brother, the late Mr. Ashton Blackburn; who added to the skill and zeal of a sportsman, the most pertinent remarks on the specimens he collected for his worthy and philosophical sister.

Pennant would thank his benefactress by naming a beautiful American warbler for her. He let another of the 101 new species he discovered in the Blackburnian collections, however, continue under the decidedly lackluster name by which it was known in its native land: "the Gray Grass-bird."

Pertinent as they may have been, the notes accompanying the skin Ashton Blackburne sent to his sister seem to have been brief. Pennant reports that the grass-bird is a permanent resident of New York, where it breeds in May, laying a clutch of five eggs in the grass. The description he provides is every bit as mundane; the specimen is gray, brown, and black above, with brown cheeks, streaked white underparts, and a dusky tail. The only color, in the bird or in Pennant's rhetoric, is the "bright bay" that distinguishes the lesser wing coverts.

When it came time to assign the new American a scientific name, Johann Friedrich Gmelin simply rendered the English "grass finch" into Latin, *Fringilla graminea*, and translated Pennant's description nearly verbatim, ending his diagnosis with the bay-colored lesser coverts.

The bird now known as the Vesper Sparrow is quickly learned by even beginning birders, who have been adjured since the late nineteenth century to identify the species in the field by "the white outer tail-feathers flashing conspicuously as the bird flies." That distinctive tail pattern was not even described, though, until 1811, when Alexander Wilson first noted that the outer vane of the outermost feather was white, "the next tipt and edged for half an inch with the same." Like his predecessors, however,

---

The chestnut lesser coverts are not always visible. The streaked breast band and flanks are subtly warmer buff than the rest of the whitish underparts. *British Columbia, June. Brian E. Small*

Wilson was more impressed by the color of the lesser secondary coverts, and he headed his account with a new English name, the Bay-winged Bunting.

While the less eloquent alternative "grass finch" survived into the twentieth century, Wilson's pleasing "Bay-winged" was adopted—if not always exclusively—by John James Audubon, Thomas Nuttall, and Spencer Baird. The name entered the more or less popular literature in Baird, Brewer, and Ridgway's *History* of 1874, but by the publication of the first edition of the AOU *Check-list* a dozen years later, it had been largely displaced by a more poetic coinage.

In the mid-nineteenth century, when the landscape of the eastern United States was still dominated by extensive cultivation, this species was a common bird east of the Alleghenies, wintering in "astonishing numbers" in the Southeast. In New York and New England, where they are now local and rare, these open-country sparrows were a familiar sight and sound, and Nuttall waxed enthusiastic in his description of the males' territorial behavior:

> They sing with a clear and agreeable note, scarcely inferior to that of the Canary . . . . Their song begins at early dawn, and is again peculiarly frequent after sun-set until dark . . . when other songsters have retired to rest.

Wilson Flagg, one of the most popular nature writers of his mid-century day, made a similar observation in the pages of the first volume of the *Atlantic Monthly*. If the Song Sparrow is "more companionable" in singing throughout the day, these buntings are silent from the end of the dawn performance "until sunset, when they repeat their concert, with still greater zeal than they chanted in the morning." That habit is the reason, says Flagg, that the bird is known familiarly as the "Vesper-bird," a name "worthy of being retained as its distinguishing *cognomen*." "Grass Finch" and, especially, "Bay-winged Bunting" continued to appear alongside the new name—probably Flagg's own neologism, in spite of his claim to have heard it first from the country folk—but both those labels soon fell out of use, replaced by the much more evocative and more memorable, if not entirely accurate, name still used today.

FIELD IDENTIFICATION

In much of eastern North America, where the Vesper Sparrow is no longer the familiar bird it was less than a hundred years ago, this is one of the most frequently overreported of the sparrows; it is simply more difficult there than in most places west of the Mississippi to gain experience in what can seem at first a subtle, even difficult identification. Happily, most potential confusion results from a reliance on outmoded field mark approaches to passerellid identification, and a more rational approach focused on structure and behavior can help even the new observer to an immediate appreciation of the characters distinguishing this streaky brown bird from others that are superficially similar.

A rather large, portly, thick-billed bird, the Vesper Sparrow differs from the two most frequently mistaken species—the Song and the Savannah Sparrows—in the length and shape of its tail and in the general blandness of its plumage. The tail is broader, proportionally somewhat shorter, and noticeably less rounded than that of any Song Sparrow, usually obviously faintly notched even on perched birds. It is at the same time broader, noticeably longer, and less pointed than that of any Savannah or Ipswich Sparrow.

More subtly and perhaps less consistently, Vesper Sparrows can seem oddly flat-backed when they are feeding on the ground. This impression is more likely due to the full breast and belly and, especially, the long wing than to any real difference in the structure of the back plumage. Both Savannah and Song Sparrows appear more nearly spherical or "domed" when perched.

Song Sparrows and, with the exception of the Ipswich, Savannah Sparrows are also significantly more contrastingly plumaged than the Vesper Sparrow, which usually strikes the field observer as conspicuously bland above and below, with little or no clear, bright white or black. Instead, Vesper Spar-

Stocky and plain, the Vesper Sparrow has a broad, white-edged tail and a swollen bill. *Maine, June. Brian E. Small*

The odd facial pattern is created by the heavy whisker, nearly surrounding the paler ear coverts, and the long extension of the pale jaw stripe behind the ear coverts. *Maine, June. Brian E. Small*

rows give the impression in most views of being darker gray-brown above and paler buff-whitish below; especially when the underparts are strongly tinged buff, the transition in color from upperparts to breast and flanks can seem quite gradual.

In closer views, the head pattern of the Vesper Sparrow is diagnostic in both its detail and its overall impression. The combination of heavily outlined ear coverts and broad creamy ear surround makes the bird look heavy-headed and "jowly," an impression heightened by the large, thick-based bill. Even at distances from which the individual plumage components are invisible, the appearance of dark cheeks with pale "mutton chops" is quite different from the fine, neat streaking and striping of the Savannah Sparrow or the overall darkness of the Song Sparrow's head and face.

If any passerellid has traditionally been assigned a single, definitive field mark, it is the Vesper Sparrow with its white-edged tail. Alexander Wilson was the first to point that character out, but not even he considered it indispensable to the bird's identification; as late as the early twentieth century, bird guides still focused on the head pattern and the color of the lesser coverts, mentioning the tail pattern only as a secondary feature. The great Ralph Hoffmann was among the first to emphasize the usefulness of the white rectrices in the field, italicizing the words "outer pair of tail-feathers mostly white" in the spe-

cies account he prepared for his *Guide to the Birds of New England and Eastern New York*. Thirty years later, Roger Tory Peterson would point to those same feathers with one of his famous arrows.

The pattern of the Vesper Sparrow's tail feathers is a useful mark, but its emphasis in the literature of field identification over the past hundred years has understated the difficulty of seeing any but the central rectrices on perched birds; even viewed from below, when the outer feathers are at their most conspicuous, the precise color of the undertail can be hard to judge, the distinction between white and pale often a difficult one in the open, brightly lit habitats this species prefers.

The traditional Petersonian comparison of Vesper Sparrows to juncos, pipits, longspurs, and even meadowlarks misleads some observers to forget that other open-country species not always thought of as white-tailed can also show considerable paleness on the tail in flight. Many Savannah Sparrows in fresh plumage have extensive white edgings on multiple rectrices, and the hopeful and the inexperienced can mistake that flash for the nearly completely white outermost rectrix of a Vesper Sparrow. The tail pattern should be considered an excellent confirming character once structure, shape, and overall plumage have been assessed.

At any distance, the song of the Vesper Sparrow can recall that of a Song Sparrow. The similarity

The most contrasting part of the wing is the even row of blackish median coverts. *Maine, June. Brian E. Small*

is greatest in the introductory whistles with which both species begin their typical song. Those notes are richer, deeper, and steadier in pitch in the Vesper than in the Song Sparrow; the whistles of the latter species tend to be thinner, shorter, and higher-pitched, often with a marked descending component. Many Song Sparrows begin with three or four such notes, while Vesper Sparrows usually utter only two or three.

The remainder of the Song Sparrow's song is typically well structured, with a discrete buzz followed by a distinct lighter trill; the fast central buzz is the highest-pitched and loudest portion of the song.

Vesper Sparrows are more given to improvisation. The introductory whistles are often the loudest part of their song, and of the two or three or more variably paced trills that follow, none is usually significantly higher-pitched than the introduction. For all its pleasing variety and richness of tone, the song often seems to play out, as it were, over a pitch range of only one or two whole steps. This melodic flatness is the most easily recognized feature of a song that can otherwise seem only vague and patternless.

Perched and in flight, Vesper Sparrows give fairly long *tseep* calls that can be surprisingly thin and high-pitched for so bulky a bird. The call is usually longer and less staccato than that of a Savannah or Lark Sparrow, but finer and shorter than that of a Song Sparrow; at close range, the call may have a faint but discernible buzz.

RANGE AND GEOGRAPHIC VARIATION

While almost two centuries ago Audubon was "inclined to look upon it as a resident of the country lying to the eastward of the range of the Alleghanies" exclusively, most American birders today think of the Vesper Sparrow as a western and midwestern bird, with the isolated remnants of the eastern populations battling urbanization and the regrowth of forest.

As a species, Vesper Sparrows breed from eastern

Summer birds can be messy, making the bill look even larger than usual. *British Columbia, June. Brian E. Small*

British Columbia and northern Alberta east across southern Saskatchewan, Manitoba, and Ontario into southern Quebec, New Brunswick, and Nova Scotia; the species occasionally breeds farther north where large tracts of boreal forest have been cleared. Vesper Sparrows, presumably of the subspecies *affinis*, formerly bred on British Columbia's Lower Mainland as well; that race currently breeds south through western Washington and Oregon to extreme northwestern California.

In the south, Vesper Sparrows breed east of the coastal mountain ranges to central California and Nevada and the northern half of Arizona and New Mexico; east of the Rockies, they are generally scarce on the Great Plains south of the Platte River. The agricultural steppes of Iowa, Illinois, and northern Indiana are a stronghold, where the species has even adapted to nesting in cornfields. Farther east, breeding is now erratic and local east to western Virginia, New York, Massachusetts, and inland of the Atlantic coastal strip north to Nova Scotia. In central and western Pennsylvania, grasslands on reclaimed strip mines are a favored local habitat.

Vesper Sparrows winter across the entire southern tier of the United States and south through Mexico to Oaxaca, with occasional cold-season records north on the coasts to British Columbia and Nova Scotia; they are rare inland north of Arizona, New Mexico, Oklahoma, Arkansas, Tennessee, and the Carolinas. The subspecies *affinis* is now largely restricted in winter to transmontane California, from Sutter County south to northwesternmost Baja California.

Finding "a good deal of difference in specimens before" him, Spencer Baird was at first uncertain whether his new genus *Pooecetes* comprised one species or two. Today, four subspecies are generally recognized, differing in size, bill shape, and overall color; observers tempted to identify migrant and wintering Vesper Sparrows to subspecies in the field would do well to heed the intimation in the two oldest subspecies names, *confinis* and *affinis*—both of which mean "closely similar."

The nominate race, represented by the Blackburne specimens and named *gramineus* by Gmelin, is the proportionally longest-tailed of the four, with gray upperparts and rather whitish underparts; it further differs from the widespread western *confinis* and *affinis* in its fairly thick bill. This subspecies breeds from the eastern prairies to the Canadian Maritimes, south to Missouri and Virginia.

Baird found that the western specimens he examined were larger, longer-winged, and thinner-billed than the eastern birds available to him. This western subspecies, which he named *confinis*, is also proportionally long-tailed, with a brownish wash to the gray upperparts and more distinctly creamy underparts. This is the widespread race of the inland west, breeding from British Columbia to westernmost Ontario and south to east-central California and the Nebraska Sandhills. This and the next race may intergrade in eastern Washington, Oregon, and California; intermediate birds are sometimes called *definitus*.

In 1888, the 19-year-old Gerrit S. Miller Jr., later a distinguished mammalogist, described the breeding Vesper Sparrows of Salem, Oregon, as smaller and more thoroughly buffy than more easterly breeders; he named the subspecies *affinis*. Breeding in coastal Washington and Oregon, this race is also proportionally short-tailed and comparatively dark above. This is the smallest of the Vesper Sparrows, the smallest individuals with a wing chord almost 25 percent shorter than that of the largest representatives of the fourth subspecies, *altus*.

That race is the largest of the four, with a proportionally long tail and the relatively thickest bill of any subspecies. The back is medium-dark, like that of *affinis*, and the underparts whitish, like those of nominate *gramineus*. This is the breeding Vesper Sparrow of the Four Corners area of Utah, Colorado, New Mexico, and Arizona, south to the San Francisco Mountains.

In winter, all three of the western subspecies can be found in the southwestern United States and northern Mexico; *gramineus* winters in the southeastern United States.

Young birds are somewhat more extensively streaked below. *Oregon, August. Brian E. Small*

This sparrow of prairie marshes has a spiky tail and colorful head with a purple-streaked nape, gray cheek, and narrow white central crown stripe. *North Dakota, June. Brian E. Small*

LECONTE SPARROW
Ammospiza leconteii

For more than half a century, explorers and natural historians in the eastern United States eagerly searched for this colorful little bird, which had first been described in 1790 by John Latham on the basis of a specimen from the interior of Georgia. "It has yet, however, escaped all our ornithologists," Thomas Nuttall observed in 1832, and gradually doubts crept in about the species' very existence; Charles Bonaparte suggested tentatively that Latham might have been describing a Grasshopper Sparrow, while Nuttall himself would later decide that Latham's *Fringilla caudacuta* must be "nearly allied if not identic [sic] with Henslow's Bunting."

Finally, in May of 1843, the collectors of John James Audubon's expedition up the Missouri River "procured" several specimens "of this pretty little Sharp-tailed Finch," among them "a fine male . . . shot by Mr. J.G. Bell, of New York." Audubon was able to show clearly that the bird was neither a Grasshopper nor a Henslow Sparrow, and as Latham's scientific name had turned out to be preoccupied, he took the liberty of renaming the species the Le Conte's Sharp-tailed Bunting, *Emberiza Le Conteii*, after his "young friend Doctor [John Lawrence] Le Conte, son of Major [John Eatton] Le Conte, so well known among naturalists, and who is, like his father, much attached to the study of natural history."

In a coincidence of the sort that history, including ornithological history, is full of, Audubon's priority and LeConte's ornithological fame were the direct result of the loss of the steamship *Assiniboine* in July of 1835, almost a decade before Audubon saw the Missouri. The boat caught fire on the river above St. Louis, and both vessel and cargo were completely destroyed. Besides the usual furs and trade goods, the *Assiniboine* was also carrying seven large crates of specimens and notes that had been taken by Prince Maximilian of Wied-Neuwied on the journey he made to the Yellowstone in 1832–1834.

The loss was discouraging, and only in January of 1858, 25 years after his expedition, did Maximilian begin the publication of "the extremely incomplete notes" he was able to reconstruct from "the only remnants of the ornithological observations made on a journey across North America, after the most interesting portions of the relevant collections, together with many written notes, were lost in the burning of a steamship on the Missouri River." Fortunately, neither Maximilian's private diary nor the skin of a certain tiny sparrow, shot in the spring of 1833, were on board the *Assiniboine.*

On May 8, 1833, Maximilian and his expeditionary team were in what is now extreme northeastern Nebraska, a short distance up the Missouri from the mouth of the Vermillion River. While the crew and traders were busy moving a load of furs from one boat to another, Maximilian led a small party to explore a burned patch now "covered with excellent, fresh young grass."

> On the ground we believed we saw a mouse running [near] the roots of a tree. We followed, but when looking closer, we found that it was a small bird that allowed us to come as close as two or three paces and then ran around the tree. We could not induce it to fly, and it was too close to shoot. We finally decided to shoot it with a very small charge and found then that it was an exceptionally nice, simply but extremely prettily colored *Fringilla acutipennis* that we saw today for the first and last time.

The epithet the prince used in his diary entry, *acutipennis*, was a slip of the pen and memory for Latham's *caudacuta*. Since that name was preoccupied, *acutipennis*—"sharp-feathered"—would have become the species' official name if Maximilian had published his account at some point in the next decade. As it was, though, by the time the 1858 list appeared in the *Journal für Ornithologie*, Audubon had long since published his name, and Maximilian duly and properly cited that authority (and Audubon's "fairly good" plate) in his own description of the only specimen he had ever obtained of *E[mberiza] Le Contei*, "der Ammer mit zugespitzten Schwanzfedern"—the bunting with acute tail feathers. Maximilian's skin, the oldest in existence, entered the collections of the American Museum of Natural History in 1870, where it remains—unlike Audubon's type, which was given to the young Spencer Baird, in whose care "it has somehow been mislaid."

Whether one dates the rediscovery of the LeConte Sparrow to 1833 or 1843, as Coues pointed out, the species "long remained an extreme rarity." Audubon's lost type was at last replaced at the Smithsonian in 1868, when Gideon E. Lincecum sent a specimen from Texas; the skin was in very poor condition when it was received—the 75-year-old Lincecum apologized to Baird for being so "very fumble-fisted" a preparator—and does not seem to be extant.

The Texas bird was still in the drawers in 1872, however, when Elliott Coues was able to examine it:

> This long-lost species, of which I, for one, never expected to see an example, believing it to have been based upon some particular condition of [the Grasshopper or the Henslow Sparrow], has at length been re-discovered . . . . Audubon says that the species is common on the Upper Missouri, where, let us hope, other specimens will be found.

That hope was fulfilled in the summer of 1873, when Coues himself "found the bird to be not uncommon" and collected seven "elegant skins," including young birds and adults, male and female, on what is now the border between North Dakota and Manitoba. Like Maximilian and like Audubon, Coues reported that the process of collection was "not without difficulty . . . the only chance was a snap shot as the birds, started at random, flitted in sight for a few seconds; while it was quite as hard to find them when killed. Several seen to fall were not recovered after diligent search."

Even as it became clear that the LeConte Sparrow was common, even abundant in parts of its breeding

---

The breast is plain dull ochre in adults, but juveniles, which may retain their first plumage well into autumn, show fine black streaks on the breast and flanks. *Minnesota, June. Brian E. Small*

and winter ranges, the species' habit of "skulking in rank herbage" where it was "difficult to flush, even with good spaniels," meant that some of the details of its life history would remain unknown until the 1880s. The first nest and eggs were collected in Manitoba in June of 1882—100 years after the first skin made its way across the Atlantic, 50 years after Prince Maximilian encountered the bird in Nebraska, and 40 years after John James Audubon named the little sparrow for a man who never saw one.

FIELD IDENTIFICATION

At all seasons, LeConte Sparrows are found in damp weedy fields, marshy edges, and prairies with little or no woody vegetation. Often furtive in the tall grasses they cleave to, these brightly colored, appealing sparrows are easily identified when seen well. Most individuals, however, are seen first— and some are seen only—in low, fluttering flapping directly away, making it important to learn what LeConte Sparrows look like both perched and in flight.

The most striking feature of LeConte Sparrows, whether seen fleeing, clinging to a grass stem, or scampering mouselike through the duff, is their overall uniform paleness. Apart from the clean white belly of adults, the plumage is thoroughly dominated by soft, warm buffy browns, lacking the much bolder black and white back pattern of the Saltmarsh and (especially) the Nelson Sparrow and the rich, dark scaling and streaking of the Grasshopper Sparrow. The short, sharply pointed tail feathers are the same brown as the dried-grass ground color of the back and rump; the outer tail feathers are much shorter than the innermost, and the pronounced wedge shape of the tail is often obvious when it is spread on take-off and landing.

The LeConte Sparrow's pale, finely streaked rump usually contrasts, subtly but noticeably, with the somewhat darker back. In a good view, the rump color appears to "flow" onto the back in long parallel stripes against the darker ground color. The pale-edged brown wing feathers sometimes create an impression of translucence in flight, unlike the darker wings of Grasshopper, Henslow, and the other sharp-tailed marsh sparrows. Even in flight, the orange head and oddly striped brown and silver nape of adults can be conspicuous.

Though the LeConte Sparrow is no less in absolute length than the similar species already mentioned, it is always conspicuously slender, small-headed, and thin-billed in the field, making it look tiny in flight or perched. The tail is not only short but obviously narrow in all views but directly from the side. The blue-gray bill is short and thin, neither as heavy as in the Grasshopper and Henslow

Sparrows nor as long and spikelike as in the Saltmarsh Sparrow.

The plumage markings are as fine as the bird's physical structure. The uppertail coverts and rump are very narrowly streaked black, and the back is marked by a series of alternating blackish and wheaten stripes. The bright pale buffy flanks and breast are streaked with fine blackish shaft streaks, surrounding a bright white oval belly in adults. The pale nape is visible from a surprising distance, conspicuous even in flight; at close range, it is pale silvery gray with very narrow rich brown streaks, in combination producing a band of an unusual dull purple color.

The overall impression of paleness extends to the adult's head pattern. There is no strong lateral throat stripe or dark whisker; the ear coverts are pale gray, bordered above and behind by only a vague, narrow darkish line. The colorful supercilium is very wide, leaving space for only a narrow dark lateral crown stripe. The median crown stripe is also narrow, but its mix of white and dull buffy orange makes it conspicuous in most views.

Unlike most other passerellid sparrows, some juvenile LeConte Sparrows do not undergo their preformative molt until they have reached the wintering range, some retaining their juvenile plumage until December. Nelson and Saltmarsh Sparrows complete their molts on the breeding grounds, in late summer or very early fall. In those areas and at those times when both juvenile Nelson and juvenile LeConte Sparrows can be present, the LeConte can be distinguished by its paler overall plumage, including the crown; the extensive fine streaking of the breast and flanks; and the bland, blank face pattern with no lateral throat stripe or whisker. Any juvenile *Ammospiza* sparrow after early October can be presumed to be a LeConte.

Migrant LeConte Sparrows tend to be relatively quiet, though when alarmed or alert they may give a series of surprisingly low-pitched, smacking notes recalling the chip of a Palm Warbler or even at times a Common Yellowthroat. They also call with a very high-pitched, short, thin *tsit*, which agitated birds may repeat in a long, fairly fast series; the single vocalization is easily mistaken for an insect sound coming from an autumn field.

The song of territorial males is soft and also insectlike. It begins with a hurried series of two to four very short, slightly buzzy notes and ends with a harsh, low-pitched buzz: *tiktetik BRZZ*. At any distance, the brief introductory phrase may be inaudible, leaving only the long, irregularly modulated buzz. From a distance, that concluding note can resemble part of the song of a Clay-colored Sparrow, but it is longer, lower-pitched, and "thicker" than the finer, briefer, repeated buzz of that species. The

The complex back pattern echoes the grassy habitats these birds prefer. *Michigan, May. Brian E. Small*

end of the LeConte Sparrow's song is also not unlike the complete song of the Nelson Sparrow; the buzz of that species is even longer, more coarsely modulated, and slightly slower, usually with a distinctive lower note at the end.

LeConte Sparrows also have an infrequently witnessed longer song, usually given in flight but sometimes delivered from the ground. This song typically is preceded by several of the short, high-pitched call notes, which are followed by a fast rising and falling whistled phrase as the bird flies up; the typical terminal buzz is given just before landing. The "aerial trill" attributed to the LeConte Sparrow in some older sources is likely not produced by this species.

RANGE AND GEOGRAPHIC VARIATION

In 1872, on the eve of his discovery of the large breeding populations along the 49th parallel, Elliott Coues described the known range of the LeConte Sparrow as "Missouri region; Texas," the collection localities for the two specimens then known to have been in existence. By the turn of the twentieth century, however, the species' breeding and wintering distributions and its migration routes were well known, especially given this sparrow's fairly secretive behavior.

The LeConte Sparrow breeds in eastern North Dakota, northern Minnesota and Wisconsin, and the upper peninsula of Michigan, but the heart of its range and the bulk of its population is in the prairie provinces of Canada, from extreme eastern-central British Columbia and the southernmost Northwest Territories across Alberta, Saskatchewan, and Manitoba. Numbers breed in northwestern and northern Ontario, especially around James Bay, and there

are isolated breeding populations in southern Quebec. There is also a disjunct breeding population in northeastern Montana and southeastern British Columbia.

Migrants move south between mid-September and early November along the eastern Great Plains and down the Mississippi Valley to the wintering areas in the south-central and, apparently to a lesser degree, the southeastern United States, from the Ohio River Valley of Illinois, Indiana, and Kentucky south to eastern Texas and east along the Gulf Coast to the Florida Panhandle. The species is much less common or more local in winter from the Florida peninsula north to the Carolinas. The northward migration typically begins in late March, with arrival on the breeding grounds no later than late May.

Vagrancy outside of what appears to be the normal range is most frequent in fall and winter, with rather few records in late spring. Since the late 1970s, the LeConte Sparrow has been a nearly annual autumn rarity in California, with occurrences north along the Pacific Coast to western British Columbia and south to Coahuila. It is significantly less frequent, or significantly less often detected, in the interior West, with only two records in Arizona, for example, and five in Utah. Some of this apparent rarity may be due to the species' retiring nature, and it is possible that the situation over much of the West resembles that in parts of the East and Northeast, where LeConte Sparrows can reasonably be regarded as regular migrants and winter visitors at extremely low densities rather than as accidental strays. This is a sparrow that can appear almost anywhere, and probably does, most often undetected.

No subspecies are recognized.

The sturdy, spike-billed Seaside Sparrow inhabits coastal marshes from New England to Texas. *New Jersey, February. Kevin Bolton*

SEASIDE SPARROW
Ammospiza maritima

One of the last new birds to be described from the United States in the twentieth century was discovered by Arthur H. Howell in February of 1918. Howell, a government biologist, was on a short collecting trip to southernmost Florida when he took specimens of a green-backed, white-breasted, sharply streaked sparrow at Cape Sable. He quickly determined that his specimens were so strikingly different from any other Florida passerellid that they should be accorded full species rank as the Cape Sable Seaside Sparrow, *Thryospiza mirabilis*.

Howell's miracle bird hardly crests most birders' mental horizons today, but its human history—the surprise discovery, the initial taxonomic certainty, and the wholesale lumping of this and other putative species 55 years later—is emblematic of the twisted routes along which scientific ornithology has approached the entire complicated complex of what is known today as the Seaside Sparrow.

The first Seaside Sparrows known to science—representing, by that very fact, what would become the nominate race *Ammospiza maritima maritima*—were taken by Alexander Wilson, probably on one of his six collecting trips to southern New Jersey. Though Wilson succeeded in shooting "a great number of individuals," his account of the species, published in 1811, dwells at evocative length on the inaccessibility of its saltmarsh habitat and the skill with which it eludes its pursuers "among the holes and interstices of the weeds and sea-wrack." The Father of American Ornithology was also the first to record an impression of the Seaside Sparrow in the kitchen:

Its flesh, . . . as was to be expected, tasted of fish, or was what is usually termed *sedgy*. [emphasis in original]

Rather little seems to have been learned about this bird over the next two decades, though John James Audubon was able to add a description of the nest and a disdainful mention of the "monotonous chirpings" that make up the species' song. Audubon was less interested, however, in adding to the knowledge about the Seaside Sparrow's biology than in seizing yet another Oedipal opportunity to point out the superiority of his own efforts to Wilson's. He agreed with his predecessor about the bird's nimble furtiveness; but its flight habit, he writes, is so straight and level that a good marksman "can easily kill them before they alight." To remind his audience once again of his superlative skill with a fowling piece, Audubon informs the reader that he one day shot a good number of Seaside Sparrows "merely for the sake of practice"—and while Wilson's culinary remarks have the air of the merely dutiful report of an experiment, Audubon had his specimens made into a pie, "which, however, could not be eaten, on account of its fishy savour." Even Audubon's failures had to be more impressively unappetizing than Wilson's.

Neither Wilson nor Audubon had a clear sense at first of these sparrows' geographic distribution, described by both simply as the seaside salt marshes of the Atlantic Coast. In 1834, though, not only was Audubon able to offer a more precise statement of these birds' range, but he also identified the sparrows of the salt marshes around Charleston, South Carolina, as a species distinct from, if allied to, the original Wilsonian Seaside Sparrow. John Bachman had sent him a dozen specimens of this newest finch, which Audubon named for his scientific and literary collaborator William MacGillivray, the coauthor of the *Ornithological Biography*. Audubon described his *Fringilla macgillivraii* as combining the dark brown-black upperparts of the Seaside Sparrow with the yellow-tinged breast sides of the Saltmarsh Sparrow, but nevertheless readily distinguishable from both. Four years later, no one had found the MacGillivray's Finch anywhere else on the eastern seaboard—but on their southwestern tour of 1837, Audubon and his party made the unexpected discovery that this uncommon bird of the Carolina marshes was "very abundant in the Texas" and the delta of the Mississippi River, 700 miles away.

Thirty-five years later, C. J. Maynard added another piece to what was quickly becoming the Seaside Sparrow puzzle with his discovery of a "very remarkable," "very striking" bird on the Indian River of southern Florida. Ridgway, who acquired several specimens from the collector, at first described this black and white sparrow as a variety of the "normal" Seaside; on Maynard's insistence, Ridgway would soon account this Dusky Seaside Sparrow a species of its own.

The recognition of these scattered, visually different populations—whether species or subspecies, races, or "varieties"—of Seaside Sparrows drove the search for new forms that might bridge the geographic and phenotypic gaps. W. E. D. Scott, who spent nearly three years collecting in southwestern Florida in the late 1880s, submitted a number of sparrow skins to Joel Asaph Allen at the American Museum of Natural History. Allen determined that Scott's birds were intermediates between the classic Seaside Sparrow and Ridgway and Maynard's Dusky, and described them as a new subspecies of the former, *peninsulae*. As a mark of gratitude to the collector, Allen assigned the newly discovered population the English name Scott's Seaside Sparrow, a gesture that seems to have been only partly satisfactory. The next year, in the pages of *The Auk*, Scott reported having collected another 67 specimens of *peninsulae*:

The Cape Sable Seaside Sparrow shares with the Florida Grasshopper Sparrow the regrettable distinction of being the most severely endangered passerine in the United States. *Florida, April. Brian E. Small*

In the light of this new material I am inclined to regard this form as a *species* rather than a *subspecies* of . . . *maritimus* . . . . In this entire series there are no individuals that could not at a glance be selected from the true *maritimus*. The bird seems quite as distinct from that species as from . . . *nigrescens*, and of equal value as a species to either of these two . . . .

Unconvinced of the specific rank of the Florida sparrow, Allen was more interested in the new population's geographic distribution. He decided that the Louisiana birds Audubon had included in *macgillivraii* with Bachman's Carolina specimens were in fact unrecognized representatives of Scott's *peninsulae*, which in Allen's new concept ranged from southwestern Florida to the mouth of the Mississippi.

Allen did not stop there. When the opportunity arose to compare Scott's Florida skins with a small series of seasidelike sparrows from the upper Texas coast, Allen declared those birds "evidently entitled to recognition" as a distinct subspecies, too. This time, it was in the scientific name, *sennetti*, that Allen thanked the collector, George B. Sennett, while as an English name he coined the more prosaic Texan Seaside Sparrow.

Frank Chapman's revision of the Seaside Sparrows at the end of the century appears to have temporarily dampened the enthusiasm of the search for new forms. But 20 years on, a new generation of systematists took up the hunt. Howell's collecting efforts on the Gulf Coast, which had resulted in the discovery of the Cape Sable Seaside Sparrow, also made available significantly more specimen material than had existed before from Florida, Mississippi, and Alabama, making it possible to draw a finer portrait of the variation and distribution of these birds.

In 1920, Ludlow Griscom and J. T. Nichols took advantage of the Howell material and information gathered by Harry C. Oberholser to lay out a new understanding of the Seaside Sparrows—an understanding that would include the naming of two new taxa. From the marshes of the northern Florida coast, Griscom and Nichols described *juncicola*, the "rush dweller," while they gave the birds of Alabama and eastern Mississippi the overdue honorific *howelli*.

Ten years later, Oberholser crowded the subspecific field even further with his descriptions of the new Atlantic Coast races *pelonota*, from northeastern Florida, and *waynei*, from Georgia and extreme southeastern South Carolina. In that same year of 1931, Harold H. Bailey named a Seaside Sparrow

Birds on the western coast of the Gulf of Mexico are grayer, with less clearly demarcated streaks beneath. *Texas, February. Brian E. Small*

from northeastern Florida *shannoni*, in honor of its first collector, W. E. Shannon. There still remains some uncertainty as to whether Bailey's *shannoni* and Oberholser's *pelonota* referred to different populations, an uncertainty unlikely ever to be resolved, as the birds—whatever their taxonomic status—have been extinct from Jacksonville, Florida, south since the 1970s.

By 1932, with twelve species and subspecies described, the pieces of the Seaside Sparrow mosaic seemed to be more or less in place: the species, or species complex, was extremely variable in plumage, and that variation was in many cases the direct opposite of clinal, with dark populations and light, heavily streaked and obscurely marked, alternating regularly from region to region. A century after Audubon described the first hints of the range of variation in this species, Ivan R. Tompkins proposed a cogent explanation for its strange zoogeographic patterns. By flooding coastal salt marshes, Tompkins suggested, tropical storms fragmented what had once been a continuous range, isolating populations that had once been in contact and permitting their divergence; winter storms, too, could wipe out entire breeding populations, as Seaside Sparrows tend to move faithfully from nesting area to wintering area. The peculiar characteristics of the survivors, or in some cases perhaps of immigrant replacements, had a "founder's effect" on the future development of each local population, resulting in much of the variety seen today.

After more than a century of collecting and study, scientific ornithology finally had a clear sense of what "the" Seaside Sparrow comprised and why the geographic picture was so varied. All that knowledge, however, was not enough to save one of the species' most distinctive local populations.

Insisting in 1875 that the Dusky Seaside Sparrow be considered a full species, C. J. Maynard claimed with equal stridency that he should be allowed to name it, whether he had scientific priority or not:

> A certain Finch which I discovered in Florida . . . I now think entitled to specific rank, and therefore propose to name it. I am perfectly aware at this time that Mr. Ridgway has already given it a name as a *variety* of *Ammodramus maritimus*. I now, however, as seems to me perfectly justifiable, it being solely a discovery of my own, baptize it a *species*. Allow me, therefore, Mr. Editor, to present to you and the public my new species, *Ammodromus melanoleucus*, the Black and White Shore Finch.

The AOU disagreed in part: for nearly a century, it recognized the Dusky Seaside Sparrow as a dis-

The very striking Dusky Seaside Sparrow, once an object of more or less polite contention among American ornithologists, has been extinct for more than 20 years. *Florida. Paul Sykes, courtesy US Fish & Wildlife Service*

tinct species, but credited Ridgway with the original and authoritative description and name. In 1973, though, the Dusky and Howell's Cape Sable Sparrow were lumped with the more widespread Atlantic Coast forms, a taxonomic decision made on purely scientific grounds but one with severe consequences for sparrow conservation.

The last Dusky Seaside Sparrow, a male, died in captivity in Florida in June 1987. Destruction of marsh habitat by drainage, impoundment, and uncontrolled fire had significantly reduced the population when the AOU determined that the birds should no longer be recognized as a distinct species, a status they had enjoyed for 100 years. In one of the starkest demonstrations ever recorded of the connection between taxonomy and the politics of conservation, the public interest and governmental effort devoted to the Dusky Seaside Sparrow were greatly reduced by the "lump." By the time a captive breeding program was initiated in 1979, the entire population, probably approaching 2,000 individuals ten years earlier, comprised six males—and no females. Five of those surviving birds were captured, and in the spring of 1980, one of the males was bred to a female Scott Seaside Sparrow; three fledged young were the result. It was planned that the two intergrade females produced in that brood would be backcrossed with Dusky males, and so on through the generations until the young were as like pure Dusky Seaside Sparrows as possible. In 1981, however, the Florida Game Commission ceded custody of the Dusky males to the United States Fish and Wildlife Service, which declined to make them available for further crossbreeding experiments.

FIELD IDENTIFICATION

Over most of its range, the Seaside Sparrow is a rather large, heavy, short-tailed, and long-billed bird, in adult plumage unlikely to be mistaken for any other denizen of the high salt marshes it inhabits; the subspecies known as the Cape Sable Sparrow shares the general shape and structure of its conspecifics, but is small and now breeds only on wet inland prairies in the Florida Everglades.

The size and generally somber plumage are usually sufficient to identify this sparrow on its spartina-clad breeding grounds. The species' large size is made the more obvious in the field by what often appears to be a proportionally longer neck and larger head than in the smaller marsh sparrows; though the bill is long, it usually seems less strikingly sharp and spikelike than that of the Saltmarsh Sparrow. The Seaside Sparrow's tail is markedly short, but especially in flight it appears obviously broader and less pointed than that of the Saltmarsh and Nelson Sparrows; in worn individuals, the tail can be decidedly ragged, but it is never as conspicuously sharp as on the other, noticeably smaller *Ammospiza* of the coastal marshes. Perched birds usually appear decidedly disheveled, especially after the breeding season, when the plumage is at its most worn.

At a distance, birds of all extant subspecies appear dark and dull-colored, though fresh-plumaged Gulf Coast birds can be conspicuously buffy below and somewhat rusty on the wing. All have a fairly bright white throat and pale jaw stripe, white on the Atlantic and yellow-tinged on the Gulf Coast; when other, more subtle characters are obscured by distance or worn into uniformity, the throat may be the only contrasting region of the entire plumage. Seaside Sparrows in the Northeast and in the mid-Atlantic states share their high-marsh habitat with Swamp Sparrows, which are also conspicuously white-throated, but are always obviously shorter-tailed, shorter-winged, and longer-billed, without

Heavily streaked beneath, the Cape Sable differs from other Seaside Sparrows in its decidedly greenish back and head. *Florida, April. Brian E. Small*

the handsome deep chestnut tones of the smaller bird.

Most American birders first become familiar with the gray Seaside Sparrows of the East, and can be momentarily confused on first encountering the more colorful birds of the Gulf Coast, from Florida to Texas. The three breeding subspecies of this region share brighter underparts and generally more distinctly marked upperparts, and are not infrequently mistaken for wintering Nelson Sparrows, which are obviously much smaller, sharper-tailed, larger-billed, and clearer orange on the breast and face. In close views, the patterns of the wings, back, breast, and head differ clearly between the two species. Gulf Coast Seaside Sparrows are relatively dark-winged and plain-backed, while the Nelson Sparrows wintering in that species' range have bland gray-brown wing feathers and contrastingly white-streaked mantles; the lead-gray ear coverts of the

Nelson Sparrow are neatly and sharply surrounded by clear orange, while the face of even the most colorful Seaside Sparrow shows a muddy mix of dull gray and buffy tan. The Seaside Sparrow's uniformly bland crown is quite unlike the neatly zoned head of the Nelson Sparrow, with its clear blue-gray median crown stripe bordered with broad blackish median stripes. Beneath, the Nelson Sparrows wintering on the Gulf Coast are neatly marked with fine streaks, where the Seaside Sparrows of the same region have only soft, blurry stripes.

A more daunting and greatly underappreciated identification challenge is posed by the juvenile Seaside Sparrow, which can wear a puzzlingly yellowish plumage through the end of its first calendar year of life. When they are not accompanied by adults, such birds can easily be confused with juvenile Saltmarsh Sparrows, which are similarly colored. In both species, the partial molt to formative—or first-winter—plumage can be protracted, but Saltmarsh Sparrows finish theirs on the breeding grounds, assuming more or less adultlike body plumage by October; Seaside Sparrows of the same age can still be wearing fully juvenile plumage in December, and some

The juvenile Seaside Sparrow is white-throated and smudgy-crowned, with a subtle pattern of scaling on the scapulars. *New Jersey, February. Kevin Bolton*

The wings and tail of Texas birds can be quite rusty, as can the blurry streaks on the flank. *Texas, February. Brian E. Small*

maritima arrive on their wintering grounds still retaining juvenile feathers.

In addition to the larger size, broader tail, and notably heavier bill of the Seaside Sparrow, there are also plumage characters that can be helpful at close range to distinguish that species from juvenile Saltmarsh Sparrows. While the Saltmarsh Sparrow's back and scapulars are streaked in all plumages, the juvenile Seaside Sparrow shows dark scapulars with fine pale tips and edgings, creating a vaguely scaled rather than streaked pattern; the streaking of the mantle feathers is smudgy and obscure. On most juvenile Seaside Sparrows, the whitest region of the underparts is the throat; the throat of the juvenile Saltmarsh Sparrow is usually strongly buffy, and does not contrast markedly with the breast. Like the adult, juvenile Saltmarsh Sparrows have a dark crown above a wide orange supercilium; the Seaside Sparrow's crown is of the same color and tone as the back, with dark streaks, and does not offer a pronounced contrast with the face.

Where the species is common, the distinctive song of the Seaside Sparrow is one of the characteristic sounds of the summer salt marsh. Sonograms and slowed-down recordings reveal a variability and complexity inaudible to human ears, including clicking tremolos, glissando phrases, warblings, and buzzes. The field observer generally hears the primary song as a fast, dry tremolo followed immediately by a loud, broad buzz. The traditional comparison to the song of a distant Red-winged Blackbird is, like so many similar memory crutches, most helpful to birders who already know the Seaside Sparrow's song.

As might be expected given its disjunct populations over a wide range, this species reveals apparently considerable variation in songs. On the Atlantic Coast, northern birds are said to sing at a lower pitch than southern birds, which in turn may add whistled song elements more typical of Gulf Coast males. The Dusky Seaside Sparrow's primary song, "very simple and mostly noise," was decidedly insectlike, exceeded in that quality only by the Cape Sable Seaside Sparrow, described by Hardy as two or three dry clicks, a very finely modulated buzz, and a thin concluding buzz quite unlike the broadly modulated, burry buzzes of coastal birds.

Like many other grass and marsh sparrows, Seaside Sparrows also have a longer, more complex song given primarily in flight. This vocalization is heard most often at the height of breeding, and is made up of a short series of introductory *tsee* notes followed by lower-pitched chips and concluding with a condensed version of the buzzing primary song.

Other than singing males, Seaside Sparrows are often quiet, but both sexes have a wide repertoire of calls. Agitated birds give a loud, low-pitched smacking *chuk*, often while perched at the top of a cordgrass stem. Flock members communicate, perched and in flight, with a short, high-pitched *tsi*, all attack with no decay; often repeated in an accelerating series of four or five notes, this call is easily overlooked as that of an insect. The most complex call is a whinny, given by both members of a pair, described as "a whirling, quavering" call often given together with a louder slurred chatter.

RANGE AND GEOGRAPHIC VARIATION

Measured in degrees of latitude, the Seaside Sparrow is an extremely widespread bird, breeding over some 1,100 miles from southernmost Maine to northeastern Florida. It is important to recall, however, that within that extensive range the species occupies only salt marsh, a decidedly discontinuous habitat only rarely more than a mile deep. Measured in acres, even the most abundant of the Seaside Sparrows, the Atlantic Coast's *maritima*, occupies a very small range.

Rare in Maine, New Hampshire, and most of the Massachusetts shore, the nominate *maritima* subspecies breeds in coastal marshes from those states south to Duval and Flagler Counties, Florida. The breeding range extends into the salt marshes of Delaware and Chesapeake Bays; Virginia's Chesapeake Bay Islands Important Bird Area (IBA) supports some 2,900 pairs.

This Atlantic Coast race is the only Seaside Sparrow to migrate. There is evidence of a northeasterly post-breeding dispersal into Maine and the Maritimes. Migrating birds are very rarely detected, and

Seaside Sparrows are very rarely seen away from their saltmarsh breeding habitat, where they clamber acrobatically through grasses, sedges, and salt-tolerant shrubs. *Texas, April. Brian E. Small*

when they are encountered, it can be impossible to distinguish them from breeding residents of the salt marshes to which even migrants are virtually restricted. Most individuals in New York and New England have withdrawn to the south by October, when they begin to appear on the wintering grounds as far south as northeastern Florida. The return flight appears to begin in April, with birds typically arriving in the northerly portions of the breeding range late that month and in May.

Given the length of some individuals' migration and the evidence that at least some take an overland route through the Carolinas, it is surprising that there are so few records of even short-distance vagrants of this subspecies. There are inland specimens or photographs from southeastern Pennsylvania, the lower Hudson Valley in New York, and eastern Massachusetts.

The resident subspecies *mirabilis*, the Cape Sable Seaside Sparrow, was thought for several years to be extinct after the hurricane of September 1935 swamped the eponymous type locality. Rediscovered in the early 1940s, it now breeds, in small and declining numbers, in Big Cypress National Preserve and Everglades National Park. These are the only noncoastal Seaside Sparrows, nesting preferentially instead in inland marsh and "mixed prairie, often with a substantial percentage of muhly grass."

The Gulf Coast subspecies breed in salt marshes from Pasco County, Florida, to the mouth of the Rio Grande in Texas. From Alabama west to just north of Corpus Christi, Texas, the breeding subspecies is *fisheri*; south of San Antonio Bay, Texas, *sennetti* is an uncommon and local breeder, rarer to the south. There are apparently no breeding records for Mexico, though *sennetti* "wanders to [the] adjacent bank" of the Rio Grande in Tamaulipas.

Inland vagrancy is even less frequent in the Gulf Coast populations than in the birds of the Atlantic salt marshes. One presumably of the subspecies *fisheri* that wintered in McLennan County, Texas, was photographed in the hand in 1974.

As even the briefest account of the taxonomic and identification history of this species reveals, the Seaside Sparrow is among the most variable of North America's passerellids. For a century or more, a clear understanding of that variability was frustrated by uncertainties about the range of sexual, individual, and age variation within populations, and by the apparent preponderance in the specimen record of migrants or wintering birds.

The Atlantic Seaside Sparrow, comprising the subspecies *maritima* and its synonyms—*waynei, shannoni, macgillivraii,* and *pelonota*—is the plainest and grayest of the populations, indistinctly streaked above and only slightly more clearly marked below. The underparts are decidedly gray, paler to the south, with only a faint overlay of buff in fresh plumage. This is also the largest of the Seaside Sparrows, on average fully half an inch longer than the Dusky Seaside.

Seaside Sparrows on the shores of the Gulf of Mexico, from the central west coast of Florida to southern Texas, are noticeably brighter and more distinctly marked above and below. The Scott Seaside Sparrow, *peninsulae* (including *juncicola*), is dark olive-gray above with clear blackish streaking; the buff-tinged gray breast is moderately streaked with black.

The Louisiana Seaside Sparrow, *fisheri*, breeds from the northwest Florida coast to eastern Texas. Quite small in comparison to either the Atlantic or the Scott, this race lacks strong gray tones on its heavily black-streaked dark olive back; the breast is decidedly buffy, with variably well-defined dark brown streaking. Woltmann et al. found that *fisheri* had interbred near Corpus Christi with the most genetically distinct of the Gulf Coast races, *sennetti*. That Texas Seaside Sparrow, another small race, is pale buffy below and on the face; the streaking of both upperparts and underparts is clear and black.

The smallest, the most distinctive, and the rarest —one of them by a factor of tragic infinity—of the Seaside Sparrows are the two "specialty" races of Florida, the imperiled *mirabilis* and the extinct *nigrescens*. The Cape Sable Seaside Sparrow, *mirabilis*, is olive-brown above with clear dark brown streaking; below, it is white-breasted with distinct blackish streaks. The dramatically patterned Dusky Seaside Sparrow, now to be seen only in the museum, was blackish above with obscure dark streaking; its white breast was heavily and neatly streaked black.

Treating all Atlantic Coast birds as members of the nominate race, and given the sedentary habit of most populations, identification to subspecies is generally simple even in the field. Only on the upper Texas coast, where the Texas and Louisiana Seaside Sparrows may overlap, and possibly at a potential contact point in the Florida Panhandle between the eastern *peninsulae* and the western *fisheri* might intergrades be expected.

Nelson Sparrows from the center of the continent are colorful and contrasting. *Texas, April. Brian E. Small*

NELSON SPARROW
Ammospiza nelsoni

"One of the keenest naturalists we have ever had," in Theodore Roosevelt's estimation, Edward W. Nelson led a life that can be described only as picaresque, driven along by coincidence and chance, luck sometimes good and often not. Following the early death of his father in the Civil War, his mother moved the two sons from New Hampshire to Chicago, where young Nelson set out to become an entomologist. On October 9, 1871, the family lost almost everything in the Great Fire—Nelson saved his insect collections from the flames, only to have them stolen on the way out of the burning city.

If clouds so dark can be said to have a silver lining, it was revealed when a family friend soothed the boy's loss with the gift of a shotgun. His natural historical efforts were thereafter directed to ornithology, and in just a few years Nelson amassed a collection of more than 3,000 bird skins and eggs. In September 1874, when he was 19, Nelson shot eight "sharp-tailed finches" along the Calumet River south of Chicago; he sent one of his specimens to Joel Asaph Allen in Cambridge, who found it "very markedly" different from the sharp-tailed sparrows

of eastern Massachusetts salt marshes, and named it *Ammodromus* [sic] *caudacutus nelsoni*. Allen, whose interest in geographic and climatic variation is commemorated in the "rule" that bears his name, determined that Nelson's bird must be the representative of a southern race of the familiar bird of the Atlantic Coast.

Later famous as the explorer of Mexico, California, and Alaska and second Chief of the Biological Survey, Nelson was certain that the sparrow Allen named for him was breeding in the marshes of south Chicago, a record no longer given credence but not beyond the pale of plausibility. Over the next 20 years, the Nelson Sparrow was also reported in the breeding season from Wisconsin, Texas, Maine, and Kansas; the first actual nesting records within what is now understood to be the species' true range were obtained in 1888 in the Dakota Territory and in 1892 in Manitoba.

Although uncertainty about the Nelson Sparrow's breeding range would persist well into the early twentieth century, it was eventually clear that the bird was "the inland representative of its strictly littoral relatives," the coastal-breeding birds we know today as the Saltmarsh Sparrows. At that point, the search began for a missing link to fill the geographic gap between the sharp-tailed sparrows of the Midwest and those of the Atlantic salt marshes. Jonathan Dwight discovered a compelling candidate on the shores of New Brunswick in July 1886. This *subvirgatus*, as he named his new subspecies, "passes gradually into *nelsoni*," just as individuals of *nelsoni* "show a gradual and complete gradation into *subvirgatus*," suggesting to Dwight the possibility that

> some inland marshes may . . . furnish a regular supply of connecting links between *nelsoni* and the new race, which is certainly more closely related to *nelsoni* than to true [i.e., nominate] *caudacutus*.

Several years later, Dwight was forced to admit that "there appears to be a wide gap between the headquarters of this form [*subvirgatus*] and those of *nelsoni*—over one thousand miles," a geographic disjunction almost without parallel among North American birds.

The puzzle of these birds' distributions and relationships grew only the more vexed in 1926 when George Sutton discovered a breeding population on the shores of James Bay. W. E. Clyde Todd described those birds as another subspecies, aptly to be named *altera*, and observed that southbound migrants of the new race—grayer, paler, and duller than *nelsoni*—were responsible "for the number of presumed intergrades between *nelsoni* and *subvirgata* recorded from" the Atlantic Coast. Now, instead of two widely separated but apparently closely related subspecies there were three, and even the visually intermediate birds of James Bay were separated by hundreds of miles from their brighter cousins of the upper midwestern prairies and their duller relatives in the southern maritimes, a confusing state of affairs that would complicate the birds' taxonomy and their identification for more than a century after the young Edward Nelson first collected his sparrow in the suburbs of Chicago.

The first suggestion that that species might in fact be two was made by Jonathan Dwight in 1896, writing that

> if it should be proved that *subvirgatus* regularly breeds on the same ground as *caudacutus*, the question of considering *nelsoni* as a separate species with *subvirgatus* as its eastern race may be seriously discussed.

Arthur Norton rose to Dwight's challenge. When he found *subvirgatus* breeding within 30 miles of nominate *caudacutus*, he proposed in 1897 that the two were indeed separate species, and that *subvirgatus* and *nelsoni* should be combined and renamed *Ammodramus nelsoni*.

The American Ornithologists' Union adopted that view in the Ninth Supplement, published in 1899, and reaffirmed it three years later. When Harry Oberholser argued that *Ammodramus* "belonged" to the Grasshopper Sparrow alone, the Nelson Sparrows were moved into C. J. Maynard's resurrected *Passerherbulus*, a genus that would now also comprise the LeConte, Seaside, Saltmarsh, and Henslow Sparrows. By 1931, when the fourth edition of the *Check-list* appeared, that grouping, too, was deemed artificial and "composite," and the Nelson Sparrow was among the members assigned to their own genus, *Ammospiza*, a name coined by Oberholser.

In addition to adding another genus to the Nelson Sparrow's extensive synonymy, the 1931 *Check-list* reversed its 22-year-old ratification of the two-species hypothesis. The committee's pronouncement was handed down without explanation (or any particular regard for English syntax):

> *Ammospiza nelsoni* being now regarded as only subspecifically different from *caudacuta*, becomes a subspecies of it, as does also *subvirgata*.

Nelson Sparrows of the population breeding on the northern coast of New England are very pale and gray. *Maine, June. Brian E. Small*

Fifty years later, Jon S. Greenlaw took up the investigation proposed by Dwight and Norton. Working at the point of contact in southwestern Maine between *subvirgata* and *caudacuta*, Greenlaw discovered consistent differences in the songs and singing behavior of the two populations, coinciding with differences in plumage and bill size. Although he recognized that gene flow continued between the two groups, he proposed a new historical explanation for their contact: according to Greenlaw, it is likely that a widespread ancestral population was split by glaciation; the two descendent populations evolved in isolation, only to meet again when the new interior group moved east and the new coastal group moved north, encountering each other in Maine. The proposal that the two, now in secondary contact, be treated as distinct species was accepted by the AOU in 1995, on the basis of Greenlaw's evidence of differences in song, morphology, and habitat, and the absence of extensive interbreeding where their ranges overlap. The unwieldy but informative English name Nelson's Sharp-tailed Sparrow for *nelsoni*, *subvirgata*, and *altera* was formally simplified in 2009 to Nelson's Sparrow.

FIELD IDENTIFICATION

As is often the case in birds displaying pronounced geographic variation, the Nelson Sparrow can be confused with different species in different parts of its range. Inland and on the Gulf Coast, it must be distinguished from the paler, more finely marked LeConte Sparrow, while on the Atlantic Coast from Maine to central Florida, it can be mistaken for the longer-billed, more clearly streaked Saltmarsh Sparrow. In the latter case, regular and probably not infrequent hybridization can make it impossible to identify certain puzzling individuals as *subvirgata* Nelson Sparrows, Saltmarsh Sparrows, hybrids, or backcrosses.

In good views, the separation in the field of adult Nelson Sparrows of the Great Plains and Hudson Bay subspecies and adult LeConte Sparrows is straightforward. Both are small, slender-billed, short-tailed, colorful sparrows of tall, moist grasslands and marshes. The adult LeConte Sparrow always gives an impression of paleness, while *nelsoni* and *altera* Nelson Sparrows are darker, more somber, and more saturated in all their colors. LeConte Sparrows often appear more uniformly straw-orange above and below; Nelson Sparrows show decidedly more contrast between the dark head, bright breast, white belly, and dark back, that last a difference that can be especially pronounced when sparrows flush from the grass to flee directly away from the observer.

The LeConte Sparrow is the larger of these two small species in absolute measurements, but in the field, its tiny bill and relatively small, round head often make that species seem very slight, while Nelson Sparrows, slightly larger-billed and bigger-

Many Nelson Sparrows from inland populations are almost as bright orange on the breast as on the face. *Texas, April. Brian E. Small*

The very bold black and white upperparts and ashy gray nape distinguish this species from the equally colorful LeConte Sparrow. *Texas, April. Brian E. Small*

parts streaking of Nelson Sparrows is typically restricted to the flanks and sides of the breast, while LeConte Sparrow juveniles show scattered streaks across most of the underparts. From September on, all Nelson Sparrows have molted into an adultlike plumage, but LeConte Sparrows can retain juvenile body feathers into early December.

In the marshes of southwestern Maine, Nelson Sparrows of the subspecies *subvirgata* overlap in the breeding season with Saltmarsh Sparrows; from October to April and even into May, all three Nelson Sparrow subspecies can be found on the coastal wintering grounds of the Saltmarsh Sparrow. The long, spikelike bill of the Saltmarsh Sparrow makes that species look front-heavy and flat-crowned. Saltmarsh Sparrows are neatly, fairly densely streaked across the dull yellowish or whitish breast, a clear distinction from the brighter, more diffusely marked underparts of Nelson Sparrows. Adults of the Saltmarsh and the inland subspecies of the Nelson Sparrow are all brightly, contrastingly marked on the back; there may be a slight tendency in the Nelson Sparrow, especially the nominate subspecies, to show deeper chestnut edges on the mantle feathers.

The breeding range of the drabbest of the Nelson Sparrows, *subvirgata*, overlaps with that of the Saltmarsh Sparrow in southwestern Maine, New Hampshire, and northeastern Massachusetts. Adults of the two are traditionally thought of as readily separated: *subvirgata* is smaller-billed, plainer-faced, duller gray above, brighter yellow below, and much less distinctly streaked on breast and flanks than its marsh mates. A recent study of birds from more than 20 marshes where both species breed has determined that hybridization and backcrossing—the successful breeding of hybrids with each other or, more usually, with a "pure" member of one of the parent species—is far more frequent than hitherto believed:

> Genetic data revealed that 52 percent of individuals sampled . . . were of mixed ancestry, and the majority of these were backcrossed.

In the field, a few apparent hybrids are obviously intermediate in appearance, combining, for example, the bright underparts of the one parent species with the neat streaking of the other. Walsh et al. discovered, however, that other hybrids—definitively identified by genetic analysis—were indistinguishable from "pure" Nelson or Saltmarsh Sparrows by plumage or morphology, posing what the authors wryly term "a challenge for accurate hybrid identification in the field." Pressing that observation to its logical conclusion, it is strictly speaking impossible—without a feather or a blood sample—to be certain

headed, usually appear blockier and more substantial at first sight. In flight, the rump of both species is dull to moderately bright orange-brown; in addition to the differences in back color, the deep bluegray nape of the Nelson Sparrow can be an obvious distinction from the paler-necked LeConte.

Perched adults of the inland races are distinguishable from LeConte Sparrows using the same characters. In a leisurely view, the head and breast patterns are also quite different. LeConte Sparrows are marked below with long, fine streaks, while Nelson Sparrows show clear streaking only on the flank, if at all, and the breast is streaked with short, blurred marks. The Nelson Sparrow's crown is dark gray with black lateral stripes; the LeConte Sparrow's crown is black with a sharply defined central stripe. In a more typical view, the color of the tarsus is often helpful—usually dull pink in the LeConte, usually drab gray in the Nelson.

Juveniles of these two prairie marsh breeders are more difficult to see and sometimes more difficult to separate. Both are yellowish above and below, with many juvenile Nelson Sparrows decidedly orange on the head and breast. While the juvenile Nelson is very dark on the wing and back, the juvenile LeConte has a more varied pattern of pale mantle stripes and whitish tips and fringes to the tertials and wing coverts; those same areas are rusty yellow

Unlike inland-breeding Nelson Sparrows, coastal birds can be very white on the breast; they lack the clear black streaks below of the Saltmarsh Sparrow. *Maine, June. Brian E. Small*

of the identity of a visually "pure" Saltmarsh or *subvirgata* Nelson Sparrow anywhere in the hybrid zone or, in winter, on the Atlantic Coast between Maine and Florida.

On the breeding grounds, song is a helpful way to distinguish Saltmarsh Sparrows from Nelson Sparrows inhabiting the same marshes. The Nelson sings a fast, two-parted buzz, both parts loud and broadly modulated. The first element of the song is higher-pitched, with a distinctive metallic jangling quality; the second element is lower-pitched and more shirring or sibilant, though still decidedly a buzz. Some songs begin with a very brief lower-pitched buzz, barely separated from the remaining elements; other songs are followed by two or three fairly faint, high-pitched ticking notes. In overall pattern, the main two-parted song can recall that of the Seaside Sparrow, but it is faster, higher-pitched, and more silvery, with a bright, breathy quality lacking in the Seaside Sparrow's more rattling buzzes. This song, *krrZZHHsshhtt*, has no parallel in the vocal repertoire of the Saltmarsh Sparrow, making it a diagnostic feature in areas where both may be present.

This song also makes up part of the Nelson Sparrow's flight display, in which the male ascends to a height of as much as 50 feet while uttering a variable series of three to ten irregularly spaced ticking or squeaking notes; the buzzing song follows these notes as the bird begins its descent. Twice over a period of ten years, Greenlaw observed Saltmarsh Sparrows uttering similar ticking notes (but without the species-specific buzz of the Nelson Sparrow) in an apparent flight display.

The Nelson Sparrow lacks the "whisper" song given by Saltmarsh Sparrows.

Nelson Sparrows in migration and winter call only infrequently, a very high-pitched, thin *teek* without noticeable attack or decay; it can be given in a fast series by birds under stress, and this call is among the notes introducing the terminal buzz of the flight song.

RANGE AND GEOGRAPHIC VARIATION

The three recognized subspecies of the Nelson Sparrow occupy three widely disjunct breeding ranges. The nominate race—reported in the nineteenth century, almost certainly erroneously, to breed as far south as Illinois and Kansas—nests in grassy marshland from northeastern South Dakota and northwestern Minnesota north and west across the prairies to central Manitoba, central Saskatchewan, and northern Alberta. The westernmost breeders are found in the Peace Lowlands and on the Kiskatinaw Plateau of northeastern British Columbia, where numbers can vary considerably with water levels. The northernmost regular breeding takes place in the southern Northwest Territories.

The James and Hudson Bay population, *altera*, breeds above the tide line in sedge- and bulrush-dominated bogs with scattered birches and willows, from the Eastmain River in Quebec west to Churchill, Manitoba. Coastal *subvirgata* nests in freshwater marshland on both shores of the lower St. Lawrence River and south through the coastal wetlands, fresh- and saltwater, of the Maritimes to southwestern Maine. The southern limit of this subspecies' current regular breeding range appears to be the Scarborough marshlands along Maine's Nonesuch River, where it encounters and sometimes hybridizes with Saltmarsh Sparrows.

Breeding hundreds of miles apart, all three subspecies can be met with at many of the same marshy localities in migration and winter. In winter as in summer, *nelsoni* is the most widespread Nelson Sparrow, spending the cold season on the Atlantic and Gulf Coasts from North Carolina west to south Texas. The very similar *altera* appears to occupy much the same winter range, though it is probably significantly less common on the Gulf Coast; there are specimens from the upper and central Texas coast, but the winter status of this subspecies west of Florida remains unclear. Less abundant and more difficult to identify—at least in areas where Saltmarsh Sparrows also winter—the subspecies *subvirgata* is known at that season chiefly from the Atlantic Coast from South Carolina to central Florida; there do not appear to be any adequately documented records of this subspecies from the Gulf Coast.

Nelson Sparrows of any subspecies are rather rarely detected on migration, a result of their small size, furtive behavior, and preference for soggy, densely vegetated habitats. The nominate race *nelsoni*, with the largest breeding range, the largest global population, and the longest migration route, is the most frequently seen. Northbound and southbound birds can be encountered anywhere in the Midwest and on the eastern Great Plains; the fall migration, which usually begins in September and continues into early November, takes place over a broader front, and some birds of this subspecies fly overland across the southeastern United States, while others appear to move east into the salt marshes of the Atlantic Coast before continuing south into the usual wintering areas.

Autumn is also the season for vagrant Nelson Sparrows, all of which are almost certainly attributable to the nominate race. There are numerous records at this season from the western Great Plains, not far from the "normal" migration route, and individuals have drifted farther west into Arizona, Utah, interior California, Washington, and British Columbia's lower mainland (eBird). In coastal California, where there were very few records between the first in 1891 and 1960, the Nelson Sparrow is now considered a rare fall migrant and rare winter visitor rather than a vagrant; it may have the same status in Baja California.

Largely because of its pronounced similarity

The center of the Nelson Sparrow's crown is slaty gray, as in the Saltmarsh Sparrow; this species' bill is somewhat less remarkably spikelike. *Texas, April. Brian E. Small*

in the field to the nominate *nelsoni*, the Hudson Bay subspecies *altera* is much less well known as a migrant. Originally thought to be largely coastal in its occurrence in the fall, there are autumn specimens from Michigan, Ohio, and western Pennsylvania, and from far inland in the southeastern United States, demonstrating that at least some individuals of this subspecies take an overland route south rather than first flying southeast to the Atlantic Coast.

The coastal *subvirgata* has the shortest route to its wintering grounds of any Nelson Sparrow. In both spring and fall, migrants stay close to the coast, though there are a few specimens suggesting that some individuals cut south across inland Massachusetts, Connecticut, and New York before returning to their coastal path.

The northbound migration in spring is less well understood, especially for *subvirgata* and *altera*. The nominate race *nelsoni* appears to avoid the

Atlantic Coast north of the Delmarva Peninsula in spring, instead choosing an overland route across the southeastern United States or, for those wintering on the western Gulf Coast, straight north to the prairie breeding grounds. The species as a whole is a late but rapid spring migrant, with many individuals lingering on the wintering grounds well into May—accounting for the nineteenth-century reports of over-summering in areas south of the currently understood breeding range. Vagrants are essentially unknown from spring; two records from southeastern California in May probably represent birds returning after a winter in coastal marshes there.

Seventy years ago, Roger Tory Peterson reported that

> Almost yearly the Linnaean Society of New York held seminars devoted to the field identification of [the subspecies of the "sharp-tailed" sparrows] so that its members could add them to their lists . . . . It is obvious that many misidentifications have been made in the past. This points up clearly the folly of being too concerned with subspecies . . . . I would recommend that a Sharp-tail be called just a Sharp-tail . . . .

That advice to the field observer remains helpful at its core. Without a thorough understanding of the geographic variation exhibited by the Nelson Sparrow as it is currently understood, however, many individuals of the coastal subspecies *subvirgata*—the Nelson Sparrow subspecies most closely resembling the other member of the old "sharp-tailed" species pair, the Saltmarsh Sparrow—will go unnoticed, unidentified, or misidentified.

The nominate *nelsoni*, breeding on the northern Great Plains, is by far the brightest and most clearly marked of the three currently recognized subspecies. This is a small and short-billed bird, strikingly so in comparison with many Saltmarsh Sparrows. The dark brown-black ground color of the back contrasts very strongly with the bold whitish stripes. The orange of the jaw stripe and supercilium extends to the sides of the nape; the upper breast and gray ear coverts are usually washed with the same color. The white belly is bordered by rich brown flank streaking, especially conspicuous in fresh plumage.

The James Bay subspecies *altera*, always identifiable by definition on the breeding grounds, can be closely similar to *nelsoni*; most wintering birds and migrants are not safely distinguished in the field from duller individuals of *nelsoni*. Any flank streaking is gray and blurry in *altera*, and the median crown stripe is broader and cleaner gray than that of *nelsoni*, without the intrusion of black spotting or streaking from the lateral stripes. The ground color of the back and scapulars is duller gray-brown than in *nelsoni*, and there is typically less contrast between the dark feather centers and paler edges.

The coastal breeder *subvirgata* is often easily distinguished from the other two subspecies of the Nelson Sparrow—indeed, with its rather long, thin bill, it is generally more likely to be mistaken for a Saltmarsh Sparrow than for either of its conspecifics. The flank streaking is blurry and gray, giving some individuals a "soft" appearance below that may even recall a Seaside Sparrow. The back is usually very dull, with a gray ground color and, in most birds, only inconspicuous whitish stripes created by the pale feather edges. The crown is paler than in either *nelsoni* or *altera*, with brownish lateral stripes contrasting only slightly with the wide gray-brown median stripe. The impression given by birds in flight or by perched birds quickly glimpsed is of a decidedly grayish sparrow with an only slightly more colorful face and breast.

SALTMARSH SPARROW
Ammospiza caudacuta

The very long bill, entirely whitish underparts, and heavy streaking across the breast are good distinctions from the Nelson Sparrow where the two overlap. *Maine, June. Brian E. Small*

In the twenty-first century, very few birds—and even fewer sparrows—have genuine English-language folk names. Birds and other wildlife are no longer part of most people's everyday world in North America, giving them no reason to coin new vernacular names or to preserve old ones; those few who go out of their quotidian way to notice and identify birds invariably use the "official" book names promulgated by the American Ornithological Society and reproduced in the field guides. The standardization of species-level English nomenclature, which dates only to the publication in 1957 of the fifth edition of the *Check-list*, has been a great and obvious benefit to clarity and understanding, but it has also flattened much of what was once a richer cultural connection between humans and birds in North America.

Historically, much of that connection was founded on the culinary role so many birds were forced to play in earlier America. Even this small species, no doubt as "sedgy" as its relative the Seaside Sparrow, would appear to have been taken for the table on occasion; it is difficult otherwise to imagine the circumstance in which Audubon would report having "seen more than forty . . . killed at one shot" in the marshes of South Carolina. On the shores of mid-nineteenth-century Long Island, gunners called it by "the name of Quail-head . . . derived from its distant resemblance to the head of the" Northern Bobwhite.

Those hunters would no doubt have been surprised to learn that a bird they knew so intimately was and would be until the end of the nineteenth century a source of taxonomic anxiety. Alexander Wilson was the first ornithologist in America to take note of this sparrow—and to admit to his puzzlement about the identity of "this new (as I apprehend it) and beautiful species." Oddly, even sloppily, although Wilson described his Sharp-tailed Finch as a new bird, he gave it a scientific name, *Fringilla caudacuta*, already assigned to what he thought was a different species—but which, more oddly still, turned out to be after all the same.

A bird of this denomination is described by Turton, Syst. p. 562; but which by no means agrees with the present. This, however, may be the fault of the describer, as it is said to be a bird of Georgia; unwilling, therefore, to multiply names unnecessarily, I have adopted his appellation. In some future part of the work I shall settle this matter with more precision,

a promise still unkept when Wilson died two years later.

William Turton had published his English translation of Johann Gmelin's *Systema naturae* in 1802. Wilson, who had borrowed a copy belonging to the great American naturalist and entomologist Thomas Say, was right to find the diagnosis of *Fringilla caudacuta* in the *General System of Nature* vague; it is, in fact, so unclear as to be unidentifiable to species. Had the American ornithologist's bibliographic effort been more thorough in this case, however, he would have found that Turton's account, badly garbled as it was, was derived ultimately from a more authoritative source, a source that describes the bird clearly and accurately—and a source, oddly enough, that Wilson cited elsewhere in his *American Ornithology*.

The chain of textual descent is long but straight, and easily traced. Turton, the English physician, translated the Latin of Gmelin, the German naturalist. Gmelin translated the English of John Latham, the English physician. And Latham tells his reader that his account and illustration of the bird he called the Sharp-tailed Oriole was taken in its entirety from Thomas Pennant, the Welsh collector.

Wilson cited Pennant 184 times in the *American Ornithology,* but somehow overlooked the original account of this "oriole" in the Welsh author's *Arctic Zoology.* Pennant's lengthy diagnosis of a New York specimen in the collections of Anna Blackburne provides a good description of today's Saltmarsh Sparrow and the pattern of its head, breast, and tail, the feathers of the last of which "slope off on each side to a point, not unlike those of a Woodpecker."

This was the description that Latham copied out and that Gmelin translated; any doubt that it applies to Wilson's bird as well is easily dispelled by the plate in Latham's *General Synopsis,* which, as Jonathan Dwight would later point out, is obviously, and attractively, a Saltmarsh Sparrow of the nominate race. In spite of his claim that his bird was the representative of a new species, Wilson did exactly (and inadvertently) the right thing taxonomically in applying Gmelin and Turton's name *caudacuta.*

The early classification of this species as an oriole reflects its colorful face and slender bill. Wilson was the first to recognize it unequivocally as a sparrow, finding it so similar to the Seaside Sparrows of the Atlantic Coast that he was "almost willing to believe they are identical." Wilson's tentativeness would be occasion 20 years later for yet another posthumous jab from Audubon's pen; Audubon agreed that the Saltmarsh Sparrow was "allied in form and habits" to the Seaside—"with which, however, it cannot possibly be confounded by any person possessing the least observation."

FIELD IDENTIFICATION

Adult Saltmarsh Sparrows of either subspecies are readily distinguished from all other sparrows but the Nelson. Both breed in the marshes of southwestern Maine, where the Nelson Sparrow is represented by the subspecies *subvirgata,* the dullest and in many ways most Saltmarsh-like of its races. From October to April and May, both subspecies of the Saltmarsh Sparrow and all three subspecies of the Nelson Sparrow overlap on coastal wintering grounds from New Jersey to Florida.

With good views, these two species are generally separable in the field; unfortunately, such views may require both considerable patience and considerable good luck. Saltmarsh Sparrows are significantly larger than Nelson Sparrows, a difference sometimes noticeable even on birds flushed from wintertime spartina marshes. Perched, Saltmarsh Sparrows look decidedly front-heavy and forward-leaning, with a flat crown accentuated by the long, spikelike bill.

The pattern of the underparts is usually clearly different in these two species. Nelson Sparrows are only indistinctly marked below, with a bright breast colored identically to the wide jaw stripe and supercilium. The breast of the Saltmarsh Sparrow is less colorful, dull yellowish or whitish, and noticeably less bright than the face. Above, the Saltmarsh Sparrow—particularly the dark southern subspecies *diversa*—and the two inland races of the Nelson Sparrow are contrastingly marked, unlike the somewhat plainer, grayer *subvirgata,* with which the Saltmarsh overlaps in the breeding season in southwestern Maine, New Hampshire, and northeastern Massachusetts.

As already noted in the discussion of the Nelson Sparrow, hybridization between Saltmarsh Sparrows and *subvirgata* Nelson Sparrows and the successful breeding of hybrids with each other or, probably more usually, with a "pure" member of one of the parent species are more frequent than hitherto believed. Some of those mixed birds are obviously intermediate in appearance; others, however, though identified as intergrades by genetic analysis, were indistinguishable from "pure" Nelson or Saltmarsh Sparrows by plumage or morphology, posing "a challenge for accurate hybrid identification in the field." Pressing that observation to its logical conclusion, it is strictly speaking impossible—without a feather or a blood sample—to be certain of the identity of a visually "pure" Saltmarsh or *subvirgata* Nelson Sparrow anywhere in the hybrid zone or, in winter, on the Atlantic Coast between Maine and Florida.

Juvenile Saltmarsh and Nelson Sparrows in the area where both species breed can be difficult to identify. Saltmarsh Sparrows of any age are larger and longer-billed than Nelson Sparrows; this is obvious when the two are seen close to each other, but extremely hard to assess in lone individuals. The Saltmarsh Sparrow has a whiter throat bordered by darker, better-defined lateral throat stripes. Perhaps the most useful mark is the extent of streaking below, which is restricted to the breast sides and upper flanks in the Nelson Sparrow but covers most of the yellow-tinged breast and whitish upper belly in the Saltmarsh.

On the breeding grounds, song is a helpful way to distinguish male Saltmarsh and Nelson Sparrows. Saltmarsh Sparrows do not sing the primary song given by Nelson Sparrows, audible to human ears as two gasping buzzes. Neither does the species normally perform the flight display commonly seen in Nelson Sparrows, in which a primary-like song is preceded by twittering tickings. Conversely, the "whisper" song of the Saltmarsh Sparrow is without a counterpart in the vocal repertoire of the Nelson. This quiet and inconspicuous song, given from a perch or in horizontal flight between perches, com-

The juvenile Saltmarsh Sparrow is dark and messy, with a strikingly long bill. *New Jersey, June. Donna Schulman*

prises a series of phrases—sometimes lasting more than a minute—made up of a ticking introduction, short trills or mordents, and a longer, lower-pitched trill; phrases may be separated by a variable number of clicking notes, but are usually performed in a continuous, run-on delivery.

Saltmarsh Sparrows call only infrequently in migration and winter, a very high-pitched, thin *teek* without noticeable attack or decay; it can be given in a fast series by birds under stress. This call is probably not distinguishable from that of the Nelson Sparrow.

RANGE AND GEOGRAPHIC VARIATION

Bound to coastal marshes, the Saltmarsh Sparrow occupies one of the narrowest breeding ranges of any North American bird. The northernmost recent breeding reports are from Knox County, Maine; the most reliable sites in that state are farther south, including Popham Beach and Scarborough Marsh, where this species overlaps and occasionally hybridizes with Nelson Sparrows of the race *subvirgata*. Breeders are found up Delaware Bay to Kent County, Delaware, and Salem County, New Jersey; they also occurred, at least formerly, as high as Baltimore on Chesapeake Bay. Saltmarsh Sparrows seem to be significantly less widespread and

less common south of Cape Charles, Virginia, with the southernmost recent breeding records from the Outer Banks of North Carolina.

Fall migration appears to be quite leisurely. Early December still finds some birds on the breeding grounds in New York and New Jersey, though more timely migrants begin to arrive in the Carolinas and Florida as early as October. True wintering birds can be found as far north as New Jersey, but most seem to spend the cold months on the coasts of the Carolinas and central Florida. Very little is known of the winter occurrence of this species on the coast of the Gulf of Mexico. All Texas records are now believed to represent misidentified Nelson Sparrows, and it would be unsurprising to find that most or all reports west of Florida have a similar origin. An individual of the northern subspecies *caudacuta* discovered in Nova Scotia in early April had likely wintered nearby, as the northbound flight in spring typically begins at the end of that month. Most male Saltmarsh Sparrows arrive on their Massachusetts breeding grounds in mid-May.

Unlike the Nelson Sparrow, the two recognized subspecies of the Saltmarsh Sparrow occupy adjacent breeding ranges, meeting in the salt marshes of New Jersey; in winter, both can be present anywhere in coastal marshes from that state south to Florida,

Like many of today's birders, Alexander Wilson first encountered the colorful Saltmarsh Sparrow (bird with orange head), in the cordgrass of coastal New Jersey. *Image from the Biodiversity Heritage Library. Digitized by the Smithsonian Libraries. | www.biodiversitylibrary.org*

while the southerly breeder, *diversa*, also spreads onto the Gulf Coast as far west as Louisiana, where it overlaps with wintering Nelson Sparrows of the races *nelsoni* and, most likely, *altera*.

The most recent study found that in the hand, more than 90 percent of Saltmarsh Sparrows in reasonably fresh plumage—in fall and early winter—could be identified to subspecies on the wintering grounds in Virginia. Louis B. Bishop described the "very distinct" southern-breeding *diversa* in 1901 from a series of February and May specimens collected on Pea and Roanoke Islands, North Carolina. The name—"different"—emphasizes the clarity of the distinctions Bishop was able to draw between his new form and birds taken in Connecticut at the same seasons.

The southern *diversa* is darker and more contrasting above than the nominate race. The scapulars have blackish, rather than brown-gray, centers, and buffy fringes; in *caudacuta*, whiter fringes are separated from the scapular centers by a broad black submarginal bar. The same pattern obtains in the tertials, which in *diversa* show no internal contrast but in *caudacuta* are visibly divided into a brown center, black submarginal bar, and buffy white fringes. The tail and rump are darker brown in *diversa*. In his original description, Bishop noted, with only moderate exaggeration, that the darkness and rich contrast of the upperparts in *diversa* more closely recalled a Nelson Sparrow than a Saltmarsh Sparrow of the nominate race.

The darkness of the upperparts in *diversa* is echoed in the crown pattern, where the division into a median and lateral stripes is obscured by dense blackish streaking and spotting throughout; in the nominate subspecies, the gray central crown is bordered by dark brown lateral stripes, and any black spots or streaks are restricted to those stripes. Fine black streaking may also intrude into the supercilium in *diversa*.

A few individuals (8 percent in the Virginia study reported by Smith) show a combination of characters; those birds are probably intergrades or backcrosses between nominate *caudacuta* and *diversa*.

BELL SPARROW
Artemisiospiza belli

The Bell Sparrow shares its long-tailed, round-bodied aspect with the very similar Sagebrush Sparrow. *California, April. Brian E. Small*

This species and its "sister," the Sagebrush Sparrow, have been a source of ontological confusion almost from the beginning. That beginning came in 1850, when John Cassin at the Academy of Natural Sciences described a new *Emberiza* from California. Sent to Philadelphia by John Graham Bell—co-discoverer a few years earlier of the Baird Sparrow—the specimens reminded Cassin to "rather a remarkable degree" of a South American tanager, the Rufous-sided Warbling-Finch (*Poospiza hypochondria*), but he observed sufficient differences in size and plumage to justify the recognition of a new species. Cassin took the opportunity to honor Bell in the name, "a gentleman possessing a very extensive knowledge of natural history, and whose attachment to the pursuit of which, induced him to make the visit to California, which resulted in the discovery of this and other interesting birds."

In 1873, the 22-year-old Robert Ridgway described a new subspecies of the Bell Sparrow, "much larger and all the colors paler" than nominate *belli*, "with very distinct streaks on the back." Over the next decade, the known range of Ridgway's *nevadensis* (today's Sagebrush Sparrow) was expanded to include much of the interior West and Southwest, and many specimens identified as of that form were collected in California, in both the breeding season and the winter, beginning in the mid-1870s.

In July 1897, Joseph Grinnell found both forms—nominate *belli* and *nevadensis*—breeding side by side in the mountains of central Los Angeles County without any intergradation. Grinnell adduced this discovery as evidence that the two were in fact separate species, an argument promptly refuted by A. K. Fisher, who had examined many California specimens "intermediate in color and size," apparent intergrades between two populations not deserving recognition at the species level.

Grinnell returned to the dilemma eight years later. Comparing a series of skins from Kern, Ventura, Los Angeles, and Riverside Counties in California with undoubted specimens of *nevadensis* from Nevada, Arizona, and southeasternmost California, Grinnell discovered that the birds he "and others [had] repeatedly recorded from Los Angeles as *nevadensis*" were in fact quite different from *both* nominate *belli* and Ridgway's *nevadensis*—and deserved a name of their own. On reexamining Fisher's "intermediates," Grinnell found that they, too, belonged to the new form, which he called *canescens*, "grayish." Grinnell observed that the

"gaps" in color and size between *canescens* and *belli* on the one hand and between *canescens* and *nevadensis* on the other were

> so definite . . . that were it not for current rulings being overwhelmingly against it, [he] should not hesitate to consider them specifically distinct. . . . Not a single intermediate is to be found between *canescens* and *belli* (or *canescens* and *nevadensis*, for that matter). . . .

At the time Grinnell was writing, *nevadensis* had been elevated to species status, and so he found himself obliged to choose between identifying *canescens* as a subspecies of that bird or as a subspecies of the Bell Sparrow in its strict sense. He settled unhappily on the latter course: "It is . . . only under protest that I use the combination *Amphispiza belli canescens*," an expression of frustration and uncertainty that more than a hundred years later still echoes through any discussion of the "sage sparrows" and their taxonomy.

The heavy, wedge-shaped lateral throat stripe of the Bell Sparrow may or may not reach the base of the bill. The long, narrow tail is held straight out and often twitched upward. *California, April. Brian E. Small*

The formal recognition in 2013 of, once again, two species concealed in the "sage complex" was effected "on the basis of differences in mitochondrial DNA, morphology, and ecology, and limited gene flow at the contact zone in eastern California." This most recent split, however, did not simply restore the taxonomic status quo of the early twentieth century. The Bell Sparrow and the Sagebrush Sparrow as understood today are constituted in nearly exactly the opposite way as they were in the years between the Fourteenth and Nineteenth Supplements to the *Check-list*. The Sagebrush Sparrow, *Artemisiospiza nevadensis*, is now monotypic; the Bell Sparrow, in its latest sense, now takes in nominate *belli*, *cinerea*, *clementeae*, and *canescens*, the pale California race that Grinnell and his contemporaries more than a century ago had "repeatedly" mistaken for *nevadensis* of the interior West.

The shuffling and the reshuffling may not be over. Grinnell's 1905 protest at having to subsume his *canescens* under either of the sage species then recognized may well have been prescient. In 2012, Carla Cicero and Michelle S. Koo published the results of a study of the mitochondrial DNA, morphology, and ecology of three populations; they found that the ecological overlap was greatest between *canescens* and *nevadensis*, but that even where the two populations bred in the same area, they remained distinguishable by phenotype and

DNA. There may yet be a third species of "sage sparrow" in the mix, the Mojave Sparrow, *Artemisiospiza canescens*.

FIELD IDENTIFICATION

Adults of the darker races *belli*, *clementeae*, and *cinerea* are unlikely to be misidentified. Juvenile Black-throated Sparrows, which lack the black throat and breast of adults of that species, somewhat resemble both adult and juvenile Bell Sparrows, but have bright white edges to the tail, heavy narrow streaking on the back and breast, a plain gray crown, and a long clear white supercilium reaching from the base of the bill nearly to the side of the nape.

Adult *canescens* can be very difficult to distinguish in the field from the larger, more heavily streaked Sagebrush Sparrow; even in the hand, measurements definitively separate the two only in the case of individuals of known sex, as male *canescens* and female Sagebrush Sparrows may overlap in wing and tail length.

The plumage characters used to differentiate these species are subtle and variable, with some overlap. The two are most readily distinguished in fresh plumage, attained in both species by a prebasic molt that concludes before the southbound migration in September and October; worn adults on the breeding grounds in late spring and summer

This apparent Mojave Bell Sparrow, *canescens*, closely resembles a Sagebrush Sparrow in back pattern, but the lateral throat stripe is heavy, blackish, and nearly reaches the bill base. *California, April. Brian E. Small*

can be expected to be even more challenging. At any season, birds should be identified using as many different plumage characters as possible; both species are fairly shy and often reluctant to give prolonged looks, and inevitably, some sightings will not result in identification beyond the genus level.

The outermost tail feather of *canescens* has most of the outer web and a portion of the inner web buffy white; some Sagebrush Sparrows, even in fresh plumage, show no obvious white in the tail in the field. Both species have brownish flanks with dark brown or black streaking; the shaft streaks may be broader and darker in *canescens* and narrower and less contrasting in Sagebrush. Though the scapulars may have fairly obvious sparse dark shaft streaks, the central back of *canescens* is so faintly streaked as to often appear solid gray-brown in the field; the dark streaks are significantly more extensive and distinct in Sagebrush Sparrows, though the upperpart streaking in that species may become more obscure in spring. The head of *canescens* may average browner than the more purely gray head of the Sagebrush, resulting in slightly greater contrast between the head and the back in fresh plumage.

The *Artemisiospiza* sparrows present one of the cases in which the precise details of head and throat pattern are essential in distinguishing the members of a very similar species pair. Sagebrush Sparrows may average more blackish streaking on the crown

and a slightly longer whitish supercilium; the crown of many Bell Sparrows shows no streaking at all in the field, and the supercilium typically does not reach behind the narrow white eye-ring. The shape, extent, and darkness of the lateral throat stripe differ subtly but consistently between Bell and Sagebrush Sparrows; the combination of back pattern and throat pattern should serve to identify most individuals—if seen well—from the completion of the prebasic molt well into spring.

Bell Sparrows have a dark, thick, and long lateral throat stripe. This is particularly noticeable in the dark resident subspecies, but even in *canescens*, which typically possesses the most poorly defined head pattern in this species, the base of the stripe is noticeably thick, even triangular. In most *canescens*, the stripe narrows rapidly as it approaches the base of the bill, but in most individuals, it is complete from the bottom of the throat to the bill; even when, as in a minority of birds, it does not quite reach the bill base, the stripe is uniformly dark gray throughout its length. In many *canescens*, the lateral throat stripe is slightly but noticeably sootier than the paler, more brown-tinged ground color of the ear coverts.

Sagebrush Sparrows in the same fresh plumage have a paler, narrow, and short lateral throat stripe. The base of the stripe is only slightly broader than the remainder, such that the width over the stripe's

The Bell Sparrow is very sparsely marked above, with the back streaking short, scattered, or apparently entirely absent in the field. *California, April. Brian E. Small*

entire length is more uniform than in the Bell Sparrow. Unlike in that species, the Sagebrush Sparrow's lateral throat stripe appears never to be complete, breaking or fading well below the base of the bill. The stripe may be slightly darker at its base and slightly paler as it ascends the side of the throat; it is paler gray throughout its length than that of the Bell Sparrow, and does not contrast with the grayish ear coverts.

In late spring and midsummer, when the plumage of breeding adults is at its oldest and most worn, Sagebrush Sparrows may be less conspicuously streaked above, and the lateral throat stripe of the Bell Sparrow may be ragged and narrow: thus, the two species may approach each other even more closely in appearance. When this is the case, or at

any time when a sparrow of this genus declines to give reasonably prolonged views, it is best to identify the bird simply as a "sage sparrow," *Artemisiospiza* sp.

Juveniles, like adults, complete the prebasic molt before migration, making it extremely unlikely that they should ever be seen without a parent nearby. In the hand, juvenile Bell Sparrows—presumably but not certainly including juvenile *canescens*—have blacker streaks than Sagebrush Sparrows of the same age.

Adult Bell Sparrows of any subspecies are moderately large, long-tailed, rather short-winged sparrows with a decidedly dark aspect, particularly in the case of the resident subspecies. The head is fairly large and square, but the bill is short, at times giving the impression of unnatural truncation. The four resident subspecies are brown-gray above, from crown to tail, with browner back and wings; the gunmetal blue seen in so many older field guide illustrations of this species is an artifact of reproduction. The tail feathers have narrow whitish edges in fresh plumage, difficult to see in the field, but lack conspicuous white vanes and tips. The most conspicuous plumage mark at any distance is the broad white jaw stripe, separated from the bright white throat by the thick dark lateral throat stripe. The streaking of the breast sides and flanks is usually difficult to see, as is the central breast spot, which varies in size, neatness, and darkness.

The Mojave Sparrow, *canescens*, is, as described above, paler and more streaked; at a distance, it is more likely to strike the observer as a brown bird with a dark tail than as a uniformly gray. The most helpful first impression is often that of the contrast between the grayer head and the brownish back, a difference usually inconspicuous in the more uniformly colored *belli* and other subspecies. The lateral throat stripe of *canescens* often seems "permeable" at first glance, letting the white jaw stripe and throat fuse vaguely in distant views.

The songs of the Bell Sparrow and the Sagebrush Sparrow are quite different. While Sagebrush Sparrows give buzzy, low-pitched, strongly cadenced songs, Bell Sparrows of the subspecies *belli* and *clementeae* (and presumably of the southern race *cinerea*) sing a bright, light, chirping and chipping warble. The thin tone and slightly stuttering rhythm may recall the song of a Horned Lark, while in its length and jumbled cheerfulness the song resembles that of a Painted Bunting.

The Mojave Sparrow, *canescens,* appears to be capable of songs resembling both *belli* and the Sagebrush Sparrow; the song repertoire of this "confusing intermediate group" may in fact "grade clinally from *nevadensis*-like songs in the northeastern part of its range to *belli*-like songs in the south and west," a circumstance that further complicates the field identification of this bird.

The flight calls of the Bell Sparrow are high and metallic, with the hint of a penetrating buzz; they are very short, with a harsh attack and virtually no audible decay. In flight, the calls may be even shorter and uttered in a quick series of up to ten notes, verging on a tremolo.

RANGE AND GEOGRAPHIC VARIATION

The Bell Sparrow is represented on San Clemente Island by the resident race *clementeae*. On the mainland, the nominate race is resident in California's northern Coast Range from Trinity and Shasta Counties and, on the coast, Marin County, south; "an isolated pocket" also occurs in the Sierra foothills between El Dorado and Mariposa Counties. The range extends south through western California to Ballenas Bay in central Baja California; *belli* yields to *cinerea* south of Santa Catarina. Vagrant individuals of *belli* are known from Southeast Farallon Island and Alameda and Orange Counties in California, and, remarkably, from Baja California Sur.

The Mojave Sparrow, *canescens*, is a breeder in the San Joaquin Valley from Merced to Kern Counties, on the slopes of the Transverse Ranges from San Luis Obispo County east, and in the western Mojave Desert from San Bernardino County north to the White Mountains of Mono County. It has also been recorded as breeding in western Nevada's Grapevine Mountains. This race encounters breeding Sagebrush Sparrows in Inyo County, California, and breeding *belli* along the eastern Coast Range.

This is the only regular migrant among the Bell Sparrows; as already noted, though, the determination of its winter range is complicated by the difficulty of distinguishing *canescens* from Sagebrush Sparrows in the field. The specimen record demonstrates regular wintering in southwestern Arizona, though recent observations make it seem unlikely that it is always "nearly as common as" the Sagebrush Sparrow there. Among the preliminary results of an ongoing study of wintering birds in southern and western Arizona is the suggestion that Sagebrush and *canescens* Bell Sparrows there use different habitats and rarely mix in the winter.

Given that all populations of the Bell Sparrow are resident or, in the case of *canescens*, migrate only short distances, it is not surprising that there are no definite records of this species as a vagrant far from its currently expected range. The possible exception is a bird discovered in Nova Scotia in November 1994; the published photograph shows an obscurely streaked brown-tinged back and a broad, blackish lateral throat stripe, both characters suggesting a Bell Sparrow of the subspecies *canescens*. As of 2018, all other identifiable vagrant *Artemisiospiza* sparrows seem to be referable not to any population of this species but to the Sagebrush Sparrow.

As currently recognized by the AOU, the Bell Sparrow comprises four subspecies. The most widespread race, nominate *belli* is small and small-billed, with indistinct dark streaking on the back; the black lateral throat stripe is well defined and usually reaches the base of the bill.

Evaluating the specimen material collected by Charles Townsend on the *Albatross* expedition, in 1898 Robert Ridgway described the resident Bell Sparrows of San Clemente Island as a distinct subspecies, *clementeae*, "exactly like *A. belli [belli]* (Cassin) in coloration, but larger and with relatively larger bill." A few years later, he retracted the claim: "the difference," he wrote, "proves too slight to warrant recognition of the alleged subspecies." In October of 1930, A. J. van Rossem and John Roy Pemberton visited San Clemente to collect Bell Sparrows. Van Rossem agreed with Ridgway that there were "no tangible differences" in body measurements between the island birds and their mainland counterparts, but he found that

> In comparison with *belli* the island birds have slightly, but noticeably, longer bills; the dorsal coloration of freshly plumaged adults is definitely paler, grayer, and more clearly streaked, and the juveniles are paler and with very much narrower ventral streaking . . . .

Charles Townsend described the race *cinerea* from the shores of Ballenas Bay, on the central Pacific Coast of Baja California. Similar to the nominate subspecies of the Bell Sparrow in body size and bill size, this bird is notably paler, with a buff-tinged, vaguely streaked back. The lateral throat stripe is gray rather than blackish, and narrower and more often incomplete than in *belli,* a character more often thought of as typical of the Sagebrush Sparrow; indeed, Townsend began his description of *cinerea* by comparing it to that population rather than to the "normal" Bell Sparrow. The subspecies *belli* and *cinerea* meet at the latitude of Santa Catarina, Baja California Sur, producing birds darker above than typical *cinerea,* with narrow but dark lateral throat stripes; the flanks of these individuals are usually pale brown with black streaks. Described by Laurence M. Huey as a distinct subspecies *xerophilus,* such birds are now considered intergrades between the two recognized races.

The fly in the ointment of *Artemisiospiza* taxonomy and identification is *canescens*, the population Joseph Grinnell described "only under protest" as a subspecies of the Bell Sparrow. Decidedly larger

Bell Sparrows of coastal populations can appear very dark in the field, the head almost steel blue. *California, April. Brian E. Small*

than other Bell Sparrows, this is the only migratory race of that species, wintering east at least as far as the Gila River Indian Reservation in south-central Arizona; the precise limits of its winter range remain obscure because of its remarkable similarity in the field to the Sagebrush Sparrow.

Known in English as the California, Mojave, or Saltbush Sparrow, *canescens* is fairly easily distinguished from all other Bell Sparrows. It is larger and "very much paler" than *belli*, with a slight but definite contrast between the gray head and the brownish, very faintly streaked back; the lateral throat stripe is clear and dark, but does not usually reach the base of the bill. The outer tail feathers have light tan-white edging; the same feathers are edged darker brown in *belli*.

Like a pastel version of the smaller, darker Bell Sparrow, this species is more heavily streaked above but less strongly marked on the throat. *New Mexico, December. Brian E. Small*

SAGEBRUSH SPARROW
Artemisiospiza nevadensis

What is now the type specimen of the Sagebrush Sparrow—the red-tagged skin that serves as physical referent for the species name—was collected by Robert Ridgway in September 1867 in the West Humboldt Mountains of Nevada. But that Nevada individual, whose origin is memorialized in the name Ridgway eventually gave the new taxon, was not the first Sagebrush Sparrow known to science.

That fatal honor went instead to a bird collected at Fort Thorn, New Mexico, in the mid-1850s by the Army surgeon Thomas Charlton Henry, "whose exertions in the investigation of the natural productions of New Mexico . . . for several years, and the formation of large collections in various departments . . . attest [to] his zeal and attachment to zoological science." Henry named two new species himself during his six years on the southwestern frontier, the Crissal Thrasher and the Red-backed Junco, and John Cassin named a subspecies of the Common Nighthawk in his honor. Neither Henry nor Spencer Baird, though, recognized the

Fort Thorn sparrow as new, and Henry's record was published without further comment in 1858 as the first Bell Sparrow from New Mexico. Not for another 15 years would Ridgway describe the birds of the inland West as representing a distinct taxon; by then, Henry, his health destroyed by his service with the Union Army in the Civil War, had retired to Kentucky, where he died in 1877, remembered still by some as "a zealous naturalist, whose untimely recall from this world's duties cut short a career which opened in full promise of usefulness and honor."

Spencer Baird was content in 1874 to accept Ridgway's designation of his Nevada Sagebrush Sparrow as a race of the Bell Sparrow, its differentiation possibly an example of the climatic effect described by Bergmann's Rule:

> The difference in size between the race of the Great Basin [our Sagebrush Sparrow] and that of the southern Pacific Province [the nominate Bell Sparrow], of this species, is quite remarkable, being much greater than in any other instance within our knowledge. This may, perhaps, be explained by the fact the former [viz.,

the Sagebrush Sparrow] . . . [inhabits] the most northern part of its range; while the California one is . . . an inhabitant of only the southern portion of the coast region, not reaching nearly so far north as the race of the interior.

Three years later, Ridgway—one of the coauthors of Baird's *History*—elevated the Nevada sparrow to full species rank, calling it *Amphispiza nevadensis*, the Artemisia Sparrow. That quickly passed, however, and by 1880 he too had restored *nevadensis* to its place within a polytypic Bell Sparrow.

Joseph Grinnell took up the problem again in 1898. Collecting breeding sparrows in central Los Angeles County, California, Grinnell was struck by the fact that *belli* Bell Sparrows and what he believed were *nevadensis* Bell Sparrows were nesting in the same locality, with no sign of hybridization or intergradation. He determined that the two forms should be treated as specifically distinct—a conclusion made no less logical by the knowledge that Grinnell's comparison was actually not with *nevadensis* but with the race *canescens*, which he would describe in 1905.

In 1908, the AOU agreed that Ridgway's *nevadensis* was a species distinct from the Bell Sparrow, and that the new species should also comprise Grinnell's newly described *canescens* and the Baja California subspecies *cinerea*, described in 1890 by Charles Townsend. The Bell Sparrow—in its new, narrow sense—was left monotypic, while the newly constituted Sage Sparrow comprised three subspecies: *Amphispiza nevadensis nevadensis*, *cinerea*, and *canescens*.

Thirty years on, Charles E. Hellmayr treated what were meanwhile six recognized populations as belonging to a single species; *Amphispiza belli*—in its old, broad sense—was back, as a polytypic species ranging from Washington to Baja California. In the Nineteenth Supplement, the AOU followed Hellmayr; the fifth edition of the *Check-list*, published in 1957, changed the English name of the entire species to "Sage Sparrow." There is no reason to believe

The Sagebrush Sparrow's lateral throat stripe may be nearly complete, but it is always thin and hardly widens at its base. *New Mexico, December. Brian E. Small*

that the twenty-first-century view of these birds as making up two species, the Bell and the Sagebrush Sparrow, will prove definitive—ornithological classification changes even more rapidly than the birds themselves, and the only thing to be counted on is that any given concept will eventually be replaced.

FIELD IDENTIFICATION

In 1964, Gale Monson wrote that

> After the last few species [namely, the very challenging and secretive sparrows of the genera *Aimophila* and *Peucaea*], the Sage Sparrow is a relief. He runs around waving his long black tail in the air above his back, so anyone can identify him who can tell a sparrow's beak from a wren's.

True as that may have been half a century ago, the (re-)split of the old Sage Sparrow into the Bell Sparrow and this species has once again complicated what was for long decades a straightforward identification. And it is precisely in Arizona, as well as those small areas in eastern California and probably western Nevada where both species breed, that the field observer is confronted with the considerable challenge of distinguishing wintering Sagebrush Sparrows from *canescens* Bell Sparrows.

Until the recent split by the AOU, the classic species to be confused with the Sagebrush Sparrow was the juvenile Black-throated Sparrow, which lacks the black throat and breast of the adult. Even when they are not accompanied by an adult, however, juvenile Black-throated Sparrows are easily identified by the long clear white supercilium, which extends from the base of the bill nearly to the side of the nape.

Now, whether *canescens* is treated—as it is by the AOU—as a subspecies of the Bell Sparrow or as a distinct species on its own, identifying the Sagebrush Sparrow in the field wherever both are possible can be extremely difficult without excellent views and some experience. Even in the hand, measurements definitively separate the two only in the case of individuals of known sex, as male *canescens* and female Sagebrush Sparrows may overlap in wing and tail length.

The plumage characters used to differentiate these species are subtle and variable, with some overlap. The two are most readily distinguished in fresh plumage, attained in both species by a prebasic molt that concludes before the southbound migration in September and October; worn adults on the breeding grounds in late spring and summer can be expected to be even more challenging. At any season, birds should be identified using as many different plumage characters as possible; both species

are fairly shy and often reluctant to give prolonged looks, and inevitably, some sightings will not result in identification beyond the genus level.

Some Sagebrush Sparrows, even in fresh plumage, show no obvious white in the tail in the field, while the outermost tail feather of *canescens* is buffy white over most of the outer web and a portion of the inner web. Both have brownish flanks with dark brown or black streaking; the shaft streaks may be broader and darker in *canescens* and narrower and less contrasting in Sagebrush. In the Sagebrush Sparrow, both the scapulars and the central back are extensively and distinctly streaked blackish, though the upperpart streaking in that species may become more obscure in spring. The head of *canescens* may average browner than the more purely gray head of the Sagebrush, resulting in slightly greater contrast between the head and the back in fresh plumage.

The *Artemisiospiza* sparrows present one of the cases in which the precise details of head and throat pattern are essential in distinguishing the members of a very similar species pair. Sagebrush Sparrows may average more blackish streaking on the crown and a slightly longer whitish supercilium; the crown of many Bell Sparrows shows no streaking at all in the field, and the supercilium typically does not reach behind the narrow white eye-ring. The shape, extent, and darkness of the lateral throat stripe differ subtly but consistently between Bell and Sagebrush Sparrows; the combination of back pattern and throat pattern should serve to identify most individuals—if seen well—from the completion of the prebasic molt well into spring.

Sagebrush Sparrows in fresh plumage have a pale, narrow, and short lateral throat stripe. The base of the stripe is only slightly broader than the remainder, such that the width over the stripe's entire length is more uniform than in the Bell Sparrow. Unlike in that species, the Sagebrush Sparrow's lateral throat stripe appears never to be complete, breaking or fading well below the base of the bill. The stripe may be slightly darker at its base and slightly paler as it ascends the side of the throat; it is paler gray throughout its length than that of the Bell Sparrow, and does not contrast with the grayish ear coverts.

In late spring and midsummer, when the plumage of breeding adults is at its oldest and most worn, Sagebrush Sparrows may be less conspicuously streaked above, and the lateral throat stripe of the Bell Sparrow may be ragged and narrow: thus, the two species may approach each other even more closely in appearance. When this is the case, or at any time when a sparrow of this genus declines to give reasonably prolonged views, it is best to identify the bird simply as a "sage sparrow," *Artemisiospiza* sp.

Juveniles, like adults, complete the prebasic molt before migration, making it extremely unlikely that they should ever be seen without a parent nearby. In the hand, juvenile Sagebrush Sparrows tend to have duller, grayer streaks than Bell Sparrows of the same age.

Sagebrush Sparrows of any subspecies are moderately large, long-tailed, rather short-winged sparrows with a decidedly grayish aspect. The head is fairly large and square, but the bill is short. The tail feathers have white edges in fresh plumage. The most conspicuous plumage mark at any distance is the contrast between the grayish head and the clearly browner, more heavily streaked back. The streaking of the breast sides and flanks is usually difficult to see, as is the central breast spot, which varies in size, neatness, and darkness.

The long, blackish tail is often twitched upward, a habit shared with the Bell and Black-throated Sparrows. *California, March. Brian E. Small*

The buzzy, low-pitched, strongly cadenced song of the Sagebrush Sparrow has moved even normally sober ornithologists to poetic flight. Robert Ridgway, best known today for his dry scientific descriptions of thousands of avian taxa, found the Sagebrush Sparrow singing as early as late February in Nevada:

> The song of this bird, although not brilliant in execution nor by any means loud, is nevertheless of such a character as to attract attention. It has a melancholy pensiveness, remarkably in accord with the dreary monotony of the surroundings, yet as a sort of compensation, is possessed of delicacy of expression and peculiar pathos—just as the fine lights and shadows on the sunlit mountains, combined with a certain vagueness in the dreamy distance, subdue the harsher features of the desert landscape.

Even observers prone to a less romantic view of their surroundings can come to think of this sweet-voiced sparrow as the *genius* of its sadly vanishing *loci*.

Unlike the chippering warble of most Bell Sparrows, almost every note of the song of the Sagebrush Sparrow has an underlying buzz; some individuals can even recall a distant Blue Grosbeak. The song is variable in length, but usually follows an easily recognized and distinctive pattern: two or three short buzzing notes are followed by a somewhat longer buzzing trill, which gives way to two or three shorter notes introducing a distinctive slurred cadence. Any of the first elements—the introductory notes, the longer trill, or the following short notes—can be repeated, delaying the cadence. The song is continuous, without discernible pauses or phrasing, *beebeebzzzbzabzazeerup*. Volume and tone remain consistent through the song, and the only obvious emphasis is on the very last note, which can be a quite sharp chip.

The Mojave Sparrow *canescens* appears to be capable of songs resembling that of the Sagebrush Sparrow; the song repertoire of that "confusing intermediate group" may in fact "grade clinally from *nevadensis*-like songs in the northeastern part of its range to *belli*-like songs in the south and west," a circumstance that further complicates the field identification of that bird.

The flight calls of the Sagebrush Sparrow are high and metallic, almost juncolike, with the hint of a penetrating buzz; they are very short, with a harsh attack and virtually no audible decay. In flight, the calls may be even shorter and uttered in a quick series of up to ten notes, verging on a tremolo.

RANGE AND GEOGRAPHIC VARIATION

A Great Basin specialty, the Sagebrush Sparrow breeds from northern Arizona, northwestern New Mexico, and far western Colorado north through the western half of Wyoming and all but the southernmost portions of Utah and Nevada to southwestern Montana, southern Idaho, eastern Oregon, and south-central Washington. In Arizona, most breeding records have come from the sagebrush flats north and east of the Little Colorado River; at least in years of more abundant moisture, however, the species also breeds west to the Arizona Strip and south to the South Rim of the Grand Canyon. On the northern edge of the species' range, Montana reports are infrequent, most from Carbon County; in central and eastern Washington, Sagebrush Sparrows are found as far north as Douglas and Lincoln Counties. The eastern boundaries of the species' regular breeding range are somewhat unclear; great areas of evidently suitable habitat appear to be unoccupied in western Colorado, and there is anecdotal evidence that the Sagebrush Sparrow has abandoned some of its historical range in eastern Wyoming. In the west, this is a breeding bird on California's Modoc Plateau and from the Mono Basin east.

The southbound migration of Sagebrush Sparrows is leisurely, with departure from the northern portions of the breeding range stretching from mid-September into early November. During this period, and less frequently in the northbound migration in April, occasional individuals are detected on the plains of eastern Wyoming and west to the Pacific Coast in British Columbia, Washington, and Oregon.

Wintering birds can be seen north as far as the southern parts of the breeding range in Nevada, Utah, Arizona, and New Mexico. Most move farther, typically arriving in southern New Mexico, southern Arizona, northern Sonora, and central Chihuahua in October. The easternmost regular wintering sites are in the Trans-Pecos of Texas.

For a bird as widespread, as common, and as decidedly migratory as this, the Sagebrush Sparrow is remarkably scarce as a vagrant outside of the range sketched above; it is far less frequent in the East, for example, than Black-throated or Golden-crowned Sparrows, a southwestern and northwestern species, respectively, with scattered records from across the continent. The Sagebrush Sparrow is very rare on the western Great Plains even adjacent to its breeding grounds in Wyoming and Colorado, and South Dakota, Nebraska, and Kansas have not a dozen recent records between them. North of the expected breeding range, there are at least 15 *Artemisiospiza* records for British Columbia, most in spring, evenly distributed between the Okanagan deserts of the south-central portion of the province and the coast of the George Depression; all that can be identified to species pertain to the Sagerush Sparrow. There is a 1992 report from Hamilton County, Ohio, and one was photographed in Warren County, Kentucky, in 2006. An *Artemisiospiza* sparrow photographed in Nova Scotia appears to have been a Bell Sparrow of the subspecies *canescens*.

In the concept currently accepted by the AOS, the Sagebrush Sparrow is monotypic. In 1946, though, Harry C. Oberholser named the northernmost population, breeding from central Oregon and southern Idaho north to central Idaho and eastern Washington, as a new race, *campicola* ("field dweller"); he described this Idaho Sage Sparrow as somewhat larger than *nevadensis*, with darker, more grayish upperparts and less buffy flanks. The differences are now generally thought to be clinal and sufficient for the recognition of a distinct subspecies; in any event, it is extremely unlikely that they should ever be discernible in the field.

CANYON TOWHEE
Melozone fusca

This very large, relatively nondescript sparrow was first encountered in the United States by Caleb Kennerly, a student and protégé of Spencer Baird. On Baird's recommendation, Kennerly had been appointed surgeon-naturalist to the Whipple expedition surveying a possible railroad route from Arkansas to California by way of northern Arizona. In February 1854, Kennerly's party was on the Bill Williams Fork, where he shot the type specimen of what Baird, on receiving the skin in Washington, would describe as a new species, *Pipilo mesoleucus*, the Cañon Towhee.

Oddly, even 20 years later the life history of this species—common and confiding over its fairly large range—was still only poorly known to orni-

One of the plainest of the American sparrows, this large towhee is paler and more uniformly colored than the White-throated, California, and Abert Towhees. *Arizona, April. Brian E. Small*

thologists. Many observers found it "restless" and "retiring," "preferring the dense bushes in the valleys." Such characterizations raise the possibility that some early sight records in fact pertain not to this tame bird of sparsely vegetated rocky slopes but to the Abert Towhee, a relatively furtive resident of dense riparian thickets. Even Elliott Coues, a careful observer if ever any was, in 1866 reported behavior quite unlike what the modern birder expects from the Canyon Towhee: it "associates freely with the" Abert Towhee, he claimed, "and inhabits the same regions; and the two have very similar habits." Henry Henshaw, later head of the United States Biological Survey, was the first ornithologist to draw clear behavioral distinctions between the brown towhees of the Southwest. Of the Canyon Towhee, which he encountered for the first time near Camp Grant, at the base of Arizona's Pinaleño Mountains, Henshaw wrote:

> In the localities it selects for its abode, its taste differs much from that of the following species [the Abert Towhee], and it was found by us in situations more congenial to the nature of the previous bird [the Spotted Towhee] than to Abert's [Towhee], with which, indeed, I believe it never associates.

Clarity came too late for Caleb Kennerly. Returning from the Northwest Boundary Survey in February 1861, Kennerly suffered a stroke and died, at the age of 32, off the coast of Baja California. Spencer Baird eulogized the young collector:

> No one of the gentlemen who have labored so zealously to extend a knowledge of the natural history of the west within the last ten or twelve years has been more successful than Dr. Kennerly. Many new species have been first described by himself or from his collections, while his contributions to the biography of American animals have been of the highest interest.

Scientific views of the relationship between this species and the California Towhee have wavered over the years. The California Towhee was briefly but cogently described as a new, distinct species *Fringilla crissalis* by Nicholas Vigors in 1839. Though Coues peremptorily declared it a subspecies of the Canyon Towhee in 1872, Robert Ridgway continued to treat the two as distinct until 1879, when he too synonymized the brown towhees of the West. In 1901, however, Ridgway returned to his original concept of two separate species, inspiring the AOU to appoint William Brewster, Jonathan

Some Canyon Towhees show a vaguely bicolored throat, with whitish chin and peach-washed lower throat. *Arizona, April. Brian E. Small*

Dwight, and Witmer Stone as a committee to investigate the matter.

A decision was reached in 1908, when the AOU removed the subspecies *crissalis* and *senicula* from the Canyon Towhee to make of them a full species, *Pipilo crissalis*. It took only 11 years for the taxonomic pendulum to swing back, with the description by Harry C. Oberholser of a new subspecies of brown towhee, *aripolius* ("purely grayish"). Oberholser found that his new bird, collected in Baja California by Edward Nelson and Edward Goldman, should be

> of considerable interest since it establishes direct and complete liaison in both geographic distribution and in characters between [populations of the Canyon and California Towhees], and shows clearly that these birds, commonly regarded as distinct species, are but sub-specifically related.

That lumping was adopted in the fourth edition of the *Check-list*, and maintained in the fifth and sixth. Just two years after the publication of the sixth edition in 1986, Robert M. Zink conducted biochem-ical and morphological studies revealing genetic differences between the California Towhee and the Canyon Towhee as great as those between most pairs of full species. Zink advocated that the two be considered distinct species, "almost certainly not each other's nearest relative." In ratifying Zink's proposal, the AOU pointed out, too, that the California and Canyon Towhees have different vocal repertoires, a factor of increasing significance today in attempts to determine the species-level relationships of many passerine birds.

FIELD IDENTIFICATION

The range of this species does not overlap with that of the California Towhee, making identification straightforward over most of those two species' range. It is barely conceivable that an ambitious Canyon Towhee might someday find its way across the bleak flats of the lower Colorado River to enter California, where it would stand out by its paler, more grayish upperparts and more contrasting crown; the Canyon Towhee's very pale central belly is also unlike the California's. The throat and upper breast are pale yellowish, surrounded by fine irregular streaks and marked by a "pendant" spot at the top of the belly. Plumage characters aside, the most obvious distinction between the two species is the common call given by each: high, sharp, and ringing in the California Towhee, lower-pitched, dull, and overlaid by a vague squeakiness in the Canyon.

In northwestern Oaxaca, this species and the closely related White-throated Towhee narrowly overlap. The more social White-throated is slightly darker and browner above than the sympatric Canyon Towhees, with faint white tips to the coverts of the uppertail and wing. The sides and occasionally center of the white breast are washed or faintly spotted gray; the throat is bright white, with an uneven band of buffy stretching back nearly to the side of the nape. Juveniles of both species have tawny wing bars; the Canyon Towhee is streaked brown below, the White-throated mottled gray-brown.

Birders visiting the range of the Canyon Towhee for the first time are sometimes briefly misled by the robust, ground-loving Rufous-crowned Sparrow, which shares its rusty crown with several populations of the towhee. The sparrow, though, is somewhat smaller, noticeably darker, and usually shyer, keeping to more heavily vegetated slopes. The rounded, blackish tail, more clearly contrasted wing and back, and heavily marked throat and face distinguish the Rufous-crowned immediately, as do its nasal call and brittle, rattling song.

Canyon Towhees are large, round, long-tailed sparrows of dry, rocky brushland, where they spend most of their time on or near the ground. At any distance, they are uniformly medium-pale, with a slightly darker tail, no strong head markings, and only a sparse, messy pattern of short streaks and spots on the breast; the somewhat brighter buff of the undertail coverts, vent, and lower flanks is rarely conspicuous in the field.

At closer range, the rather large, blocky head and swollen bill are obvious. The poorly defined buffy throat is irregularly framed by short streaks; in most, the upper edge of the whitish belly is marked with a smudgy blackish spot. Even in *mesoleucus* and other rusty-capped subspecies, the reddish crown usually shows only a light contrast to the rest of the head. More noticeable, and entirely distinctive, is the bland, feckless appearance of the face, resulting from the lack of strong black and the narrow buffy eye-ring; Canyon Towhees invariably make a distinctive gentle, innocent impression in the field, an impression heightened by their tameness and homely chirping calls.

Juveniles are plainer-headed still, in all subspecies lacking or nearly lacking rusty on the crown. The underparts are messily streaked, and the wing coverts are usually faintly tipped buffy, creating two narrow wing bars.

The most commonly heard vocalization is a loud, low-pitched *djump*, calling to mind the less abruptly

The Canyon Towhee's throat is often paler than the rest of the head, and can be largely plain or set off by a vague spotted necklace. *New Mexico, December. Brian E. Small*

explosive, higher-pitched *chimp* of the Song Sparrow, but with a decidedly squeaky overtone, especially at the beginning of the note. At times it is drawn out into two stuttering syllables, *sd-jump*. This call is often given by feeding birds, on the ground or in low vegetation, and is sometimes the first clue to the presence of the species. Individuals hiding in denser brush usually switch to a long, hissing *tseee*, similar to the corresponding call of the Abert and California Towhees but without the discordant buzz of the note given by the Spotted Towhee.

Like the Abert and California Towhees, this species also has a reunion greeting, a very loud, accelerating series of squealing chips and rattles given "dozens of times" each day in affirmation of the bond joining a mated pair, typically uttered when the pair moves from feeding site to feeding site on its forage beat. Harsh and grating, the duet is not usually given in perfect synchrony, but with one bird's performance set off by a split second from the other's, thus doubling all of the notes. Each bird's part begins with a few abrupt hissing notes, which are followed by a hoarse, rasping cackle: *tee-tee-ckackacka*.

The song of the Canyon Towhee is a simple and cheerful series of sweet slurred whistles, often introduced by one or two *djump* notes. The slurred series is often broken into two phrases, one—whether the first or the second—faster than the other. The effect can be surprisingly cardinal-like, especially when it is the latter half of the song that is sung faster. Shorter songs without a change in tempo can closely recall the song of a Pyrrhuloxia. A single individual male is capable of songs of both types.

RANGE AND GEOGRAPHIC VARIATION

The Canyon Towhee is a permanent resident of the southwestern United States and much of Mexico. In Colorado, it is found from El Paso County south to Baca County; there are also records from Montezuma and Mesa Counties in the southwest. A historical population in Boulder County, the northernmost known, is apparently extirpated, though there is a record from the summer of 2011 (eBird).

The species is scarce in extreme southwestern Kansas, northwestern Oklahoma, and the southern tier of Utah counties. In Arizona, it is a common resident in appropriate habitat northwest of and south of the Mogollon Rim, west to the Sierra Pinta and disjunctly in the Kofa Mountains. It is much less common and more local north and east of the Rim. The Canyon Towhee is common nearly throughout New Mexico, less frequent and less common only on the eastern plains.

Canyon Towhees are resident in Texas from the

Dark-crowned birds may recall the smaller, darker Rufous-crowned Sparrow, which lacks the buffy or pale orange undertail of the towhee. *Texas, April. Brian E. Small*

Trans-Pecos east to the Rolling Plains and Edwards Plateau, rare and local on the eastern edges of those two regions; wintertime wanderers have been recorded east to southern Texas. In the south, this species occurs in Sonora and northern Sinaloa on Mexico's Pacific Slope, and from Chihuahua and Coahuila to northern Oaxaca; see below for the distribution of each subspecies within that vast range.

Records of this species outside its known breeding range are scant, and appear to be limited to southeastern Utah and the western Great Plains in Colorado, southwest Kansas, and southern Texas. A 1975 report from Nebraska is nearly undocumented; one or two more recent occurrences there are rumored to involve birds transported from Arizona (W. Flack in litt.), a plausible scenario given this species' habit of investigating vehicle interiors.

As of 2017, ten geographic races of the Canyon Towhee are widely recognized. The subspecies most familiar to birders in the United States is Baird's *mesoleuca* (grammatically feminine to agree with the genus name *Melozone*), found across the lowlands of central and eastern Arizona, New Mexico, and western Texas; it reaches its southern limit in northern Sonora and Chihuahua. This is a rather large, long-tailed and large-billed race; the decidedly rusty crown contrasts with the dark gray-brown upperparts.

The northernmost race, *mesata* ("most intermediate"), is large, relatively short-tailed and thick-billed, with a paler, less rufous crown and paler, grayer upperparts than *mesoleuca*; the flanks are also lighter. Originally believed to occur only in southern Colorado, this is also the resident race of northeastern New Mexico and western Oklahoma, and may be a local resident in the northwestern Panhandle of Texas.

The Canyon Towhee of northwestern Coahuila and central and southwestern Texas is *texana*. This is the easternmost race of the species in the United States. It is small, like *mesoleuca*, but shorter-tailed, with a thinner bill; the upperparts are darker and grayer, and the dull crown lacks or nearly lacks rufous. In Texas, this race is resident in the southern Trans-Pecos and Edwards Plateau, north to the southeastern portions of the Panhandle; to the west and north, it intergrades with *mesoleuca*, which is resident from El Paso County east to Culberson County.

South of the western range of *mesoleuca*, the subspecies *intermedia* inhabits the southern half of Sonora and northern Sinaloa. This long-tailed race is smaller than *mesoleuca*, with darker, grayer upperparts and crown, the head without or nearly without rufous; beneath, the flanks are slightly darker than in *mesoleuca*. The race *jamesi*, named for Arthur Curtis James, the railroad investor

This species is often quite tame, feeding in picnic areas and parking lots and readily perching up. *Texas, April. Brian E. Small*

and trustee of the American Museum, is entirely restricted to Sonora's Tiburón Island. It is smaller and paler than its mainland counterpart *intermedia*, with the upperparts much grayer and the crown lighter rufous; it is also less tawny on the vent and undertail coverts.

To the east, the "desert pocket" between the central highlands and the Sierra Madre in Chihuahua, Durango, and western Zacatecas is inhabited by the subspecies *perpallida*. Grayer and—as its name suggests—paler than all but *jamesi*, it is smaller and shorter-tailed than *mesoleuca*, and lighter gray and redder-crowned than *potosina*. That last subspecies is very widespread on Mexico's central plateau, from Coahuila, Nuevo León, and western Tamaulipas, south to northern Jalisco, Guanajuato, Querétaro, and Aguascalientes; first collected by the French entomologist Auguste Sallé in the mid-1850s, it was not recognized as a distinct subspecies until 1899, when Robert Ridgway described it on the basis of a male specimen collected by Alfredo Dugès in Guanajuato. It differs from the nominate subspecies in its larger size and paler, grayer plumage with a pale rusty crown and dull yellowish undertail coverts.

The nominate *Melozone fusca fusca*, described by Swainson in 1827, is resident across the southern Mexican plateau from Colima east to northern Guerrero and the state of Mexico. It is uniformly medium-pale above, from rump to crown, with a prominent whitish area on the lower breast and belly. The wing is relatively long and the tail relatively short, creating a wing-to-tail ratio of nearly 1:1. The Hidalgo Canyon Towhee, *campoi*, is a pale, small subspecies, with wing and tail approximately equal in length; the back and crown are gray. It is paler below and darker above than *fusca*, with a pale buffy throat and a whiter breast. This race was named by Robert T. Moore for Rafael Martín del Campo, curator of ornithology at the Instituto de Biología in Mexico City.

The southernmost Canyon Towhees, found from Tlaxcala and western Veracruz south through Puebla to Oaxaca, are members of the large race *toroi*, the palest of the species, lighter brown than either of the neighboring subspecies *fusca* or *campoi*. Unlike *fusca*, birds of this subspecies have pale gray upperparts without brown; there is only a trace of brownish tinge on the crown. They are cold gray-brown beneath, with purer white on the center of the breast and belly and no mottling or spotting across the upper breast. Moore's name commemorates the professional collector Mario del Toro Avilés.

The differences among these many described subspecies tend to be subtle in the field, where the assessment of size, tail-to-wing ratios, and precise shades of gray and brown can be difficult indeed; even in the museum drawer, it is a challenge to assign every specimen to a geographic area. Where subspecies approach each other geographically, of course, intergradation is—virtually by definition—frequent. Migration in this species, though, is restricted to local movements, making it possible to presumptively identify to race birds in the heart of their range.

CALIFORNIA TOWHEE
Melozone crissalis

The discovery and naming of birds, as basic, even as trivial as it may seem, has more than once been the source of contention, collegial most of the time, but occasionally more controversial. In 1855, the German museum man Gustav Hartlaub published his diagnosis of an "undescribed" bird in the collections of the natural history museum in Leiden. He concluded his description with a gentle jab at a much more famous contemporary, Charles Bonaparte. Bonaparte, Hartlaub sniffed, also knew the Leiden specimen, and in fact had published on it himself five years earlier—but "quite feebly, characterizing it with the single word *'variegatus'*."

The response was swift and withering. Bonaparte offered a series of corrections to his earlier account: the specimen had come from Russian America, not from Siberia; it was clearly a member not of the genus to which he had preliminarily assigned it but rather of Swainson's genus *Pipilo*; "*variegatus*" —"heavily marked"—was an inexplicable printer's error; and on reflection, Bonaparte concluded, the bird seemed to be not new at all. Rather than let any of these errors move him to contrition, however, Bonaparte testily reproached his colleague: "How did Hartlaub not notice all this after describing the bird in such detail?"

The failure to "notice" persisted for a quarter of a century after this chilly exchange. The bird we know as the California Towhee had in fact been discovered and named 15 years before Bonaparte and Hartlaub started sorting the sparrows in Leiden. Working from specimens collected on Frederick William Beechey's 1825 Pacific voyage aboard the *Blossom*, the Irish ornithologist Nicholas Aylward Vigors in 1839 described a new species of heavy-billed finch,

> above dusky brown, below dull whitish,
> with a rusty eye line, throat, and undertail.

Between 1850 and 1856, Bonaparte came to recognize that the Leiden skin he and Hartlaub had squabbled over was in fact identical to Vigors's *Fringilla crissalis*, making its correct name—with an update to the genus to reflect the incipient dismantling of *Fringilla—Pipilo crissalis*.

English-language ornithology, however, was slow to identify Vigors's *crissalis* with the dark towhees shipped east by collectors in California. For 30 years, and in a few cases beyond, scientists in the United States and Britain simply assigned those specimens to the broadly conceived species *fuscus*,

The California Towhee's lore, forehead, and often throat are dull orange, darker than the rest of the head. *California, June. Brian E. Small*

the Canyon Towhee, and Vigors's *crissalis* was essentially forgotten, Bonaparte's attempt to revive it notwithstanding. In 1870, George Robert Gray at last gave Vigors credit and the birds a separate identity in the *Hand-list*, where he assigned the specimens in the British Museum the name *Pipilo crissalis*.

This time, the name stuck. There is some irony, though, in the fact that it should have been Gray, of all people, who finally restored it to widespread ornithological usage. Beechey, in the course of whose expedition the type specimen was collected, had solicited Gray's assistance in preparing the zoological report of the voyage. Beechey reports the result:

> The publication has suffered so much by delay in consequence of his having been connected with it, that it is a matter of the greatest regret to me that I ever acceded

The throat pattern is variable, but there is often a neat band of dark spots across the lower throat and upper breast. *California, October.* Brian E. Small

to his offer to engage in it. This delay has from various causes been extended over a period of eight years . . . it has been occasioned entirely by Mr. Gray's failing to furnish his part.

The existence of the California Towhee, and its right to the name *crissalis*, was by then well established. Over the next century or more, the focus of the debate shifted to the relationships of *crissalis* and its status as a species of its own. Happily, that debate was conducted with less overt acrimony.

Nicholas Vigors had described his *Fringilla crissalis* with no hint that it could be any but a distinct species. It was Elliott Coues who in 1872 lumped Vigors's bird with the Canyon Towhee, identifying it as the "dark coast form" of that species; Coues was also the first to explain the towhees' relationship by analogy to that between the somberly colored, coastally distributed California Thrasher and the paler, inland-dwelling LeConte Thrasher.

Coues's concept prevailed without contradiction for more than a quarter of a century, with the exception of a few years in the 1870s when Robert Ridgway again recognized the brown towhees as separate species. Single-species orthodoxy had been restored by the time of the publication in 1886 of the first edition of the AOU *Check-list*, in which the two subspecies of the California Towhee then known,

Baja California's *albigula* and Upper California's *crissalis*, were treated as races of an enlarged brown towhee species. Two more subspecies, *senicula* of southern California and *carolae* of the state's central interior, were added in 1895 and 1901, respectively, both, of course, as additional races of the enlarged species *Pipilo fuscus*.

Ridgway's treatment of the brown towhees in the *Birds of North and Middle America* appeared in 1901, spurring the AOU to reexamine the lines hitherto drawn within the group. The eminent subcommittee responsible—William Brewster, Jonathan Dwight, and Witmer Stone—narrowed its focus to three questions: whether *albigula* should be considered a distinct species of its own; whether *crissalis*, *senicula*, and *carolae* should be split (in the case of *crissalis*, again) from the other brown towhees; and whether *carolae* merited recognition at all.

The last question was easily, if not definitively, answered the next year: the "alleged form" *carolae* was "to be eliminated, as indistinguishable from *P. fuscus crissalis.*" The other issues were resolved in 1908, when the populations known as *crissalis* and *senicula* were removed to their own species, *Pipilo crissalis*; the southern *albigula*, on the other hand, would continue to be classed as a subspecies of the other brown towhee, *Pipilo fuscus albigula*.

In 1919, however, Harry Oberholser described a new population of brown towhees from central Baja California, *aripolius*, geographically and morphologically intermediate between *senicula*—considered by the AOU a member of the species *crissalis*—and *albigula*—considered by the AOU a member of the species *fuscus*. Oberholser's *aripolius*, he wrote,

> establishes direct and complete liaison in both geographic distribution and in characters between *Pipilo fuscus senicula* and *Pipilo fuscus albigulus* [sic], and shows clearly that these birds, commonly regarded as distinct species, are but sub-specifically related. Since *Pipilo fuscus albigulus* intergrades individually with *Pipilo fuscus mesoleucus* of northwestern Mexico and Arizona, it follows that *Pipilo crissalis* and its races are all subspecies of *Pipilo fuscus*.

The AOU agreed, and what had been split asunder was again joined together on publication of the fourth edition of the *Check-list*. Moreover, the new race *petulans*, described in 1926, was added to the already long list of recognized subspecies of *Pipilo fuscus*, and *carolae*, summarily "eliminated" in 1904, was resurrected. The lump into a single species was ratified by Davis's morphological study of all known brown towhees, and was maintained by the AOU through the sixth, 1983 edition of the *Check-list.*

Just two years after the publication of that edition, Robert M. Zink conducted biochemical and morphological studies revealing genetic differences between the California Towhee and the Canyon Towhee as great as those between most pairs of full species. Zink advocated that the two should properly be considered distinct species, "almost certainly not each other's nearest relative." In adopting Zink's proposal, the AOU pointed out, too, that the California and Canyon Towhees have different vocal repertoires, as keen-eared birders have always known.

FIELD IDENTIFICATION

The range of this species does not overlap with that of the Canyon Towhee, making identification straightforward over most of the two species' range. Plumage characters aside, the most obvious distinction between the Canyon and the California Towhees species is the common call given by each; low-pitched and flat in the Canyon, it is high, sharp, and ringing in the California Towhee, reminding many observers of a breathless, squeaky Black Phoebe. The call is similar to the single note of an Abert Towhee, a species differing from the California Towhee in its even larger size, unstreaked upper breast, and black chin and lore; the Abert and California Towhees do not normally overlap, though the Salton Sea record of the latter comes from well within the usual range of the former.

California Towhees are large, round, long-tailed sparrows of usually moist, patchy brushland and chaparral, where they spend most of their time on or near the ground, darting nervously into cover at the least disturbance. At any distance, birds of the more northerly subspecies are uniformly dark, with little contrast between upperparts and underparts or tail and back. The head is even plainer than in the Canyon Towhee, and the breast markings are inconspicuous, sometimes invisible, against the general darkness of the breast and throat.

At closer range, the rather large, blocky head and swollen bill are obvious. In northern birds, the deep yellowish buffy throat is irregularly framed by short coarse streaks, which intrude as flecks onto the throat; there is usually no breast spot in these subspecies. The crown varies geographically and individually from somewhat reddish to nearly concolorous with the rest of the upperparts. The face shows no black, but because the lore, ear coverts, and eye-ring are dark yellowish brown, the bird's expression is noticeably "stern," unlike the gentle, open face of the Canyon Towhee.

In southern Baja California, the California Towhee subspecies *albigula* is—as its name suggests —noticeably different from its more northerly relatives. Pale above and below, with an extensive pale rufous crown and nape, this San Lucas Towhee resembles the Canyon Towhees of the Mexican mainland. The throat is pale buffy white, bordered by dusky streaks and, in at least some birds, a Canyon-like breast spot. The flanks and sides of the breast are light grayish brown, setting off the whitish central belly, which grades into the tawny buff of the vent and undertail coverts.

Juveniles of all subspecies of the California Towhee are browner above, the feathers of the wing and back faintly edged brighter brown. The crown is essentially the same color as the back, even in those subspecies with rusty-crowned adults. The uniformly dim brown underparts are messily streaked, and the wing coverts are usually faintly tipped buffy or cinnamon, creating two narrow, ill-defined wing bars.

The most commonly heard vocalization is a loud, sharp, high-pitched *tjeek*, calling to mind the chip of a Blue Grosbeak. The tone is decidedly metallic, with a noticeable squeakiness in the attack. This call

The undertail coverts are darker than in the Canyon Towhee and usually as deep russet as in the Abert Towhee. *California, January. Brian E. Small*

The long dark tail is clearly graduated, the outer feathers shorter than the inner. *California, January. Brian E. Small*

can be given singly or in a long series, exceptionally for as long as 25 minutes; it is often given by feeding birds, on the ground or in low vegetation, and is sometimes the first clue to the presence of the species. Individuals hiding in denser brush may switch to a long, hissing *tseee*, similar to the corresponding call of the Abert and Canyon Towhees but without the discordant buzz of the note given by the Spotted Towhee.

Like the Abert and Canyon Towhees, this species also has a reunion greeting, a loud, accelerating series of squeals and hoarse squeaks given in affirmation of the bond joining a mated pair. The duet is not usually given in perfect synchrony, but with one bird's performance set off by a split second from the other's, thus doubling all of the notes. Each bird's part begins with a few bright, metallic ticking notes, which are followed by a faster rattle: *teep-teep-claclacla*.

The song is often surprisingly soft for a bird of this size. The typical series of metallic chips begins three or four call-like notes, which are followed by a ragged accelerating trill on the same pitch; some songs end with a soft, lower-pitched chortle: *teepteeptptptptchur*.

RANGE AND GEOGRAPHIC VARIATION

Apart from the disjunct populations in southwest Oregon and northern California and in Inyo County, California, the range of this widespread species extends from Humboldt County south to southernmost Baja California; it is absent from California's southern Central Valley and the most arid portions of northeastern Baja California. Over this large range, California Towhees occupy brushy habitats including chaparral, riparian and woodland thickets, hedgerows, and vegetated suburban landscapes, including parks and yards; the small Inyo County population is dependent on dense streamside willow thickets. Foraging birds happily venture into more open woods, though rarely far from heavy cover.

Like the Canyon Towhee, this species is resident throughout its range, moving down from higher elevations only locally in winter and fall. As expected, apparent vagrants are virtually unheard of; strays have been recorded in Douglas County, Oregon; at the Salton Sea; and on Todos Santos Island. The single report from the Russian Far East almost certainly derives from a mislabeled specimen.

Given the subtle and gradual variation exhibited by this species over most—but not all—of its range, it is most practical from the point of view of the field observer to be conservative in identifying and naming discrete geographic populations. It may not be strictly speaking true that "too many [!] subspecies of the California Towhee have been described north of the Mexican border," but it is indisputable that many of the races named over the decades are as difficult to distinguish in the museum drawer as they are in the field.

Thus, it is a source of some relief to find that Benedict et al. reduce the number of recognized subspecies from eight or more to five, following the revision carried out by Patten et al.

The subspecies of northern California's coast, *petulans*, is larger than the nominate race, with a proportionally larger bill; its brownish plumage is reddish in tone, showing "the extreme of ruddiness seen in all the brown towhees." The deep chestnut of the crown extends farther to the rear and down the nape than in other races. This subspecies is resident from Humboldt to Santa Cruz Counties, California, intergrading at the southern edge of its range with the nominate subspecies *crissalis*.

As defined by Patten et al., *crissalis* includes the populations originally described under the names *carolae*, *bullata*, *eremophila*, and *kernensis*. Birds of this race are paler and grayer than *petulans*, above and below, with the crown dull rust; they are also smaller and smaller-billed (Pyle). This is a fairly common resident of chaparral habitats in southern Oregon's Rogue, Applegate, and Illinois Valleys, and occurs disjunctly in extreme north-central California and from the interior of Humboldt County—east of the range of *petulans*—south to the coasts of northern Monterey County and east to Kern and Ventura Counties; the disjunct population of the Argus Mountains in Inyo County, originally described as a distinct subspecies, *eremophila*, occupies a range equivalent to a circle of no more than eleven miles in diameter, most of it on the China Lake Naval Air Weapons Center.

Smaller still than *crissalis*, the subspecies *senicula* (its epithet translated from one of the species' Spanish names, *viejecita*) is a coastal resident from Los Angeles County, California, south to southern Baja California Norte; there is a single December specimen from the Salton Sea. These small, small-billed birds are darker above than any other California Towhee, with somber gray tones overlying the dark brown and a low-contrast rusty crown.

The rest of the Baja California peninsula is occupied by Oberholser's *aripolia* ("thoroughly gray") and Baird's *albigula* ("white-throated"). Both are paler than the more northerly *senicula*. From Playa Santa Maria to Guajademi, *aripolia* shows a paler belly and throat and longer bill than *senicula*. It is darker and grayer than *albigula*, which, with a rufous crown, pale buff throat, whitish belly, and at least occasionally a breast spot, recalls a Canyon Towhee more than any other California Towhee. Indeed, that San Lucas Towhee has been assigned at different times to both species; Zink found continuous "gradation in several phenotypic characteristics" from California to southernmost Baja California, leading him to retain it among the subspecies of *crissalis*, a treatment followed in the major subsequent lists.

The juvenile, messy like all juvenile towhees, is variably streaked beneath but otherwise resembles its parents. *California, August. Marjorie Kibby*

ABERT TOWHEE
Melozone aberti

The Abert Towhee is a very large, long-tailed, thick-footed sparrow with rusty undertail coverts and black lore and chin. *Arizona, October. Brian E. Small*

The military parties exploring the American West in the mid-nineteenth century included a startlingly disproportionate number of polymath geniuses. James William Abert, watercolorist, engineer, naturalist, soldier, and literature professor, was among them, accompanying Fremont, Emory, and Stansbury to the Southwest on expeditions beginning in 1845. Among the fruits of his efforts on the Stansbury exploration, Abert sent back east a "small but exceedingly interesting collection of birds and mammals, procured . . . in New Mexico." On examining the specimens at the Smithsonian, Spencer Baird found two undescribed species, a mammal and a bird. Baird's *Vulpes macroura* was later reidentified as a local variant of the red fox, the most widely distributed carnivorous mammal in the world. The bird's "characteristic features," however, have stood the test of taxonomic time, and Baird's description of the specimen as a new species has never been questioned. He named the bird for "its accomplished discoverer": *Pipilo aberti*.

Abert would be utterly and unjustly forgotten today if not for the name of this huge, somber-colored towhee. Unfortunately, the honor Baird bestowed on the young naturalist was as misplaced as it was well meant. There is no record of Abert's ever having entered the range of this species, and nowhere in his own writings does he claim to have seen or to have collected the bird. Instead, the "mutilated" specimen was probably slipped into Abert's shipment by a colleague, perhaps William Emory, who had explored west along the Gila River and on to the Colorado.

Badly shot, badly prepared, or badly preserved, Baird's type specimen exhibited only a trace of one of this species' most characteristic marks, the black chin and lore. Nevertheless, it was clear from the beginning that this southwestern sparrow was distinct from any other in plumage and in size, and unlike its congeners, it has never been seriously proposed that it might represent a "variety" of another brown towhee species.

In 2009, a study of mitochondrial DNA data from all species in the old genus *Pipilo* determined that the traditional groups in the genus, the "brown" and the "rufous-sided" towhees, were "highly divergent and not each other's closest relatives" and thus not properly considered congeneric. The brown towhees, including the Abert Towhee, were found to be more closely related to the *Melozone* ground sparrows, particularly to the Rusty-crowned Ground Sparrow. To reflect the greater proximity of the brown towhees to that last species than to the other *Melozone*, DaCosta et al. suggested assigning the towhees and the Rusty-crowned to a genus of their own, *Pyrgisoma*, the genus in which the Rusty-crowned Ground Sparrow was first described. On the strength of this study, the next year the AOU split the genus *Pipilo*. Rather than emphasize the different degrees of relatedness between the various *Melozone* ground sparrows and the brown towhees, however, the committee took "a more conservative approach" intended to reflect the general relatedness of the entire complex, assigning them all to a newly expanded *Melozone*, where they remain.

FIELD IDENTIFICATION

Few are the passerellid sparrows that can be said to be instantly and definitively identifiable, but the adult of this heavy, dark, long-tailed bird is one of them. Even at a distance, the elongated shape, the odd pinkish brown of the body plumage, the dark chin, and the dull pink bill are distinctive, eliminating any possible confusion with the paler Canyon Towhee, the chunky and silver-billed Blue Grosbeak, or any other species of similar size.

The notably long, dark brown tail and very short, rounded wings speak to this species' fondness for dark thickets; if disturbed, Abert's Towhees dart into cover with impressive speed, twisting around twigs and branches in search of an impregnable perch. Seen well, they show a uniformly pink-brown body with contrasting darker tail and black chin and lore; the deep orange undertail coverts, vent, and lower belly can be difficult to see in the dim, tangled habitats they frequent, but are usually conspicuous on birds feeding on open ground or at feeders.

Juvenile Abert Towhees closely resemble their parents, which they follow for a month or more before becoming independent. Juveniles have variable faint rusty edges on the tertials and rusty tips to the greater and median coverts, forming weak wing bars. The underparts tend to be paler than in adults; the flanks and sides of the breast are marked with blurry brownish streaks. The lore is black, but the black chin patch is smaller than in most adults; a faint black lateral stripe borders the throat on most birds.

It is important to recall that not every towhee-sized streaked juvenile begging from a towhee pair is itself a towhee. This species is a common foster parent for both Bronzed and, especially, Brown-headed Cowbirds; juveniles of both cowbird species are more heavily marked, with scalloping on the wing and back and longer, darker bills.

Abert and California Towhees are more conspicuously vocal than the Canyon Towhee. The most frequently heard call of the Abert Towhee is a sharp, metallic, high-pitched chirp, with an explosive attack and very little decay; very similar to the corresponding note given by the California Towhee, this call is unlike the lower, slower, squeakier chip of the Canyon Towhee, and may recall the chirp of a rock squirrel or the loud *teep* produced by the tail feathers of a displaying Anna Hummingbird. Like the other brown towhees, the Abert also has a long, slightly trilling *seee* call, given by birds perched deep in cover.

The frequently heard reunion duet of this species is loud and harsh. It begins with three or four hesitant chirp calls, which are followed by a slow rattling chatter, subtly decelerating and clearly descending in pitch: *teep teep teep rha-rho-rhi*. This vocalization tends to be slower, more raucous, and more "tropical sounding" than the duets of the Canyon Towhee.

Male Abert Towhees sing rather infrequently. The close and durable, often years-long bond between the members of a pair makes "advertising" necessary only for unmated birds; most songs are presumably given to defend the pair's territory against intruders. The usual song is a series of rapidly accelerating chips, very similar in tone and pitch to the usual chirping call; the last notes can be very short,

nearly running together into a tremolo, and in some individuals rapidly fading in volume. In one common variant, the faster notes are separated from the introduction by a pause, and drop noticeably in pitch.

RANGE AND GEOGRAPHIC VARIATION

As a species, the Abert Towhee is resident from southwestern Utah and southeastern Nevada to extreme northeastern Baja California and north-westernmost Sonora. The range extends east along the entire course of the Gila River into New Mexico, and then southeast along the San Pedro and Santa Cruz Rivers just across the border into northern Sonora. This is a pronouncedly riparian species, occurring only in thickets along desert rivers, washes, and creeks; it barely ranges above 5,000 feet in elevation anywhere in its range.

In Utah, Abert Towhees occur along the Virgin River, and have been recorded as far northeast as the mouth of Zion Canyon. The species follows the Virgin River into Nevada, where it is also resident along the Muddy and Colorado Rivers and in the Las Vegas area. In California, it occupies suitable habitats along the Colorado from San Bernadino, Riverside, and Imperial Counties, occurring as far west as Palm Springs. The Gila River provides a habitat corridor for this species to enter New Mexico in Hidalgo and Grant Counties.

One of the quintessential Arizona specialties, more than 80 percent of the Abert Towhee's global range is within that state, where it occurs in appropriate riparian habitat south of the Mogollon Rim and north along the Colorado River to Fort Mohave;

The large size, strongly graduated tail, and brush-loving habits can suggest a desert thrasher. *Arizona, October. Brian E. Small*

The forehead, neck, and throat are often tinged orange-buff, revealing this species' close relationship to the California Towhee. *Arizona, October. Brian E. Small*

the Virgin River population extends into the north-westernmost corner of the state.

Almost entirely restricted to the United States, with only a tiny Mexican range, the Abert Towhee is the mirror opposite of many sought-after birds of the Southwest. With the decline of river flow and the subsequent loss of habitat, it is now rather scarce in the old delta of the Colorado River in western Sonora and adjacent Baja California Norte. In central Sonora, it barely enters the country along the upper San Pedro and Santa Cruz Rivers; it has also recently become resident in the washes on the border at Sasabe.

Living out their lives in small stretches of riparian habitat surrounded by desert, Abert Towhees are unlikely vagrants anywhere outside their normal range. Reports from Nebraska and Colorado are almost certainly incorrect.

The plumage differences visible between birds in the eastern and in the northern and western portions of this species' range are so subtle as to be indiscernible to many eyes even when specimens lie next to each other on a table; the two currently recognized subspecies are separable in the field only presumptively by range.

The subspecies *dumeticola*, described in 1946 from extreme southeastern California, is slightly paler above, with "more russet and cinnamon tones" than the "more grayish and pinkish" nominate race. It is also proportionally shorter-tailed and slightly thinner-billed than *aberti*. This is the race resident in Utah, Nevada, California, Baja California, and western Sonora, and along the Williams and Colorado Rivers in Arizona.

Allan Phillips also recognized two subspecies, but—inevitably—defined them and their ranges differently. Phillips considered the bird named and described by Baird a representative of the western population; he thus found himself obliged to describe the grayer, darker birds of southeastern Arizona and New Mexico as a new race, *vorhiesi*, occupying a range stretching from southern Pima County north to the mouth of the San Pedro River.

Juvenile towhees show dull eyes, the hint of a pale fleshy gape, and lax and disheveled plumage. *Arizona, June. Candace Porth*

This brown tohwee species differs from the more widespread Canyon in its white-spotted wings and buffy throat band. *David Krueper*

WHITE-THROATED TOWHEE
Melozone albicollis

It was gold and rumors of gold that brought the earliest European explorers to the New World. By the nineteenth century, though, some had entered into a profitable sideline, collecting exotic tropical birds and their parts for shipping back to Europe, where museums, curiosity dealers, and private amateurs made up an eager market. Brilliant toucans, hummingbirds, parrots, and cotingas were among the obvious favorites, but more sophisticated collectors had their eye out for novelty, colorful or not.

Ferdinand Deppe was one of the first collectors to systematically supply ornithologists with birds from Mexico. Together with his botanist colleague Wilhelm Schiede, Deppe settled in Xalapa in 1828, from where they would sell plant and animal specimens to museums and natural history dealers in Europe. Unfortunately, what had been Deppe's best customers, the Berlin museum and its director Hinrich Lichtenstein, found themselves in a state of financial embarrassment, and Deppe and Schiede "were soon disappointed." By the late summer of 1830, the young entrepreneurs had abandoned their Mexican scheme, and eventually they advertised their remaining stock—mammals, birds,

amphibians, fish, and crustaceans—in a pamphlet printed in Berlin by Deppe's brother Wilhelm, using names supplied, and in not a few cases coined, by Lichtenstein.

Number 70 in the catalog of the bird skins sold by Deppe and Schiede was a certain *Tanagra rutila*, a male priced at two and a half Prussian thalers, the equivalent of half a week's rent in middle-class Berlin. The price list describes the bird only as "brown-gray above, the same below but paler," surely among the least eloquent diagnoses ever given a new bird species.

Lichtenstein continued to use the name *Tangara rutila* on specimen labels in the Berlin museum, labels he attached to skins and mounts of the species we now know as the White-throated Towhee. None of those birds, however, is the type specimen collected by Deppe, which cannot now be located, in all likelihood discarded sometime after its sale in 1830. Thus, *Tangara rutila* is defined only by the brief description provided in the price list. That description, however, includes no character suggesting the identification of Lichtenstein's *rutila* with the White-throated Towhee. Accordingly, in 1954, Erwin Stresemann, sitting in Lichtenstein's chair in Berlin, proclaimed *Tangara rutila* a synonym of *Pipilo fuscus*, the Canyon Towhee; other authorities treat the name as a *nomen dubium*, a label without unequivocal application to any species.

A clear description of the White-throated Towhee, neatly distinguishing it from the Canyon, California, and Abert Towhees, was finally provided in 1858 by Philip Lutley Sclater, based on a specimen taken in Oaxaca the year before. Quite rightly, Sclater did not associate his bird with Lichtenstein's *Tangara rutila*, and he gave it the name *Pipilo albicollis*. Lichtenstein's name remained oddly persistent, however, even after Sclater's was adopted—with some uncertainty about its status as a full species—in Baird, Brewer, and Ridgway's influential *History of North American Birds*. Indeed, Ridgway himself would later assign *rutilus* priority over Sclater's new name, and it remains so familiar an alternative that even the most recent edition of the AOU *Check-list* offers a warning:

> *Pipilo rutilus* W. Deppe, 1830, sometimes used for this species [namely, the White-throated Towhee], is now regarded as a synonym of *P. fuscus*.

FIELD IDENTIFICATION

In southern Puebla and extreme northern Oaxaca, the White-throated Towhee overlaps with Canyon Towhees of the rather pale grayish subspecies *toroi*. Observers familiar only with more northerly Canyon Towhees, especially those occurring in the United States, may be momentarily misled by the plain crown and whitish, relatively unspotted underparts shown by *toroi*, but the Canyon Towhee never exhibits the white-tipped wing and uppertail coverts of the White-throated.

The wing pattern is "the most positive character," or at least the most readily visible, of the White-throated Towhee in hand or in the field, but the color and marking of the underparts are equally distinctive. The dull rufous-yellow of the lower belly contrasts with the whitish upper belly and the grayish flanks and sides of the breast, especially in the colder-colored northern race *marshalli*. The center of the breast is white, with a variable loose necklace of gray barring; some individuals are said to have a blackish central breast spot. The throat and chin are the brightest areas of the underparts, the clear white contrasting with the gray breast sides and necklace. The upper throat is crossed by a diffuse band of tawny orange, similar in color to the lower belly; the band's width and extent are variable, but it often reaches through the jaw stripe to the side of the neck. The white lore and eye-ring are often conspicuous, and differ strikingly from the smudgy brown shown by Canyon Towhees.

Juvenile White-throated Towhees show the species' eponymous white throat, with heavy, irregular spotting elsewhere below, usually concentrated on the lower breast; the crown is unstreaked, and the back is faintly spotted or barred.

The vocabulary of this noisy and gregarious species closely resembles that of the other brown towhees. The chirp note is squeaky like that of the Canyon Towhee, but often more extended and higher-pitched, with a less breathy tone; there is also a high-pitched hissing *see*. Individuals in flocks frequently give a loud, raucous chatter, often running into a fast, almost wrenlike trill.

Perhaps as a result of this species' more sociable nature, its reunion duet appears to be less stereotyped than in other brown towhees and less distinct from the rest of its vocal repertoire. The paired rattles are fast, high-pitched, and thin, without, or at least not invariably with, the concluding descent in pitch characteristic of the corresponding vocalizations of related species.

The White-throated Towhee's song resembles that of its congeners in pattern. The introductory one to four notes can resemble the usual chirping calls, or they can be clearer, sharper downslurred whistles. In either case, the introduction is followed by two or more loud trills on different pitches, in some individuals recalling the song of an outsized House Wren.

RANGE AND GEOGRAPHIC VARIATION

Like the other brown towhees, this species is a sedentary bird of more or less dry brushy habitats. Fairly common within its limited range, it is restricted to the highlands of eastern Guerrero, Oaxaca, and southern Puebla; the southeastern-most localities lie just north of Oaxaca City.

As its name indicates, the subspecies *parvirostris* of the White-throated Towhee was distinguished largely by its shorter bill. The describer of that race later suggested that variation in bill length in this species might be a matter of seasonal change, and *parvirostris* is no longer recognized as a valid subspecies.

In 1974, Kenneth C. Parkes distinguished the White-throated Towhees found in Guerrero, Puebla, and northernmost Oaxaca as a new subspecies *marshalli*, named for the ornithologist and harpsichord builder Joe T. Marshall. These birds are colder gray above than the nominate race, with duller, less rufous flanks; the tips of the greater and median coverts are more whitish and less orange, as are the tips of the uppertail coverts. The orange throat band averages paler and less extensive.

Shy but colorful, the Rusty-crowned Ground Sparrow is a forest bird of western and south-central Mexico. *Chuck Slusarczyk Jr.*

RUSTY-CROWNED GROUND SPARROW
Melozone kieneri

The last major expedition of the Smithsonian collector Pierre Louis Jouy took him across Mexico. Accompanied by his wife, Jouy traveled by train, canoe, and horse from Tampico on the Gulf of Mexico to Jalisco and Colima, then up the Pacific Coast to southern Sonora. His favorite locality among all those he visited was Jalisco's Barranca Ibarra, half a day's journey north of Gaudalajara. Here, on May 11 and May 13, 1892, Jouy found the only Rusty-crowned Ground Sparrows of his trip, "almost exclusively around the head of the barranca on the bare hillsides." He found the species "not exactly rare," but "a very shy bird and difficult to get."

Jouy was neither the first nor the last to complain about the difficulty of obtaining specimens of this retiring, ground-hugging sparrow. Charles Lucien Bonaparte described the species in July 1850 from the sole example available in Paris's Natural History Museum, naming it for that institution's distinguished curator, Louis Charles Kiener. Bonaparte's description was brief and, as it developed, not entirely accurate, and he was able to specify the bird's range only as "western America." Thirty-five years later, Bonaparte's type—collected by the later admiral Charles Jaurès in 1843 in the course of the circumnavigation of the globe by the French warship *Danaïde*—was still deemed one of a kind, as it was as late as 1901, when Robert Ridgway could only speculate that the sparrow had been taken in Mexico.

Less than a year after Bonaparte dedicated his bird to Kiener, the German cataloger and ornithologist Jean Cabanis described a specimen in the Berlin museum. Cabanis published the bird, represented by a single skin collected by Ferdinant Deppe in Puebla, Mexico, under the name first given it by Lichtenstein, *Atlapetes rubricatus* ("painted"). In 1860, Cabanis expressed some doubt about the bird's specific status, musing in print that the skin—still to his knowledge unique in European collections—might represent the female of another species, the tropical Prevost Ground Sparrow. Six years later, and fully 15 years after describing the Berlin sparrow, Cabanis finally had the second specimen he needed, a male; this time he confidently asserted that the birds from Puebla were "clearly different" from any known species.

Meanwhile, in June 1863, the Hungarian-American collector John Xantus collected yet another would-be new sparrow in Colima. George Newbold Lawrence published a description of the bird four years later, naming it for the collector and distinguishing it carefully from Bonaparte's *kieneri*; Lawrence lamented that he had no specimen of the Prevost Ground Sparrow with which to compare the new *Pyrgisoma xantusii*. He would stake out a different position in 1874, quietly deciding that his *xantusii* was after all identical to Cabanis's *rubricatus*—but even well afterward, the lack of complete series of skins from across the species' range or ranges precluded certainty.

Instead, ornithologists were forced to rely on verbal descriptions and paintings of the scattered few specimens. Edward Nelson, eponym of a sparrow species of his own, gathered the evidence in 1898. He determined that *xantusi* (Nelson's spelling of Lawrence's *xantusii*) differed from Cabanis's sparrow only "as a well-marked geographical race," and furthermore that it so closely resembled the engraving of Bonaparte's type in Sclater and Salvin's *Exotic Ornithology* as to inspire in him "a strong suspicion of their identity," making *kieneri*, *rubricatus*, and *xantusii* three names for one species.

It took some time for Nelson's "strong suspicion" to be confirmed: as one frustrated worker noted in 1901, the specimen record of this—or these—species was still so slender that

> the series of true *P. rubricatum* that I have been able to examine is much too small to enable me to state the characters . . . and I also labor under the disadvantage of not being able to compare at one time specimens from different parts of its range,

a lament that would have been less surprising had it not come from the curator of ornithology at no less an institution than the Smithsonian.

The significant plumage differences between this species and the Prevost Ground Sparrow were pointed out unequivocally in the mid-1860s by John Cassin and by Jean Cabanis. Nevertheless, the ghost of conspecificity rose again and again even as late as the 1930s, when Charles E. Hellmayr, though acknowledging the "trenchant features" separating them, lumped those two and the Cabanis Ground Sparrow, *Melozone cabanisi*, into a single species. Since no later than 1955, though, the Prevost and the Rusty-crowned Ground Sparrows have

been universally treated as distinct at the species level.

The more difficult issue, as suggested above, was the relationship between Bonaparte's *kieneri* and Cabanis's *rubricatum*. A. J. van Rossem put the problem clearly in 1933:

> until the type of *kieneri* can be examined by someone familiar with the problem, there is no occasion to change the names in current use . . . .

That summer, van Rossem visited Paris to examine the "rumpled and stained" bird, then more than 90 years old. He quickly confirmed that *rubricatum* and *xantusii* were at most subspecifically distinct from *kieneri*, and that that name, coined before any other, was the correct scientific epithet for the species.

Settling that venerable nomenclatural question, of course, had no bearing on the assignment of the Rusty-crowned Ground Sparrow to the proper genus. Bonaparte in 1850 included this species in the genus *Pyrgisoma* ("sparrow-bodied bird"); Cabanis, describing a few months later what would turn out to be the same species, placed it instead in *Atlapetes* ("reluctant flutterer"), the name coined originally by Johann Georg Wagler for the Chestnut-capped Brush Finch. Wagler's characterization of his new genus reflects the contemporary uncertainty about the affinities of the tropical sparrows:

> the genus combines characters of the finches, manakins, euphonias, Old World warblers, and fairy wrens. The bill is entirely like that of our Eurasian Tree Sparrow . . . the wings like those of the White-bearded Manakin (so short, round, and concave that the bird can surely do no more than flutter), the feet . . . like those of a fairy wren, the rather long tail and plumage of an Old World warbler, especially of the Blackcap!

A slightly less extravagant perception of the Rusty-crowned and other ground sparrows as somehow composite would lead George Robert Gray to allocate them to the genus *Embernagra*, a portmanteau coined by René-Primevère Lesson combining *Emberiza*—the buntings—and *Tanagra*—the tropical tanagers.

Sclater and Salvin returned the ground sparrows to *Pyrgisoma* in 1868—unaware that seven weeks before Bonaparte published that name, Ludwig Reichenbach had erected the genus *Melozone* for the Prevost Ground Sparrow, making Bonaparte's name a synonym, only very scantily junior to

Melozone but nevertheless invalid. The American Ornithological Society currently assigns all of the brown towhees and ground sparrows to *Melozone*.

FIELD IDENTIFICATION

If seen well, adults of this rather large, dark sparrow are unlikely to be mistaken for anything else; there is no overlap in range with the bright white-faced Prevost Ground Sparrow of Central America. Usually seen singly or in apparent pairs, the Rusty-crowned Ground Sparrow is moderately short-tailed, with very short, rounded wings. The belly and breast are full and round; the square head, often with crest slightly raised, and thick, dark bill give the bird an unbalanced look on the ground, as if there were too much sparrow in front of the legs. Above, the entire bird is cold olive-gray from tail to nape, the flanks and breast sides barely paler. The rusty head plumage is of variable extent toward the forehead; to the rear, it extends low across most of the nape and curls forward onto the sides of the neck below the ear coverts, just reaching the bottom of the white jaw stripe. In poor light or at a distance, the most conspicuous feature of the head is the broad white eye-ring, broken behind and sometimes in front of the eye; the lore is dark, with a messy whitish spot above it.

The head pattern, large dark spot on the lower breast, and the variably yellow to tawny undertail coverts and vent call to mind an especially neatly marked brown towhee. All of those birds are much larger, however, and the Ground Sparrow's bright white throat, jaw stripe, and upper breast are entirely distinctive.

Juveniles are even less frequently observed than the adults. They are rich dark brown above with dark streaks on the back and mantle; there is a pale spot above the lore, and the dull yellowish throat is bordered by a dark lateral stripe. The undertail coverts and vent are pale yellowish brown; the belly and lower breast are vaguely streaked slaty brown.

Rusty-crowned Ground Sparrows give high, thin, short *tsik* calls, almost like the flight note of a Savannah Sparrow. They also have a buzzy, descending, rather towheelike *dzeer*, thinner and softer than what is apparently the corresponding call of the Canyon Towhee. The male's homely song is a slow, well-articulated series of notes of different lengths, pitches, and qualities; it often begins with two or three *tsik* notes, followed by three or four lower, slurred liquid notes *djiu djiu*. Some songs end with a sputtering labial trill, somewhat like that of a Botteri Sparrow.

RANGE AND GEOGRAPHIC VARIATION

This colorful but retiring sparrow is endemic to thorn forest and dry brush in the Sierra of western and south-central Mexico, from southern Sonora

Truly unmistakable sparrows are few, but this species qualifies when it is seen well. *Chuck Slusarczyk Jr.*

and Sinaloa to west-central Oaxaca; it appears to be uncommon, or at least infrequently detected, over much of that range. In the north, recent sightings are concentrated along the Durango Highway of southern Sinaloa and northern Durango in winter, no doubt reflecting the distribution of birders as much as of birds. It is regularly—though again, not frequently—recorded on the Alamos, Sonora, Christmas Bird Count; there is also a late autumn record from southwestern Chihuahua. The Rusty-crowned Ground Sparrow is said to be among the commonest passerines in the state of Morelos. In the southwest, it has frequently been observed on the flanks of the Volcán de Colima, and west nearly to the Pacific. The southeasternmost recent reports are from Oaxaca's Sierra de Miahuatlán (eBird).

The Rusty-crowned Ground Sparrow is at present understood to comprise three subspecies. (Lawrence's *xantusii,* Allan Phillips's *obscurior,* and Pierce Brodkorb's "very poor race" *hartwegi* are not generally recognized.) The nominate race, Bonaparte's *kieneri,* occurs from extreme southern Sonora south to western Jalisco and Colima. It is large, with a sturdier bill and tarsus than the others; the rusty of the crown reaches farther toward the forehead.

Cabanis's *rubricata* (grammatically feminine to agree with *Melozone*) is smaller, with a smaller, more slender bill and tarsus; the rusty of the crown is brighter but slightly more restricted than in *kieneri*. It is resident from Guanajuato and Michoacán to Morelos, Puebla, and western Oaxaca. Occupying a small corner of eastern Sonora and northern Sinaloa, Van Rossem's *grisior,* as its name suggests, is paler and grayer above and on the flanks; the rusty of the head is brighter cinnamon than in the nominate race, as are the undertail coverts.

The subspecific differences are subtle and minor overall, but because this species is nonmigratory across its distribution, individuals can be identified presumptively in the field by range.

Joseph Smit's handsome depiction of the type specimen of the Rusty-crowned Ground Sparrow reveals the artist's skill at revivifying even poorly prepared or poorly preserved skins. *Image from the Biodiversity Heritage Library. Digitized by the Smithsonian Libraries. | www.biodiversitylibrary.org*

Larger, larger-billed, and more heavily marked on the head than the Rufous-crowned, the Rusty Sparrow is found in northwestern Mexico not far south of the United States border. *Sonora, June. Amy McAndrews*

RUSTY SPARROW
Aimophila rufescens

On his return from southeastern Mexico in 1856, the "indefatigable explorer" Auguste Sallé brought with him "a magnificent collection of birds . . . around two hundred species . . . in a most perfect state of preservation." Among them, Charles Bonaparte found

> a finch so remarkable that it has convinced us to found a new genus, comprising this Mexican species, which wc will name *melanotis* ["black-eared"], and a Colombian species that we have already named *Passerculus geospizopsis* in our notes on the collection of Delattre . . . . We will name the genus *Geospizopsis* ["like a ground finch"].

Bonaparte went on to describe his new Mexican bird as blackish, with each feather completely edged rufous, a very wide whitish supercilium, and black cheeks and ear coverts. Beneath, it was pale fawn-colored, with a plain throat and densely streaked breast; the short tail feathers were sharply pointed.

Philip Lutley Sclater, examining the specimen for his formal catalog of the Sallé collection, agreed that it was "certainly a remarkable bird . . . . I hardly know what to make of it." Unlike Bonaparte, though, Sclater recognized that the skin was of a juvenile—and that it had "nothing to do with" the only distantly related Plumbeous Sierra Finch. Two years later, Sclater had narrowed down the possibilities: with a second specimen in hand, he determined that the Mexican sparrow was the young of either the Striped Sparrow or, more likely, the Rusty Sparrow, described by William Swainson 30 years earlier. That identification made Bonaparte's *Geospizopsis* inapplicable, and Sclater pronounced himself especially pleased "that this ugly generic name may be altogether cancelled as useless."

Swainson had described the adult Rusty Sparrow from a specimen sent back to England by the William Bullocks, father and son. Before allowing their Mexican birds to be "dispersed by the hammer of the auctioneer," the elder Bullock asked Swainson to prepare a synopsis of the collection, thus "securing the honour of these discoveries to Mr. Bullock and his son." This first description of the Rusty Sparrow was brief but adequate; a dozen years later, Swainson had acquired a specimen of his own, and he would provide a more detailed account in the appendix to his *Animals in Menageries*.

Birds in the southern and eastern portions of the species' range are especially colorful, with bright rusty wings and warm buff flanks and undertail. *Veracruz. David Krueper*

Swainson's original description of the Rusty Sparrow assigned the new species the telling name *Pipilo rufescens*, grouping the large, heavy-billed, ground-loving bird with the Spotted and Canyon Towhees. In 1837, however, in his *Menageries* he assigned the bird to a new genus, *Aimophila* ("thicket-loving"), along with the Striped Sparrow.

That genus name has proved durable over nearly 200 years, even as species have been moved in and out of it with startling frequency. Its spelling, though, would be the subject of long disagreement. In 1850, Ludwig Reichenbach "corrected" Swainson's original spelling to *Haemophila*, changing these inoffensive tropical sparrows from thicket dwellers to "blood lovers." The emended spelling prevailed in the European literature up to the very end of the nineteenth century, while most American authors properly retained Swainson's *Aimophila*. The single notable exception was Elliott Coues, who in the last, posthumous edition of the *Key* reproached both Swainson and the American Ornithologists' Union for having "misspelled" the name of the genus—and at the same time queried in print "what application" the hypercorrection to "blood lover" might possibly have to the birds in question.

FIELD IDENTIFICATION

Large to very large—longer and heavier on average than the Green-tailed Towhee—big-headed, and long-tailed, over most of its range the Rusty

Sparrow is unlikely to be confused with any other species. In the northwestern reaches, however, where the smaller and grayer subspecies *mcleodii* and *antonensis* are resident, this species and the Rufous-crowned Sparrow overlap. Even in these areas, however, the Rusty Sparrow is still larger and heavier, with a decidedly large bill; the eye line of the Rufous-crowned Sparrow is less distinct and usually browner than the neat black line behind the eye of the Rusty Sparrow. In upperparts color, these northern birds are remarkably similar in the field to Rufous-crowned Sparrows, but all Rusty Sparrows have broad reddish edges to the secondaries and tertials, creating a conspicuously contrasting wing panel always absent in the smaller, more slender-billed Rufous-crowned.

Habitat can also be a useful clue in northwest Mexico. Though in the more southerly portions of its range the Rusty Sparrow is widespread in lowland habitats, in the north it is more pronouncedly a bird of open pine and oak forests in mid-elevation canyons and slopes, on average found higher than the Rufous-crowned, which prefers more arid, more sparsely vegetated rocky habitat.

There is some uncertainty as to whether the Rusty and the Oaxaca Sparrows overlap in range. The Oaxaca Sparrow is smaller, and its bill is black above and below, unlike the two-toned bill of the Rusty Sparrow. The Oaxaca Sparrow's face pattern is clearer, with the eye-ring unbroken in front.

All Rusty Sparrows are obviously large and bulky. The full belly and breast, very short wing, and square, heavily marked head make them look brutish in comparison to the slighter, finer-billed Rufous-crowned; the Rusty is also a noisier, more energetic forager, scratching and rustling towheelike in the leaves. In all subspecies, the tail and uppertail coverts contrast in color with the duller rump and back: blackish against brown-gray in the northern races, bright rust against colder rust in the south. In flight, the tail is obviously graduated, the central rectrices noticeably longer than the outer tail feathers.

Southern and eastern Rusty Sparrows fully live up to their name, with bright rufous tail, uppertail coverts, and wings; the back is often only slightly duller, with short blackish streaks or splotches forming a visually "rough" pattern. The brightest part of the bird in the field is often the tertials, which have very broad bright rufous edges, merging with the rusty secondaries into a large square panel; there is usually no hint of a wing bar.

In Sonora, Chihuahua, and northern Sinaloa, birds of the races *antonensis* and *mcleodii* are grayer overall, with colder tones to the upperparts; the tertials, though, are broadly edged rufous. The back is only faintly marked with short, dull, sooty streaks or blotches.

Rusty Sparrows of the southern races are also more distinctly marked on the head and throat, with a bold black eye line, large black lore patch, and variable black lateral crown stripe; the front of the supercilium is often clean, bright white. In southern birds, the lateral throat stripe is finer, and the lore and eye line are sooty rather than black; the supercilium is more evenly colored, pale gray rather than white at the front, and there is usually little or no black at the side of the crown.

Juvenile Rusty Sparrows resemble their parents in structure and plumage tone, but they are slightly more conspicuously streaked on the back. The faintly streaked crown has a poorly defined pale buffy median stripe. The underparts show a "unique . . . strong yellow wash," with blackish streaks or chevrons across the breast.

Rusty Sparrows have a dry, chattering call; loud and fast, it often changes pitch or tempo in the middle, calling to mind the rattle of a distant Belted Kingfisher. Mated pairs also perform a "reunion duet," in which one member gives a chattering vocalization while the other gives a sustained trill.

There is also a hissing *tzee* note, which may introduce the duet.

The variable song is a cheerful phrase with a rich, chirping quality. One common form begins with a hesitant rising chirp, then continues with three or four higher-pitched, slightly slurred notes before concluding with a strongly downslurred flourish; the effect of many songs is surprisingly Dickcissel-like, though the Rusty Sparrow's song is never as buzzy.

RANGE AND GEOGRAPHIC VARIATION

The Rusty Sparrow occurs in western Mexico from about 30 degrees north in central and eastern Sonora and western Chihuahua, and in eastern Mexico from about 23 degrees north in Tamaulipas, south to Oaxaca and Chiapas. It is also resident south through Honduras, central El Salvador, and northern Nicaragua. In northwestern Costa Rica, it is restricted to the Pacific Slope of the Guanacaste Range.

West Mexican birds occupy relatively open and grassy oak and pine-oak habitats on mountain

Even relatively dull individuals of this large, heavy-billed sparrow are obviously more coarsely marked than the smaller, plainer Rufous-crowned Sparrow. *Sonora. David Krueper*

slopes and in canyons; in the east and south, this species also occurs in grassy lowlands with sparse pines, cactus, or shrubby growth.

Though Rusty Sparrows breed in Sonora not more than 100 miles south of Arizona, they have not yet been documented to occur in the United States.

The Rusty Sparrow has been divided into a dozen subspecies over the past nearly 200 years. Of the seven generally recognized today, four occur north of the Trans-Mexican Volcanic Belt; the other three are resident from Chiapas to El Salvador (*pectoralis*), in Honduras and Nicaragua (*discolor*), and in Costa Rica's Guanacaste (*hypaethra*).

The nominate subspecies, *Aimophila rufescens rufescens*, is found from southern Sinaloa south to western Chiapas and southern Puebla. It is moderately streaked above, with pale gray-buff to brown-buff underparts; the rusty crown shows a gray median crown stripe and narrow blackish streaking above the gray supercilium.

The subspecies *mcleodii*, named for the Maine collector Richard Randall McLeod, was described tentatively as a new species by William Brewster in 1888; in the same publication, Brewster also described another possible new species, *Aimophila cahooni*, named for Brewster's collector John C. Cahoon, who three years later would fall from a cliff and die while taking nests and eggs in Newfoundland.

Brewster suspected from the beginning that both *cahooni* and *mcleodii* might be only subspecifically distinct from the nominate *rufescens*. That supposition would prove to be correct, but not before *mcleodii* was again named and described as new, in 1889 by Osbert Salvin and Frederick DuCane Godman (as *Peucaea megarhyncha*). All are now known as *mcleodii*, the northernmost of the Rusty Sparrows; this subspecies is resident from east-central Sonora and adjacent Chihuahua south to northern Sinaloa and Durango. Birds of this race are slightly smaller than *rufescens*, overall duller brown and less conspicuously streaked above; the rusty crown lacks central and median stripes. The northwesternmost Rusty Sparrows, discovered in the Sierra San Antonio some 95 miles south of the United States border, are of the paler subspecies *antonensis*; the breast, sides of breast flanks, undertail coverts, and edgings of the wing feathers are grayer than in *mcleodii*.

La Fresnaye's *pyrgitoides* is an eastern subspecies, resident from southern Tamaulipas to northern Oaxaca and thence south to El Salvador and northwestern Nicaragua. This race is buffier above and below than the nominate race, with a distinctly streaked back.

Because Rusty Sparrows appear to be resident throughout their range, individuals can generally be identified presumptively to subspecies.

Juveniles are messily streaked above and below. They are plainer-headed than adults but do share with them the large size and outsized bill. Where the two overlap, this Middle American species occurs at higher elevations and in more wooded habitat than the Rufous-crowned Sparrow. *Sonora. David Krueper*

Uniformly colored, even plain at a distance, this large terrestrial sparrow has a distinctive head pattern and an equally distinctive song. *Texas, March. Brian E. Small*

RUFOUS-CROWNED SPARROW
Aimophila ruficeps

If it is sometimes challenging to trace the career of a bird through the scientific literature, it is too often impossible to re-create the tradition of any particular bit of more popular birding lore. The Rufous-crowned Sparrow, a familiar and fairly common resident of southwestern North America, offers a happy exception to the rule.

Every birder who encounters this species in the field becomes instantly familiar with its nasal, petulant call, the famous "dear, dear" that makes the birds sound so worried as they forage beneath the vegetation of rocky slopes. The transcription is so memorable as to seem natural, even inevitable—but it is not, and someone had to introduce the sparrow's "dear, dear" into the culture of North American birding. Astonishingly, we know who that was.

In his *Birds of the Pacific States* of 1927, the great Ralph Hoffmann credits the now standard characterization of the Rufous-crowned's call to "Myers." Harriet Williams Myers (1867–1950) was an influential figure on the early California conservation scene. A founder of the California Audubon Society, she was also an early bird photographer and a mod-

erately successful author, writing "for the needs of many amateur bird students of California and other western states." The most-read of her works was the simply titled *Western Birds*, an introduction to the identification and life history of some 200 common species. In California, Myers writes,

> these Sparrows are locally common in the foothills, where one may hear their plaintive, yet liquid, *dear, dear, dear*, as they flit about on the weed stalks or low bushes . . . .

It was on the afternoon of April 10, 1909, that Myers appears to have coined the familiar verbalization of the "plaintive" call. A friend had led Myers to the nest of a pair of Rufous-crowneds, and they watched as the female flew back and forth from the nest in front them, expressing her "distress" with a repeated call that Myers transcribed as "dear, dear." Nine days later, Myers took what is likely the first photograph of living Rufous-crowned Sparrows ever obtained, an adult feeding two newly hatched nestlings.

As a species, the Rufous-crowned Sparrow had been discovered more than half a century before Harriet Myers had her inspiration. The first specimens reached Philadelphia in 1852; collected in

California by Adolphus L. Heermann, they reminded John Cassin so much of the Swamp Sparrow that he named the new species "the Western Swamp Sparrow" and guessed that it preferred the same habitats—"the margins of streams of fresh water"—as its eastern cousin. James Graham Cooper, following Spencer Baird, discerned a different resemblance: the birds he saw on California's Catalina Island recalled the Bachman Sparrow, and he concluded that "their favorite resort, like that of the Eastern species, may, perhaps, be pine woods." Even in 1874, very little was known of the Rufous-crowned Sparrow's life history and habitat. Five years later, though, Charles A. Allen—the California collector and eponym of the hummingbird—was finally able to provide William Brewster with the first definitive notes on the bird's "chosen haunts . . . destitute of forests, [where] the exceedingly steep, rocky sides are abundantly clothed with 'wild oats' . . . dry and barren to a degree," an accurate assessment very different indeed from the earlier speculations.

FIELD IDENTIFICATION

This chunky, round-tailed and big-headed sparrow can be confused only with the larger, darker Rusty Sparrow. The dull-colored northwestern race of that species, *mcleodii*, is the most similar to the Rufous-crowned, but can always be distinguished by its size, heavy bill, darker tail and brighter tertials, and stronger face and throat patterns.

The eye-ring and very dark lateral throat stripe are usually easy to see. *Texas, March. Brian E. Small*

Generally first detected feeding quietly beneath the vegetation on rocky slopes, Rufous-crowned Sparrows give an impression of uniform somber darkness, with little contrast between the tail and back or even between the upperparts and underparts, distantly recalling a bland towhee at first glance. The flight feathers of the wing and tail often show faint narrow edges of brighter brown, but the tertials are never as broadly striped chestnut as in the Rusty Sparrow; the central tail feathers may be faintly barred in close views. The brown, sooty, or reddish streaking of the back and scapulars varies both geographically and individually, but at least some is always visible. The underparts vary from dull gray to dull whitish brown, brightest on the lower belly and especially the throat; at any distance, the whitish throat and fine but strong black lateral throat stripe are usually the most striking characters seen in the field.

At closer range, the pattern of the head is striking and distinctive. The line behind the eye and most of the crown are brown, varying from dull rust to bright rufous from subspecies to subspecies and individual to individual. In many Rufous-crowned Sparrows, the forehead is marked by a thin white line, similar to that seen in the Chipping Sparrow, and the sides of the crown are bounded with very narrow black lateral stripes. There is often a smudgy black patch on the lore, bordered above by a gray supercilium, which tends to white as it approaches the bill.

Unlike the Rusty Sparrow, the eye-ring in this species is usually broken only behind the eye; in the field, it often appears to be fully complete, giving the Rufous-crowned Sparrow a startled, alert look. The heavy blue-gray bill distinguishes it immediately from the superficially similar Rufous-winged Sparrow.

Juveniles closely resemble juvenile Rufous-winged Sparrows, but can be identified by the presence of dark brown shaft streaks and the absence of buff fringes on the back feathers; the underparts of the Rufous-crowned Sparrow are brownish, while those of the juvenile Rufous-winged are white (Pyle). The Rufous-crowned's crown is less rufous at this age than that of the Rufous-winged, with only obscure dark streaks.

The most characteristic and easily recognized of the Rufous-crowned Sparrow's vocalizations is the famous nasal "dear, dear" call, given singly or in an irregular series; given more often by males, it "is thought to convey mild alertness to danger." Pairs maintain contact with a soft, slurred *dzeep*, and also have a reunion duet analogous to that of the brown towhees. Agitated birds frequently give a loose, liquid chatter, similar to the rattle contributed by the female to the reunion duet.

The rusty crown and monotone body can recall the larger Canyon Towhee, which is more confiding than this often rather secretive species. *Texas, March. Brian E. Small*

Once learned, the singing of the Rufous-crowned Sparrow becomes a characteristic spring and summer sound of rocky hillsides in southwestern North America. Typical songs are rather quiet but sustained, a series of rapidly delivered chips and lisps like a handful of tiny pebbles tossed down the slope. The song is often compared to that of northern House Wrens, sharing its wide pitch range and chattering quality, but higher-pitched and less musical; the frequent doubling of notes may distantly recall the song of the Indigo Bunting. An infrequent variant concludes with a long trill.

RANGE AND GEOGRAPHIC VARIATION

On the Pacific Slope, Rufous-crowned Sparrows are found in foothills and valleys from Mendocino County, California, south to northern Baja California; farther east, this species is resident in the Cascades foothills of Shasta County and south to Kern, San Bernardino, and Sutter Counties. Populations also survive on Santa Cruz and Anacapa Islands.

Extremely local in southernmost Nevada and Utah, Rufous-crowned Sparrows breed in Clark County, Nevada, and in Zion National Park. In Arizona, they are resident in the southeast and along the Gila River drainage into New Mexico, where the range extends northeast to the southeastern counties of Colorado and the western Oklahoma Panhandle. The species breeds more widely in central Oklahoma and over most of the western half of Texas, east to the Edwards Plateau and south to Starr County. The northernmost birds have been

The reddish back streaking can be faint, especially on worn individuals; the tail and wing are cold brown, unlike in the noticeably larger Rusty Sparrow. *Texas, April. Brian E. Small*

Juveniles share their parents' stocky build, long tail, and thick bill, but are variably streaked beneath, with a less colorful head. *Luke Tiller*

seen in north-central Kansas, and the easternmost regular breeders are in Logan County, Arkansas.

On the Mexican mainland, the Rufous-crowned Sparrow is found from the northern borders of Sonora and Chihuahua and from Coahuila south to west-central Oaxaca; there is a small disjunct population in eastern Tamaulipas.

No regular migration is known for this species, though there is probably some elevational movement after breeding. The farthest-flung individuals have been a few scattered birds recorded on the Gulf of Mexico coast of Texas.

As many as 18 subspecies of this widespread bird are recognized by various authorities; most are restricted to Mexico. The treatment below of the northern subspecies is from Pyle, while southern populations are described from John P. Hubbard's 1975 review.

The nominate race *Aimophila ruficeps ruficeps* is resident in the mountains of central California; it is small and small-billed, with dark buff-brown streaking above and pale underparts. Resident from northern Los Angeles County, California, south to northern Baja California, *canescens* is slightly larger than *ruficeps*; both the back streaking and the underparts have, as the name suggests, a grayish tone. Santa Cruz, Anacapa, and Santa Catalina Islands are—or in the case of the last, possibly were—occupied by *obscura*, which is darker above and below than either of its mainland counterparts, with dark buffy gray streaking on the back.

To the east, the large, medium-billed *scottii* has reddish-washed gray upperparts and pale, slightly gray-tinged underparts; this subspecies is resident from the deserts of eastern California, Nevada, Utah,

northern Sonora, northern Chihuahua, and Arizona east to extreme western Texas and north to southeastern Colorado and western Oklahoma. The Great Plains subspecies *eremoeca*, more olive and less densely streaked, with a paler rufous crown, occurs from central Oklahoma and central Texas south to northern Chihuahua and central Coahuila; this is presumably the race represented in north-central Kansas and western Arkansas. These two northeastern races intergrade where they meet in southeastern New Mexico and Texas.

In northern Mexico's *pallidissima*, the back is darker and more extensively streaked than in *eremoeca*, and the underparts are darker and in some individuals buffier; it ranges from southern Coahuila and southwestern Tamaulipas to northern San Luis Potosí and Nuevo León. This race intergrades with *boucardi* at the southern edges of its range; *boucardi*, which occurs from southern San Luis Potosí to northern Michoacán and central Puebla, is a dark and brown bird, with more conspicuous back streaking than its northern neighbor.

The darkest and grayest of all Rufous-crowned Sparrows make up the subspecies *duponti*; these birds are somberly colored, with less reddish than any other race. This subspecies occurs only in the state and Federal District of Mexico. The pale and grayish *laybournae*, less distinctly streaked than adjacent populations, is found from central Veracruz and southeastern Puebla to northern Oaxaca. Central Oaxaca is occupied by the decidedly reddish *australis*, which is buffier below. From Guerrero and adjacent Puebla south to southern Oaxaca, the species is represented by the "variable race" *extima*, which is usually browner than *australis* and *laybournae*, with broader brown streaks on the back.

The aptly named *fusca* is dark, more reddish and more heavily streaked than *boucardi*, without that subspecies' dominant brown tones; it occurs in northern Michoacán and eastern Jalisco. In comparison to *fusca*, *suttoni*, found in eastern Nayarit, northern Jalisco, and Colima, is pale above and below, with more restricted brownish streaking on the back.

Still known only from southeastern Sinaloa, *phillipsi* is very dark and brown, with extensive streaking above. The well-named *simulans*, resident from eastern Sonora south to northern Jalisco, resembles the widespread northern *scottii*, but is darker overall, with a shorter wing and tail; it is paler than *phillipsi* and more reddish than *boucardi*.

Most of these differences, gradual and subtle from one race to the next, are not likely to be visible in the field, but because this species is almost entirely resident throughout its range, individuals at the heart of their breeding distribution can be identified presumptively on the basis of geography.

Not all sparrows are small, and not all sparrows are brown. The tropical brush finches, like this Rufous-capped, are very attractively colored, but share with some other passerellids a furtiveness that can make them hard to appreciate. *Amy McAndrews*

RUFOUS-CAPPED BRUSH FINCH
Atlapetes pileatus

A year before his accidental death at the age of 31, the young director of the Munich natural history museum, Johann Georg Wagler, received a box of specimens from Mexico, donated by the Bavarian collector and explorer F. W. Keerl. Among the skins, Wagler found "a number of new species, even new genera," one of which he named "Flatterfink," the "fluttering finch." Wagler, of course, had never seen the bird alive, but he found the wings of his specimen

> so short, so blunt-tipped, and so concave that surely the bird can do no more than flutter on them.

The scientific name Wagler coined for the genus refers just as clearly to this tropical sparrow's aerial challenge: *Atlapetes*, a syncopated form of *Atlantopes*, means "reluctant flier."

In founding his new, "very distinctive" genus, Wagler described it as combining characteristics otherwise typical of finches, manakins, tanagers, Old World warblers, and even fairy wrens, but identical to none of them. Philip Lutley Sclater would add *Embernagra* to the list of genera recalled by this short-billed, short-winged sparrow; Sclater found the brush finch otherwise "quite different." But in 1870, George Robert Gray assigned this species to a greatly expanded *Embernagra*, which also included the *Aimophila* sparrows, the brown towhees, and the Prevost and Rusty-crowned Ground Sparrows. Christoph Giebel adopted this same, extremely catholic generic concept in his *Thesaurus*, but all succeeding authorities have recognized *Atlapetes* as distinct from *Embernagra*.

FIELD IDENTIFICATION

This fairly large, chunky but long-tailed sparrow is distinctive when seen well. In a brief glimpse, the dull greenish upperparts and bright yellow throat can recall the similarly sized Yellow-breasted Chat; the chat generally occurs at lower elevations and in damper habitats than the sparrow. The greenish wings and tail and rusty cap are rather like those of a Green-tailed Towhee, from which this species is immediately distinguished by the yellow throat and breast.

Often first detected by their sharp chipping calls, adult Rufous-capped Brush Finches in a good view are easily identified by the combination of dull green back, somber gray ear coverts and supercilium, blackish half-mask, gleaming yellow throat, and dark rusty crown.

The juvenile is plainer but equally distinctive. In this plumage, Rufous-capped Brush Finches are uniformly green-gray above, including the nape and crown; the gray ear coverts and blackish lore resemble those of the adult. The underparts are unstreaked pale lemon yellow, lighter than in adults. At least some individuals are said to show pale brown tips to the wing coverts.

The common calls of this species include a long, thin *tsee-up*, with a leisurely attack and sustained, slightly ascending decay. Pairs and family groups give an incessant series of squeaky, chattering chips, the individual notes sometimes doubled and often with an odd "electronic" quality. There is also an apparent "reunion duet," more clearly structured and more obviously descending than the usual chatter.

The simple but variable song usually begins with one or two thin notes resembling the *tsee-up* call; that introduction is followed by four or five squeaky chips or, at times, a faster trill. In some birds, the entire song is composed of squeaky chips, while in others, the concluding notes are clear and ringing, somewhat like the introductory notes of a Louisiana Waterthrush.

RANGE AND GEOGRAPHIC VARIATION

Nominate *pileatus* occurs on the Mexican Plateau and in the southern mountains from Sinaloa south to Michoacán, Guanajuato, Hidalgo, Veracruz, Guerrero, Puebla, and Oaxaca; resident farther north and east, *dilutus* occurs in southwestern Chihuahua, Durango, Coahuila, Nuevo León, Tamaulipas, and San Luis Potosí. This species has been recorded numerous times since 1993 in the Yécora region of east-central Sonora (eBird); the subspecies involved there is probably *dilutus*. There is also a February report from just south of Santa Ana, Sonora (eBird); a scant 125 miles south of the Arizona border, this lowland site is devoid of the high-elevation pine and oak forests the Rufous-capped Brush Finch is bound to elsewhere.

The nominate race is larger and larger-billed than the grayer, paler subspecies *dilutus*; "these differences are easily distinguished but of no great magnitude."

This dramatically colorful tropical sparrow barely enters our region at Mexico's central latitudes. *Veracruz, April. David Krueper*

WHITE-NAPED BRUSH FINCH
Atlapetes albinucha

Antoine Marie Ferdinand de Maussion de Candé would serve with distinction as a counter admiral in the French navy and as the colonial governor of Martinique. In the mid-1830s, however, de Candé was still a lieutenant and aide de camp aboard the *Didon* on its exploration of southern Mexico and northeastern South America. On his temporary return to France in 1838, he brought with him specimens representing 17 species of neotropical birds. Among the four new species was a bird labeled as having been collected at Cartagena, Colombia; Frédéric de La Fresnaye and Alcide d'Orbigny— himself recently returned from seven years in the New World tropics—described the exotic new species as *Embernagra albinucha*, comparing it in structure and plumage ("and probably in habits as well") to the Pectoral Sparrow of northeastern South America and to the Great Pampa Finch of southeastern South America.

At very nearly the same time as Maussion de Candé's stay in France, the Bordeaux physician Grégoire Abeillé acquired a specimen from Mexico of the same bird for his extensive collection. Unaware, or perhaps unconcerned, that the species had already been described and named, René Primevère Lesson introduced it to science again in the next volume of the same journal, naming it this time *Embernagra Mexicana*.

The priority of the name assigned Maussion de Candé's specimen was quickly established. It would take more than a century, though, to determine that while that individual made it into European print first, the label Maussion de Candé had prepared for it was wrong: the bird almost certainly did not come from anywhere near Cartagena, but must instead have been collected on one of the *Didon*'s stops in Mexico. The error might seem trivial, but in fact it would have serious implications for the ornithological assessment of the relationship between this bird and another, the Yellow-throated Brush Finch of central America and Colombia, described a few years later, this one too by La Fresnaye.

The original assignment of this species to the

genus *Embernagra,* along with La Fresnaye and d'Orbigny's comparison to two such different species, indicates ornithology's longstanding uncertainty about whether these large, heavy-billed birds should be classed as sparrows or as tanagers; indeed, when Lesson erected the genus in 1831, he identified it in the vernacular as comprising the *tangaras-bruants,* "bunting-tanagers."

In 1851, Jean Cabanis determined that this species was better considered a finch, and he assigned it and its close relatives to the new genus *Atlapetes.* Nevertheless, other authoritative taxonomists cast the group's lot with the tanagers, a circumstance that did not change after the bird was moved to the genus *Buarremon* by Charles Bonaparte. *Buarremon* was still accounted a tanager as late as 1886, but Robert Ridgway definitively reassigned that genus and the newly revived *Atlapetes,* including the White-naped Brush Finch, to the sparrows in 1901.

Most birders in the northern parts of North America flatter themselves that they know a sparrow when they see it, but many tropical species can still throw the observer unprepared for just how different some members of the family can seem at first acquaintance.

FIELD IDENTIFICATION

Large, rotund, and broad-tailed, White-naped Brush Finches of the northern subspecies are easily identified by the olive-gray back, black head and nape, and yellow underparts. The much smaller Rufous-capped Brush Finch lacks the eponymous brown crown in juvenile plumage, but the head is dull olive-gray with darker ear coverts and lores. The southern *griseipectus* overlaps with neither the nominate race nor any of the Yellow-throated Brush Finches.

The call is a high, thin, decidedly downslurred *dswee,* with a soft attack and long, trailing decay. The song is a well-cadenced but variable series of loud, slurred whistles, often beginning with an emphatic slur followed by a pause and a descending phrase of two or three sweet whistles: *teeu—tee weeree tee.*

RANGE AND GEOGRAPHIC VARIATION

Nominate *albinucha* is resident at elevations of 2,000 to 8,000 feet on the Atlantic slope of Mexico from southeastern San Luis Potosí south to eastern Chiapas. The southerly race *griseipectus* is found from southern Chiapas into Guatemala and western El Salvador.

When considered specifically distinct from the Yellow-throated Brush Finch, the White-naped Brush Finch comprises two subspecies. The northern, nominate race is darker and grayer above than *griseipectus,* with the underparts yellow from throat to vent. The latter, very distinctive subspecies has the color of the throat "lighter yellow and more extensive than in any" of the Yellow-throated Brush Finch forms, and the breast is clear gray.

The adult male is almost glossy black, the eyes in northern birds a deep, demonic red. *New Jersey, January. Eric C. Reuter*

EASTERN TOWHEE
Pipilo erythrophthalmus

This familiar and colorful bird shares with the Slate-colored Junco the distinction of being the first of the American sparrows ever depicted by a European painter. John White, an English "gentleman artist," was commissioned by Sir Walter Raleigh to document the human inhabitants and wildlife of Virginia and North Carolina; the 75 watercolors White produced and sent back to England between 1584 and 1590 are the earliest surviving visual records of North America as it was at the time of European settlement.

White painted both the male and (probably) the female Eastern Towhee, labeling the former with the Native name "Chúwhweeo." Though the originals are now lost, both watercolors were copied, first in the early seventeenth century by Edward Topsell and then a hundred years later for Hans Sloane. The paintings—and the birds—were then forgotten, and only when Mark Catesby published his account of the "Towhe-bird" in 1731 did European science become more broadly aware of the existence of the species. Catesby characterized the bird as "*Passer niger, oculis rubris*"—"a black sparrow with red eyes"—a diagnosis that Linnaeus in 1758 simply translated into Greek, giving the species the scientific binomial *Fringilla erythrophthalma*. Even today, the northern populations are sometimes known in English as "red-eyed towhees."

This species has a strikingly different eye color in its southeastern range. The collectors John Abbot and Steven Elliot were the first to report this phenomenon to Alexander Wilson, who confirmed that he himself had

> examined a great number of these birds in the month of March, in Georgia, every one of which had the iris of the eye *white*. Mr. Abbot of Savannah assured me, that at this season, every one of these birds he shot had the iris white, while at other times it was red; and Mr. Elliot, of Beaufort, a judicious naturalist, informed me, that in the month of February he killed a Towhe Bunting with one eye red and the other white!

Wilson wondered, logically enough, whether the brown eye of juveniles turned red before finally becoming white in the adult. Sixty years later, Gilbert Burling—probably a relative of the well-known American painter, rather than the artist himself—proposed a different interpretation, suggesting that the towhee's eye color varied not with age but seasonally: the

> color of the iris . . . with the summer dress is red, and with the winter white—the change taking place sometimes first in one eye, and then in the other, during the process of the autumn moulting.

Joel Asaph Allen, working from specimens collected in Florida by C. J. Maynard, answered the riddle of the towhee's eye color in 1871, when he determined that it was in fact a matter of geographic variation, and that the white-eyed birds of the Southeast were "evidently . . . a local race" of the Eastern Towhee. Allen, ever a great opponent of the unnecessary multiplication of taxa, declined to name the bird, a lapse made up later that year by Elliott Coues, who described "the curious little *Pipilo*" as a full species, *Pipilo alleni*, the species epithet a tribute "to our esteemed friend."

Coues almost immediately revised his opinion of the status of the white-eyed birds to accord with Allen's, treating it in all subsequent publications as a subspecies of the Eastern Towhee; indeed, as early as 1874, he went so far as to misquote his own nomenclatural act in the attempt to tacitly deny ever having assigned it full species rank. There was one taxonomic holdout, however. Maynard, who could justly claim to be the modern rediscoverer of the white-eyed towhees, not only continued to believe that his birds constituted a distinct species, but he also, with a presumptuousness matched only by his characteristic prolixity, maintained his right to name them.

> A few words of explanation concerning my use of the name of *Pipilo leucopis* for a species which had already been described by Dr. Elliott Coues as *Pipilo Alleni*, may be necessary. In the winter of 1868, I discovered the White-eyed Towhee . . . and took many specimens. Arriving in Jacksonville later in the season, I met my friend, Mr. J. A. Allen . . . . I called his attention to some living specimens which were exposed for sale in a cage in the city market. After examining them, we walked down Bay Street and going a short distance, met Mr. George A. Boardman who was at that time unacquainted with the bird in question . . . . Up to that time, be it noted, no one, excepting myself, had ever observed that there was a Tow-

This large and handsome sparrow has eyes varying from red in the north to yellow or even white in the extreme southeastern portion of its range. *Florida, February. Brian E. Small*

hee in Florida having white eyes. Upon my return home, I sent some specimens to the Smithsonian . . . upon condition that they should not be described until I had decided respecting their specific rank . . . . [Later] I wrote Prof. Baird and other friends, from Florida, stating that I had decided that the *Pipilo* was new and that I should describe it upon my return. In fact, my name was in manuscript when Dr. Coues' *Pipilo Alleni* appeared . . . . Dr. Coues was at the time, however, unaware that I was about to describe the bird. Influenced by the circumstances, I concluded not to discard my name [for the bird, *Pipilo leucopis*] and have since used it in the body of the work.

The ornithological establishment's exasperation with Maynard was largely restricted to private criticism of what Allen in a letter to Ridgway called his "absurd classifications" and "newly-coined names of groups, constructed in accordance with the rules of no known language." In public, Maynard's colleagues were more delicate. Henry Henshaw, writing specifically about "the case of *Pipilo alleni*," remarked quietly on Maynard's "desire for originality" and excused the renaming of the sparrow "simply as an expression of his claim to the discovery of the form" that "can do no special harm," gently defusing what could easily have become—given the persons and personalities involved—a genuine taxonomic feud.

A more drawn-out and more significant argument was kindled in the 1870s by the repeated discovery of apparent Eastern Towhees with white spotting and streaking where there should be none, from the East Coast to the prairies; a number of Minnesota specimens, representing both sexes, were found to be marked with "a great or less number of minute white spots more or less distinctly indicated on the portion where the large white spots of the western forms are located." Elliott Coues, Thomas Brewer, and Joel Asaph Allen all found in these birds evidence that the "black" towhees of the eastern United States and those of the West were better considered members of a single variable species rather than as distinct species.

In spite of those early suspicions, the Eastern Towhee—typically under the English name "red-eyed towhee" or simply "towhee"—and the Spotted Towhees of the West were maintained by the AOU and other authorities as distinct at the species level until the middle of the twentieth century, when Charles G. Sibley, in the course of his studies of the Spotted Towhees of Mexico, suggested that those forms and the Eastern Towhee be considered conspecific, a recommendation greeted as "particularly pleasurable" and "long . . . indicated." The American Ornithologists' Union ratified Sibley's view in 1954, with the result that the fifth, 1957 edition of the *Check-list* included only a single extremely far-flung species, the Rufous-sided Towhee, comprising no fewer than 16 subspecies in the area covered.

Sibley continued his study of the towhees, extending it to the northern Great Plains beginning in 1953. More than 500 specimens were collected in Colorado, Nebraska, the Dakotas, and Manitoba; it was determined that the eastern and western populations were in secondary contact only in narrow strips of riparian habitat, chiefly along the Platte and Niobrara Rivers, where there was found to be a "gradient" in the amount of back spotting and the darkness of female plumages.

Nevertheless, visually "pure" individuals were in the majority in much of the hybrid zone, and even along the Platte River, where hybrids were numerically most dominant, fully a fifth of the birds were deemed pure representatives of the eastern or western type, circumstances that "strongly suggest assortative mating," in which birds seek out individuals like themselves for breeding and avoid hybridization.

There is poetic justice in the fact that it was molecular techniques pioneered by Sibley in the 1960s that eventually led to the formal resplitting of the towhees. Study of mitochondrial DNA in specimens from eastern and western populations discovered "marked mtDNA phylogeographic differentiation" and "strongly suggest[ed] a considerable genetic distinction." Indeed, the difference between eastern and western Rufous-sided Towhees exceeded that separating such species pairs as King and Clapper Rails and Mallard and Mottled Ducks, and considerably greater than any geographic divergence observed within the study's Mourning Doves, Downy Woodpeckers, Song Sparrows, and Brown-headed Cowbirds. Armed with that biochemical information and the well-known and conspicuous differences between eastern and western towhee populations in vocalizations, plumage, and extent of sexual dimorphism, the American Ornithologists' Union resplit them in 1995, reviving the former English name Spotted Towhee for the western species and coining the new English name Eastern Towhee for the other.

FIELD IDENTIFICATION

With its large size, colorful plumage, and characteristic vocalizations, the Eastern Towhee is distinctive and immediately recognizable over much of its extensive range; if it were less secretive and slinky, this species could well displace such familiar birds as robins, mockingbirds, and blue jays in the popular imagination and on the lists of state and provincial emblems.

Juvenile Eastern Towhees, brown and coarsely streaked, provide a signal example of how misleading plumage characters can occasionally be for less experienced observers. The smudgy streaks, plain head, and dark tail could recall juvenile Brown-headed Cowbirds—themselves commonly seen following towhee foster parents—and the thick, conical bill can reinforce that misapprehension. First impressions are readily corrected, though, by a glance at the sparrow's distinctive posture, with head and neck held low and long, narrow tail raised; unlike cowbirds and other blackbirds, the towhee hops and kicks on slender pale feet, rather than walking and waddling on coarsely scaled, gnarled dark tarsi and toes. Juvenile towhees vocalize almost continuously when soliciting their parents for food, with a high-pitched, fingernails-on-the-chalkboard *tititi*; cowbirds are less obstreperous in their insistence.

All other streaked passerellid sparrows within the breeding range of the Eastern Towhee are much smaller, smaller-billed, and less lavishly long-tailed. None shares the extensively white tail tips (varying geographically as described above, largest in the north and most restricted in the south) or the white-tipped tertials and white-edged primaries of the towhee.

Towhee identification, whether of adults or of juveniles, becomes significantly more challenging on the eastern and central Great Plains. Most breeding birds in the lower Missouri River Valley, north to a short distance above the mouth of the Platte River, are apparently "pure" Eastern Towhees in plumage and in voice, but to the north and west, the black towhees are a perplexing mix of hybrids and backcrosses. Classically intermediate birds (which are not necessarily first-generation hybrids) combine the white primary bases of the Eastern Towhee with the spotted wing coverts and scapulars of the Spotted Towhee; along the central Platte River, however, a very wide range of plumage phenotypes can be observed, including individuals that may appear nearly "pure" in the field but in the hand reveal obscure spotting or reduced white in the primary bases. It is very important in these areas to listen carefully to any towhees observed; birds otherwise closely resembling Eastern Towhees often give calls and songs typically associated with Spotted Towhees, in many cases probably the result of introgression. Interestingly, visually "purish" Spotted Towhees in these same areas appear to vocalize like Easterns far less often.

Adding to the complexity is the fact, known for nearly a century and a half, that occasional Eastern Towhees hatched far from any zone of hybridization can have white spots or bars on the scapulars and wing coverts. This phenomenon is often assumed to be an atavism, "due to 'ancestral' genes," or a simple mutation.

Many birders on the Great Plains use the system created by Sibley and West for "scoring" the plumage of adult towhees. A bird scored 0 lacks spots on the wing coverts, scapulars, and back, but has a full-sized white patch at the primary bases; such individuals are visually pure Eastern Towhees. A score of 4 indicates a bird visually indistinguishable from a Spotted Towhee of the very heavily marked race *arcticus*, with fully spotted upperparts and no white at the base of the primaries. Birds scored 1 and 3, respectively, combine a trace of spotting with a reduced white primary patch or a trace of white primary spotting with reduced upperparts spotting; a score of 2 is assigned to towhees with abundant spotting and a primary patch approximately half as extensive as in a "pure" Eastern.

The Eastern Towhee is one of few passerellid sparrows to exhibit obvious sexual dimorphism in plumage; in Spotted Towhees, the sexes resemble each other much more closely, sometimes to the extent that it can be impossible in poor light to be sure whether a given individual is more likely a male or a female. The ground color of the head and upperparts of the female Spotted Towhees of the Great Plains is a cold juncolike gray-brown, while all Eastern Towhee females are a rich chocolate brown. In apparent hybrids and backcrosses, the head and back are "muddied brown" or brownish gray.

Juvenile hybrids and intergrades can tentatively be recognized by a white primary patch that is clearly present but, in comparison with "pure" Eastern Towhees, reduced in extent. The tail spots do not differ sufficiently between the two species to reliably distinguish Eastern and *arcticus* Spotted Towhees in the field; "pure" individuals show an average difference in the relative lengths of wing and tail, but intergradation is almost certain to elide that difference.

Eastern Towhees, whether the products of miscegenation or not, are large, long-tailed, thick-billed sparrows; the solid patches of plumage color in adults are unique among the passerellids of eastern North America. On perched or feeding birds, the tail is usually held closed, creating a straight edge and, from above, a solid black or chocolate-brown appearance; only from beneath or in flight are the large white tail spots conspicuous. The belly and breast are full and rounded, and the large, domed head makes the bird seem foreshortened and neckless, especially when the head is raised.

Most towhees are first detected by ear. The sound of a tiny chain gang hard at work in the leaf litter often resolves into the vigorous double-scratching of an Eastern Towhee, throwing its feet forward and then scraping abruptly back with all its might.

This species does not generally vocalize when it is actively feeding, but individuals may pause to give a long, soft, slightly scratchy *dsee,* said to be a contact note used within family groups or wintertime flocks. This call is usually much less heavily modulated than the buzzier *dzrr* given by most Spotted Towhees. Alarmed birds give a sharp, almost cardinal-like *tipk,* running into a series that accelerates with increasing excitement.

The most familiar call—source of such vernacular names as "towhee," "touit," "chewink," "joree," and "chúwhweeo"—varies over this species' wide range. In the northern, nominate subspecies, this call is usually metallic and clearly two-syllabled, with a low-pitched, lingering introduction that slurs slowly upward into an explosive conclusion. The corresponding note given along the central Gulf of Mexico coast is "somewhat hoarser, more nasal." Florida's *alleni* has a markedly different analogue, sweetly whistled and evenly upslurred over its entire length, calling to mind an outsized American Goldfinch.

The famous "drink your tea" song of male (and rarely of female) Eastern Towhees is more variable than any of the call notes within and among populations. It is normally two-parted, the introduction ("drink your") made up of two to four more or less discrete notes. Many northern birds start with a loud, ringing note similar to the beginning of the "towhee" or "chewink" call, which they follow with a lower-pitched, slightly gulping click; the introductory series can be simpler and more uniform in the song of *alleni.* These notes are followed by a tremolo ("tea") of varying speed and tone, occasionally repeated. In some birds, this concluding phrase is musical and liquid, in others hurried and rattling. The tremolo is usually the highest-pitched portion of the song in northern birds, but in some individuals, perhaps especially of southern populations, it barely differs from the introductory notes, thus rather resembling some Spotted Towhee songs.

An infrequently heard "complex quiet song" is described by Greenlaw as "muted . . . highly variable, disjointed muttering without any apparent temporal or syntactic pattern," incorporating calls, trills, and other notes and phrases; given from the ground or a low perch, this vocalization, apparently given only by males, can continue for up to 20 minutes.

RANGE AND GEOGRAPHIC VARIATION

Though less widespread than the Spotted Towhee, the Eastern Towhee is one of the most widely distributed birds in the southeastern quarter of North America. Breeding birds occur in warm brushy habitats as far north and west as southeastern Saskatchewan and North Dakota's Turtle Mountains. The

The female is a warm chocolate brown above, with soft rusty sides. *New Jersey, January. Eric C. Reuter*

species breeds across the southern quarter of Manitoba, south and east through northern Minnesota, Wisconsin, Michigan, extreme southern Ontario, extreme southern Quebec, Vermont, New Hampshire, and southernmost Maine; it may also nest as far east as central Nova Scotia. Eastern Towhees breed south to the southern tip of Florida and thence west to western Louisiana and possibly the Piney Woods of Texas.

The western breeding limits of this species are blurred by hybridization with the Spotted Towhee. "Pure" individuals occur across eastern Arkansas and Missouri, extreme eastern Kansas, most of Iowa, southeastern South Dakota, and eastern Minnesota. In Nebraska, genuine Eastern Towhees nest in the southeastern corner of the state, most commonly along the Missouri and lower Platte Rivers, with recent breeding-season reports from as far west along the North Platte River as Scotts Bluff County; north of Douglas County and west of Hall County, visually apparent "Eastern" Towhees should be listened to carefully for signs of hybridization and intergradation. The species is said to be a rare to uncommon breeding resident in extreme northeastern and southeastern Colorado as well.

Eastern Towhees of northerly populations are short-distance migrants; the more southern subspecies *canaster* and *rileyi* are largely resident, and Florida's *alleni* appears to be entirely nonmigratory. Even northern *erythrophthalmus* is quite hardy, and some individuals may not leave the breeding range at all until forced to do so by frozen or snow-covered ground. Wintering is regular north to New Jersey and the Ohio River Valley. Most autumn departures appear to be from mid-September to mid-October, with birds arriving on the wintering grounds in October and November. Most leave the southern range in April, arriving in the breeding areas in April or early May. Individuals seen at feeders before late March are more likely to be undetected winterers than arriving breeders.

Like most short-range migrants, the Eastern Towhee has an only limited record of vagrancy. The species regularly overshoots in spring north to Nova Scotia, and is even more frequent there in autumn, not infrequently wintering at feeders. There are also several records for the French islands of St. Pierre and Miquelon. Even farther north and east, an adult female reached England—possibly under its own steam—in June 1966; the bird was captured and banded on Lundy Island.

Wanderers to the west and south pose a greater identification challenge. In Colorado, Eastern Towhees are considered resident along the South Platte and Arkansas Rivers; there are a few records, from all seasons, west to the foothills of the Rockies, with accidental occurrences in the mountains themselves and in the western valleys. There are also a few infrequent records from New Mexico, Montana, Idaho, and Arizona; most such birds are detected at feeders in winter and spring. An early autumn report from California almost certainly refers to a misidentified Spotted Towhee, probably to an individual of the very lightly marked subspecies *oregonus*.

Four subspecies of the Eastern Towhee are generally recognized. Though bright-eyed birds of the southernmost race *alleni* are identifiable in their Florida range, it is important to remember that eye color varies by age and individually in this species. Most birds are difficult or impossible to identify to subspecies in the field away from the breeding grounds, and all four occur in winter in Florida.

The nominate, northernmost, and most highly migratory subspecies is large and small-billed, with high-contrast plumage (especially in males) and, in adults, bright red eyes. This subspecies has the most deeply colored flanks of any. Females are decidedly rusty-backed. The white tail spots are large, cover-

ing more than one-third of the outermost rectrix in both sexes; more than half of all males have white on four or, in a very few individuals, five pairs of tail feathers.

The southernmost race is *alleni*—Maynard's would-be *leucopis*—which is resident in southern peninsular Florida. Birds of this subspecies are small and medium- to large-billed, with pale straw-white eyes in the adult. Females of this race and of *rileyi* are often notably gray-backed; the flanks of both male and female *alleni* are much paler brown than those of other races. There is noticeably little white in the tail, the spot of the outermost rectrix in both sexes barely half as extensive as in the nominate race; more than half show white in only three pairs of tail feathers, and in fully a third, white is present on only the outer two rectrices, while a small number of females have white limited to the outermost tail feathers alone.

Breeding from Louisiana and western Tennessee east to the Florida Panhandle and South Carolina, the subspecies *canaster* ("grayish") is large, large-billed, and pale; the adult's eye ranges from red to pale orange or yellow. The long tail shows slightly less white than in the nominate race, but considerably more than in *alleni*. In most individuals—three-quarters of males and 80 percent of females—white is restricted to only three pairs of tail feathers.

Paler-eyed and darker-tailed than *canaster*, *rileyi*—named for the Smithsonian ornithologist Joseph Harvey Riley—can resemble *alleni* in eye color; it is medium-sized and large-billed, with rather less white in the tail than either *canaster* or, especially, *erythrophthalmus*. With *canaster*, this is the longest-tailed of the Eastern Towhees; though most individuals have white on three pairs of rectrices, a notable percentage—one of twelve males and one of five females—have spots on only two pairs. The flanks, especially of males, are noticeably darker than in *alleni*. This largely coastal race is essentially resident from the north-central Florida Panhandle and eastern Alabama north through Georgia and the Carolinas to southern Virginia.

SPOTTED TOWHEE
Pipilo maculatus

Elliott Coues, in the second and all subsequent editions of his *Key*, drew a telling—even chilling—comparison:

> The black series of *Pipilo* offers a case nearly parallel with those of *Melospiza, Passerella,* and *Junco . . . .*

The similarity Coues refers to between the towhees, the Song Sparrows, the fox sparrows, and the juncos has nothing to do with the birds' plumage, their behavior, or their range; instead, the parallel is the inscrutability of the relationships among the populations in each group. William Brewster warned that "the North American towhees of the *maculatus* group are . . . involved in much confusion," and that certain names had been coined in a sort of nomenclatural helplessness as "a convenient receptacle" for individual towhees that could not be fitted into any other of the arbitrary categories devised for the species. In vain did the ornithologists of the nineteenth century look for "constant tangible differences between" named populations, as the line between species and "strongly marked . . . varieties" of these birds blurred ever more. Jean Cabanis, with one of Europe's most extensive collections at hand, raised a white flag when he essentially despaired of untangling the relationships among these towhees, synonymizing some and considering the status of others irrecoverable.

By the 1880s, there was general agreement that the Spotted Towhees represented "one of the worst muddles which at present exist to confuse the students of North American birds." Most of the confusion was then, and remains today, based in the genuine complexities presented by the complex. The source of some of the obfuscation, however, was William Swainson's descriptions of the three towhees he described and named in the 1820s and 1830s. All—*Pipilo macronyx, Pipilo maculata,* and *Pipilo arctica*—are now considered subspecies of the Spotted Towhee in its modern sense, but Swainson treated all three "ground-finches" as distinct species, even though in the case of the latter two "we might, at first sight, be tempted to think they were the same."

The new birds brought the tally of "typical" *Pipilo* species to five—a number of crucial significance in the "quinary" scheme of classification to which Swainson then subscribed. "Aesthetically pleasing and . . . reconcilable with the Linnean and Natural Theology traditions," the quinary scheme

The male Spotted Towhee is deep, even glossy black on the back. In many parts of the range, the female can be nearly as dark. *California, April. Brian E. Small*

claimed to discover a regularity in natural relationships based on groups of five, which could then be organized into a series of circles; the point at which separate circles are tangent is occupied by a species showing characteristics of both groups of five. In triumphantly completing the set of "typical" *Pipilo* species, Swainson hoped to answer the question of the affinities of one of Linnaeus's mysterious finch descriptions: although *Pipilo,* he wrote,

> was supposed to consist but of one example, we have now characterised no less than five additional species, four of which are typical. The group appears confined to America, and seems to be the Rasorial type of the true Sparrows . . . if so, it will consequently touch the circle of the

Tanagrinae at that point which brings *Pipilo* into junction with such a form as . . . the *Fringilla Xena* of Linnaeus, a rare and interesting bird.

Those words must have been as mystifying to the contemporary reader as they are today, and elaborate sets of circles and epicycles were, happily, soon abandoned as taxonomic tools. William Swainson's need to impose system on systematics, though, contributed to a multiplication of towhee species that would continue through the century, further complicating the long task of unraveling the relationships of one—or some—of the most perplexing of the sparrows.

The changing nomenclature of this species over the past nearly two centuries could be read as virtually encapsulating the modern history of ornithological taxonomy. The early nineteenth century was a period of exciting discovery and gnawing uncertainty, when the question to be answered was largely and simply what was to be found "out there" on new collecting grounds. For a good century thereafter, much of ornithology was concerned with discerning and describing different "kinds"—species and then, increasingly, subspecies—among the new birds from the Americas, Africa, Asia, and Australia; especially among taxonomists in the United States, the description and naming of new races—"splitting"—was a growth industry well into the middle of the twentieth century. The reaction, tardy but inevitable, came in the 1960s and 1970s, when geographic variation, or at least the use of rigid subspecies names to describe it, fell decidedly out of favor, and even "good" genera and species of long standing came to be synonymized, a tendency epitomized in the 1973 Supplement to the AOU *Check-list,* promulgating the widest-ranging set of "lumps" in the history of American ornithology. The creation in the 1960s of biochemical techniques for genetic study permitted in many cases a more measured view of the relationships between bird populations, and such investigations continue today to force the reevaluation of many taxonomic actions taken in the past.

In 1827, in the span of a single page, William Swainson had described the Spotted Towhee twice, as two separate species—*Pipilo macronyx* and *Pipilo maculata* (Swainson treated the genus name as grammatically feminine). Misinformed about the localities where his specimens had been collected, Swainson believed that the two were resident in the same range and thus could not possibly be anything but specifically distinct; their status as full species was affirmed through the nineteenth century. Not until the middle of the twentieth century did Charles G. Sibley subject the early accounts

of the two taxa to a critical review; his conclusion was that they should be considered conspecific and that each, along with all the other spotted towhees, should be treated as "merely a subspecific unit within" an enlarged species embracing the eastern, plain-backed towhees as well.

Probably because so much of the geographic variation in the Spotted Towhees is concentrated south of the United States border (D. Donsker in litt.), it took considerable time for order to be imposed on their taxonomy. Of the twenty-one currently recognized subspecies, seven—a full third—were originally described as full species, several of them more than once.

The New York taxidermist and collector John Bell accompanied Audubon on his Missouri River expedition of 1843. No doubt out of respect for his mentor, Bell waited until the year after Audubon's death to point out in print that both the great ornithologist and Thomas Nuttall had "confounded" two distinct towhee taxa. At least one of the specimens Audubon painted for the *Birds of America,* there assigned the name *arcticus*—properly given to birds from the Great Plains—was, Bell wrote, in fact "procured in the Oregon territory," and that Oregon bird exhibits "many strongly marked characters . . . which distinguish it from the *Pipilo arcticus* of Mr. Swainson," the Great Plains towhees. Bell accordingly named the bird on Audubon's plate *Pipilo Oregonus,* the first new member of the group to be described since Swainson's day.

In the first edition of his *Key,* published in 1872, Elliott Coues determined that Bell's *oregonus* "shades into" the "typical" *maculatus,* of which he considered the West Coast bird a geographic variety, a subspecies. The bird entered the AOU *Check-list* as such in 1886, and has never been seriously treated as specifically distinct since.

Spencer Baird was more hesitant in 1858 when he named a new *Pipilo* from the southwestern United States, finding that

This form, if not a distinct species, constitutes so strongly marked a variety as to be worthy of particular description.

His caution notwithstanding, Baird named the bird as a full species, *Pipilo megalonyx,* the scientific name signaling its "enormously large claws" and heavy tarsi. In 1866, Coues expressed his doubts, finding "it difficult to discern constant and tangible differences between *arcticus* and *megalonyx.*" The *Key* treated them as conspecific, differing only at the subspecies level, a treatment adopted and maintained by the AOU.

The Socorro Towhee, too, is among those Spotted Towhees named more than once. The second time was in a paper by George Lawrence, in which he described a number of specimens collected by Andrew Jackson Grayson on Socorro Island; "several of the species," Lawrence wrote, had "manuscript names given them by Prof. Baird, which in all such cases have been retained." Baird, and consequently Lawrence, named the "diminutive species" *Pipilo carmani,* after the collector's friend Dr. B. F. Carman of Mazatlan. Lawrence reconsidered not long thereafter, reducing *carmani* to subspecific rank, but Robert Ridgway restored it to the dignity of a species in 1887, a position it continued to hold in his 1901 *Birds of North and Middle America.*

Opinions about this bird continued to vary in the twentieth century. Charles Hellmayr lumped *carmani* as not "anything but [a] strongly marked insular race" of the Spotted Towhee, differing from other representatives of the species "only [in] degree." Alden H. Miller et al. disagreed, and considered the Socorro Towhee a full biological species:

> This dwarf form is so strongly characterized and its modifications are so great that there is real doubt it would interbreed with its mainland relative.

The Socorro Towhee somehow failed to crest the horizon of the AOU *Check-list* until the sixth edition, of 1983, when it was at last included as "a derivative of" the Spotted Towhee (which was at that time itself still lumped with the eastern birds under the name Rufous-sided Towhee). Nevertheless, well into this century such authoritative lists as *The Clements Checklist* continued to list the Socorro Towhee as a full species before acquiescing in what is now largely the conventional taxonomic wisdom.

In March 1897, Joseph Grinnell collected the type specimen of the last Spotted Towhee to be initially described as a full species, the San Clemente Towhee, *Pipilo clementae.* Grinnell warned that "more material may relegate *Pipilo clementae* to subspecific rank," a caution borne out when the AOU added the island bird to the *Check-list* as a pale, large-billed subspecies of the Spotted Towhee.

The taxonomic career of the Guadalupe Towhee has been—or rather was—particularly tumultuous. Robert Ridgway described this "not abundant" bird in 1876 on the basis of eight specimens, which he found "somewhat intermediate" between the Oregon Towhee and the Socorro Towhee; he accordingly named it as a race *Pipilo maculatus consobrinus,* the subspecific epithet—"cousin"—emphasizing its similarity and close relationship to the other Spotted Towhees. A year later, Ridgway treated the towhee as a distinct species, though with a "close affinity" to the Spotted Towhees of the Mexican mainland.

Richard Bowdler Sharpe demurred, tentatively: since "most American naturalists admit that the

This formative male Spotted Towhee's wing will be blacker after its prebasic molt into definitive plumage. *California, June. Brian E. Small*

different races of *Pipilo maculatus* merge into each other, it is questionable whether they should not all be united under a single heading"; Sharpe did just that, lumping *consobrinus* with eight other races as components of an expanded *Pipilo maculatus*. Nonetheless, American ornithology followed Ridgway in maintaining the distinctness of the Guadalupe Towhee until 1944, when the AOU ratified Hellmayr's observation that, like Mexico's other island towhee, it was only another "strongly marked" subspecies of the Spotted Towhee.

Further oscillation can be expected as time goes on, but in the early twenty-first century, most authorities consider all of these towhees Spotted Towhees, *Pipilo maculatus*. It remains unclear how the name *maculatus* came to be exclusively preferred over *macronyx* for the species as a whole; both names were coined by Swainson simultaneously in 1827 and published on the same page. Somehow there was tacit and universal agreement that *maculatus* should serve as the species name, a nomenclatural convention finally made explicit in 1981.

FIELD IDENTIFICATION

Large, square-headed, lavishly long-tailed, and strikingly colorful, adult Spotted Towhees of any subspecies pose relatively few identification challenges. In the far west, from British Columbia south to central California, lightly spotted birds of the race *oregonus* can call to mind an Eastern Towhee. Conversely, in the East, Eastern Towhees with sparse white marks on the scapulars or wing coverts are easily confused with Spotted Towhees; it has also happened that inexperienced observers have been misled on first noticing the broad white tertial edges of Eastern Towhees, mistaking those for the streaks and spots belonging to the other species.

The first plumage feature to confirm on any puzzling towhee is the pattern of the primaries; happily, given the high-contrast patterns and colors of adults, those feathers are usually easily seen in the field even on perched birds. In Spotted Towhees, the bases of the primaries are the same black or slaty as the remainder of the feather. Eastern Towhees, on the other hand, have rectangular white patches at the base of the outer web of several primaries, creating a conspicuous "comma" in flight and a neat, straight-edged panel in the folded wing. The presence of that "pocket handkerchief" eliminates the possibility that a given bird is a pure Spotted Towhee, and its absence rules out identification as a pure Eastern Towhee.

The presence of white spots on the wings or scapulars of Eastern Towhees from the eastern part of that species' range is difficult to explain. It may be simply an atavism, the expression of a primitive character now not normally shown by the species, or it may reflect the results of interbreeding between the two species where their ranges meet on the central Great Plains.

Many birders on the Great Plains use the system created by Sibley and West for "scoring" the plumage of adult towhees. A bird scored 0 lacks spots on the wing coverts, scapulars, and back, but has a full-sized white patch at the primary bases; such individuals are visually pure Eastern Towhees. A score of 4 indicates a bird visually indistinguishable from a Spotted Towhee of the very heavily marked race *arcticus*, with fully spotted upperparts and no white at the base of the primaries. Birds scored 1 and 3, respectively, combine a trace of spotting with a reduced white primary patch, or a trace of white primary spotting with reduced upperparts spotting; a score of 2 is assigned to towhees with abundant spotting and a primary patch approximately half as extensive as in a "pure" Eastern. It should be recalled that visual "purity" is not in every case identical to genetic "purity," and many birds of mixed ancestry probably resemble one or the other species so closely as to be unrecognizable on plumage characters alone.

Fewer observers are as familiar with another identification difficulty caused by interbreeding, in this case between Spotted Towhees and Collared Towhees in the central highlands of Mexico. The two are sympatric in Oaxaca, but do not interbreed there. On Mount Orizaba, in Puebla, however, both species forage and nest in a zone at around 10,000 feet of elevation, where agricultural development has created thickets suitable to both species; Charles G. Sibley found that approximately 16 percent of the towhees in that belt of overlap were neither Spotted nor Collared, but the results of hybridization. At the summit of Jalisco's Cerro Viejo, where *Pipilo* towhees were "exceptionally abundant, being the commonest [birds] present," the entire population was determined to be "a freely interbreeding one," with no individuals among the specimens collected considered visually pure and the enormous variability in the population as a whole "at least fifteen times" higher than in a genetically pure population of either parent species would be.

The situation is more complex across Mexico's central Plateau, in which the apparent extent of interbreeding ranges along a gradient from the stable Spotted Towhee population in Jalisco to the stable Collared Towhee population of southern Hidalgo. The hybrid populations known between those two areas of "purity" are more Spotted-like at Tancítaro and Pátzcuaro in west-central Michoacán, and more Collared-like on Mount Popocatépetl and on the western side of the Valley of México; the appearance of the towhees of Mil Cumbres, in eastern Michoacán,

The back spotting is largest and whitest on birds in the eastern part of the species' range. *Texas, March. Brian E. Small*

falls very close to the intermediate point . . . . The Mil Cumbres population occupies the curious position of being the transitional population between . . . two species.

The offspring of mixed towhee pairs in Mexico are extremely variable; on the Cerro Viejo, Sibley found that "no two specimens in the 77 examined were alike." Faced with an odd-looking towhee in the zones of hybridization, the observer should carefully note the crown color (chestnut, black, or mixed), the spotting of wing and scapulars (absent, extensive, or intermediate), the back color (green, black, or green with black streaks or bars), the throat color (solid white, solid black, or a combination), the flank color (brown-gray, rufous, or mixed), and the spotting of the tail (absent, extensive, or intermediate). These visual characters appear to vary independently of one another, such that any given hybrid or backcross can combine features more like those of a Spotted Towhee with other features more like those of a Collared Towhee.

Juvenile Spotted Towhees are dark brown, heavily streaked sparrows with a hint of pinkish rust on the flanks and vent; they share the adult's pattern of tail spots, and the scapulars have variable fine buffy streaks. The greater and median coverts are tipped with pale buff, creating jagged dotted wing bars that distinguish them from juvenile Eastern Towhees. Juvenile hybrids and intergrades, whether with Eastern or with Collared Towhees, are generally identifiable only if they show a conspicuous combination of characters typical of both species.

Thomas Nuttall appears to have been the first European ornithologist to draw attention to the most distinctive vocalization of the Spotted Towhee, writing to Audubon that the

> note (*towee*) so continually repeated by our humble and familiar Ground Robin [that is to say, the Eastern Towhee] is never heard in the western wilds, our present species uttering in its stead the common complaint, and almost mew, of the Cat Bird.

That hoarse, whining, scratchy call, given by both sexes in states of both calm and alarm, typically rises, then falls, slowing toward its end. Shared in similar form among all Spotted Towhee populations in Canada, the United States, and northern Mexico, this call is apparently not given by the towhees of central and southern Mexico and Guatemala, which instead utter a clearer, consistently rising "ch-wee," recalling the analogous notes of the Eastern Towhee.

Molting Spotted Towhees can be notably unkempt. *Arizona. Laurens Halsey*

All populations also have a hissing, lisping *tzeet* call, less buzzy than the corresponding note of the Eastern Towhee; this vocalization "often occurs in social contexts involving mated pair."

As in the Eastern Towhee, the songs of the Spotted Towhee are rather variable. The usual pattern consists of a series of short separate notes followed by a faster, drier trill; the two parts are similar in tone and delivery, without the extravagant slurring and pitch changes of the Eastern Towhee's song. Some songs, even within a single individual's repertoire, are relatively leisurely, others so fast as to end in less a trill than a rattle or buzz. In the northern portion of the species' range, inland birds tend to sing complete songs, while those on the Pacific Coast more frequently omit the introductory notes, reducing the song to a trill or buzz.

RANGE AND GEOGRAPHIC VARIATION

Few indeed are the American passerines that breed in both British Columbia and Guatemala. At the northwestern edge of its range in summer, the Spotted Towhee is commonest along the Georgia Depression, nesting from sea level to nearly 4,000 feet and present year-round. Inland in western Canada, it breeds north to Riske Creek, British Columbia; Big Valley, Alberta; and Regina, Saskatchewan.

On the northern Great Plains, breeders are found east to the central Dakotas and most of Nebraska in suitable habitat along and north of the Platte River, where hybridization with the Eastern Towhee is frequent; breeding also takes place in northwestern Kansas and eastern Colorado.

The breeding range along the Pacific Coast reaches from Vancouver Island, British Columbia, south to northwestern Baja California, with one December record of *umbraticola* from El Rosario; *magnirostris* is resident in a distant and disjunct range in the mountains of that state's Cape district, occasionally descending to lower elevations.

As the brief subspecies accounts above suggest, this species' populations become more pronouncedly sedentary to the south. Even most of the more northerly races appear to be merely altitudinal migrants or "half hardy," withdrawing short distances when snowfall makes feeding on the ground more difficult. The most clearly migratory populations are those breeding in the northern Rockies and on the Great Plains, which regularly move south and east in September and October after breeding. Winterers on the southern Great Plains and at lower elevations in the southwestern United States typically arrive in October or early November, and linger on the wintering grounds into March and early April.

In general, of the migrant populations, *arcticus* winters farthest east, south to southern New Mexico, Texas, Tamaulipas, and Nuevo León; *montanus* occurs from Arizona, Sonora, and Chihuahua east to central Texas, while *curtatus* moves into California and southern Arizona. The northern limits of the migratory races' winter range are variable and poorly defined, but it is unsurprising to find the occasional bird lingering in or immediately south of the breeding range well into the cold season; as a rule of thumb, winterers are regular and more or less common in most years north to at least the fortieth parallel.

The regular wintering range extends reliably east into the breeding range of the Eastern Towhee in eastern South Dakota, Nebraska, Kansas, and Oklahoma, and in western Iowa and Missouri. Wintering densities decrease rapidly farther east, and the species is generally considered notable in winter at any considerable distance east of the Mississippi River. Small numbers reach the Atlantic Coast every year, where they are most usually detected as single birds wintering at feeders; the southeasternmost records, all presumably representing the highly migratory and geographically most proximate race *arcticus*, are from the Florida Panhandle in the period from December to late April.

Bizarrely, a male Spotted Towhee spent the fall and early winter of 1975–1976 on the Yorkshire coast of the North Sea. This bird was examined in the hand three times over its long stay, beginning on September 5, but, although it "showed characters of one of the western races," including "white spots in the scapulars," it was not identified more precisely to subspecies, leaving its more precise geographic origin, or the geographic origin of its ancestors, unknown. The record was initially accepted by the *British Birds* Rarities Committee, but has since been widely assumed to pertain to an escape from captivity. If the towhee was in fact a representative of one of the nonmigratory or less migratory subspecies, it seems quite certain that it did not arrive in England under its own short-winged steam. The published photograph of this individual, however, revealing very coarse scapular markings and well-defined back streaks, at least does not rule out an identification of the bird as a member of the more mobile race *arcticus*, which is a not entirely inconceivable source of genuine wild trans-Atlantic vagrants.

With a range extending from southeastern Alaska to Guatemala, it is unsurprising that the Spotted Towhee should be one of the most visually and vocally variable birds in North America. Subspecies vary in size, back color, extent of spotting, tail pattern, bill size, and toe length; most of these differences "occur in a mosaic pattern across the species' range," with little indication of the gradual passing of one character into another in adjacent subspecies.

The northwesternmost subspecies, John Bell's *oregonus*, is short-tailed and small-footed, with considerably reduced white spotting on the scapulars and wing coverts; the white tail spots are small in both sexes. Adult males are deep shiny black above and on the head, females dark blackish brown. This is the breeding towhee from southern coastal British Columbia south to Oregon, showing some southward movement in winter as far as central coastal California.

Breeding farther inland, in relatively cool, high-elevation desert from southern British Columbia to central California and Nevada, *curtatus* was named for its rather short tail and wing. It differs from the more easterly *montanus* in the darker black of the male's upperparts and head; females are dark slaty gray. It is also darker than the Great Plains race *arcticus*, which is slightly shorter-toed and shorter-tailed. The extensive white spots of wing and tail in *curtatus* are "moderate" in size (Pyle), of much greater extent than in coastal California's *megalonyx*. These Great Basin breeders move south in winter as far as southern California and Arizona; in Arizona, they are said to be as common in winter as the state's breeding race, *montanus*.

Resident inland from the southern edge of the range of *oregonus* into south-central California, *falcinellus*—the "little scytheman"—is slightly longer-tailed and longer-toed than that more northerly

Spotted Towhees on the northern Pacific Coast are very modestly marked above. *British Columbia, January. Brian E. Small*

The back spotting can be virtually invisible in some representatives of the subspecies *oregonus* from some angles. *British Columbia, January. Brian E. Small*

subspecies. The head and back of both sexes are slightly paler, and females may show a more pronounced brown tone than in *oregonus*; the rump in both sexes is grayish. The extensive white spots in wing and tail are moderate in size (Pyle).

The coastal race *falcifer*, the "scytheman," is resident from northernmost coastal California south to Santa Cruz County. Both its tail and its toes are slightly longer than in *oregonus*, which it otherwise resembles in plumage; the white spotting of tail and wing in *falcifer* is somewhat more extensive, and the spots themselves slightly larger, than in *oregonus*.

The common towhee of the southwestern mountains in the United States, *montanus*, occupies a much larger range, breeding from Inyo County, California, across the eastern two-thirds of Nevada and Utah into southeastern Idaho and southwestern Wyoming; this is also the breeding Spotted Towhee of high elevations in Colorado, most of New Mexico, Arizona, northeastern Sonora, and northwestern Chihuahua. Wintering birds occur east to western Texas. This is a long-tailed race, with medium-short toes; the upperparts are black in males, blackish in females. The spotting of the wing and tail is fairly extensive in both sexes.

The easternmost race in the United States and Canada, breeding across a range as large as that of *montanus*, is Swainson's *arcticus*. This widespread race is medium-tailed and small-footed, with slightly more sexual dimorphism in plumage than other races: the male's head and upperparts are dull black, while the female is noticeably dusky brown. Both sexes have a faint dull greenish tone to the rump. The wings and scapulars are extensively spotted with white, the spots of moderate size; the tail spots are also moderately large. Breeding birds occur from southeastern British Columbia east to southeastern Saskatchewan, south through eastern Idaho, eastern Wyoming, the western half of the Dakotas, and most of northern and western Nebraska. Most birds of this subspecies move south and southeast after breeding, regularly occurring to the eastern edge of the Great Plains in Iowa, Missouri, and Kansas; they winter commonly as far south as New Mexico and Texas, where they are variably common in most of the western two-thirds of the state. There are winter specimens from eastern Arizona and Utah, and it is almost certainly this subspecies that accounts for the sparse records of Spotted Towhees as far east as Florida and Nova Scotia.

As currently understood, the subspecies *megalonyx* is resident on the central and southern California coast from Monterey County south; it also occurs on Santa Cruz Island, and is sometimes said to be the towhee of Santa Rosa Island as well. As its name suggests—"huge nail"—this is a large-footed, long-toed race, with a medium-long tail. As in *oregonus* and *falcifer*, males are noticeably glossy black above, while females are matte slate on the upperparts and head. The size and shape of the white spots on the wings and scapulars varies from small to large; the tail spots are somewhat larger than in *oregonus* and *falcifer*, but smaller than in *montanus*.

The largest-footed of the Spotted Towhees is the subspecies *clementae*, named for its range, which includes Santa Catalina, San Clemente, and—according to some authorities, including Pyle—Santa Rosa Islands off southern California. The bill

The large-billed subspecies *magnirostris* inhabits southernmost Baja California. This juvenile shows the typical heavy streaking of *Pipilo* towhees. *Baja California Sur. David Krueper*

is also large, the tail medium-short. The plumage of both sexes is duller than that of *megalonyx*, the upperparts and head dark brown or slate in females and black in males; the rump is gray-tinged rather than black. The spotting of wings and scapulars is moderately extensive, the tail spots slightly smaller than in *megalonyx*.

The range of *megalonyx* is adjoined in northwestern Baja California by the subspecies *umbraticola* ("shadow-dweller"). Both sexes are notably black above and on the wing and tail, the female slatier than that of *megalonyx* or *falcifer*, without a brown tinge; at the same time, the white markings of the wing and tail are large and extensive. The rather small bill and feet average "decidedly blacker" than in any of the other West Coast races.

Restricted to a very small range in the mountains of southernmost Baja California Sur, *magnirostris* also resembles *megalonyx*, but is paler rufous on the breast sides and flanks, with larger white tail spots. Both the bill and the feet are larger than in *umbraticola*, to which it is not "bridged by any connecting links" in plumage or morphology.

The breeding range of *montanus* is bordered to the south by the Mexican subspecies *griseipygius* and to the southeast by *gaigei*. A resident of the Sierra Madre from central Chihuahua south to southern Durango and northern Nayarit, *griseipy-*

gius has a gray rump and uppertail coverts, the male with more olive on the back than in the otherwise similar *montanus*.

The mountains of eastern New Mexico, southwestern Texas, and northern Coahuila are the home of the race *gaigei*, named for Frederick M. Gaige, who was director of the University of Michigan's Museum of Zoology. This subspecies differs from *montanus*, which breeds to its north and west, and from *arcticus*, which winters in its resident range, in having paler but more extensive rufous on the breast sides, less white in the tertials and wing coverts, and a shorter wing and tail; the upperparts and head of the female are much blacker. The ranges of *gaigei* and *griseipygius* are not adjacent, though the two resemble each other; in *gaigei*, the male's uppertail coverts, lower back, and rump are more extensively black, and the breast sides are paler.

The northernmost subspecies in eastern Mexico is the appropriately named *orientalis*, which occupies the Sierra Madre Oriental from the Mesa de Chipinque in Coahuila east to southwestern Tamaulipas and south to eastern San Luis Potosí. It resembles *gaigei*, but the back is brighter black and the flanks and breast sides darker; the uppertail coverts and remainder of the upperparts are darker than in *griseipygius*. Males lack the olive tones above of *maculatus*, and differ from *montanus* in having the rump gray with considerable black flecking; *orientalis* is also less extensively spotted white above.

The highlands of central Mexico are inhabited by three subspecies with restricted ranges. Nominate *maculatus*, found in pines, oaks, and firs high in the Sierra Madre of Hidalgo and Puebla, is olive-toned above, with no black in the rump; the undertail, flanks, and breast sides are rather pale, and the spots and streaks of the wing and scapulars tend to whitish rather than yellowish.

A century and a quarter after Swainson described *macronyx* as a distinct species sympatric with *maculatus*, Charles G. Sibley determined that the type specimens of *macronyx* had in fact come from a different locality, which he formally restricted to "the western slope of the Volcán de Toluca, state of México." This bird, named for its long hind toe and nail, is resident in the alpine forests of the western border of the Valley of Mexico, west to the border with Michoacán; it is the only Spotted Towhee in its range. This subspecies differs from *vulcanorum*, resident immediately to its south, in its greener back with finer, less extensive black shaft streaking; the restricted spotting of the wing, back, and tail is pale yellow.

On the southeastern side of the Valley, especially on the volcanoes of Popocatépetl and Ixtaccíhuatl, *vulcanorum* is olive-greenish above, with more prominent black shaft streaks on the back. The spots of the wing, back, and tail are more prominent than in *macronyx*, "slightly clouded with yellowish." The greater amount of black above in *vulcanorum* "caus[es] head to merge gradually with dorsal coloration rather than to be sharply delimited as in *macronyx*." The two races are in contact at the southern edge of the Valley of Mexico, where they intergrade to produce individuals of intermediate appearance.

Three Spotted Towhee subspecies are currently recognized from Middle America south of Mexico's Transvolcanic Belt. Restricted to the highlands of Oaxaca, the "surprisingly well-marked race" *oaxacae* resembles the larger, longer-winged nominate *maculatus* but is paler and browner above, especially on the rump; the white markings above are more extensive, particularly the white tertial edges.

The abundant Spotted Towhee of pine-oak forests in north-central Chiapas, *chiapensis*, is darker than either *oaxacae* to the north or *repetens* to the south. The upperparts are buffy, and the white markings are "heavily clouded" with brown. The southernmost and easternmost populations of the species, *repetens*, occupy the Pacific highlands of Guatemala and Volcán Tacaná in Chiapas; this subspecies resembles the more northerly and westerly *chiapensis*, but is paler below and less buffy above, with clearer white spotting, while it is darker above than *oaxacae*.

The island races *consobrinus* and *socorroensis* are described in separate accounts below.

This Guadalupe Towhee, a male, was collected by W. E. Bryant in January 1886, hardly more than a decade before the population was completely wiped out by feral cats and other introduced mammals. *Baja California Norte, January. Courtesy American Museum of Natural History*

GUADALUPE TOWHEE
Pipilo maculatus consobrinus

When Robert Ridgway returned to Washington from a visit to his family in November 1875, he found Elliott Coues busily working up a number of birds sent to the Smithsonian by the botanist and naturalist Edward Palmer, who had spent the first months of that year collecting birds and plants on Guadalupe, a volcanic island 150 miles west of Baja California. Ridgway was incensed, and immediately wrote to ask Palmer to confirm to Spencer Baird that he had already assigned Ridgway "the first right" to describe and name any new forms among Palmer's birds.

It appears that Palmer, who had disliked and mistrusted Coues for the better part of a decade, wrote to Baird as requested. Ridgway—not Coues—published full descriptions of Palmer's "very interesting" Guadalupe birds the next year, finding "that every one of the resident species is distinct from any found on the neighboring main-land." Among the birds Ridgway described to science were two sparrows, the Guadalupe Junco and the Guadalupe Towhee.

The eye of living Guadalupe Towhees was red; it is unclear whether this specimen's glass eye has faded or was always bluish white. *Baja California Norte, January. Courtesy American Museum of Natural History*

Tactlessly, Ridgway made a point in his paper of complaining that Palmer's "notes accompanying the specimens are so meagre." The 25-year-old ornithologist should instead have counted himself lucky that Palmer and his birds returned from the Guadalupe expedition at all. The collecting party was to have been picked up by a schooner six weeks

Male Guadalupe Towhees tended to be sooty above rather than pure black; there was relatively little difference in plumage color between the sexes. *Baja California Norte, January. Courtesy American Museum of Natural History*

after arriving in February, but no ship arrived, and by May, Palmer and the others had "nothing to eat but old beans and goat meat [and] mustard leaves." By the time they were finally rescued, on May 15, the men were weak and sick, and they found it "a hard task to make the journey to the beach" to go on board. Palmer reached San Diego on May 20, "weighing 125 pounds; going on the Island, weighing one hundred and sixty." Palmer survived, but the Guadalupe Towhee would not be so fortunate.

Over a ten-day visit to Guadalupe in the early summer of 1897, members of the United States Commission on Fur Seal Investigations encountered a single Guadalupe Towhee, which they collected. It seems unlikely that that bird was the final survivor of its subspecies, but it was the last ever seen by scientists. Even by island standards, Guadalupe suffered dramatically as a result of the introduction of goats and housecats, and by the middle of the twentieth century, observers were no longer as much surprised by the towhees' extinction as by the fact that the bird had ever been able to exist on a piece of land now so thoroughly devoid of "shrubs or understory of any kind." Whether the last towhee fell to habitat loss or the claws of a feral cat will never be known, but it is certain that the Guadalupe Towhee has been extinct for a century or more.

FIELD IDENTIFICATION

The Guadalupe Towhee was a lightly spotted Spotted Towhee, the male plumage most closely resembling the Oregon Towhee *Pipilo maculatus oregonus*. Females were notably dark, thus more malelike than the females of other Spotted Towhee subspecies. The most striking distinctions were structural: both wing and tail were shorter than in mainland Spotted Towhees, and the nail of the hind toe longer.

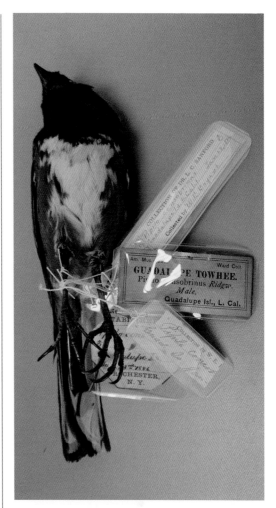

The surviving labels record this towhee's path from Guadalupe Island to Walter E. Bryant's collection in California to the commercial stock of Ward's Scientific in Rochester, New York, to the holdings of Leonard Cutler Sanford, thanks to whose generosity this and many other rare birds entered the collections of the American Museum. *Baja California Norte, January. Courtesy American Museum of Natural History*

It can be assumed that the vocalizations of birds in this population shared the whining, scratching quality of the songs and calls of other Spotted Towhees. Palmer, the first collector to encounter the island birds, told Ridgway that "when startled, they emit a short whistle of three or four syllables," a description difficult to allocate to any Spotted Towhee notes.

RANGE AND GEOGRAPHIC VARIATION

As its English name suggests, this sparrow was found only on Mexico's Guadalupe Island, 140 miles west of Baja California.

Decidedly olive above and brownish overall, with extremely limited back markings, the Socorro Towhee is unmistakable in its tiny island range. *Manuel Grosselet*

SOCORRO TOWHEE
Pipilo maculatus socorroensis

On May 19, 1867, the sloop carrying Andrew Jackson Grayson, his 16-year-old son Edward, and their 14-year-old assistant Cristobal put in at Socorro Island, 425 miles from their launching point in Mazatlán. The cove where they hoped to land was surrounded by steep rocky cliffs. On dropping anchor, the captain expressed "a great deal of uneasiness," and indeed, the boat was soon dashed onto the rocks, where its keel broke and the rest was threatened with disintegration by the wind and waves.

With only a small skiff intact, the little expedition was stranded. Grayson and his companions were able to get their guns, ammunition, and other collecting supplies ashore, and the manifest abundance of fish assured the group that they would not starve. But, Grayson later wrote,

> the contemplation of the hardships, toil and intense suffering in searching for water in a locality where it seemed extremely doubtful of success, filled my mind with the greatest anxiety . . . I felt pretty certain that the preservation of our lives depended on it.

Happily, Cristobal, "with demonstrations of the most lively joy," soon called Grayson over to a pile of rocks on the shore, which concealed a tiny spring of warm—but fresh—water.

Cristobal's discovery saved the expedition members' lives. But the credit was not his alone: Grayson wrote that "it was by observing the habits of" a "very abundant . . . and remarkably tame" finchlike bird "that water . . . was discovered; but for this incident we might have suffered greatly . . . had we not seen our little friend in the act of drinking." "For this providential service he was a welcome visitor and a privileged character" about camp.

The same welcome and privilege were not extended to a half dozen of the feathered benefactor's conspecifics, who were summarily collected by snaring them with a noose on the end of a long stick. On his return to the mainland, Grayson the next year described the specimens as representatives of a new sparrow species, which he named *Pipilo Socorroensis*. Grayson explained:

> I have given it the provincial name of the Island upon which it is found, meaning in the Spanish language, Succor, and which is the most appropriate name that could be given to it, from the fact that this bird led to the discovery of the only water found upon the Island . . . thus preserving the lives of the whole party.

Grayson gave his new species two alternative English names: the "Socorro Tawhee [sic] Finch" and "The Water Finder."

Grayson's party was unexpectedly rescued on May 28, when the crew of the *A. A. Eldridge*, bound for Chile, saw the smoke from a fire Grayson had set. Even shipwrecked, Grayson hesitated to leave the island after so short a visit, but climbed reluctantly aboard the *Eldridge* when he found his young son eager to go. That eagerness, sadly, would cost Edward his life: he was murdered "by the rude hand of some unknown assassin" on arriving in Mazatlán, an event that surely hastened Grayson's own death just two years later, on the eve of his fiftieth birthday.

Later nineteenth-century visitors found this bird "abundant" or "very common" across the island. In the 1990s, it was still said to be common to fairly common, and has recently been reported as common in forested areas at the island's higher elevations (eBird). Like many other oceanic islands, Socorro has suffered from the introduction of non-native vertebrates, including sheep, housecats, lizards, and rodents. Grayson himself had abandoned a pair of domestic hogs on his first visit to the island in 1865, and was "gratified" in 1867 to find the sow "very fat, and far advanced in pregnancy." By 2010, thanks to eradication efforts, the only invasive introduced mammals remaining on Socorro were the housecat and the house mouse. The immediate impetus for habitat restoration on the island is the scheme to reintroduce the severely endangered Socorro Dove (a species, now extinct in the wild, named *Zenaida graysoni* in memory of Grayson's son), but the removal of introduced predators and grazing ruminants is likely to benefit native ground-feeding birds such as the towhee as well.

FIELD IDENTIFICATION

The Socorro Towhee can be mistaken for none of the other nine native land birds present on the island. Like the Oregon Spotted Towhee and the extinct Guadalupe Towhee, it is only modestly streaked and spotted with white, the tail spots barely one-third as extensive as in the heavily marked *arcticus*, for example; Grayson remarked on their similarity in "general appearance" to the Eastern Towhee rather than to the western species.

The calls of the Socorro Towhee include a thin, hissing *tzee* and a harsh, rising and falling *tzhur-rEEa*. A clearer, more musical *doREE* recalls the contact notes of some Eastern Towhees from the southern part of that species' range.

The Socorro Towhee's songs, like those of other *Pipilo* towhees, are variable. The song typically begins with one to three call-like notes, harsh or sweet, and concludes with a louder, higher-pitched buzz or rattle.

RANGE AND GEOGRAPHIC VARIATION

The Socorro Towhee is found only on the eponymous island, a member of the volcanic Revillagigedo archipelago lying 250 miles southwest of the southern tip of Baja California.

Ground-nesting birds are tragically susceptible to habitat destruction by livestock and depredations by introduced housecats, reptiles, and rodents. Eradication efforts on Socorro should help the towhee and the island's other birds. *Manuel Grosselet*

The broad breast band, bright upperparts, and gray or whitish supercilium help distinguish this species from the smaller Chestnut-capped Brush Finch. *David Krueper*

COLLARED TOWHEE
Pipilo ocai

In 1847, Bernard du Bus de Gisignies, director of the new Royal Belgian Museum of Natural Sciences, commenced publication of a number of birds in that collection that he deemed new to science. The first lot of 15 new species, all collected in the New World tropics, included a Mexican bird du Bus named *Pipilo torquatus* ("collared"). The description presented an exotically handsome bird:

> A towhee with cheeks, sides of the head, and broad breast band black; crown rusty marked with black; throat and narrow supercilium white; nape and upper back olive spotted with black; lower back and rump rufous-olive; wings and tail yellow-ish olive above, the two outer tail feathers on each side spotted with yellow at the tip; center of lower breast and belly whitish; vent and undertail coverts rufous. Bill dark horn, feet yellowish.

Four years later, du Bus issued a colored engraving of the bird, as attractive as his Latin diagnosis had suggested.

Du Bus's Mexican towhee remained scarce in collections, however, and little known to European ornithologists. Osbert Salvin, the great British cataloger of Central America's nature, at first confused the towhee with another species, the White-eared Ground Sparrow (*Melozone leucotis*); Salvin even shipped mislabeled skins of the ground sparrow to the Smithsonian, where the error was at first overlooked and the specimens apparently deemed true representatives of the Collared Towhee.

Those misidentified skins were consulted by

George Newbold Lawrence when, in 1865, he received a male sparrow from the vicinity of Xalapa. Collected by the naturalist, author, and painter Rafael Montes de Oca, the specimen resembled neither the mislabeled birds supplied by Salvin nor the species described and depicted by du Bus. Lawrence, logically enough, described Montes de Oca's bird as new, naming it for its discoverer. A decade later, Salvin recognized that Lawrence had been misled by his misidentified ground sparrows, and he declared Lawrence's name *ocai* invalid, a synonym, younger by almost two decades, of du Bus's original *Pipilo torquatus*.

There matters rested until A. J. van Rossem reexamined the matter. Van Rossem observed that du Bus's

> original diagnosis and plate by no means describe or picture the collared towhee which for so many years has passed as *torquatus*; in fact they are so at variance that it is difficult to understand why they were ever confused.

On the eve of World War II, van Rossem traveled to Brussels to examine du Bus's type specimen, which he declared "an obvious hybrid between what has currently been called *torquatus* and some other *Pipilo*," probably a Spotted Towhee. Lawrence's type, on the other hand, was a true, "typical" Collared Towhee, without signs of hybridization or intergradation. Lawrence had been right to describe his Xalapa bird as new, and right to honor a distinguished Mexican colleague in the name *ocai*, which remains the species' official name today.

FIELD IDENTIFICATION

The visual resemblance between this species and the Chestnut-capped Brush Finch is startling, and there is essentially no portion of the towhee's range that is not also at least potentially occupied by the brush finch. Both species occur in dense brush and in damp, pine-oak and pine-evergreen forest, though the brush finch ranges significantly lower; it can be found locally down to 1,500 feet, while Collared Towhees prefer elevations above 3,000 feet.

The Collared Towhee is both larger and proportionally longer-tailed than the brush finch, with brighter olive-green upperparts, wings, and tail; the towhee's flanks are also paler and less contrasting, though this feature varies by subspecies as described below. The dark breast band is wider in Collared Towhees, especially of the races *ocai* and *alticola*; the black forehead is also more extensive,

Hybrids with the Spotted Towhee are frequent. This very striking individual combines the head and underpart pattern of a Spotted with the back and crown pattern of a Collared Towhee. *Veracruz, June. Amy McAndrews*

meeting the rufous of the top of the head on the forecrown. Except in the dark headed *nigrescens* and some *alticola*, the Collared Towhee shows a narrow but bright white or silvery gray supercilium extending from the bill base to the nape, neatly setting the black sides of the head off from the rufous crown. In the Chestnut-capped Brush Finch, the supercilium is shorter and duller yellowish green, fading into the nape; it is separated from the small white loral spot by a fine black spot or line intruding from the forehead.

Short-winged and long-tailed, with a sturdy neck and a stout, dark bill, the Collared Towhee is otherwise distinctive in its range. The principal identification difficulty once brush finches have been eliminated from consideration is posed by the existence in some areas of hybrids and intergrades with Spotted Towhees. A visually "pure" Collared Towhee will show an extensive chestnut cap, a solid greenish back without white spots or dark shaft streaks, a solid white throat (except in *nigrescens* and some *alticola*), brown or gray flanks, and no white in the tail. Presumed hybrids and intergrades may have a reduced crown patch; pale or white spots and edges on the back, wing coverts, and scapulars; noticeable dark shaft streaks on the back; an admixture of rufous on the flanks; and small to large white tail spots. Individual hybrids and backcrosses can show almost any combination of these features, each to a variable extent; some more closely resemble "pure" Spotted Towhees, others "pure" Collared Towhees, and it can be assumed that many of the descendants of hybrids are indistinguishable from one or the other of the parental types.

Adding to the complexity, hybridization and intergradation do not occur uniformly across the Collared Towhee's range. On the Cerro San Felipe of Oaxaca, Collared and Spotted Towhees commonly occur together without "the slightest hint of any crossing."

Occasional individuals that appear to be hybrids have been found on Mount Orizaba, straddling the border of Veracruz and Puebla; Sibley's 1950 study found signs of interbreeding in approximately 16 percent of the specimens collected there. Over the years, hybrids or backcrosses from this area of Puebla have repeatedly been named as new species, beginning with du Bus's *Pipilo torquata.* Robert Ridgway described two other specimens as *Pipilo submaculatus* and *Pipilo complexus,* respectively; he later synonymized both with the nominate race of the Spotted Towhee, but it is clear in each case that the specimens before him were of mixed ancestry. The type of Ridgway's *complexus* was "so near the precise intermediate point . . . that it may represent an example" of a first-generation hybrid, the offspring of "pure" Collared and Spotted Towhee par-

ents. Ridgway described yet another hybrid in 1894 as *Pipilo orizabae,* determining later that it, too, was an aberrant Spotted Towhee.

In 1948, Sibley found towhees "exceptionally abundant . . . the commonest species present" on the northern slope of Jalisco's Cerro Viejo. Of the specimens collected there, none was considered a truly "pure" representative of either species, and there was no indication of any preferential mating system in which Spotted-like or Collared-like individuals paired more frequently with similar birds. Instead, the entire population was deemed "a freely interbreeding one," exhibiting nearly the full range of plumage variation between the two species. This variability is explained by the "insular" nature of the Cerro Viejo population, into which "occasional immigrants" of either species inject new genetic material.

The situation is at its most complicated in those areas where the two species display a "gradient," in which a wide variety of visual hybrid types occupy a "long, narrow belt of suitable habitat between two large areas occupied by the parental species." This is especially striking in the *Pipilo* towhees of the long band of suitable habitat stretching across the Mexican Transvolcanic Belt from eastern Jalisco to western Puebla and Tlaxcala. The intermediate populations resident over these 300 miles begin in the east with Spotted-like birds and become more and more Collared-like to the west. Obviously, it is impossible to assign any of these populations to either of the "pure" species approached at the gradient's end points, but Sibley suggests that "for nomenclatural convenience" birds east of Morelia in Michoacán can be called Spotted and those west of that city Collared—though all are the products of intergradation and crossbreeding. Other, similar gradients, though on a smaller geographic scale, can be observed in eastern Jalisco and northwestern Michoacáan, from the Spotted-like birds of the Mesa de los Puercos south to the Collared-like birds at Mazamitla; and in Michoacán from the Cerro El Fraile to Las Joyas and Lake Pátzcuaro.

Whatever their genetic heritage, Collared Towhees resemble their Spotted and Eastern relatives in their secretive, ground-loving behavior. Keeping close to cover in the brushy thickets, fencerows, and coniferous forests they prefer, Collared Towhees are often first detected by their scratching and jumping through the leaf litter in search of seeds and insects and other invertebrates.

The calls of the Collared Towhee have been described as very similar to those of adjacent populations of Spotted Towhees; Sibley described the Collared Towhee's typical calls as shorter and sharper than the *zhree* or *jor-ee* of the Spotted. Calls recorded in Jalisco include a low, grumbling chatter; a very thin, slightly grating *dzeer*; and a thin descending whistle. That last call is likely a variant of what Howell and Webb note as a distinctive "high, clear, usually ascending whistle, *pseeeeeu* or *teeeeeu*."

Male Collared Towhees utter a wide variety of songs, all similar in tone and pattern to those of Spotted Towhees. One simple variant begins with a descending whistled call-like note, followed by a higher-pitched, rather musical trill; in a more assertive version, the introductory note is loud, brassy, and ascending, recalling the *chewink* of an Eastern Towhee. Longer, more complex songs are also frequently given, in which the trill is preceded or, less often, followed by several chips at different pitches and of different tones; a given song type may be repeated for several minutes before the singer switches to another.

RANGE AND GEOGRAPHIC VARIATION

The Collared Towhee is an endemic resident of the mountains of central and southern Mexico, where its range is fairly large but broken into discrete segments. As a whole, the species occurs from northwestern Jalisco and northeasternmost Colima to west-central Veracruz and southern Oaxaca, through central Michoacán, the Sierra Madre del Sur of Guerrero, and eastern Puebla.

Variation in this species is clouded by the prevalence in some parts of its range of hybrids with the Spotted Towhee. That circumstance notwithstanding, five subspecies are recognized.

The nominate race *ocai* occupies the mountains of eastern Puebla and west-central Veracruz, from Teziutlán south to Zoquitlán. It has grayer flanks and darker undertail coverts than the similar subspecies *brunnescens* and *guerrerensis*; it differs from *alticola* and the very distinctive *nigrescens* in the presence of a thin white line dividing the black forehead.

The high mountains of Oaxaca are occupied by the race *brunnescens*, with more brown and less gray in the flanks. The forehead stripe is well developed. The range of this subspecies is entirely disjunct, nowhere adjoining the range of any other Collared Towhees.

The flanks of *guerrerensis*, resident in the Sierra Madre of Guerrero west of Chilpancingo, are browner still; the undertail coverts are paler than in either *ocai* or *brunnescens*. The breast band is narrow but fully developed.

The race *alticola*, first collected at high elevation in the Sierra Nevada of Colima, was originally described by Salvin and Godman as a distinct species. Ridgway was the first to classify it as a dark subspecies of the Collared Towhee. It differs from the three paler races in lacking the forehead stripe and having only a very narrow supercilium; the breast band is very wide.

The completely white throat and extensively rufous crown distinguish *alticola* from *nigrescens*, which was also initially considered a separate species; subsequently thought variously to be a "good" species or a hybrid population, its true relationship to the other Collared Towhees was not determined for nearly 60 years after its discovery. This subspecies is restricted to Michoacán, from Cerro San Andrés to Mount Tancítaro. The white throat patch in *nigrescens* is reduced, vestigial, or even absent, and there is no white supercilium. The crown sometimes lacks rufous; the flanks are tinged rusty, and the tail sometimes shows weak terminal spots. These features are generally attributed to the "effects of infiltration of" Spotted Towhee genes, and indeed, Sibley eventually considered *nigrescens* simply an extreme in the hybrid "gradient" connecting the two species across the northern end of the Mexican Plateau.

Few sparrows are as well named as this colorful species of dry brushland; in the field, however, the white throat is often more immediately conspicuous than the yellow-olive of the tail. *California, June. Brian E. Small*

GREEN-TAILED TOWHEE
Pipilo chlorurus

This colorful little towhee, justly praised by John Cassin as "the handsomest bird of the family of Sparrows yet discovered in the United States," is a familiar breeding bird of the brushlands of the Great Basin and a common winterer across southwestern North America. In Cassin's mid-nineteenth-century day, however, this exotically beautiful sparrow was little known indeed, with no more than four adult specimens available to scientists and natural historians in the eastern United States.

The first was a male shot in central Arizona in September 1842 by the 19-year-old William Gambel; on his return to Philadelphia, Gambel named his "new and singularly marked species" *Fringilla Blandingiana,* in honor of William Blanding, owner of one of the largest private natural history collections in the United States and an important benefactor of the Academy of Natural Sciences. This was the individual pictured for Gambel by J. H. Colen four years later in the first published illustration of the species; almost certainly, the same specimen

served as the model for George White's drawing lithographed for Cassin's *Illustrations.*

The next Green-tailed Towhee was not collected until early April of 1851, when Samuel Washington Woodhouse took one near San Antonio, Texas. He later found the species "quite abundant" in the dry mountains of western New Mexico, but for some reason declined to collect more. Serving as naturalist to the Williamson Railroad Survey of 1853, Gambel's Philadelphia acquaintance Adolphus Lewis Heermann expanded the specimen record west to California when he shot a single individual from a mixed flock of sparrows in the Tejon Valley; he had earlier secured a specimen in the Sacramento Valley.

In his 1847 account of the species, Gambel raised for the first time the possibility that the bird he had described as a new species four years earlier might be identical to a sparrow collected by John Townsend in the Rocky Mountains in July 1834. That specimen, the only one of its kind Townsend ever encountered, was described by Audubon in the appendix to his *Ornithological Biography,* where he named it *Fringilla chlorura,* the Green-tailed Sparrow. Both Heermann and Woodhouse shared Gambel's suspicion. While Woodhouse included Audubon's name—with a question mark—in the synonymy for *blandingiana,* Heermann asserted their identity without comment.

John Cassin, though, would have none of it. In his *Illustrations* of western birds, Cassin rejected the possibility that the Green-tailed Sparrow and Blanding's Finch were the same species. He admitted that the two resembled each other "in some respects," but proceeded to claim that they were not even of the same genus. Townsend's description, as relayed by Audubon, he deemed insufficiently precise; oddly, Cassin also asserts that Townsend's was purely a sight record and that he had "procured no specimens."

Townsend had indeed shot his bird, as Audubon's account states explicitly. It was not, however, among the Townsend specimens that ended up in the collections of the Academy of Natural Sciences in Philadelphia, and by 1847 that specimen was clearly unavailable for study—at least to Cassin, Gambel, Woodhouse, and Heermann, all of whom cite Audubon's published description as their only source of information about the bird.

The skin finally reemerged, in the collection of Spencer Baird, who had presumably received it as a gift from Audubon; Baird, too, seems to have forgotten its existence, as he himself was still using the name *blandingiana* as late as 1852. Sometime between then and 1858, though, Baird relocated the Townsend specimen, which he was able to compare with 13 other skins that had meanwhile entered the collections of the Smithsonian. One of those 13, collected in August 1856, was a male beginning the molt from its juvenile plumage; here at last was the missing link between Townsend's very young bird and the adult specimens known to the Philadelphians, and Baird declared the Blanding finch and the green-tailed sparrow "unmistakeably" the same species, settling "the question in favor of the priority of the name *chlorurus,*" which Audubon had given it a quarter century before.

FIELD IDENTIFICATION

The adult Green-tailed Towhee is easily identified once seen well; even in the briefest glimpse, the greenish wings and tail, deep rusty crown, and dazzling white throat reveal the identity of this reclusive, ground-loving sparrow.

That combination of characters occurs in no other bird north of the Mexico border. In the dim light so often favored by this species, it might conceivably be mistaken for a Rufous-crowned Sparrow, but that bird is slightly smaller, darker, and more uniformly somber above and below. The towhee's crown and throat pattern can also distantly recall that of a Swamp Sparrow; each is rare in the other species' normal range, and Green-tailed Towhees are as infrequently encountered in cattails

and wet ditches as are Swamp Sparrows in brushy mesquite bosques and chaparral. The Olive Sparrow shares the towhee's greenish upperparts, but is smaller and much plainer headed, with little contrast between the throat, breast, and cheek.

In its Mexican range, the Green-tailed Towhee may overlap with the Collared Towhee or the Chestnut-capped Brush Finch. These greenish sparrows share the towhee's rusty cap, but both have very different face patterns, with strongly contrasting black sides of the head. The brush finch is roughly the same size as the Green-tailed Towhee, while the Collared Towhee is considerably larger than either.

Green-tailed Towhees are medium-large sparrows with long, rather narrow tails and fairly short wings; a close view of a perched bird reveals that the dusky primary tips extend beyond the greenish secondaries. The belly is deep and the breast full; combined with the smoothly rounded back, the body sometimes gives the impression of an overstuffed oval with a long tail. The rather large head is noticeably square unless the crown feathers are raised in alarm or curiosity; the dark bill is relatively long and slender.

Juvenile Green-tailed Towhees are very different from their parents. Held through August, the juvenile plumage is largely brown, with only a slight greenish tinge to the secondaries and tail feathers. The upperparts, head, and most of the underparts are finely and extensively streaked with black, setting off a narrow white eye-ring. The center of the belly and, especially, the throat are whiter than the rest of the underparts and less conspicuously streaked.

Juvenile Green-tailed Towhees undergo a partial preformative molt before leaving the breeding grounds in autumn. They typically retain the brownish tertials and tail feathers of the juvenile plumage, however, until late the following summer, making it possible in good views to age some birds in their first year of life by the contrast between the new, greener secondary coverts and the retained brownish tertials (Pyle).

The best-known vocalization of the Green-tailed Tohwee, given at any time of year, from the ground, from a low perch, or even in flight, is a "loud and distinct *mew-wée.*" Traditionally compared to a cat's mewling, this note is a clear, ascending squeal, very like some of the flight calls of the Franklin Gull. The other frequently heard call is a long, decidedly buzzy *dzeee,* thinner and less strongly modulated than the corresponding call of the Spotted Towhee, longer and more strongly modulated than the corresponding calls of White-crowned and White-throated Sparrows. Nesting adults also give a sharp *tick* alarm note and an abrupt, slightly popping *poitt.*

The song, given by males beginning in spring migration, is extremely variable, with some local populations exhibiting more than 50 distinguishable song types. The song is commonly three-parted, beginning with a variable introduction of two to four sweet sliding notes; this is usually followed by a slower, variably buzzy or trilling note. The song typically ends with a faster, rasping buzz, often recalling the end of a Spotted Towhee's song. Some songs or singers omit the central portion, and a few may replace the warbled introduction with a buzzing trill.

RANGE AND GEOGRAPHIC VARIATION

Breeding Green-tailed Towhees are found in a wide range of brushy, semi-open habitats over most of the western United States, at altitudes from 3,000 to 12,000 feet. Decidedly a bird of dry, but not barren, places, the species breeds from southeast Washington, eastern Oregon, southern Idaho, and southwestern Montana south through most of Wyoming, western Colorado, and Utah to the northern half of New Mexico, the mountains of northern and central Arizona, and the central and eastern mountains of California. The northernmost breeders are in the Blue Mountains of Washington and northern Oregon and in south-central Montana. The southernmost occur in San Diego County, California; the Sierra de San Pedro Mártir of Baja California Norte; the White Mountains of Arizona; the Sacramento and Mogollon Mountains of New Mexico; and, at least occasionally, the Guadalupe and Chisos Mountains of west Texas. The eastern edge of the species' breeding range essentially follows the eastern foothills of the southern Rocky Mountains from Blaine County, Montana, south to central New Mexico; there are regular summer records from the Wyoming Black Hills, and the species may also breed or have bred in the South Dakota Black Hills.

The fall migration begins early, with some birds moving to higher elevations before beginning the southbound flight. By the first of August, Green-tailed Towhees are being seen out of their breeding areas in the desert Southwest. Most birds have left the breeding areas by the end of September; arrival on the Mexican wintering grounds typically extends from August to October. Most migrants appear to take an essentially straight north-south route in the autumn, though a few occur in September on the western Great Plains slightly to the east of the breeding range.

The wintering distribution of this species can show marked local variation from year to year, presumably depending on the productivity of amaranth and other favored winter seeds. Sensitive to cold and snow cover, Green-tailed Towhees winter north irregularly and in small numbers to the Great Basin sagebrush flats of southern Nevada, Utah, and southwestern Colorado. Numbers in California are typically highest in coastal San Diego County and in the southeastern deserts. This species is fairly common from September to March most years throughout Arizona south of the Mogollon Rim and in southern and southwestern New Mexico. Texas winterers are most frequent in the western Panhandle, Edwards Plateau, and the brush country of south Texas, but in some years can be found virtually throughout the state, east and south to the mouth of the Rio Grande.

The greater part of the Green-tailed Towhee's winter range is in Mexico, including nearly all of the country north of the Transvolcanic Belt; north of Oaxaca, it is regularly absent only from southeastern Tamaulipas, Veracruz, and the coasts of Guerrero, Michoacán, and Jalisco.

At close range the head pattern is quite complex, and the bill rather small for a towhee. *Arizona, April. Brian E. Small*

Outside of the usual winter range, there are scattered winter records of the Green-tailed Towhee from much of the United States and southern Canada. Successful wintering has been recorded at feeders as far northwest as Jasper, Alberta (eBird). There are also records from November and December for southern Manitoba, northern North Dakota, northern Michigan, southern Quebec, central Maine, and New Brunswick (eBird); even at well-stocked feeders, most vagrant individuals at these latitudes can be expected to die or withdraw as the weather grows colder and the snow deeper, while the rare birds that appear in early winter from the eastern Great Plains to southern New England and the mid-Atlantic states are more likely to survive into February and beyond.

Small numbers of Green-tailed Towhees winter irregularly on the Texas coast, more abundantly in "irruption" years. Reports are significantly fewer and less frequent eastward on the coast of the Gulf of Mexico. Florida has at least ten records of the species between October and April, a remarkable six of them from the single winter of 2011–2012. The southeasternmost records for the species are from Cuba and Cayo Coco in the Greater Antilles.

The return migration in spring is apparent in southern wintering areas as early as March, with most birds arriving on the breeding grounds in late April and May; some linger in the winter range in the southwestern United States and northwestern Mexico into May or even early June. Apparently overshooting birds occur north to northwestern Washington and southern British Columbia and Alberta. Springtime records east of the Great Plains that do not obviously pertain to overwintering towhees are infrequent, but include three May and June individuals from Cape Sable and Sable Island, Nova Scotia, a decided stretch of the concept of the "overshoot."

In 1930, Elizabeth Beardsley Bingham Blossom funded a collecting trip to south-central Oregon to benefit the Cleveland Museum of Natural History. Among the specimens sent back from the Warner Valley were 29 Green-tailed Towhees of both sexes. Harry Oberholser determined that those birds differed from the Green-tailed Towhees of the Rocky Mountains in their grayer back, flanks, and breast, and named them as a distinct subspecies, *Oberholseria chlorura zapolia* ("quite gray"), breeding north to central Oregon, east to northeastern Nevada, and south to southwestern Nevada and the San Jacinto Mountains of California.

Alden H. Miller determined that those differences were more likely due to seasonal wear than to geographic variation, and Oberholser's *zapolia* was omitted from the 1957 edition of the AOU *Check-list* and the 1970 volume of Peters and Paynter's *Check-list of Birds of the World*. The Green-tailed Towhee is now considered monotypic by all authorities.

AMERICAN TREE SPARROW
Spizelloides arborea

This handsome, cheerful-voiced bird is familiar at one season or another across much of North America. As common and widespread as it is, however, it took science the better part of a half century to unravel the identity of the American Tree Sparrow.

Modern sources tend to dismiss the name "tree sparrow" as "a misnomer," and to blame its coinage on "early European settlers" who discovered in the American bird a "superficial resemblance" to the Eurasian Tree Sparrow. In fact, however, the name "tree sparrow," in its application to both birds, is entirely a book name, created not by colonial farmers clearing forests and brush but by European natural historians at work in their specimen cabinets.

In 1760, George Edwards published an account of what he named the Mountain Sparrow, illustrating it with an attractive colored engraving of two birds—one a male, the other, he supposed, the female. The male, clearly recognizable as what we know today as the Eurasian Tree Sparrow, had been shot in England. The putative female, though, had been sent from Pennsylvania by William Bartram, the Philadelphia writer and collector with whom Edwards and so many other British naturalists were in correspondence. Edwards's painting of this second bird is a remarkably good depiction of an American Tree Sparrow, down to the details of bill color and tail pattern. But Edwards, in what William Swainson would later call "one of the very few mistakes" ever made by that very careful ornithologist, identified both birds as individuals of his Mountain Sparrow, *Fringilla montana*, not suspecting that the specimens before him in fact represented two quite distinct species.

For a dozen years thereafter, Edwards's colleagues agreed. Writing in that same year of 1760, the French ornithologist Mathurin Brisson too asserted that the sexes of the Eurasian Tree Sparrow—Edwards's Mountain Sparrow—differed, just as do the sexes of the similar and closely related House Sparrow, signally in the absence of black on the female's head and throat. Johann Reinhold Forster was the first European scientist to express doubts about the identity of the two sparrows in Edwards's plate. On examining a specimen sent to him from Hudson Bay, Forster reluctantly identified it as the female of Edwards's Mountain Sparrow, "as it has no black under the throat and eyes, and no white collar." But he also observed "many differences" between the Canadian skin and the Mountain Sparrow, and admitted that he had been at first "inclined to make this bird a new species"; he forbore only because "the

An American Tree Sparrow gazes lovingly on her "mate," a Eurasian Tree Sparrow. *George Edwards,* Gleanings of Natural History. *Image from the Biodiversity Heritage Library. Digitized by National Library Board, Singapore.*

specimen sent over was not in the best order, and might be a female." In 1785, Thomas Pennant, later praised by Alexander Wilson as "that judicious and excellent naturalist," had the opportunity to look closely at specimens from Canada and the United States in the collections of Anna Blackburne and Ashton Lever; he found that all "agreed in marks and colors," and declared the American bird a species distinct from Edwards's male Mountain Sparrow. Pennant had already renamed the European bird the Tree Sparrow noting that it was more often found in woody vegetation than around houses, and now he rechristened the New World species, rather unhelpfully, the Tree Finch.

In 1810, Alexander Wilson Americanized Pennant's Tree Finch into the Tree Sparrow, and translated the name into scientific Latin as *Fringilla arborea*, justifying it by the tendency of wintering birds to "if disturbed take to trees . . . contrary to the habit of most of the others, who are inclined rather to dive into thickets." Wilson explicitly corrected Edwards's 50-year-old error in classification, but pointed out at the same time that United States natural historians had not entirely escaped confusion themselves:

> By some of our own naturalists this species has been confounded with the Chipping Sparrow, which it very much resembles; but is larger and handsomer; and is never found with us in summer. The [Chipping Sparrow] departs for the south about the time that the [American Tree Sparrow] arrives from the north; and from this circumstance, and their general resemblance, has arisen the mistake,

a mistake far more understandable than the one made 50 years before when the American Tree Sparrow was thought to be the female of a visually dissimilar Eurasian species.

FIELD IDENTIFICATION

The American Tree Sparrow is distinctive, especially by the standards of passerellid sparrows, and unlikely to be mistaken by experienced observers for any other brown bird. As the historical record shows, however, those who have not had the opportunity to become familiar with the species—from eighteenth-century museum ornithologists to twenty-first-century beginning birders—may on first acquaintance find it confusingly reminiscent of a number of other brown birds.

Few observers today are likely to be as badly misled as was Buffon in the mid-eighteenth century, who found in the American Tree Sparrow the diminished, degenerate reflex of Europe's Rock Petronia. But new birders and birders who do not see the species regularly continue to struggle to distinguish this species from other ruddy-capped brown birds.

Most of that confusion is the result of the stubborn birding tradition of identification head-first. Formative-plumaged White-crowned Sparrows—Brisson's *canadensis*—have broad rusty lateral crown stripes, but the tail and wingtip are both brownish tan, unlike the slaty wing and tail feathers of the American Tree Sparrow. Field Sparrows, with a pale rusty crown pattern and grayish neck, are much smaller and longer-tailed, with an overall pinkish orange aspect quite unlike the elegant chestnut and gray elegance of the significantly more gregarious Tree Sparrow.

The sparrow most frequently confused with this species is the tiny, slim, fine-billed Chipping Sparrow. As early as 1810, Alexander Wilson observed that "some of our own naturalists . . . confounded" the two species, though the American Tree Sparrow "is larger and handsomer." Historically, over much

of their shared range, the Tree Sparrow arrived in autumn just as the last of the Chipping Sparrows were leaving for their wintering grounds farther south; this has changed in recent decades with the expansion of the Chipping Sparrow's winter range to the north, but the two are still sometimes thought of as seasonal replacements, a notion long encouraged by the alternative name "winter chippy" for the American Tree Sparrow.

The only similarity between the two species is the chestnut crown, which is retained year-round by the Tree Sparrow but replaced in the winter Chipping Sparrow by a dull chestnut or brown crown with fine blackish streaks. At any season, the Chipping Sparrow's tail is longer, narrower, and more deeply forked, and its rump is silvery gray or bluish, quite unlike the buffy rump of the American Tree Spar-

The rich chestnut back and wing contrast with the adult American Tree Sparrow's dove gray underparts. *Alaska, June. Brian E. Small*

The American Tree Sparrow's long, rather narrow tail is blackish with narrow white edges to the outer tail feathers. *Alaska, June. Brian E. Small*

row. The Chipping Sparrow's back is cold brown with coarse black streaks, and its underparts are uniformly gray or pale buffy, without the contrast shown in the Tree Sparrow between the gray belly and warm peach-buff flanks and breast sides. The American Tree Sparrow has a complete but very narrow whitish eye-ring, behind which a narrow reddish eye line extends nearly to the nape; the winter Chipping Sparrow's eye-ring is broken by a coarse blackish eye line reaching from bill base to nape. The Tree Sparrow's rather stout bill is "swollen" at the base and always dark above; in winter, the Chipping Sparrow's thin, evenly pointed black bill is dull horn-pink on both upper and lower mandibles.

The Swamp Sparrow shares with the American Tree Sparrow a gray nape, rusty crown, buffy sides, and yellow-based lower mandible; Swamp Sparrows may also show a central breast spot. They are noticeably chunky, however, with a shorter, broader tail

that is usually rounded or square at the tip, unlike the long, slender, deeply notched tail of Tree Sparrows. Swamp Sparrows are considerably less gregarious than Tree Sparrows, maintaining a certain individual distance even when large numbers are present; both species may frequent the same brushy ditches in fall and winter, but while American Tree Sparrows fly to an open, conspicuous perch when disturbed, a Swamp Sparrow will more typically seek shelter in the dark depths of the vegetation. Tree Sparrows, like the *Spizella* sparrows, are usually vocal when flushed, unlike Swamp Sparrows, which rarely call except from a perch.

Certain of the tropical *Peucaea* sparrows, especially the Rufous-winged Sparrow, might recall an American Tree Sparrow with their rusty crowns and eye lines and thick, yellow-based bills. Those birds have grayer backs and tails, with plain wings and underparts. There is at present no range overlap between any of these birds and the American Tree Sparrow.

In the field, American Tree Sparrows usually appear medium-sized, their true bulk belied by the long, narrow tail. Perched, they show a full belly and breast, rather rounded back, and long, pointed wingtip. The head is large and smoothly domed, though birds sometimes raise the crown feathers into a puffy, rounded crest.

This is a richly colored, contrastingly marked bird. At any distance, the tail is conspicuously dark and the head conspicuously gray. The upperparts and wings are deep bright chestnut, darker in eastern breeders. In winter, American Tree Sparrows frequently droop the scapulars over the inner wing, concealing the median coverts and leaving visible only a single bright white wing bar. The bright orange-buff of the flanks and breast sides is broadest at the top, bulging toward the center of the underparts to create an irregularly shaped partial breast band; that pattern is both more readily visible and more useful to identification than the sooty gray spot on the center of the breast.

The long blackish tail is a useful character when American Tree Sparrows flush from the ground. The outermost tail feather is edged in buff or white; this can be very noticeable in some birds, especially those of the western subspecies *ochracea*.

At closer range, the simple but appealing head pattern can be appreciated. Largely gray from nape to lore, the head is marked with a fine rufous eye line reaching back to separate the gray ear coverts from the very broad gray supercilium. The crown is dull reddish rust, browner than the decidedly red crown of an alternate-plumaged Chipping Sparrow; in winter, many American Tree Sparrows show a gray central crown stripe, usually narrow but occasionally covering as much as a quarter of the crown.

American Tree Sparrows are quite vocal, even in winter, when the members of a flock often join in a rollicking chant of two- or three-syllable phrases *tlee-dee, tleedle-dee*. The alarm call, given by birds perched or in flight, is short and harsh, with an emphatic attack and very fast decay; it resembles the short flight calls of the Chipping Sparrow, but is briefer, with a more emphatic beginning and less prolonged ending.

Male American Tree Sparrows regularly sing on warm, calm days in late winter, and April migrants may sing persistently. Though the song is variable, individual males sing only a single type. All song types include a slow introductory series of sweet, clear notes, lower-pitched but otherwise reminiscent of a Brown Creeper's song; most songs also have a fuller-voiced, accelerating second phrase, the final note often the highest-pitched of the song. In some birds, that second phrase ends with a sweet, slow tremolo followed by a high note. The clear, measured introduction is nearly distinctive, but can be confused with the imperfectly heard song of a distant fox sparrow.

RANGE AND GEOGRAPHIC VARIATION

The American Tree Sparrow breeds across Alaska and the northern Canadian territories and provinces north to the tree line. In Alaska, the species nests from the Brooks Range south to the coasts of Bristol Bay, the Gulf of Alaska, and the northern Panhandle. Breeding birds are found in British Columbia south to the southern edge of the boreal mountains, just south of the Stikine River. American Tree Sparrows nest throughout the Yukon and Northwest Territories and mainland Nunavut. There is also a "small, isolated population" in north-central Alberta, and the species breeds fairly commonly in northeastern Saskatchewan and far northern Manitoba. In the East, American Tree Sparrows breed in the Hudson Bay lowlands of extreme northern Ontario and in Quebec south to approximately the latitude of Lake Mistassini. The easternmost breeders are in Labrador, primarily on the central coast but also in appropriate habitats to the north and south.

The southbound migration begins by the end of July, and the northernmost breeding areas have been abandoned entirely by mid- or late October, which is also a typical arrival date across the wintering grounds. Rare individuals and small flocks have been seen in August and September west of the breeding areas and the normal migration route on Alaska's St. Lawrence Island. This species is very social in autumn and winter, occurring in single-species or mixed flocks of dozens or hundreds of often very vocal individuals.

Wintering birds are most abundant on the central and eastern Great Plains, where many Christmas Bird Counts regularly tally more than a thousand individuals. The species is less common west of the Rocky Mountains, and is notably scarce at most locations on the Pacific Coast.

The southern limits of the winter range vary from year to year, depending on snow cover and the availability of food. In California, winterers are regular in the Klamath Basin and south to Lassen County; they are "very rare and erratic" elsewhere in the state, especially in the Central Valley, the southern deserts, and in the southeast. Arizona winterers are almost entirely restricted to areas north and northeast of the Mogollon Rim, though there are recent November and December records from the White

The colorful upperparts are neatly set off from the gray and rusty head. This summer individual's pale tail is probably the result of plumage wear. *Alaska, June. Brian E. Small*

Mountains (eBird). New Mexico records are concentrated in the northern two-thirds of the state, with a few early-winter reports from as far south as Eddy County; a mid-March record from Doña Ana County is exceptional, and probably pertains to a bird that wintered locally undetected (eBird).

Local and uncommon in the Texas Panhandle, American Tree Sparrows are now casual or accidental elsewhere in the state, with records documented as far south as Cameron and Nueces Counties, the southernmost records of the species anywhere. The regularly wintering race is *ochracea*; *arborea* has also been documented in the northeastern region, west to Denton and Dallas Counties.

There are relatively few records of this species from Louisiana to South Carolina, though American Tree Sparrows are regular winterers across the band of states just to the north, from Arkansas to North Carolina. The half dozen recorded from Florida include single individuals in Wakulla, Escambia, and Brevard Counties; there is also a January report from Okaloosa County (eBird).

The northern extent of the wintering range is equally variable, but regular wintering takes place from southeastern British Columbia across the southern tier of Canadian provinces to Quebec, New Brunswick, and Nova Scotia.

The spring migration begins in March, with the largest number of migrants on the move in early or mid-April in most areas. Arrival on the breeding grounds appears to be timed to coincide with the spring thaw, when food resources and nesting sites become available again as snow cover recedes. Overshoots have occurred on St. Lawrence Island and on the Chukchi Peninsula and Wrangel Island in Russia.

American Tree Sparrows from the western portions of the breeding range are larger and paler than those that breed farther east. Although this variation is "weak and probably clinal" (Pyle), two subspecies are generally recognized.

The nominate, eastern race, *Spizelloides arborea arborea* (formerly *monticola monticola*), has less conspicuous pale edging to the outermost tail feathers; that edging is typically gray or buffy, rather than whitish or even bright white as in the western subspecies, *Spizelloides arborea ochracea*. The back of eastern breeders averages grayer and more coarsely streaked than in *ochracea*.

The back and rump of *ochracea* tend to be a paler sandy brown; the back streaking is more sharply defined, and the black streaks show no chestnut edges. The nape is paler than in *arborea*, in some individuals more clearly tinged brownish than gray. The rufous of the crown in *ochracea* is said to be paler, invaded by gray at the rear and in the center, though the color and pattern of the crown appear to vary with age, sex, and season in both subspecies. The flanks, belly, and sides of the breast may average browner or yellower in *ochracea*.

The smallest female *arborea* measure nearly three-quarters of an inch shorter than the largest male *ochracea,* but the average size difference between the two subspecies is only slight, and unlikely to be discernible in the field.

The western subspecies was first collected on the wintering grounds, by Charles Bendire. Breeding *ochracea* are found in Alaska, the Yukon, and norhtwestern British Columbia; the rest of the breeding range, from the eastern Mackenzie River basin east to Labrador, is occupied by the nominate subspecies. The wintering ranges of the two populations are sometimes said to be divided by the Rocky Mountains, but *ochracea* is known in winter on the eastern Great Plains from Saskatchewan to Texas and even east to the Mississippi; in Nebraska, *ochracea* and apparent intergrades are said to be significantly more abundant than are pure individuals of the eastern race. The uncertainty likely reflects nothing more than the subtlety required to distinguish the two subspecies—if they are reliably distinguishable at all.

The rusty rump and tail and bold, discrete red striping on the back and flank offer a striking contrast to the gray lower back and heavily marked gray head. *Alaska, June. Brian E. Small*

RED FOX SPARROW
Passerella iliaca

In 1786, Blasius Merrem, professor of mathematics and physical sciences at Duisburg and later the first professor of zoology at the University of Marburg, issued the second fascicle of a volume of portraits and descriptions of rare and little-known birds. Most of the colorful tropical rarities had been acquired from the itinerant English dealer Dolmer, but a few specimens had been sent directly to Merrem by his contacts in the field. A farmer near Göttingen furnished the professor with a White-tailed Eagle, while a large, brightly colored sparrow came from much farther away: "This new species of finch was sent from North America by a Hessian soldier and transferred into my collection."

Their reputation as brutish, bloodthirsty mercenaries notwithstanding, certain of the German soldiers who fought for the British during the American Revolution ultimately made significant contributions to American natural history. Friedrich von Wangenheim made a thorough study of the new continent's trees, while Johann David Schoepf stayed on after the war to retrace the naturalizing travels of Mark Catesby. And at least one was collecting birds for European ornithologists.

Unfortunately, we do not know the name of Merrem's correspondent, or even the locality where he collected the new finch. Charles Richmond believed, for reasons unstated, that Merrem's bird "was a winter specimen taken somewhere between New York and Washington." In 1946, Harry C. Oberholser restricted the type locality to Quebec, not on the basis of any new historical insight but because the plate and description provided by Merrem "seem to indicate that they were derived from a specimen of this species from eastern North America" rather than from a site farther to the west.

Though the Canadian sparrow still bears the name Merrem gave it—*iliaca*, "spot-flanked"—the species had not gone entirely unnoticed earlier. Before 1760, the curate of Saint-Louis en l'Isle in Paris, Jean-Thomas Aubry, had added to his famous natural history cabinet a large bunting from Canada, which Mathurin Brisson would later describe:

> It is the size of our Corn Bunting . . . . The upper part of the head is marked chestnut and brown . . . . The nape and back, along with the scapulars and coverts of the upper wing, are marked with the same

The Red Fox Sparrow depicted here, shipped home by an anonymous German mercenary, is the "type" of the species. *Image from the Biodiversity Heritage Library. Digitized by Missouri Botanical Garden.*

colors, mixed with a bit of gray . . . . The throat, the lower neck, the breast, the belly, the sides, and the thigh feathers are off-white, marked with chestnut spots, more sparsely on the belly than elsewhere. The flight feathers of the wing are brown, their outer edges gray shading to chestnut. The twelve tail feathers are of the same color as the wings.

The accompanying engraving, by François Nicolas Martinet, shows a long-tailed rusty and gray sparrow with heavily marked underparts, one that well merits the vernacular name Buffon would give it in 1778: *cul-rousset*, the red-tail. Characteristically, neither French ornithologist assigned the bird a Linnaean binomial, leaving it to Merrem—and, indirectly, to George III—to formally introduce the Red Fox Sparrow to science.

FIELD IDENTIFICATION

This rotund, long-tailed, and colorful sparrow can hardly be mistaken for any other when seen well; indeed, it is hard to suppress the suspicion that most of the identification confusion about the Red Fox Sparrow has been manufactured by field guide authors eager to fill the rubric "Similar Species."

Only at the edges of its range is the Red Fox Sparrow likely to be taken for a Song Sparrow—and even then only by observers unaware of the variation exhibited by that latter species. The very large, long-billed Song Sparrows of the Alaska islands, recalling a fox sparrow in their impressive bulk, are somberly colored and diffusely marked, entirely unlike the bright rust, gray, and white of the Red Fox Sparrow. The more pronouncedly rusty Song Sparrows of the Pacific coast of British Columbia, Washington, Oregon, and northern California are likewise dark, with slender dark bills and blurry underparts patterns, again quite dissimilar to the crisp breast markings of Red Fox Sparrows.

In the Southwest, birders focused exclusively on field marks can be momentarily led astray by the pale rusty *fallax* Song Sparrow, which has a coppery tail, reddish-streaked gray back, gray and rusty head, and whitish underparts with well-spaced rusty streaks—all plumage characters that can be cited for the Red Fox Sparrow as well. The Song Sparrow, however, is much smaller, more narrow-tailed, and less portly than the fox sparrow, with a slender bill and a habit of jumping and running with its tail cocked through the marshes, wet ditches, and desert ponds it prefers over the dark, woodland thickets preferred by the much more retiring Red Fox Sparrow.

Like their congeners, Red Fox Sparrows are noticeably large sparrows, with long, lavishly broad tails and rather long wings. The belly and breast are smoothly curved, the neck long but thick, and the large head decidedly square. Depending on the viewer's angle and the bird's boldness, the most striking plumage feature on first sighting is either the orange rusty tail and wings or the clear, bright off-white ground color of the underparts; neither is shared by any other sparrow, and neither truly recalls a Hermit Thrush, with its duller russet tail and subtle gray breast.

In flight or perched, the contrast between the Red Fox Sparrow's colorful wings and tail and the dove gray of the lower back and much of the head is eye-catching and distinctive. The white flanks and breast sides are heavily and coarsely marked with long, thick chestnut streaks, which grow denser on the breast, almost always forming a very large, ragged splotch in the center of the breast. The chestnut markings of the lower breast and belly often, but not invariably, take the form of discrete check marks or chevrons, leaving the center and tip of the feather white; the presence of both blurry flank streaks and neat, fine inverted Vs on the underparts is a further distinction from Song Sparrows, whose breast and flanks are more uniformly marked with streaks or spots.

The Red Fox Sparrow's head is noticeably pale, the dove-gray ground color only sparsely marked with soft chestnut on the crown and messily surrounding the ear coverts. The white eye-ring is nar-

The loud, melodious song of the Red Fox Sparrow is given both in migration and on the breeding grounds; it can be heard, too, on bright, warm winter days when most other birds are silent. It typically comprises four phrases: the first is a hesitant introduction of short, usually well-separated notes; the second is a series of two or three higher-pitched whistles; the third is a parallel series of lower-pitched whistles; and the concluding phrase is a glissando cadence, sometimes terminating in a sharp *tick*. The "sliding" quality of the last phrase is distinctive. The entire *tee-teet toota toota weewee teeup* song, audible from a great distance on calm days in open woodlands, can last as long as four seconds. Both males and females are known to sing.

RANGE AND GEOGRAPHIC VARIATION

The Red Fox Sparrow is one of the most widespread breeding birds of the boreal forest, breeding from northern and western Alaska east to Newfoundland and Prince Edward Island by way of the Yukon and

The pale greater covert tips are sometimes small, but generally form a clear white wing bar on the rusty wing. *Alaska, June. Brian E. Small*

The bright rusty underpart markings are heaviest and blurriest on the breast sides, usually running into the broad rusty lateral throat stripe; the marks are finer and sharper on the lower breast and belly. *Brian E. Small*

row and broken both in front and rear, giving the bird a squinting aspect at times.

The duller breeding race of central British Columbia and Alberta, *altivagans*, can be more challenging. Apparently "pure" individuals differ from the Slate-colored Fox Sparrow in their rust-streaked back and brighter underparts markings; hybrids and intergrades cloud identification to the point that many apparently intermediate individuals must be left undetermined.

In its feeding behavior, the Red Fox Sparrow recalls a towhee or a White-throated Sparrow, kicking and scratching vigorously in or within easy reach of dark cover, often well within a woodland. Disturbed in their thickety fastness, birds often give a square, harsh, anxious-sounding *tchak*, louder and lower-pitched than the *tek* note of a Lincoln Sparrow or junco and higher-pitched and faintly squeakier than the very similar *tchuk* of the Brown Thrasher. A prolonged, sweet *seeb* note, with slow attack and decay, is given from a perch or in flight; it is fuller, longer, and more clearly ascending than the corresponding, slightly wavering call of the White-throated Sparrow.

Seasonal and individual variation in fox sparrows' bill color is still poorly known. The bill on this Red Fox Sparrow is unusually pale and pink. *Steve Hampton*

Northwest Territories, southernmost Nunavut, and the northern portions of the Canadian provinces. The breeding range reaches south in the West to British Columbia's Peace River region, in the East to northwestern Maine, New Brunswick, and Prince Edward Island.

This is the most strongly migratory of the fox sparrows, with some birds covering more than 1,200 miles between the breeding and the wintering grounds. Red Fox Sparrows are late autumn migrants, typically not reaching the northern United States and southern Ontario and Quebec until late September or even October; on the southern edge of the wintering grounds in Texas, fall migrants arrive between mid-October and the end of November. Of the two Greenland records, one is from October, and there is an early November record from Iceland, along with an early December record from Estonia.

This late and leisurely departure from the nesting areas, along with the pattern of autumn vagrancy, reflects the species' overall hardiness. Wintering birds are regularly found as far north as the Canadian Maritimes in the East and southern Minnesota and South Dakota in the West. There are also rare winter records north of the normal breeding range, to northernmost Alaska and Nunavut.

At the southern edges of the wintering range, Red Fox Sparrows of the subspecies *altivagans* have their principal wintering grounds in the foothills of California's Sierra Nevada and in northern Baja California. Birds representing *altivagans* or *zaboria* are encountered annually in California, southern Nevada, and southern, especially southeastern, Arizona, and along the Rio Grande in New Mexico (eBird); they probably occur more frequently in Chihuahua and northern Sonora than the few published records suggest.

The species is locally common from November to March in the eastern two-thirds of Texas, west as far as the eastern Panhandle and the central Edwards Plateau and south on the coast as far as San Patricio County; most wintering individuals are of the northwestern race *zaboria,* but there are, somewhat surprisingly, records of the nominate race from the state as well. Red Fox Sparrows are common winterers in the states bordering the Gulf of Mexico east to northern Florida. The southernmost Florida records are from Marion and Orange Counties; the bird is rare in most winters anywhere in the state, but can be uncommon in severe weather from November to March in the Panhandle and northern third of the Florida peninsula.

Just as the autumn migration is late, the northbound movement of the Red Fox Sparrow begins early in spring; the beginning of this migration can be difficult to detect, especially on the northern wintering grounds, as local birds depart and are replaced by individuals from slightly farther south. By early March, though, obvious migrants have reached the Great Lakes and New England, and the species is present in Newfoundland by early April; famously, it was this species' song that "predominated" when James Fisher and Roger Tory Peterson met at Gander to begin their cross-continent journey in April 1953. By late April, fox sparrows have become noticeably scarce in most of the eastern United States, and May finds all but the odd stragglers on the northern breeding grounds. Wayward spring migrants have been recorded in April, May, and June in Ireland and Germany.

Writing at the turn of the twentieth century, Robert Ridgway observed that specimens of the Red Fox Sparrow from Alaska "average larger than eastern examples . . . and, in view of the fact that apparently the grayest examples only occur there, it may eventually become necessary to separate them."

While Ridgway did not have sufficient specimen material to recognize the western birds as a distinct subspecies, Harry Oberholser was able to determine that the Alaskan breeders were consistently overall darker, more grayish sooty and less rusty than eastern birds, with duller streaking above and less numerous but blacker markings on the breast and sides. He named the new subspecies *zaboria*, "far northern." This race and the nominate *iliaca* come into contact, and presumably intergrade, in the breeding areas of northeastern Manitoba. In winter, the range of *zaboria* extends east to the southeastern United States, overlapping with the nominate race from the Great Lakes states to Georgia (Pyle).

The affinities of a third population of fox sparrows, identified and named by J. H. Riley in 1911, are much less clear. Riley named *Passerella iliaca altivagans* ("wandering the heights") from two "slightly immature birds" from Alberta and British Columbia; he also assigned to it four apparent migrants from Montana, Oregon, and California. Leaving aside the question of the appropriateness of naming subspecies on the basis of non-adult, nonbreeding specimens, Riley described *altivagans* in comparison not with the Red Fox Sparrow but

with the "typical" Slate-colored Fox Sparrow, from which it differs, he wrote, in the brown rather than gray back and the redder wings and tail—the same features that distinguish Red Fox Sparrows from the Slate-colored, Sooty, and Thick-billed.

Harry Swarth allocated the "relatively bright" *altivagans* firmly to the Red Fox Sparrow, though he also saw in it "a step from the Iliaca [Red] toward the Schistacea [Slate-colored]" group of taxa. Today, *altivagans* is variously considered a Red (for example, by Pyle) or a Slate-colored Fox Sparrow (for example, in the *Birds of North America* account). The most Solomonic solution is that offered in 2003 by Beadle and Rising, who listed *altivagans* as a subspecies of both the Slate-colored Fox Sparrow *and* the Red Fox Sparrow.

Biochemical study of the relationships and evolution of the fox sparrows suggests that *altivagans* may in fact be a Slate-colored Fox Sparrow. Field observers confronted with an individual of this taxon, however, are often more likely to identify it visually as a Red Fox Sparrow; given the scientific uncertainty, it can do little harm here and may be more instructive to number the bird among the subspecies of that species.

Whatever it "is," *altivagans* is slightly smaller and very slightly longer-billed than the other two Red subspecies. It is dark, somewhat duller reddish above, with variably dense and distinct sooty to rusty marking beneath. Paler, brighter birds are probably intergrades with *zaboria*, while darker, duller birds are probably hybrids with Sooty Fox Sparrows; this subspecies also hybridizes with the Slate-colored Fox Sparrow. "Pure" *altivagans* breeds in central and southeastern interior British Columbia and southern Alberta; this is probably the commoner of the two Red Fox Sparrow subspecies that occur—both decidedly scarce—in the Southwest in winter, though the specimen record in Arizona is split more or less evenly between *altivagans* and *zaboria*.

The attempt to identify Red Fox Sparrows in the field beyond the species level is bold, if not foolhardy, away from the breeding grounds. Plumage variability (especially in *altivagans*), intergradation and hybridization, and the possibility that two or even three subspecies are present at a given site make it impossible to assign any individual a racial identity with any confidence.

Extremely variable across their wide range, Sooty Fox Sparrows are very heavily marked beneath, the wedge-shaped spots on the underparts often fusing into a "breast shield" and solid brown flanks. *California, October. Brian E. Small*

SOOTY FOX SPARROW
Passerella unalaschcensis

Thanks to his mother's political connections, the German natural historian and illustrator Friedrich Wilhelm Heinrich, Baron von Kittlitz, was able to spend the years 1826–1829 as naturalist to the fourth Russian circumnavigation of the globe, under the command of Fyodor Litke. Near Sitka, Alaska—then the Russian town of New Archangel—Kittlitz discovered a large sparrow singing from the blackberry thickets. The song, he wrote, was

> the most musical of any of the local birds' . . . it ends with a cadence not unlike that of the Chaffinch, but the tone resembles that of the Nightingale.

He shot two of the singers. Only later did he attempt to identify them: finding that they were "extremely close" to the Red Fox Sparrow presented by Alexander Wilson, differing almost exclusively in their much darker plumage, Kittlitz considered them at first a climatic variant of that bird. On reflection, though, he came to believe that it was "not unlikely" that his Alaskan birds represented a distinct species, which he provisionally named *Emberiza rufina*.

Kittlitz was unaware that that name was preoccupied, having already been applied to a Song Sparrow. Neither was he aware, more significantly,

that this same Sitka sparrow had been discovered, collected, described, and named almost 50 years before his own Alaska sojourn. Brought back from the northeast Pacific by the naturalists of Captain Cook's last, third voyage, the bird found a home in the rich collections of Joseph Banks. On examining the specimen, John Latham named it the Aoon-alashka Bunting. Five years later, Gmelin took the bird over into his edition of Linnaeus's *Systema*, assigning it the binomial *Emberiza unalaschensis*.

While Kittlitz was in Alaska, the British ornithologist Nicholas Aylward Vigors was serving as naturalist on F. W. Beechey's voyage of exploration aboard the *Blossom*. In Monterey, California, he encountered a large sparrow, "brownish above, with rusty wings, rump, and undertail; white beneath, marked with rusty-brownish." Not "recogniz[ing] it among those hitherto described," he named his new bird *Fringilla meruloides*, the thrushlike finch.

Even nearly a decade after Kittlitz's find in Sitka and Vigors's in California, the scarcity of specimens and the inadequacy of published descriptions made it possible to "discover" this species yet again. In February 1836, John Townsend killed a large, heavily marked sparrow on the shores of the Columbia River; Audubon named it for the collector, *Fringilla Townsendi*, and noted, like Kittlitz, its "considerable resemblance" to the Red Fox Sparrow, from which it differed signally in its darker plumage, shorter wing, much longer and slenderer claws, and unmarked wing coverts.

The first step toward a provisional resolution of

these discoveries and rediscoveries was not taken until 1858, when Spencer Baird lumped Vigors's bird with Audubon's and suggested that Gmelin's *unalaschensis*—based on the specimen from the last Cook expedition—"probably has some relation to" the birds collected by Townsend and Vigors. That suspicion was finally affirmed in 1872 by Otto Finsch, who gathered together Gmelin's, Vigors's, Kittlitz's, and Townsend's sparrows under a single scientific name, *Passerella unalaschensis*, noting that he "could find no reason not to resume use of Gmelin's, the oldest name, given that this was the species that he had in mind, however brief the description."

The nineteenth- and twentieth-century view of the relationships between and among the fox sparrows was heavily influenced by Henry Henshaw's determination that all of the known "forms are but modifications of a single species, brought about through the agency of the laws of Geographical Variation." Henshaw's conclusion would go largely unquestioned until the 1980s and 1990s, when biochemical studies and changing species concepts once again suggested that the genus *Passerella* comprised more than one species.

FIELD IDENTIFICATION

The southern-breeding subspecies of this chunky, dark, rather long- and broad-tailed bird are—by fox sparrow standards—readily identified, distinguished from other streaked sparrows by their uniformly dark and virtually unstreaked upperparts, quite obscurely marked head, and heavy and coarse underparts streaking. That streaking, varying individually and geographically from blackish to sooty brown, nearly or entirely coalesces on the sides of the belly to form a conspicuous blurry

In some Sooty Fox Sparrows, the wing and tail are browner than the gray-brown back and head, but they are no rustier than the underpart markings. *British Columbia, January. Brian E. Small*

Some Sooty Fox Sparrows are more uniformly brown above, from crown to tail. *British Columbia, January. Brian E. Small*

flank patch, and on the breast to form a very large, poorly defined breast splotch; that splotch is often so extensive as to form an apron or shield covering much of the breast. While the color of the upperparts also differs among subspecies, there is usually only little difference in tone between the tail and the back, without the pronounced contrast created by the warmer, more reddish tail and uppertail coverts of Red Fox Sparrows; the contrast is greatest in the darkest individuals, which are unlikely in any event to be confused. Sooty Fox Sparrows lack the whitish wing bars and abundant back striping of Red Fox Sparrows, and birds of the southern subspecies have distinctively dark, only subtly patterned heads.

The northerly races of the Sooty Fox Sparrow are less well known to most observers and more variable in plumage, and can be mistaken on the wintering grounds for Slate-colored or for Red Fox Sparrows. These northern-breeding birds, representing the subspecies *sinuosa, insularis,* and, especially, nominate *unalaschcensis,* are significantly paler and grayer than their southern relatives, with much more strikingly patterned upperparts and head. The tail, uppertail coverts, and in some birds the outer undertail coverts are dull rusty, contrasting with the decidedly gray-tinged back; the greater coverts are often bright as well, forming a patch on the folded wing. The nape and supercilium are extensively soft gray, the crown overlain by a dull brownish wash; the effect in many birds is of a gray collar reaching onto the neck sides and curling narrowly above the ear coverts. The eye-ring is narrow but clear and bright, broken at the rear by a fine blackish eye line; the lore is gray, frequently surmounted by an oblong whitish spot. These patterns are extremely similar to those shown—in more vibrant tones—by the Red Fox Sparrow.

The yellow lower mandible is brighter in spring, duller in winter. British Columbia, January. Brian E. Small

Above, these gray-collared Sooty Fox Sparrows differ from the Red by their more muted colors, nearly or entirely unstreaked backs, and browner wings, without the strongly patterned tertials and narrow, jagged white wing bars of the Red Fox Sparrow. Beneath, they are less similar to Red Fox Sparrows, with denser, less even spotting on a somewhat less blindingly white ground; the spots tend to be dull reddish brown rather than bright chestnut, and as in other Sooty Fox Sparrows, they come together to form an elongated dark flank patch, visually continuous with the dark rump when the wing is lifted. The breast splotch is variable in extent, often smaller than in other Sooty Fox Sparrows. The densest spotting is typically high on the upper breast, extending up into very thick, irregularly shaped lateral throat stripes; as a result, the white throat patch is usually small, but still more conspicuous than in the darker Sooty Fox Sparrows.

The Red Fox Sparrow subspecies *altivagans* can closely resemble some of the brighter Sooty Fox Sparrows. Unlike the Sooty, that bird is always streaked above, even if faintly, and the spotting beneath is less crowded and more regular; the spotting continues onto the flanks without fusing into an extensive patch. The greater wing coverts usually have ragged white tips; Sooty Fox Sparrows do not show a wing bar.

With their gray napes, plain backs, and rusty tails, many northern Sooty Fox Sparrows may recall the Slate-colored Fox Sparrow. Rather than appearing gray-collared like the Sooty, however, Slate-coloreds, with their clean gray backs, give the impression of being swaddled in a gray cloak; the back shows more contrast with the decidedly rusty wings and tail than in any Sooty Fox Sparrow.

Hybridization between the Sooty and the Red Fox Sparrows, the Sooty and the Slate-colored Fox Sparrows, and the Sooty and the Thick-billed Fox Sparrows is believed to be "extensive" in the respective zones of range overlap. Birds of intermediate appearance should be not identified but described, with special attention to the tail and back color, flank pattern, back pattern, breast pattern, and head pattern.

Some such birds will be visually closer to one presumed ancestral species than to another, or they can be almost exactly intermediate—though intermediacy is difficult to assess in species that even in their "pure" form can be distinguished only on subtle characters. More revealing, and more humbling for the ambitious field observer, is the fact that individual birds exhibiting plumage colors and patterns belonging to one species have been found to have genetic combinations unique to another species—in other words, birds that even the most cautious observer would confidently identify to species in the field could in fact be hybrids or intergrades. One study found that where breeding Red, Sooty, and Slate-colored Fox Sparrows meet in British

Columbia, Yukon, Alberta, and Alaska, "most combinations of hybridization were observed, and 9 out of 139 (6.5%) individuals had the 'wrong'" genetic combinations for their apparent species.

Fortunately, juveniles add fairly little to the complexities of fox sparrow identification. In the Sooty as in the other species, birds in their first plumage closely resemble adults, with fewer reddish tones above and a buffier ground color below than their parents. The markings of the underparts are usually finer, sparser, and less well organized; in the case of the juvenile Sooty Fox Sparrow, this can result in a less heavy flank patch and a less shieldlike breast patch than in adults, causing some individuals to more closely resemble an *altivagans* Red Fox Sparrow. Juveniles molt into their first adultlike plumage before leaving the breeding grounds, and are indistinguishable in the field, and usually in the hand, from adults in winter.

The Sooty Fox Sparrow is the boldest of its genus, fearlessly emerging from tangled cover to kick and scratch vigorously through grass and leaf litter in yards and parks. If disturbed, birds hop or fly in low swooping flight back into brambles or other low dense vegetation, where they may perch in sight as they give a square, harsh, anxious-sounding *tchak*, louder and lower-pitched than the *tek* note of a Lincoln Sparrow or junco; this call is very similar to the corresponding note given by the Red and Slate-colored Fox Sparrows. A prolonged, fairly sweet *seeb* note, with slow attack and decay, is given from a perch or in flight; it is fuller, longer, and more clearly ascending than the corresponding, slightly wavering call of the White-throated Sparrow.

The loud, melodious song of the Sooty Fox Sparrow is given both in migration and on the breeding grounds; it can be heard, too, on bright, warm winter days when most other birds are silent. It often comprises four phrases: the first is a hesitant introduction of short, sharp, usually well-separated notes; the second is a series of two or three higher-pitched whistles; the third is a parallel series of lower-pitched whistles with a buzzy undertone; and the concluding phrase is variable, sometimes a descending slur and sometimes a dry trill. The entire song seems lower-pitched, fuller, and hoarser than the sweeter, more musical phrases of the Red Fox Sparrow, though the race *zaboria* of the latter species may sing more harshly than Red Fox Sparrows breeding farther east.

RANGE AND GEOGRAPHIC VARIATION

As a species, the Sooty Fox Sparrow breeds from the eastern Aleutians south, mostly along the Pacific Coast, to northwest Washington. The nominate subspecies occupies the Aleutians east of Unalaska and the Alaska Peninsula; *sinuosa* breeds to the

The upperpart color of fox sparrows can be hard to assess given their predilection for shade. This Sooty, probably of the widespread subspecies *sinuosa*, might look brighter in different light. *California, January. Steve Hampton*

Likely an *annectens* Sooty Fox Sparrow, this bird shows a bright contrasting reddish tail and rump. Note the clear grayish surround to the red-streaked ear coverts. *California, January. Steve Hampton*

southeast of that range on Middleton Island and in the area of Prince William Sound and the Kenai Peninsula, while *insularis* is restricted to Kodiak Island. Linking the breeding ranges of *sinuosa* and *townsendi*, *annectens* breeds coastally from Alaska's Cross Sound to the north shore of Yakutat Bay. Audubon's *townsendi* has the most extensive breeding range of any of the subspecies, from Glacier Bay in Alaska south to British Columbia's Haida Gwaii; if *chilcatensis* is recognized as racially distinct, it occupies the portion of that area between the Chilkat River and the vicinity of Stewart, British Columbia.

In southwestern British Columbia, *fuliginosa* breeds over large areas of Vancouver Island and the nearby mainland, north to approximately Hopetown and east through the Fraser lowlands; it also nests on a number of smaller offshore islands. This is also the breeding fox sparrow of northwest Washington and the Olympic Peninsula.

The migration and wintering distribution of some Sooty Fox Sparrows is often adduced as the best example of a phenomenon known as "leap-frog migration," in which

> northern breeding populations leap-frog their southern neighbors to produce a mirror image of their respective breeding distributions in winter.

This is essentially the pattern exhibited by the three eastern races of this species, *annectens, townsendi,* and *fuliginosa*. The last subspecies undertakes the shortest migratory movements, while the first—the northernmost breeder—flies farthest in spring and fall. Thus, among these three populations, the one that breeds farthest north winters farthest south, after an autumn flight along the Pacific Coast.

The western Alaska subspecies, *unalaschensis, insularis,* and *sinuosa,* all of which breed farther north than *annectens,* spend the colder season in southern California, northern Baja California, and on the Channel Islands, and so as a group they "leap-frog" all three of the eastern races. Many fly over the Pacific, though apparent *unalaschcensis* birds can be numerous in winter in many coastal areas of British Columbia, Washington, and Oregon. Wintering birds can arrive in California as early as mid-September, with peak autumn movements recorded in early October. Fall migrants regularly occur in California's eastern deserts, and there is a late November specimen of *townsendi* from southeast Arizona. Two or perhaps three individuals were recorded in September and early October 1990 on the Chukchi Peninsula.

As a whole, the Sooty Fox Sparrow winters regularly from central British Columbia south along the coast to northern Baja California; there are occasional winter records from as far west as St. Lawrence Island and as far north as Point Barrow. In California, Sooty Fox Sparrows are most common on the northern and central coast and in the valley west of the Cascades and Sierra; in the south, they winter coastally from San Luis Obispo to Los Angeles Counties and more sparingly beyond. The nominate subspecies is "numerous and widespread" in San Diego County from late September to April.

There are at least five autumn and winter records of birds of unknown subspecies from the vicinity of the Salton Sea. Unsurprisingly, in view of the long migrations undergone by some populations in this species, there are also infrequent records from east of the breeding range, including one that wintered in Ward County, North Dakota. A bird that survived from December to late March at a Strafford County, New Hampshire, bird feeder was identified by that state's records committee as a Sooty Fox Sparrow of the nominate race. To the west of the breeding range, the species is accidental in winter in Japan.

The northward movement in spring has birds arriving on the breeding grounds in late April or early May. There are no fewer than six June and July records from the Russian Far East, presumably representing birds that strayed west across the Pacific during the northbound migration. Startling to the point of inexplicable was a single Sooty Fox Sparrow photographed in New York City's Central Park in mid-May 2010; this bird was described as "gray" by many observers, suggesting that it might have been a member of the subspecies *unalaschcensis.*

Nominate *unalaschcensis* is relatively large and large-billed, with less reddish plumage than any other Sooty Fox Sparrow. Swarth discerned two "categories" among winter specimens of *unalaschcensis* taken in California: the first pale brownish, "rather ashy in general tone," with a long bill; the second darker, "rather plumbeous in tone," with a short, heavy bill. In spite of his tentative suggestion that the two might occupy different breeding ranges and thus be properly considered two separate subspecies, Swarth found that "absolute proof" of that possibility would be provided only by further study of specimens from the breeding grounds. The matter remains unsettled.

Breeding from Unalaska, Shumigan, and Semidi Islands to the adjacent Alaskan Peninsula, *unalaschcensis* probably intergrades with *sinuosa* and with *insularis* where their ranges approach. Originally described from Drier Bay, Knight Island, in Prince William Sound, *sinuosa* differs from the nominate race in its slenderer bill and "rather more reddish" plumage. It is larger-billed than *annectens, townsendi,* and *fuliginosa,* and less reddish above than *insularis, annectens,* and *townsendi*. Its plumage tone is intermediate between the slatier *unalaschcensis* and more reddish *annectens,* the bill smaller than that of the former but larger than that of the latter.

Kodiak Island is the only breeding site for the aptly named race *insularis*. This subspecies is paler, slightly larger, and, in the museum drawer, notably larger-billed than *annectens, townsendi,* and *fuliginosa;* the plumage is brighter reddish than either *sinuosa* or *unalaschcensis*, both of which tend to the grayish above. This subspecies is also known as *ridgwayi*, but that usage is correct only when the fox sparrows are included in an enlarged genus *Zonotrichia*.

Another well-named subspecies, *annectens* ("linking") breeds in the area between the breeding ranges of *sinuosa* to the northwest and *townsendi* to the south. It is smaller and smaller-billed than *sinuosa* and the other two races breeding to its north and west, *insularis* and *unalaschcensis*; it is slightly larger than *townsendi*. In plumage, *annectens* is brighter reddish than *sinuosa* and paler than *townsendi*; its breast spots are not as heavy as in that latter race. It is possible, but not certain, that Beechey's *meruloides*, collected on the wintering grounds in northern California, was of this subspecies.

Said by Swarth to be "one of the least known of the subspecies of fox sparrow," *fuliginosa* may now be the Sooty Fox Sparrow most familiar to many observers in the Pacific Northwest, given that its breeding range includes the heavily populated areas of southwestern British Columbia and northwestern Washington; it is found in winter as far south as northern California. Darkest of all the fox sparrows, this subspecies is darker and less reddish than *townsendi*; the upperparts are sooty brown, and the underparts are marked with large dark sooty spots, less distinctly separated than in *townsendi*.

Among winter specimens of apparent *fuliginosa* from Vancouver Island and California, Swarth discovered individuals that he considered "non-typical," with notably short bills, duller upperparts, more whitish undertail coverts, and less "crowded" streaking beneath. Birds of this type were described in 1983 as a distinct subspecies, *chilcatensis*, differing from *townsendi* in the less reddish, more blackish upperparts and breast spotting and from *fuliginosa* (in its new, narrower sense); the breeding range of *chilcatensis* extends from the Chilkat River area in British Columbia and Alaska southeast to the area of Stewart, British Columbia, equivalent to approximately the northern one-third of the range of *townsendi* as originally understood. This subspecies is not recognized by all authorities.

By definition, individuals on their breeding grounds are identifiable to subspecies in the field. In winter, however, when more or less discrete breeding populations mingle, identification can rarely be certain without the bird in hand. As a practical matter in the field, large, pale, large-billed, and relatively gray-headed individuals can be reasonably suspected of representing the nominate race *unalaschcensis*; the other subspecies are generally so similar, differing only slightly in size, bill size, and upperparts coloration, that even in the hand many specimens can be difficult to determine with any confidence.

The most widespread of the western fox sparrows, this species is nearly unmarked above, with brown or chestnut streaks below and a smaller bill than the Thick-billed Fox Sparrow. *California, October. Brian E. Small*

SLATE-COLORED FOX SPARROW
Passerella schistacea

All of the fox sparrows raise important ontological questions about the existence of categories, about the nature of difference and similarity, about identity—questions that affect the way we have learned to look at these birds in the field. The Slate-colored Fox Sparrow poses a far more basic conundrum: where did the first specimen come from?

The Smithsonian's National Museum of Natural History still holds a bird collected on July 19, 1856, by William S. Wood Jr., the naturalist assigned to Lieutenant Francis T. Bryan's road-surveying expedition from Fort Riley, Kansas, to Bridger's Pass, Wyoming. This female skin, now USNM A 5718, was described two years later by Spencer Baird as a new species, "readily distinguished" from the Red Fox Sparrow "by the slate back and spots on the breast, without any streaks above." Baird's account is marred by typographical errors, one of which appears to affect the indication of the type locality: in the prose text, Wood's skin is said to have been secured at "the head of the Platte," while the tabular list of specimens assigns it to the "Platte river, K.T.," an obvious lapsus for "N.T.," Nebraska Territory.

At the time of the Bryan expedition, Nebraska Territory took in most of what is now the western Dakotas, Montana, and Wyoming, along with a generous slab of northeastern Colorado. Bryan's party left Fort Riley, Kansas, in late June, then moved north to follow the South Platte in modern-day Nebraska; they eventually left the Platte to follow Lodgepole Creek upstream, arriving in what is now Laramie County, Wyoming, on July 24 or 25.

There is no doubt that the Bryan expedition was still in what is now Nebraska on July 19, when Wood shot the new sparrow. T. S. Palmer was the first to determine the type locality more precisely, informing Wells Cooke in 1897 "that the specimen in question was taken by Lieutenant Bryan's party July 19, 1859 [sic], in Nebraska, about 20 miles east of the Colorado line." When a few years later the American Ornithologists' Union began to include type localities in its *Check-list*, the site was described slightly differently, as the "south fork of the Platte River, about 25 miles east of northeastern corner of Colorado, Nebraska." That locality continued to be published through the 1957 edition, the last version of the *Check-list* to accord the Slate-colored Fox Sparrow full treatment.

Palmer's cartographic exercise yielded an accurate, if certainly too precise, answer as to the surveying party's whereabouts on July 19, 1856. Bryan described the habitat:

> The country here, and for some miles further up [Lodgepole Creek], is a high, dry prairie—a dead, flat, burned up piece of ground . . . . The country is extremely barren and burnt up; nothing green to be seen except the willows and grass immediately along the banks. The higher ground is covered with buffalo grass, which is now burned dry. Scarcely have we seen anything resembling a tree . . . .

Harry Swarth, aware of the absence of suitable habitat for any July fox sparrow in the southern Nebraska Panhandle, suggested that "it may be that this bird had strayed down" Lodgepole Creek, "and then down the South Platte"—a possible but unlikely scenario, especially for an individual he described as having been taken "in badly molting condition." Especially given that there has not been another documented occurrence in the state of Nebraska over all the intervening 160 years, it seems most plausible that the bird was mislabeled or that the date was misread, and that the Slate-colored Fox Sparrow was discovered not on Nebraska's short-grass prairies but on the brush-clad slopes of Wyoming's Rocky Mountains, where it still breeds today.

FIELD IDENTIFICATION

This species is immediately recognizable as a fox sparrow thanks to its large size, long tail, heavily marked underparts, and red-tinged wings and tail. As both the English and scientific names suggest, it is grayer, especially on the back and largely unmarked head, than either the Sooty or the Red Fox Sparrow, and all but the heftiest-billed *canescens* Slate-coloreds have slenderer bills than the Thick-billed Fox Sparrow. While the ground color of the back varies geographically from olive-gray to slate gray, it is not streaked and always contrasts well with the reddish wing and tail (rusty streaking on a browner back is a characteristic of the Red Fox Sparrow subspecies *altivagans*, sometimes assigned to this species instead).

The Slate-colored is a small but long-tailed fox sparrow; even in the field, the bill can be noticeably small and slender. The wings are rather long, as in other fox sparrows. The belly and breast are smoothly curved, the neck long but thick, and the large head decidedly square. The very narrow whitish eye-ring, blurry pale lore spot, and small bill combine to give the bird an odd look of disapproval.

In flight or perched, the contrast between the Slate-colored Fox Sparrow's deep rusty wings and tail and the dove gray of the back and head is eye-catching and distinctive. The white flanks and breast sides are fairly sparsely marked with long blackish streaks, which grow denser on the breast to form a moderately large, ragged splotch in the center of the breast. The markings of the lower breast and belly take the form of discrete checkmarks or chevrons, leaving the center and tip of the feather white; the markings of the central belly can be especially sparse.

The head is noticeably gray, blending smoothly into the gray of the nape and back. The facial markings are faint, comprising a thin dusky line behind the eye, a narrow broken eye-ring, and usually a paler spot above the lore. As in the other fox sparrows, the whitish jaw stripe is irregularly shaped and often speckled dark, separated from the throat by a poorly defined, broken lateral throat stripe.

In its feeding behavior, the Slate-colored Fox Sparrow recalls a Spotted or Green-tailed Towhee, kicking and scratching vigorously in or within easy reach of dark cover. Disturbed in their thickety fastness, birds often give a square, harsh, anxious-sounding *tchak*, louder and lower-pitched than the *tek* note of a Lincoln Sparrow or junco and lower-pitched and less metallic than the *tchink* of the Thick-billed Fox Sparrow. A prolonged, faintly trilled *seeb* note, with slow attack and decay, is given from a perch or in flight.

The loud, melodious song of the Slate-colored Fox Sparrow is given both in migration and on the breeding grounds; it can be heard, too, on bright, warm winter days when most other birds are silent. After an introduction of two or three short, well-separated notes, the two or three phrases can be thought of as comprising triplet groups with the emphasis on the second note. Some notes are buzzy, others more sweetly slurred. Individuals may sing several different song types, switching regularly from one to the other.

RANGE AND GEOGRAPHIC VARIATION

A relatively short-distance migrant taking apparently straightforward north-south routes in spring and fall, the Slate-colored Fox Sparrow occupies a well-defined breeding range in the Northwest and Great Basin and a rather compact wintering range in California and the Southwest. There are very few records from even as far east as the Great Plains—and none since 1856 from Nebraska, where the type specimen is said to have been collected.

Two subspecies of the Slate-colored Fox Sparrow breed in the interior of British Columbia. *Olivacea*

This California winterer combines the streaked rusty back and neat wing bars of a Red Fox Sparrow with the head pattern of a Slate-colored. Away from the breeding grounds, it may be impossible to assign such *altivagans*-like birds to species, much less to subspecies. *California, January. Steve Hampton*

The Slate-colored Fox Sparrow's smooth gray head and back contrast handsomely with the rusty wings and tail. The neat streaking of the underparts is black, without red tones. *California, March. Steve Hampton*

breeds at high elevations in the southwest and south-central areas; this is presumably the subspecies that has recently begun to breed in clear-cut habitats below 5,000 feet elevation in the Okanagan Valley as well. The breeding fox sparrow of the

Rocky Mountains in extreme southeastern British Columbia is *schistacea*, which occupies breeding grounds from southwest Alberta south through the Great Basin to central and eastern Oregon, western Montana, southwest Wyoming, western Colorado (less common in eastern Colorado mountains), and northern Nevada. In central Nevada and the White Mountains of east-central California, the species is represented by *canescens*, while the dark gray *swarthi* breeds from Idaho's Bannock and Bear Lake Counties south to Sanpete County, Utah. Throughout their range, breeding Slate-colored Fox Sparrows prefer creekside thickets of alder, birch, and willows.

As they pass for the most part through regions with relatively little ornithological coverage, the migratory behavior and timing of Slate-colored Fox Sparrows is not well known. The southbound flight appears to begin very early. Departures are noticeable in southern British Columbia by September, and some are already arriving on the Colorado River in Arizona and Nevada in late August and in northeastern Arizona in early September. The nominate race has been recorded in October on the eastern plains of Colorado and in late September in El Paso County, Texas.

Slate-colored Fox Sparrows winter in brush and chaparral habitats south and southwest of the breeding areas; some populations may be resident or undertake only altitudinal movements. They are common in California's foothills and lowlands away from the deserts of the southeast, where they occur less than annually between October and late March. Most of the records from the lower Colorado River region of California and Arizona are of the nominate *schistacea*, though there is at least one specimen representing *olivacea*. Slate-colored Fox Sparrows are common winter residents in northern Baja California, south to Concepción in the Sierra de San Pedro Mártir; as in southern California, most are *schistacea*, with rare records of *olivacea*. There are only a few records for this species in Sonora and Chihuahua, but its true abundance is likely masked by the relatively small number of observers; if the situation there is like that in southernmost Arizona, the species is a rare but regular winter visitor, probably slightly less common and slightly less frequent than the Red Fox Sparrow.

Farther north in Arizona, Slate-colored Fox Sparrows of the nominate race are locally common to uncommon in thickets and chaparral from September to March, with rare individuals lingering into May. There are scattered winter records in central New Mexico as far north as Albuquerque, and in Colorado north to Garfield County, with one individual photographed in January 2007 exceptionally far east in Yuma County (eBird).

This striking bird may be an unusually large-billed Slate-colored Fox Sparrow or a Thick-billed Fox Sparrow somewhat to the north of its usual range. *Washington, May. Steve Hampton*

The migration north to the breeding grounds begins in late winter, with arrival on the breeding grounds starting in late March.

Whether it is treated as a distinct species or a subspecies or subspecies group within an enlarged Fox Sparrow, the Slate-colored Fox Sparrow and its regional variation remain a subject of debate. Zink's analyses of protein variations and of skin and skeletal characters "failed to confirm the existence of traditional subspecies" within the Slate-colored Fox Sparrow. Other authorities recognize between four and six distinct races, sometimes including one or both of the confusing subspecies *altivagans*—here treated as a Red Fox Sparrow—and *fulva*—here treated as a Thick-billed Fox Sparrow.

Of the four generally accepted Slate-colored subspecies, the smallest and thickest-billed is *canescens*, described by Harry Swarth in 1918 from Inyo County, California. Breeding in the White and Inyo Mountains of eastern California and in central Nevada from Esmeralda County to White Pine County, this race is found in winter in southern California and northern Baja California. As its name indicates, *canescens* is much more uniformly gray on the head and back than the nominate race *schistacea*, from which it differs also in its more sparsely spotted underparts (Pyle).

The slightly larger race *swarthi*, named in 1951 in memory of Harry Swarth, is darker and "still grayer, so much so that *canescens* looks brown in comparison." The bill is slender, and the markings of the underparts are heavy and brown, less reddish than in *schistacea*. Breeding in southern Idaho and north-central Utah, *swarthi* is at its darkest, grayest, and "purest" in the Wasatch Mountains, and apparently undergoes a "transition" toward the browner nominate *schistacea* to the east, in the Uintas; to the

northwest, in the Raft River Mountains; and to the west, in northern Nevada. Apparent intermediates between *swarthi* and the paler *canescens* have also been found in east-central Nevada. The wintering range remains unknown.

Nominate *schistacea* is the most widespread of the Slate-colored Fox Sparrows at any season, breeding over much of the interior Northwest from southeastern British Columbia south to eastern Oregon and from Alberta south to western Colorado; it winters in California, both along the coast and in the Sierra Nevada, east in low densities across Arizona and New Mexico to southwest Texas (Pyle). The bill is slender, the white underparts heavily marked with dark brown spots and streaks on the breast and belly. The crown and back are a brown-washed medium gray (Pyle), not pure gray as in the small, pale *canescens* or the dark *swarthi*.

The Slate-colored Fox Sparrow breeding in eastern Washington and south-central British Columbia, north to at least Kootenay Lake and south to Oregon's Blue Mountains, is the subspecies *olivacea*, described in 1943 from the lower slopes of Mount Rainier. It is darker and, logically, more olivaceous above than *schistacea*, with a reddish tinge; the flight feathers of the wing and tail have brown rather than cinnamon edges.

This last subspecies comes into contact and intergrades with *schistacea* near the Wallula Gap of south-central Washington. It may also interbreed with Thick-billed Fox Sparrows of the race *fulva* in Crook and Harney Counties, Oregon, and with Red Fox Sparrows of the race *altivagans* "in some unknown region of southern British Columbia"; both of those taxa have been considered by some authorities to be Slate-colored Fox Sparrows themselves.

Aptly named, the Thick-billed Fox Sparrow recalls the Slate-colored Fox Sparrow in plumage, but the extraordinary bill identifies it. *California, October. Brian E. Small*

THICK-BILLED FOX SPARROW
Passerella megarhyncha

John Xantus—if that was his name—arrived in the United States in 1851. Four years later, he joined the United States Army as a private, and was promptly assigned to the frontier at Fort Riley, where he met William A. Hammond. Hammond, whose name Xantus would later give to a flycatcher, introduced the Hungarian to Spencer Baird of the Smithsonian, who used his extensive connections to have Xantus transferred to the richer collecting grounds of Fort Tejon, 70 miles north of Los Angeles.

There, in spite of his constitutional incapacity for peaceful relations with his colleagues and commanders, Xantus quickly proved himself one of the best of the Smithsonian's far-flung crew of collectors. Baird wrote, flatteringly but with little exaggeration, that

> all unhesitatingly concur [sic, for "concede"] you the first place as a collector, and are unanimous in saying that there is but one Xantus, unapproachable, and not to be equalled however much the second person may do.

Whether "unapproachable" here refers to Xantus's prickly personality or to his attainments in the field, it is clear that Baird was pleased with his contributions to American natural history. Xantus on his part made a point of regularly reminding his mentor in Washington how costly such contributions could be:

> I undertook a trip to the Tejon peak, the highest mountain we have around here . . . . It was too craggy, & thorny brush filled the whole desolate region, that I could not persecute anything, considering fortunate to have a sure foothold at times.

A few months later, he informed Baird that as much as 15 feet of snow lay on the road from Tejon to Los Angeles—but that he had managed in spite of it to dispatch yet another shipment of hundreds of specimens to Washington.

In the middle of that snowy winter, in January 1858, Xantus collected and sent to Baird a small series of fox sparrows. Baird, puzzled, at first assigned them to his new Slate-colored Fox Sparrow, apparently in consultation with Xantus, who in an October letter to Washington went so far as to claim to have himself "solved satisfactorily [the] difference as to species" between the Sooty and the Slate-colored.

It was not long before Baird reconsidered. Comparing Xantus's Fort Tejon specimens with 11 skins collected in the Wyoming Rockies, he found that the Wyoming sparrows, like the type specimen of the Slate-colored Fox Sparrow, all had much smaller bills than the California birds. In the appendix to the 1858 reports on the natural history of the railroad surveys, Baird proposed that "the bird of Fort Tejon may be called *P. megarhynchus*," one more species owed to the collecting skill of John Xantus.

FIELD IDENTIFICATION

Medium-large and front-heavy, with fairly long wings and tail, smaller-billed individuals of the Thick-billed Fox Sparrow can be impossible to distinguish from Slate-colored Fox Sparrows away from the breeding grounds—and even there, as in parts of Nevada and eastern California, it is not always clear which species "should" be in possession of a given site. Overlap has been detected in the mitochondrial DNA of individuals in the White and Warner Mountains and at Steens Mountain, Mono Lake, Woodfords, and Yakima, all areas "where individuals of each [species] co-occur and almost certainly interbreed."

The bill of most Thick-billed Fox Sparrows is noticeably deep at the base and broad across the top, creating a "frog-faced" impression especially when viewed from above. Confronted with a slaty-headed, gray-backed, rusty-winged fox sparrow that is not unequivocally huge-billed, an observer should note the color of the lower mandible; it is usually yellow year-round in Slate-colored Fox Sparrows, sometimes dull blue-gray in breeding Thick-billed Fox Sparrows.

A more promising distinction lies in the contact call given by the Thick-billed Fox Sparrow. Unlike the low-pitched, smacking notes of the other fox sparrows, this species has a light, bright *tdink*, higher-pitched and more musical, closer in tone to a California Towhee's or, especially, a White-crowned Sparrow's chip. An alarmed fox sparrow of any species may also give a series of high-pitched *tsip* notes, "spectrographically distinct from the 'tink' calls of Thick-billed Fox Sparrows, but [which] can be quite difficult to distinguish by ear." Waiting, if possible, for the agitation to subside offers the best chance that a bird will return to issuing its typical and distinctive contact note.

> The song of the Thick-billed Fox Sparrow in Nevada was praised by Robert Ridgway as second to none . . . though in variety, sprightliness, and continuity, and also in passionate emotional character, its song is not equal to that of the [Lark Sparrow], yet it is far superior in power and richness of tone.

Unlike many Sooty or Slate-colored Fox Sparrows, Thick-billed Fox Sparrows often show pale whitish tips to the greater coverts, forming a single wing bar. *California, October. Brian E. Small*

A loud, melodic series of phrases, it resembles the songs of the Sooty and Slate-colored Fox Sparrows in the insertion of rich buzzy notes within and between phrases; some Thick-billed appear to sing a slightly longer song than most Slate-colored Fox Sparrows, but it is not known whether this difference is consistent at the level of populations and thus useful for field identification.

RANGE AND GEOGRAPHIC VARIATION

Long thought to occur no farther north than central Oregon, in this century the Thick-billed Fox Sparrow has been reported in the breeding season in Yakima, Skamania, and Klickitat Counties in southern Washington, with the first fully documented record from Yakima County in July 2013. The species is a fairly common breeder in the northern and central Cascades of Oregon, represented there by *fulva* (or the nominate *megarhyncha*, depending on whether *fulva* is recognized as distinct).

In California, birds identified as *fulva* breed in Modoc and Lassen Counties, while *megarhyncha* in the narrower sense breeds south to the Siskiyou Mountains and the Sierra Nevada as far south as Kearsarge Pass in Inyo County. Swarth's *brevicauda* occurs in the northern coast ranges, from Trinity to Colusa Counties. The nominate subspecies grades into *stephensi* at the southern edge of its range; *stephensi* itself breeds in the southern Sierra Nevada through the high elevations of southern California to the Sierra de San Pedro Mártir in northern Baja California. The Yosemite form *mariposae* is not recognized by all authorities as distinct from *megarhyncha*.

The breeding Thick-billed Fox Sparrows of the eastern flank of the central Sierra Nevada and east to Mineral County, Nevada, are those that have been distinguished as *monoensis*; in west-central Nevada, "the situation is messier," and the division between the Thick-billed and the Slate-colored Fox Sparrows there remains unclear, though there are recent reliable sight records from Humboldt County (eBird).

The Thick-billed Fox Sparrow is an early autumn migrant, moving south in August and September the short distance to the wintering grounds, which extend from central and southern California into northern Baja California; *megarhyncha* (including *monoensis*) is a common winter visitor to Baja

The rusty wing and tail of the Thick-billed Sparrow are brighter or duller in the field, depending on light and shadow. *California, October. Steve Hampton*

California south to the Sierra de San Pedro Mártir, while *fulva* is apparently rare at the same season. The Thick-billed is the commonest of the fox sparrows in San Diego County, where *stephensi* breeds and birds of more northerly populations linger from October to April. It is unclear whether Arizona specimens from October and December represented tardy autumn wanderers or birds intending to winter.

Even more than the other fox sparrows, the Thick-billed Fox Sparrow has been subdivided on the basis of extremely slight differences in plumage color and, more cogently, in bill size. The nominate subspecies, named by Baird on the basis of Xantus's Fort Tejon specimens, breeds from southern Oregon south through California's Sierra Nevada to Inyo County. It is smaller and, its name notwithstanding, smaller-billed than the very heavy-billed *stephensi*, which breeds from Baja California's Sierra de San Pedro Mártir north at high elevations to the range of the *megarhynchus*. The back of *stephensi* is somewhat darker and grayer than that of the nominate race.

Browner and narrower-billed than *stephensi*, the race *brevicauda* breeds in northern California's coastal mountains south of the Trinity River, wintering farther south along the coast. The scientific name, coined by Joseph Mailliard in 1918, refers to "a minor characteristic," the relatively shorter tail, a distinction hardly visible in the hand and certainly not in the field; Mailliard explains that

> a geographic designation would have been preferable, but the most applicable one, that of the type locality, is too clumsy to latinize, though possible to use unaltered for the vernacular name.

Mailliard's recommendation notwithstanding, his Yolla Bolly Fox Sparrow has more commonly been known in English as the Trinity Fox Sparrow, commemorating the county, rather than the mountain range, in which the type specimen was collected in August 1913.

Two other subspecies named from their California breeding ranges are probably not to be distinguished from the nominate *megarhyncha* (Pyle). First encountered at the Mono Lake post office in Mono County, the Mono Fox Sparrow *monoensis* was described as "closely" resembling the nominate race but slighter-billed and faintly paler gray above. The subspecies *mariposae*, named for its type locality near Yosemite, has been distinguished from *megarhyncha* on the basis of its relatively long, long-tipped bill, intermediate in size and shape between that of the nominate subspecies and the truly huge-billed *stephensi*; like *monoensis*, it is also said to be more grayish above.

The uninformatively named subspecies *fulva*, "brown," presents a puzzle that is inscrutable even by the standards of the genus. Harry Swarth described this race on the basis of a large number of skins from the Warner Mountains in Modoc County, California. The bill was described as more slender than that of *monoensis*, the back color more brownish than in that race or in *mariposae*.

Significantly, Swarth was at pains to distinguish his *fulva* not just from the other Thick-billed subspecies but from the Slate-colored Fox Sparrow as well, which it resembled in coloration and in size. Indeed, *fulva* is so closely similar visually to the Slate-colored that it is today sometimes allocated to that species, while other authorities consider it a Thick-billed Fox Sparrow, even lumping it with the nominate race of that species.

The pale-headed, pale-flanked Guadalupe Junco is less colorful and contrasting than the Oregon Junco, an occasional autumn stray to the island. *Baja California. Paul Sweet*

GUADALUPE JUNCO
Junco insularis

The birds of remote oceanic islands are famous for their tameness, a quality that historically has endeared them as much as it has exposed them.

In July 1896, the 19-year-old Pasadena naturalist Horace Amidon Gaylord joined a collecting party led by the San Diego Museum of Natural History's A. W. Anthony. The expedition landed on Guadalupe Island on September 17, and promptly set about securing specimens of the island's birds; finally, on their last evening on the island, Gaylord himself managed to shoot one of the "three or four individuals" that were then "probably the only remaining representatives" of the Gualadupe Caracara, a species now—unsurprisingly—extinct.

The other island rarities were gratifyingly less elusive. Gaylord found the Rock Wrens impressively trusting, but he was most impressed by the "remarkable confidence" of the Guadalupe Juncos:

Imagine a bird so rare in collections as *Junco insularis* baffling an attempt to collect it by trying to alight on the end of the gun barrel. In September 1896 I saw this happen, and only regretted that this almost perfect confidence could not be rewarded by the bird being allowed to live.

Twenty years earlier, the Anglo-American botanist Edward Palmer had become the first scientist to note the juncos of Guadalupe Island.

They are the most abundant birds of the island, and are so tame that they may be killed with a stick, or captured in a butterfly net. When I was looking for insects under stones and logs, these birds would sometimes join in the search, and hop almost into my hands . . . . At times, they even enter the houses, picking up anything edible they can find. Numbers boarded the schooner as we neared the island, and made themselves perfectly at home, roaming over every part of the vessel . . . .

Palmer made good use of his sticks and nets, collecting a dozen specimens, males and females, adults and juveniles.

When Walter E. Bryant visited the island in 1885 and early 1886—an unexpectedly protracted expedition on which he "nearly starved to death"—he was able to nearly double the specimen record for the species, collecting thirteen birds along with nests and egg sets. Bryant found the juncos less abundant and less confiding than he had been led to expect; he noted that one locality where he had shot specimens in January 1885 hosted "the following year not more than two or three." Bryant also observed that no tree on the island had been left unscarred by the two thousand feral goats that browsed the forests and groves, and he lost at least one nest to the suspected depredations of feral housecats. Goats, cats, and quite possibly excessive collecting would eventually reduce the junco population to a number so low that the species is still considered critically endangered today.

FIELD IDENTIFICATION

Including the extinct Guadalupe Towhee, ten species of passerellid sparrow have occurred on Guadalupe. The only one resembling the island's resident junco is the Oregon Junco, "probably not uncommon" as an autumn stray. Those migrant juncos can be expected to be dark-headed and contrastingly plumaged, unlike the pale-headed, pastel-flanked Guadalupe Junco.

The plumages of the Pink-sided Junco (and of the nonmigratory Baird Junco, resident in Baja California) more closely recall those of the Guadalupe Junco, but even should a Pink-sided somehow make its way to the island, it would be immediately distinguishable by its smaller, paler, more typically juncolike bill. Baird Juncos, very unlikely strays to Guadalupe Island, also differ in their yellow eyes.

The Guadalupe Junco is a sturdy, medium-sized sparrow with a moderately long tail, short wing, round belly, and round head with a long, dark, almost spikelike bill. The adult is unstreaked, with a soft gray head and breast, dark lores, brown back and wings, and pale cinnamon flanks; there is virtually no difference between the sexes in plumage color and pattern. The outer tail feathers are less extensively white than in Oregon Juncos, a character visible only in flight or when birds spread the tail to preen or flirt it in anxiety; in most, the outermost tail feather is not entirely white, and two-thirds of Miller's specimens showed some black in that feather's outer web. Nearly all have black on the outer web of the next feather in, rectrix 5, while rectrix 4 is mostly or entirely black in 98 percent. The outer webs of rectrices 5 and 6 are significantly whiter in the Oregon Junco race *pinosus*, the subspecies judged by Miller to have the tail pattern closest to the Guadalupe Junco's.

Juvenile Guadalupe Juncos are largely brownish gray, indistinctly streaked dusky everywhere but the center of the whitish belly. Their wings may show faint brown wing bars created by the paler tips of the greater and median coverts.

The Guadalupe Junco typically shows less white in the tail than the other juncos in our area. *Baja California. Paul Sweet*

Bryant claimed that the song of the Guadalupe Junco was, like that of most of the dark-eyed juncos, a simple loose trill resembling that of a Chipping Sparrow. In fact, as recordings online demonstrate, the island bird's song is considerably more complex and considerably more variable; one song type may be repeated several times before the singer switches to a second, finally returning to the first. The songs comprise several different syllable types uttered in varying combinations; they may include a trill, a buzz, and a selection of bright, whistled syllables. One simple song type begins with a cheerful series of slurred notes, *chiptewee*, followed by a medium-slow trill.

RANGE AND GEOGRAPHIC VARIATION

The Guadalupe Junco is a permanent resident of the pine and oak woods in the northern parts of the eponymous island, which lies 150 miles west of Baja California. In 2013, those habitats still covered only about 500 acres of Guadalupe's total land area of 94 square miles.

The species does not migrate, and there are no records of vagrants away from the island. There are no recognized subspecies.

The odd olive-brown of the Baird Junco's back contrasts with the dove gray head and colorful pastel underparts. *Baja California Sur. David Krueper*

BAIRD JUNCO
Junco bairdi

Lyman Belding's collecting tour of southernmost Baja California in the winter of 1882–1883 had two objectives. He spent considerable time over his four-month stay with the archaeological anthropologist Herman ten Kate, excavating middens and graves in search of traces left by the Pericú people, who had been exterminated a century earlier. Of the skeletons they discovered in the burial caves, Belding sent one to the Smithsonian, while ten Kate provided six to his colleagues in France.

When they were not busy desecrating native tombs, the two scientists turned their attention to natural history, the second focus of Belding's trip. In early February, Belding set off alone for La Laguna, in the mountains of Baja's Sierra de la Victoria. He found "the trail leading to Laguna . . . the longest, highest, and possibly the worst" in these mountains, "which were probably never previously explored by any collector." The effort paid off handsomely, however, when, on reaching the lower edge of the pines, Belding encountered "a beautiful new Snowbird,"

which he dispatched and sent to Robert Ridgway at the Smithsonian for description, specifying that the new bird was to be named for Spencer Baird, "in consideration of [his] valuable ornithological services . . . in field and office, not the least of such services being his original, full, and accurate descriptions of so many North American birds." Ridgway, finding the bird "pretty and very distinct," obliged, concluding his formal description with the observation that the Baird Junco "is so markedly distinct . . . from all its congeners as to really need no comparison with any of them."

FIELD IDENTIFICATION

Purely casual in Baja California Sur, and unexpected at the high elevations preferred by the Baird, the Oregon Junco is unlikely to be confused with the local resident species. In addition to the obvious yellow eye, the Baird Junco differs from even the most colorful of Oregon Juncos in the rather long bill's yellowish lower mandible and the pale throat, quite unlike the Oregon Juncos' conical pink bill and striking dark hood.

Alden H. Miller saw in the Baird Junco "a pale, dwarfed representative" of the Yellow-eyed Juncos of southern Mexico and Guatemala, from which it is separated by the Gulf of California and the western slope of the Sierra Madre. The pastel cinnamon back, broad cinnamon flanks, and lack of abrupt color changes on the blended upperparts distinguish the Baird, as does its shorter wing, tail, and foot.

The Baird Junco is a sturdy, medium-sized sparrow with a moderately long tail, short wing, round belly, and round head with a fairly long, bicolored bill. The adult is unstreaked, with a pale gray head and paler breast and throat, dark lores, soft cinnamon back and wings, and extensively orange-cinnamon flanks; the color of the breast sides bulges inward, in some individuals creating a nearly complete but ill-defined cinnamon breast band. Most birds show a buffy patch on the nape.

The outermost tail feathers average more extensively white than in Mexican Yellow-eyed Juncos, a character visible only in flight or when birds spread the tail to preen or flirt it in anxiety; in most, the

The hot, dry mountain forests inhabited by this species are quite unlike the cooler habitats favored by its more northerly congeners. *Baja California Sur, January. Rick Wright*

The bicolored bill of the Baird Junco is noticeably long in many individuals. *Baja California Sur. David Krueper*

Like other juncos, the juvenile Baird Junco is coarsely streaked dull dusky beneath. *Baja California Sur. David Krueper*

outer web of the outermost tail feather is entirely white, the inner web variably marked with black. Rectrix 5, the next tail feather in, is more variable, but nearly half of Baird Juncos have the outer web of that feather all or nearly all white, unlike the Yellow-eyed Junco. In males, there is a slight tendency for the Baird Junco to show more white spotting on the outer web of rectrix 4.

Juvenile Baird Juncos are largely buffy gray, streaked dusky everywhere but the center of the grayish belly.

The song of the Baird Junco is characterized by great variety, with a wide range of syllable and phrase types and shorter, less frequent trills than in the Yellow-eyed Junco; the high number of unique syllables and the low number of repeated syllables is diagnostic of this species. The song has been compared to that of a House Wren, a *Passerina* bunting, or a Rufous-crowned Sparrow. A light, cheerful melody without pronounced cadence, the song creates an impression of blithe aimlessness, high and low notes mixed into a relaxed twittering warble. When a trill is included, it tends to be high-pitched, wiry, and short, unlike the more rolling, lower-pitched trills that make up the Dark-eyed Junco's song.

The common call is a low-pitched and harsh *chak*, often doubled or given in a rapid accelerating series that ends in rasping notes somewhat recalling those of a Canyon Towhee. There is also a very short, high-pitched *tseet*, insectlike and often faint, closely resembling the corresponding note of mainland Yellow-eyed Juncos.

RANGE AND GEOGRAPHIC VARIATION

The Baird Junco is a strongly sedentary resident of a 50-mile stretch of the Victoria Mountains, which stretch north from nearly the southern tip of the peninsula to El Triunfo. Most records are from oak, pinyon pine, and madrone forests at elevations above 4,000 feet, in an area between Miraflores and the type locality, La Laguna; the juncos remain on their breeding grounds all year, only very rarely descending as low as 1,800 feet.

The uniform underparts, lack of a conspicuous hood, and extensive red on the wings and back are every bit as suggestive as this species' eye color. *Arizona, December. Brian E. Small*

YELLOW-EYED JUNCO
Junco phaeonotus

It is likely that any given type specimen of a North American bird has been examined fewer than ten times since it was collected years, decades, or even centuries ago. The average, however, is very much higher, largely because of the efforts and activities of William Bullock Sr.

In 1823, Bullock and his son left England for a newly independent Mexico, hoping to make their fortune in silver. They quickly bought mining rights at Temiscaltepec, 90 miles from Mexico City—only to find that the "mine" had been salted and they had been duped. The elder Bullock returned to London, arriving in November of that same year, bringing with him treasures of a different sort: hundreds of artifacts and natural history objects, including specimens of no fewer than 200 species of birds. Six months later, in April 1824, Bullock opened the doors of a lavish Mexican exhibition in his Egyptian Museum; 50,000 of London's curious visited the show before it closed in September 1825 and Bullock returned to Mexico, his son, and their silver mine.

No individual or institution in England was interested in purchasing Bullock's Mexican collections in full, and so they were disposed of in private sales and at auction. Before the birds were dispersed, though, Bullock made the specimens available to William Swainson, who published an exhaustive catalog before returning them to their owner for liquidation.

Among the specimens was a sparrow from the mesas of Temiscaltepec, ashy gray above and whitish below, with a rusty back and wing coverts and a broad, white-edged tail. Recognizing the bird as new to science, Swainson named it *Fringilla cinerea*, the Ash-gray Finch. The specimen went back to Bullock, and was presumably included in "the smaller bird collection," which brought at auction the sum of £11, 11s.

Nearly all of the birds from Bullock's Mexican exhibition have been lost or destroyed over the past nearly two centuries, among them the type specimen of *Fringilla cinerea*, known today as the Yellow-eyed Junco. But before it vanished, that tiny gray bird had been seen by vastly more eyes than thousands of other types combined, lying quiet and long unexamined in the museum drawers.

FIELD IDENTIFICATION

Conspicuous, tame, and often abundant, this soft-gray sparrow with the uncanny face is easily identified in the cool, shady mountain forests it inhabits. Except in winter, when a few Yellow-eyed

The upper mandible is steely gray, the lower yellowish. Yellow-eyed Juncos are as likely to sing from a concealed lower branch as from a treetop. *Arizona, May. Brian E. Small*

Juncos descend to mid-elevation canyon feeders frequented by flocks of Gray-headed, Oregon, and Pink-sided Juncos, the ranges of the Yellow-eyed and Dark-eyed do not overlap.

Breeding birds of the two species approach each other most closely in Arizona, where the low valley of the Salt River separates the Yellow-eyed Juncos of the Pinal Mountains from the Red-backed Juncos of the Sierra Ancha, a scant 40 miles away. In September 2004, an adult junco was discovered in the high Pinals that may well have been a hybrid between the two:

> Its plumage was most like that of Yellow-eyed Junco but the eye was a dark brown, and the call note was somewhat intermediate.

In the published photograph, the eye of that bird appears to be not dark brown but dull yellowish brown, paler than the deep red-brown eye of the Red-backed but darker than the staring orange-yellow of the Yellow-eyed Junco. The tertials and wing coverts and inner secondaries are extensively reddish, extremely so for a typical Red-backed Junco. Intriguingly, the Pinal junco very closely resembles Robert Ridgway's illustration accompanying the account of the discovery of the Yellow-eyed Junco in Arizona in 1874; like the putative hybrid from the Pinals, the bird in the plate combines red greater coverts and secondaries with a vaguely yellow-tinged muddy brown eye.

Possible hybrids aside, adult Yellow-eyed Juncos are easily recognized by their two-toned bills, bright wings, eerily bright eyes, and weird waddling limp on the ground. Perhaps because of their tendency to come very close to humans in their hopeful search for crumbs, they often seem larger than they are. That tameness means that they are not often seen in flight, but the swooping habit, white outer tail, and tinkling metallic calls make them unmistakable even then.

Juveniles may be less immediately recognizable on first view. They are streaked above and on the head, speckled and spotted below, with a dull, pale brown eye and dark gray bill. They share the adult's extensively rusty wing, bright back, and white outer tail feathers, and like their parents, shuffle and scoot along the ground, a gait very different from the sprightly jumps and hops of Dark-eyed Juncos.

The calls of this species are unmistakably junco-like, including dry *tek* notes and higher, sharper *tick*s and *tseet*s; in breeding season, quiet mouse-like metallic notes can be heard even at night, calling to mind the light tinkling of a loose zipper pull. The Yellow-eyed Junco's *tek* calls are softer and less "square" than the corresponding vocalizations of the Dark-eyed, and the *tseet* is shorter and less obviously trilled.

While it requires some experience to distinguish the calls of the two species (and much more to determine that an individual is giving "intermediate" calls), the song of the Yellow-eyed Junco is very different from that of the Dark-eyed Juncos breeding in the United States and Canada. Rather than the loose, colorless trill of their relatives, male Yellow-eyeds sing a bright, varied melody, including fast and slow trills and warbling or slurred phrases. Vaguely finchlike, but with the thin tone and high pitch of some wood warblers, the song is usually given from a concealed perch at mid-height in a ponderosa pine, even more misleading to the observer who has grown used to the bird's normally terrestrial habit. Many songs begin with an introduction of two to four down-slurred notes, followed by a jumbled trill and a chattering ending: *tiu tiu tiu drdeedrdee chipadee*. In other, shorter songs the introduction is faster, followed by a slower trill at a lower pitch, with no cadence, resembling a truncated House Wren's song. The most puzzling versions, capable of confounding even the most experienced observer, are two-parted, dispensing with the trill such that a reeling introduction is followed by a higher-pitched phrase with an aggressive, rolling cadence: *pleasedpleasedpleased tomeetcha!* Songs collected in New Mexico's Big Burro Mountains "varied from being most similar to Bewick's Wrens on one extreme to Spotted Towhee on the other." (D. Griffin, in litt.)

RANGE AND GEOGRAPHIC VARIATION

Represented in the southernmost parts of its range by the race *alticola*, the Yellow-eyed Junco is a common permanent resident of humid pine, oak, and spruce forests and agricultural habitats in the high western mountains of Guatemala and extreme southeastern Chiapas, usually at elevations above 8,000 feet. The Mexican range of this subspecies, known in English as the Guatemala Junco, is restricted to the east of the Sierra Madre southwest of the Rio Grande of Chiapas.

The dark bill, black lore, and startlingly yellow eye give the adult of this species a sinister appearance. *Arizona, May. Brian E. Small*

The ranges of *alticola* and *fulvescens* are disjunct, separated by some 40 miles of unsuitable lowland habitat on both sides of the border between Guatemala and the Mexican state of Chiapas. This Chiapas Junco, as *fulvescens* is known, is limited to the pine forests in the mountains between the Rio Grande and the Rio Jataté, from San Cristóbal to Teopisca; these mountains, attaining elevations of only 8,000 feet, are lower and drier than those used by *alticola*.

The nominate subspecies is widespread in the mountains of southern Mexico, from southern Jalisco across Michoacán, the state of México, and Hidalgo to Veracruz, thence south to Oaxaca; it is a common permanent resident of mixed and coniferous forests from about 4,000 feet in Veracruz to 14,000 feet in southern Jalisco. Two isolated populations occupy the Sierra Nevada de Colima in Jalisco and the Sierra Madre in Guerrero; both were originally described as distinct subspecies (*colimae* and *australis*, respectively), but neither is currently considered separable from nominate *phaeonotus*.

The pale northern subspecies *palliatus* intergrades with *phaeonotus* from Nayarit and Nuevo León south, with intermediates thought to be most frequent in the Sierra Madre of Durango and southern Coahuila. The shift in plumage from paler, gray-sided birds to darker, buffy-sided birds "constitutes an even gradient between extremes"; as a "purely arbitrary boundary of races," Miller proposed the border between Durango and Zacatecas.

North of Sonora and Chihuahua, *palliatus* is a common and characteristic resident of the Madrean "sky islands" of southeastern Arizona, easily found in shady mountain forests and well-wooded canyons above 5,000 feet in elevation; they are most abundant at elevations higher than 6,000 feet. The surveys conducted in the last eight years of the twentieth century for the Arizona breeding bird atlas found the species throughout the Pinaleño, Santa Catalina, Rincon, Santa Rita, and Chiricahua Mountains at appropriate elevation; they were less ubiquitous in the drier Huachuca and Mule Mountains. The most northerly Yellow-eyed Juncos in the world breed in the Pinal Mountains, their only outpost north of the lands acquired by the United States in the Gadsden Purchase. In New Mexico, the species is found breeding in the Animas and Big Burro Mountains.

The Yellow-eyed Junco is largely resident through its range, with only weak seasonal movements taking birds to lower elevations. Nevertheless, individuals occur infrequently in autumn and winter in New Mexico and Arizona mountain ranges where they are not known to breed, and as of 2015, there are seven accepted records from west Texas, spread through the seasons. There are several recent

The juvenile is heavily but finely streaked, with rich chestnut upperparts and wings. The lower mandible and the eye, both very dark in newly fledged birds, lighten quickly to yellow. *Arizona, August. Laurens Halsey*

August and September records from the San Mateo, Mogollon, and Manzano Mountains of New Mexico. (D. Griffin, in litt.)

Two recognizable subspecies of the Yellow-eyed Junco occur north of Mexico's Transvolcanic Belt. The nominate race, *phaeonotus*, resident across central and much of southern Mexico, is darker gray above and often on the breast than *palliatus* ("cloaked"), which breeds from southern Arizona and New Mexico south to the range of *phaeonotus*. The rump of *palliatus* is light olive-gray, that of *phaeonotus* deeper olive; *phaeonotus* also has buffy flanks. On average, *palliatus* shows more white on the outer three pairs of tail feathers, though there is considerable overlap in this character.

To the south, isolated in the interior mountains of Chiapas, the race *fulvescens* has a duller brown back and rump than *phaeonotus*, with which its range does not overlap. The bill of *fulvescens* is longer and much thicker at the base, while the wing and tail are shorter than those of *phaeonotus*.

The southernmost subspecies of the Yellow-eyed Junco, *alticola*, is resident in the highlands of western Guatemala and southern Chiapas; its range does not overlap with the isolated, more northerly distribution of *fulvescens*. This Guatemala Junco is browner on the sides and much paler on the underparts than the buffy-flanked, olive-breasted *fulvescens*; its bill is shorter but even deeper at the base than that of *fulvescens*.

The Red-backed Junco combines the blackish lore and bright red back of the Gray-headed Junco with the dark bill of the Yellow-eyed Junco, with which some individuals share a red "bleed" onto the wing. *New Mexico, May. Michael Retter*

RED-BACKED JUNCO
Junco hyemalis dorsalis

For all their training and prestige, army surgeons on the southwestern frontier were subject to the same mundane hardships undergone by every soldier. Nevertheless, many military doctors stationed in such exotic locations as Arizona and New Mexico voluntarily lengthened their already strenuous days to collect and preserve natural history specimens, whether in cooperation with an eastern museum or simply in pursuit of private diversion.

In the years he spent in New Mexico, from 1851 to 1856, Thomas Charlton Henry devoted nearly all the time not dedicated to his official duties to ornithology, his "favorite pursuit," collecting birds both for the Academy of Natural Sciences in Philadelphia and for his own eager interest. Reassigned to Fort Fillmore in 1853, Henry reported that on the nearly 400-mile march from Fort Webster

> among the timber generally are to be found many curious birds, peculiar to the country, some specimens of which are yet undescribed . . . . I procured a large num-

> ber of fine birds (new, many of them) on the route, and prepared them while traveling rapidly through the country,

regularly staying up past midnight to skin the day's proceeds.

Most of Henry's genuine discoveries were published by others, but in 1858, he formally introduced to science two species he had collected near Fort Stanton in southwestern New Mexico: the Crissal Thrasher and the Red-backed Junco, which Henry named *Junco dorsalis*. Baird affirmed the specific distinctness of the bird that same year, and the species was illustrated in print for the first time in 1860—with a carefully prepared image depicting Henry's New Mexico type specimen with yellow eyes.

Henry's belief that he had discovered a new species was uniformly shared until the early 1870s, when Henry Henshaw, on comparing the Red-backed Junco of Arizona and New Mexico with the Gray-headed and the Yellow-eyed Juncos, determined that the three were probably best considered "only separable as varieties rather than as distinct species." Spencer Baird, Thomas Mayo Brewer, and Robert Ridgway speculated briefly that the intermediate appearance of the Red-backed might be due to its status as a "hybrid between *caniceps* and *cinereus*" and not a discrete taxon at all. Elliott Coues advocated an even more extreme solution a decade later, eschewing the practice of

> almost all late writers [of] shuffling them about in the vain attempt to decide which are "species" and which "varieties." All are either, or both, as we may elect to consider them; for the degree of difference between almost any two of the nearest related ones is about the same . . . . Upon this understanding the recognizable styles of *Junco* may all be treated alike, lumped into a single highly variable species ranging from Alaska to Mexico.

The American Ornithologists' Union adopted Henshaw's view in its first *Check-list*, published in 1886. Altering the species name to *phaeonotus* after it was pointed out that *cinereus* was preoccupied, the AOU continued to list the Red-backed as a northern, dark-eyed race of the Yellow-eyed Junco for the next 60 years.

In his monographic study of the juncos, Alden H. Miller rejected the traditional view of a close relationship between the Red-backed and the Yellow-eyed Juncos to instead treat the Gray-headed and the Red-backed as conspecific, given that

> they replace each other geographically and interbreed freely in certain isolated areas that are intermediate geographically.

Miller's classification was adopted by the AOU in 1945. Though the species, assigned the confusing English name Gray-headed, survived the lumping of all other brown-eyed juncos in 1973, it, too, was ultimately reduced to subspecific status not long thereafter, the Red-backed and Gray-headed Juncos together constituting a subspecies group within the Dark-eyed Junco, the uncomfortable status they still enjoy today.

FIELD IDENTIFICATION

Leaving aside the question of intergrades with other Dark-eyed Junco subspecies and hybrids with the Yellow-eyed Junco, the Red-backed Junco poses few real identification challenges, and recognition is made even more straightforward by its sedentary habits.

This is a fairly large, plump, pale gray junco; the dark mask and bill create a stern, even sinister expression. Unlike the Yellow-eyed Junco, this (like all Dark-eyed Juncos) moves with spry hops and jumps rather than slow-motion creeping.

The Red-backed shares with the closely related Gray-headed Junco a dark brown eye, gray underparts, and, most signally, a sharply defined, brightly contrasting red back, in the field often appearing as a colorful triangle. The rather long, square or shallowly notched tail is mostly or entirely white on the outer two feathers; nearly all also show white on rectrix 4 as well. Like other juncos, ground-feeding Red-backeds often flirt and flash the tail to reveal the white patterning.

Most Red-backed Juncos have gray scapulars and tertials, though in approximately 20 percent of birds, the outer webs of those feathers are edged, more or less broadly, with bright buff; similar markings occur in the same proportion in the Gray-headed Junco. Unlike in Gray-headed Juncos, in a small minority of Red-backed Juncos the red of the back spills over onto the tertials, scapulars, and greater coverts, creating a "shawled" impression like that of the Yellow-eyed Junco. Though in some cases that wing pattern may be the result of hybridization, as described above, it also occurs far to the north of any areas where the two might be expected to come into contact in the breeding season.

Beneath, the Red-backed Junco is pale gray, with little contrast between the gray flanks, belly, and lower breast. The upper breast and throat are soft whitish gray, sometimes obviously contrasting in the field with the slightly darker gray of the remainder of the underparts. The darker, duller gray throat of the Gray-headed Junco contributes to that bird's more uniformly hooded appearance, while Red-backed Juncos show a greater contrast between the throat and the slightly darker gray sides of the head. As in the Yellow-eyed Junco, the crown is almost always gray, with only a very few individuals showing reddish spots or streaks.

Some Gray-headed Juncos, most of them perhaps birds in their first winter, show a very faint buffy overlay on the flanks, while that same area is purer, paler gray in Red-backeds. Especially bright-sided birds otherwise identical to Gray-headed Juncos are probably hybrids or backcrosses with the Pink-sided Junco, a very attractive combination known as the Ridgway Junco.

The bill of most Red-backed Juncos appears obviously long and heavy in the field, an effect resulting from the combination of its actual great size and its odd colors. Rather than the light pinkish of both mandibles of the Gray-headed Junco, "pure" Red-backeds have a distinctly bicolored bill, dark gray or blackish above and medium-dark pearly gray beneath. The culmen and tip of the upper mandible are always dark, with any white or pink limited to a small spot or two at the very base, either in the center or on the edges; even when they are present, such pale areas are often invisible in the field unless the bird can be viewed from directly above.

The underparts are very uniform, with no marked change from breast to belly. *New Mexico, December. Brian E. Small*

Intergrades with the Gray-headed Junco can be extremely difficult to distinguish from either of the parental types. Miller found that the Gray-headed and the Red-backed "interbreed freely" on the Kaibab Plateau and in New Mexico's Zuni Mountains, with a more or less continuous gradation in head and bill color, tail length, foot length, and bill depth. The darkest, smallest, palest-billed birds in such zones of contact are identical to typical Gray-headed Juncos, while typical Red-backed Juncos are paler beneath and on the head, larger, and darker-billed. An apparently "pure" Red-backed Junco should have a culmen that is entirely black or shows just a small whitish spot where the culmen meets the forehead; the culmen of a "pure" Gray-headed Junco will be pinkish or whitish from the forehead to or very nearly to the bill tip, which may show a very small spot of dark. Intermediates show upper mandible patterns ranging from nearly black with an extensively pale culmen and bill edges to nearly pink with an extensive dark tip and patches or blotches at the bill base. Every effort should be made to clearly assess the bill pattern of any suspected Red-backed Junco out of the subspecies' expected range.

Juvenile Red-backed Juncos show the same contrasting pattern above as adults, though the gray of the upperparts is browner or yellower and the back triangle duller brown; the crown, nape, and back are marked with scattered narrow blackish streaks. The flight feathers, especially the tertials, may have broad brown edges on the outer webs. Beneath, the flanks are dull buffy with brown streaks, the throat and breast gray or grayish buff with coarse streaks often forming a loose band across the breast. The eye is dull muddy brown, and the bill is dark but less neatly patterned than that of the adult. Juveniles are usually in the company of adults; they molt into an adultlike formative plumage in late summer, and thus cannot be expected to be seen except on the breeding grounds.

The calls of the Red-backed Junco include ticking notes and high, slender lisps, similar or identical to the corresponding vocalizations of other Dark-eyed Junco subspecies. The most frequently heard calls from perched birds include a loud descending *tu*, with an abrupt attack and lingering decay; and a "square," dry *tek* sometimes repeated three or four times; and a medium-length, rather low-pitched, slightly buzzy *dsee*. Startled birds take flight with a panicked series of tinkling *tsit-tsit-tsit* notes.

While the calls are unexceptional, this subspecies' song is unlike that of more northern races of the Dark-eyed Junco, in its length and complexity sometimes even recalling the trilling and warbling performance of the Yellow-eyed Junco. The song is usually composed of two phrases, one a slow series of musical whistles, slurred up or down, and the other a faster trill; either phrase can begin the song. Some birds instead sing two canary-like trills, one slower and lower-pitched than the other, the cheerful tone and rapid transition between phrases bringing to mind a truncated goldfinch song.

RANGE AND GEOGRAPHIC VARIATION

Almost all of the Red-backed Juncos in the world occupy a well-circumscribed range in central and northern Arizona and western New Mexico, where

The thick but long bill is dull silvery pink below and variably dusky above, including at the base. *New Mexico, December. Brian E. Small*

this is a common and characteristic species of high mountain and canyon forests. In Arizona, breeders are found in ponderosa pine habitats between 6,000 and 11,000 feet, from the New Mexico border in the White Mountains west and north to the south rim of the Grand Canyon and Mingus Mountain in the north-central part of the state; the range extends across the Natanes, Mogollon, and Coconino Plateaus and into the Mazatzal, Sierra Ancha, and Bradshaw Mountains. There are also recently discovered populations nearly as far north as the Utah border in the Virgin Mountains, and as far west as Mount Dellenbaugh, the Uinkaret Mountains, and the Aubrey Cliffs. On the northern edge of its range along the Kaibab Plateau, this subspecies encounters and intergrades with the Gray-headed Junco. Just south of its normal breeding range, the Red-backed Junco has apparently hybridized with the Yellow-eyed Junco in the Pinal Mountains, producing a bird with "plumage . . . most like that of Yellow-eyed Junco but the eye . . . dark brown, and the call note . . . somewhat intermediate."

In New Mexico, the Red-backed Junco breeds in the Capitan, Sierra Blanca, Sacramento, Guadalupe, Magdalena, and Mogollon Mountains, in the Mimbres Mountains of the Black Range, and in the Pinos Altos Range; it also probably nests in the Datil and San Mateo Mountains. Red-backed Juncos were first noted in the Big Burro Mountains in 1995, when immature birds were discovered on the north side of Burro Peak. (D. Griffin, in litt.) Since then,

hybridization with the Yellow-eyed Junco has been regularly recorded in the Big Burro Mountains. (D. Griffin, in litt.) Just south of the New Mexico border, the Red-backed Junco is also an uncommon resident of high elevations in the Guadalupe Mountains of Texas.

Red-backed Juncos were first discovered in New Mexico's Big Burro Mountains at the end of the twentieth century; Yellow-eyed Juncos were also documented in that range beginning in 2003, and in 2006 mixed pairs, apparent hybrids, and birds with intermediate eye color and plumage patterns were observed. The population of both species in this range remains very small, with no more than 25 pairs as of 2016. (D. Griffin, in litt.)

Relatively sedentary at any time of year, Red-backed Juncos move only short distances in the winter, most spending the cold season in nearby valleys downslope from the breeding grounds. Birds of this subspecies are extremely scarce south to Pinal, Pima, Santa Cruz, and Cochise Counties in Arizona, and it does not occur—or is not observed—every winter in even the most heavily birded localities there. Migrants and winterers are also recorded from the Big Hatchet Mountains in New Mexico. They are more frequently observed, though in very small numbers, from October to March in the Davis and Chisos Mountains of west Texas, which appear to be regular wintering grounds for at least a few individuals, presumably from the breeding population in the Guadalupe Mountains just to the north.

Miller did not accept several nineteenth-century winter records of this subspecies from northwestern Mexico. There is one recent late October sighting from Chihuahua (eBird).

Where their breeding ranges meet in northern Arizona and New Mexico, Red-backed and Gray-headed Juncos intergrade, producing individuals clearly assignable to neither subspecies—and almost certainly, birds indistinguishable from one or the other "pure" type. Such birds are occasionally detected outside of the breeding range, as far north as the eastern foothills of the Rocky Mountains in central Colorado. Observers fortunate to encounter an apparent Red-backed Junco out of range should pay particular attention to the wing pattern, flank color, throat color, bill pattern, and bill size; even then, many birds must be recorded simply as "Red-backed or Gray-headed Junco or intergrade."

In spite of the existence of isolated populations in widely separated mountain ranges, the Red-backed Junco does not exhibit any plumage or morphological variations that can consistently be correlated with geography.

The red back, dark lore, and only slight contrast between breast and belly identify this junco; a hint of pink on the flanks may suggest some Pink-sided Junco ancestry. *New Mexico, January. Brian E. Small*

GRAY-HEADED JUNCO
Junco hyemalis caniceps

The exploration of the American West in the mid-nineteenth century led to the discovery of a surprising and often bewildering range of new birds, all carefully described—and formally named—by the collectors themselves or by the scientists who received the specimens in the great museums back East. It would be some decades before books and journals routinely reproduced color photographs, but many of the contemporary descriptions were illustrated with appealing lithographs, often depicting the very specimen discussed in the text.

Unfortunately, the southwestern juncos benefited from this practice rather less than did other new and possibly new birds. The first published image of the northern *palliatus* race of the Yellow-eyed Junco showed the bird with a decidedly brown iris; when the type specimen of the Red-backed Junco was portrayed for the first time, it was with a yellow eye and a largely light-colored bill. And the Gray-headed Junco first appeared in print with a bright white belly and a slender, dark-tipped yellow bill.

Such inaccuracies, major and minor, should usually not be laid at the door of the artist, working in the best case from well-prepared but eyeless skins, and in the worst case from fragmentary specimens or even only descriptions, themselves often no more complete. This was the case for the Gray-headed Junco, described as a new species in 1853 on the basis of birds collected in Mexico, Texas, and Arizona and shipped to the Academy of Natural Sciences in Philadelphia. It was John Cassin who first noticed that those specimens were different from the other juncos in the drawers, but he reserved the right of naming to Samuel Woodhouse, who had secured one of the birds, a female, in northern Arizona's San Francisco Peaks.

Although Woodhouse's description of the bird he named *Struthus caniceps* clearly refers to the bird known today as the Gray-headed Junco, it presents a signal anomaly in its analysis of the type specimens' bill color. Like that of the more northerly races of the Dark-eyed Junco, the bill of the Gray-headed is soft pink above and below, with or without a minute dark spot at the tip. Woodhouse, however, says that the upper mandible is "dark brown, almost black," a character proper to the Red-backed and Yellow-eyed Juncos but not to the present form, a fact that would lead C. Hart Merriam to believe that Woodhouse had unwittingly based his new species in part on skins of the Yellow-eyed Junco.

Conveniently, the only surviving specimen of those Woodhouse described is the female he collected in the San Francisco Peaks in October 1851; the other, "male birds have now vanished without a trace." As the Yellow-eyed Junco does not occur in north-central Arizona, the status of Woodhouse's name and its relation to any junco taxon remained discomfitingly mysterious until Alden H. Miller reexamined the Arizona skin in 1933. Bill color is notoriously transitory in specimens, and it would have been unreasonable after more than 80 years to expect to find the original flesh-pink preserved. What Miller found, however, was the solution to the puzzle of Woodhouse's anomalous description. The bird was badly shot: the bill tip was broken, and the central portions of the upper mandible were in fact dark, "probably due to blood stain."

Woodhouse published the description of the Gray-headed Junco nearly two years after he shot the Arizona female—a bird he had obviously not himself considered particularly notable when it was collected. Not until 1858, when Spencer Baird published his review of the four Gray-headed Junco skins available to him, was Woodhouse's erroneous description of bill color corrected; but even then, post mortem changes in the specimens misled the Smithsonian scientist into describing the birds' bills as "yellowish, black at the tip"—exactly as it would be depicted two years later when the species was first illustrated in print.

FIELD IDENTIFICATION

The Gray-headed Junco is a very distinctive sparrow, readily identified by the striking contrast between its rather dark gray head, breast, and flanks and its rich reddish back. This race differs from its apparent closest relative, the Red-backed Junco, in its pink, sometimes dark-tipped bill; in adults, the red of the upperparts is restricted to the back rather than spilling into the wing coverts and tertials, as it sometimes does in the Red-backed Junco. Where Red-backed Juncos show contrast between the pale, even whitish throat and the darker gray sides of the head, Gray-headed Juncos are more evenly gray on the head, throat, and breast.

Intergrades with the Red-backed Junco occur in northern Arizona and New Mexico. These intermediate birds may be slightly larger than "pure" Gray-headed Juncos, with a paler gray head and throat and shorter, more slender bill. While typi-

The Gray-headed Junco's confusing early publication history notwithstanding, in life this junco's bill is all pink or pink with restricted dark at the tip or base. *New Mexico, January. Brian E. Small*

cal Gray-headed Juncos show black only as small spots at the very base and, sometimes, at the tip of the upper mandible, the bill tips of intergrades are at least one-quarter black, and the black at the base of the upper mandible is often continuous across the culmen.

Junco populations in several mountain ranges in California and Nevada appear to represent the product of the interbreeding of Gray-headed Juncos with Oregon Juncos. In September 1930, A. J. van Rossem took a short collecting excursion in the Charleston and Sheep Mountains of southern Nevada. The 31 juncos shot there had the red back and gray head of the Gray-headed Junco, but they were darker on the head and breast, with more contrast between the breast and belly than shown by the smoothly blended underparts of the Gray-headed; their flanks were "more or less pinkish tinted."

Van Rossem described these birds as a new subspecies of the Oregon Junco, which he named *mutabilis*. Miller viewed this population instead as one of hybrids and backcrosses. Van Rossem had found that "the uniformly red back in both sexes," with no individuals showing an intermediate color there, "rather militates against this supposition," but Miller pointed out that in fact some birds, especially those with Oregon-like head color and breast pattern, did show "some mixture" of red and brown on the back; the predominance of bright, Gray-headed–like back colors was explainable simply as genetic dominance and the historically "great proportion of *caniceps* individuals entering into the parentage" of the hybrid population.

Similar intermediates are also found in the Panamint, Inyo, White, and Walker River Mountains, where Oregon-like plumage characters predominate over characters reflecting Gray-headed parentage. Miller speculated that this population had been produced by interbreeding between pure Oregon Juncos and "some birds of mixed blood from a mountain range such as the Charlestons," or that continued arrival of Oregon Juncos from the west had "in large part swamped" the evidence of historical Gray-headed influence. The range of visual intergradation is shifted slightly in Nevada's Grapevine Mountains, where among the intermediate specimens Miller also discovered individuals closely resembling "pure" Gray-headed Juncos; one female, a Gray-headed type but with some mixture of brown in the red back, was found to be mated with a male showing the black head, brown sides, and yellowish brown back of the Oregon Junco:

> The general aspect of the pair was that of opposite extremes, and if juncos take account of such matters, they must have appeared to each other as foreign species.

In the Pine Forest Mountains of Humboldt County, Nevada, Miller discovered an even more complex situation when he collected a mated pair of juncos, the female of which was visually a pale, dull Oregon Junco and the male a hybrid between the Gray-headed and Pink-sided Juncos. He suggested that similar or more straightforward interbreeding might be responsible for the fact that Nevada's Gray-headed Juncos include "a larger proportion of individuals with darker type of head," the result of occasional injections of Oregon Junco genes.

The most colorful of the juncos seen in the United States and Canada is the Ridgway Junco, a hybrid between the Gray-headed Junco with its bright back and the Pink-sided Junco with its pastel underparts. The taxonomic history of this form is complex, involving

> the only case in which the person for whom a species was named . . . diagnosed that taxon's illegitimacy . . . and then named one of the hybrid's parental taxa . . . after the very colleague who had tried to do him the same favor.

In 1858, Spencer Baird noted among "a large collection of *Junco* from Fort Bridger" in southwest Wyoming a "remarkable" individual "having the sides reddish . . . although with the dorsal features of *caniceps*." He tentatively identified the specimen as a hybrid between the Gray-headed and Oregon Juncos. A dozen years later, however, he reconsidered, noting that the combination of back and flank color was "very constant," without the variety of additional intermediate forms indicative of hybridism; Baird did not entirely reject the possibility of hybrid origin, but nevertheless determined that the bird was "entitled to a provisional appellation, even if a hybrid," and named it *Junco annectens*—only to recant in 1874, when he, Robert Ridgway, and Thomas Mayo Brewer again listed it as a hybrid.

In the first edition of the *Check-list*, the American Ornithologists' Union redefined Baird's *annectens* to apply not to the hybrid but to the "pure" Pink-sided Junco. Four years later, in 1890, Edgar Mearns reported his discovery in northern Arizona of a bird "above similar to *J. caniceps*; below indistinguishable from *J. annectens*," that latter name used in the sense of the AOU *Check-list*, in reference to the Pink-sided Junco. Not immediately recognizing the junco as noteworthy in the field, Mearns "preserved but a single specimen," and wrote that he could not

> now refrain from smiling at the recollection of my misdirected zeal in garnering series of specimens of Flickers, Long-crested [Steller] Jays, Black-headed

Grosbeaks and other conspicuous but well-known birds, while these Juncos . . . were almost ignored.

Apparently not so much as suspecting that "this handsome junco" was in fact Baird's old hybrid, Mearns named it *Junco ridgwayi*, the name under which it would enter the second, 1895 edition of the AOU *Check-list*.

The beneficiary of Mearns's nomenclatural gift, Robert Ridgway, raised the matter anew in 1897. Ridgway confirmed that Baird's *annectens* had in fact been described from a hybrid, and that Mearns had simply redescribed the same hybrid combination under the name *ridgwayi*. The name *annectens*, therefore, being based on a hybrid, was not properly available for application to the Pink-sided Junco, and so Ridgway, finding that form

> without a name, [took] pleasure in bestowing upon it the name *Junco mearnsi*, in compliment to my friend, Dr. Edgar A. Mearns, U.S.A.

Formally named or not, the hybrid still popularly known as the Ridgway Junco is the product of interbreeding where the ranges of the Gray-headed and the Pink-sided Juncos adjoin in northeast Nevada, southern Idaho, northeastern Utah, and southern Wyoming. It would no doubt be reported far more frequently, especially on the wintering grounds in Arizona and New Mexico, if it were known to more observers.

Mearns's description of this junco as a Gray-headed above and a Pink-sided below is apt, but neither he, Ridgway, nor Baird was aware of the variability of plumage exhibited by individual representatives of this cross. Ridgway Juncos can show any combination of the parent species' head color, flank color, and back color. Thus, they can resemble a Gray-headed Junco with an abnormally dark head, or a Pink-sided Junco with an unusually bright back; especially birds of the former aspect no doubt go unnoticed in the field most of the time.

Collecting in Utah's Wasatch Mountains, Miller was able to glimpse some of the social mechanisms that make interbreeding so frequent. On shooting one of a pair of juncos attending young, Miller found that it was a male hybrid with the back and flank patterns of the Pink-sided but with a paler, intermediate head. The female, an apparently normal Pink-sided Junco, was spared. Five hours after her mate had been collected, a new male had arrived, courting her with song and tail flitting. Miller shot this second male, a bird with pink flanks, intermediate head color, and a mixed back color. An hour later, a third male had attached itself

Originally described as a separate species, intergrades between the Gray-headed and Pink-sided Juncos can be very colorful, combining characters of both these junco kinds. *Colorado. Steven Mlodinow*

to the now twice-widowed female; the newcomer was quickly dispatched and found to be visually a Pink-sided Junco but with intermediate head color. By noon, a fourth male had given his life for science, victim to his interest in the bereaved female; this bird had the back of a Pink-sided, the flanks of a Gray-headed, and the head color of an intermediate junco. Miller wrote:

> I am doubtful that these males were all unattached previous to their interest in female X . . . . There was no doubt of the attraction of the female for all of them, however . . . . No intolerance was evidenced by the female. Some of the males gathered food for the young. This indicates disregard on the part of the junco for differences in colors of sides and backs.

The vocalizations of the Gray-headed Junco are indistinguishable in the field from those of other northerly Dark-eyed Junco subspecies. The calls include a dry *dek*, a bubbling *dewdewdew*, and a series of panicked *tzit* notes when disturbed. "The acoustic properties of song [and its] syntax [are] also uniform across dark-eyed junco subspecies," comprising a short tremolo of often rather loud, musical notes, sweeter than the song of most Chipping Sparrows and less vague and rambling than the song of the Orange-crowned Warbler, which it can otherwise resemble quite closely. Both male and female Dark-eyed Juncos also have a longer, more varied "short-range" song, including "short whistles, trills, warbles, call notes . . . and other sounds."

RANGE AND GEOGRAPHIC VARIATION

The most widespread of the breeding juncos of the southwestern United States, the Gray-headed Junco nests in coniferous mountain forests from southern Idaho, central and southern Nevada, Utah, and southern Wyoming south to central Colorado, northern Arizona, and northern New Mexico. In California, this subspecies breeds in the White and Inyo Mountains, with one summer record of a pair on San Diego County's Volcan Mountain. It intergrades with the Oregon Junco in the central, eastern, and northern mountains of Nevada, and with the Red-backed Junco on Arizona's Kaibab Plateau and in the Zuni Mountains of New Mexico. Interbreeding with Pink-sided Juncos in northeastern Nevada, southern Idaho, northeastern Utah, and southern Wyoming produces the colorful bird originally described as *Junco annectens*, now known to most observers as the Ridgway Junco.

The relatively short southbound migration takes place in September and especially October, with birds arriving on the wintering grounds in southern Arizona beginning the end of September. By late October, wintering birds are in place from the Texas Panhandle and Big Bend west across southern New Mexico and Arizona and northern Chihuahua and Sonora to southern California and northern Durango (eBird).

Autumn and winter strays occur regularly but infrequently as far northwest as Puget Sound, and as far west as California's Channel Islands. In San Diego County, California, Gray-headed Juncos are rare winterers on the immediate coast, but more common in the mountains. Recent records in Baja California Norte reach south to Ensenada and the Constitution 1857 National Park; there is also a record from southern Baja California Sur (eBird). Gray-headed Juncos are said to be scarce winterers and fall migrants in the Salton Sea area, with records concentrated in November, and individuals join wintering flocks of Oregon Juncos along the lower Colorado River in California and Arizona.

Similarly rare over the western third of Arizona, the Gray-headed Junco is very common in flocks wintering in the mountains and mid-elevation canyons of central and eastern Arizona. It winters commonly in the mountains of Utah, Colorado, and New Mexico as well, and is locally common or uncommon in the Trans-Pecos and Panhandle regions of Texas.

Gray-headed Juncos are scarce, low-density wintering birds elsewhere on the Great Plains, from North Dakota to central Texas. They are very rare and probably overreported east of the Missouri River, with reliable records east to Illinois.

Spring migration appears to begin in March, with most birds on the breeding grounds in April and May. Records of birds east of the normal range are much less frequent in spring than in autumn and winter, though there are Ontario records from May (eBird).

The Gray-headed Junco is currently treated as a single discrete subspecies of the Dark-eyed Junco, *Junco hyemalis caniceps*. Some males in Nevada are darker-headed than normal, a variation "ascribed to occasional hybridization in the past with [the Oregon Junco subspecies] *thurberi*, perhaps also currently."

A classic Pink-sided Junco like this one shows a pale, almost bluish head with a well-defined black lore and an orange-pink back and flanks, with the color bulging toward the center of the breast. *New Mexico, January. Brian E. Small*

PINK-SIDED JUNCO
Junco hyemalis mearnsi

Spencer Baird's corps of Smithsonian collectors brought together an impressive number of professionals and refined amateurs, from army officers to physicians, writers, and artists. But Baird also recognized skill and industry in the less educated, rewarding even the nearly illiterate with opportunities to contribute to the vast enterprise of nineteenth-century American natural history.

Of German immigrant heritage, Constantin Charles Drexler was a hospital orderly in Washington, D.C. He also worked as a taxidermist at the Smithsonian, where Baird described him as

> not very polished in his manners, . . . yet well behaved, intelligent, and obedient to instructions, and . . . highly accomplished in all that relates to his business.

Drexler would eventually be sent to arctic Canada in search of birds, mammals, and insects, but his first commission on behalf of the Smithsonian was to serve as assistant to James Graham Cooper, surgeon to the South Pass Wagon Road expedition. Cooper, Drexler, and the rest of their party arrived in Wyoming's Wind River Range in September 1857, too late in the season to continue into Utah; Drexler was among those who wintered on the Popo Agie River.

The expedition was underway again the following March, reaching Fort Bridger by the end of the month. There, on April 12, 1858, Drexler secured a sparrow that was shipped back to Washington, mounted, and duly labeled as an Oregon Junco. Almost 40 years later, Robert Ridgway pointed out that this specimen was an example of the "pure" Pink-sided Junco—in fact, the earliest such specimen known, the others originally used to describe the species having been reidentified as hybrids. Drexler, the coarse, unschooled menial laborer, had collected the type of a species unknown to science, more than fair recompense for Spencer Baird's faith in him.

FIELD IDENTIFICATION

It is no accident that when this handsome western sparrow is discussed, its name is so often preceded by adjectives such as "true," "real," and "genuine"—a tacit and salutary reminder that the identification of

Pink-sided Juncos, as much as that of any sparrow, requires caution, close observation, and, ideally, experience.

The Pink-sided is a rather large junco, its greater size often noticeable in the field when it is in company with the smaller, neater Oregon Juncos. The first impression is of a pale but colorful bird, without deep rust or saturated black tones anywhere. Instead, the back is dull brown with a yellowish pink overlay, not nearly as bright as the Gray-headed Junco's chestnut triangle but quite similar to the color shown by many female Oregon Juncos. On Pink-sideds, the brown of the back spills onto the innermost one or two greater coverts and the tertial edges.

The underparts are softly, smoothly blended. The undertail coverts and center of the lower belly are very pale whitish gray; the vent and some undertail coverts are often washed faintly with pinkish, sometimes forming a diffuse band across the vent. The eponymous color of the flanks and sides of the breast is an unusual pinkish buff or pastel cinnamon, without obvious yellow, brown, or gray. More distinctive than the quality of the color is its extent. The pink of the flanks reaches far toward the center of the belly, often making the bird look entirely pink-bellied. The same color typically bulges from the sides toward the center of the lower breast; in a significant minority, the pink reaches nearly or entirely across the lower breast, forming an almost bluebirdlike band beneath the gray of the upper breast and hood.

The lower edge of that hood meets the whitish gray or pink of the lower breast convexly—that is to say, the hood intrudes into and truncates the pale portion of the underparts, rather than marking one neatly rounded end of a pale ellipse.

The nape, crown, neck, ear coverts, and upper breast of adult Pink-sided Juncos are concolorous light gray, paler and bluer than in any Oregon Junco and usually slightly paler than in the Gray-headed Junco. As in that latter bird, the lores and supercilium in front of the Pink-sided's eye are contrastingly dark, satiny black in males and very dark slaty in females and young birds, which may have sparse brownish streaking on the nape and crown as well.

Pink-sided Juncos, especially the males, also average more white in the tail than do either Oregon or Gray-headed Juncos. Roughly one-quarter of males show some white on rectrix 3, and a very few have a trace of white even on rectrix 2, the next feather out from the central pair; rectrix 4 may be entirely white, giving some Pink-sided Juncos essentially as much white as black in the tail. Digital photographs of birds spreading the tail in takeoff or on landing can make it possible to assess this character, which is otherwise invisible in the field.

Before gaining a certain familiarity with Pink-sided Juncos, birders might find it difficult to distinguish this bird from many female Oregon Juncos,

The back and scapulars are only slightly duller than the breast sides, without deep chestnut tones. *New Mexico, December. Brian E. Small*

Particularly pale birds are less colorful above and below, but the contrasting lore and extensive and bulging pink of the breast sides are consistent. *New Mexico, December. Brian E. Small*

Cassiar Juncos, or even Slate-colored Juncos. That difficulty was not lessened by the poor, even inaccurate illustrations of the Pink-sided in virtually all of the standard field guides published in the twentieth century. Careful attention to back color, flank and breast pattern, and, especially, the darkness of the head and the contrast presented by the lores should permit the identification of all typical, genuine, real, and true Pink-sided Juncos.

Apparent hybrids between the Pink-sided Junco and adjacent subspecies complicate matters. Any potential Pink-sided Junco with a richly colored reddish back, grayish or mixed gray and pink flanks, or an unusually dark head with low-contrast lores should be considered critically; many such birds will be ultimately unidentifiable.

One relatively common hybrid combination is the bird known as the Ridgway Junco, discussed in greater detail under the Gray-headed Junco, above. Essentially a Pink-sided Junco below and a Gray-headed Junco above, this bird combines the rich rusty back of the latter with the pale head and pastel underparts of the former. Ridgway Juncos are most frequently recorded on the wintering grounds, especially in Arizona and eastern Colorado, though rare winter wanderers have been observed as far east as the Missouri River in Nebraska (P. Swanson, in litt.). Typical representatives of this hybrid exhibit very broad, bright-colored flanks and breast sides, though others may approach the pattern of pure Gray-headed Juncos in formative plumage, which may show buffy tones on the rear flanks.

Miller reports four "conclusive" hybrids between the Pink-sided and White-winged Juncos, two forms whose normal breeding ranges do not overlap today; he explained the interbreeding as the result of "occasional *mearnsi* wander[ing] eastward from their breeding range in the Big Horn Mountains of Montana and Wyoming to breed with *aikeni*" on the Pine Ridge or in the Black Hills. One specimen taken on the wintering grounds in Colorado combined the typical back color of the Pink-sided with the concave breast pattern of the White-winged; the flanks and breast sides were mixed pinkish and light gray. Though the wing coverts had prominent gray edgings rather than white tips, this individual's large size and the light gray of the flank satisfactorily ruled out the Slate-colored Junco as one of the parent species.

Other birds of apparent mixed ancestry showed a combination of brown and slaty on the back or of gray and pink on the flanks. The most subtly marked of Miller's hybrids was visually "chiefly" a White-winged Junco, but the hood met the breast in a convex rather than concave curve, and the dark head and upper breast contrasted with the pale slate breast sides. Similar birds would be easily overlooked, and should be sought and carefully described and photographed in flocks of White-winged and other juncos on the breeding and wintering grounds.

In south-central Idaho, Miller found a gradient between the Oregon Junco of the subspecies *shufeldti* and the Pink-sided Junco. Juncos breeding

Hybrids and intergrades with the White-winged Junco are regularly produced where those two junco kinds' breeding ranges overlap. *Colorado, December. Steven Mlodinow*

This apparent hybrid or backcross combines the pastel tones of a Pink-sided with the large size, large bill, and white-tipped coverts of a White-winged Junco. *Colorado, December. Steve Mlodinow.*

on the state's western border represented "normal" *shufeldti* Oregon Juncos, while more than half of the birds in the southeastern Sawtooth Mountains were closer to the Pink-sided Junco. Some 120 miles to the northeast, in the Lemhi Mountains, typical Pink-sideds replaced the Oregon Juncos entirely.

Between those areas, Miller evaluated the Idaho birds on head color—darker in the Oregon, paler in the Pink-sided Junco—and flank and breast sides pattern and color—moderately extensive brownish to fawn in the Oregon, yellowish or pinkish cinnamon and extending toward the center of the breast in the Pink-sided. The slight differences in back color are probably not discernible in the field, and indeed are barely visible in the hand in many cases. Birds intermediate in one or the other of these characters occurred from Payette Lake to Alturas Lake in the Sawtooths, where the visual average shifted from Oregon-like to Pink-sided-like before the latter form asserted itself definitively in the Lemhis.

The vocalizations of the Pink-sided Junco are indistinguishable in the field from those of other northerly Dark-eyed Junco subspecies. The calls include a dry *tek,* a bubbling *ditditdit,* and a series of panicked *tzit* notes when disturbed. "The acoustic properties of song [and its] syntax [are] also uniform across dark-eyed junco subspecies," comprising a short tremolo of often rather loud, musical notes, sweeter than the song of most Chipping Sparrows and less vague and rambling than the song of the Orange-crowned Warbler, which it can otherwise resemble quite closely. Both male and female Dark-eyed Juncos also have a longer, more varied "short-range" song, including "short whistles, trills, warbles, call notes . . . and other sounds."

RANGE AND GEOGRAPHIC VARIATION

The Pink-sided Junco breeds in the Rocky Mountains from Banff National Park and the Cypress Hills of Alberta and southwestern Saskatchewan south to Montana, from the Continental Divide east to Big Horn, Yellowstone, and Blaine Counties, and in eastern Idaho and the Teton and Wind River Mountains of northeastern Wyoming. It has been suggested that Pink-headed Juncos "are the most common birds in Yellowstone."

Autumn flocks form on the breeding grounds as early as August, as family groups merge. At this season, many young birds are still in juvenile plumage, but they undergo their preformative molt before leaving the breeding range to move south. The first flocks arrive on the wintering grounds in October.

The Pink-sided is a decidedly western junco, even in winter, when its range covers much of Utah, Colorado, New Mexico, eastern Arizona, the Trans-Pecos and High Plains of Texas, and northern

This winter bird appears to combine the large size, long bill, and white-tipped wing coverts of the White-winged with a clear hint of the back and flank color of the Pink-sided Junco. *Colorado, December.* *Steven Mlodinow*

Sonora and Chihuahua. This is the most common winter junco on the plains of eastern Colorado and is common in New Mexico, west Texas, the Oklahoma Panhandle, and westernmost Kansas, but its abundance decreases dramatically to the east. The most abundant wintering form in most of the Nebraska Panhandle, Pink-sideds are uncommon in the central part of that state and barely annual east to the Missouri River.

Nearly all reports east of the Great Plains probably pertain to misidentified juncos of other races. There are photographs from late fall and winter 1999 of a Pink-sided Junco in Northampton County, Virginia, and a specimen was secured in the upper peninsula of Michigan in 2001; New

York's first was discovered in late December 1999. The early winter of 2000–2001 brought one to a feeder in Pulaski County, Kentucky, where it was well photographed (eBird). Two of four reports in Nova Scotia were accepted by McLaren, while "photographs of [the other] two of these suggest otherwise." The difficulty, however, of evaluating photographs of claimed Pink-sided Juncos is made clear by published images of New York's second record, a female netted in May 2002 on Fire Island; the written account of the capture, by two cautious and authoritative observers, argues cogently that this was a Pink-sided Junco, but in the accompanying photographs, the head is too dark, the back too brown, and the sides too muddy for a Pink-sided.

To the west of the heart of the southwestern wintering range, the Pink-sided Junco is scarce but regular across western Arizona and southern California to the mountains of San Diego County, moving rarely as far as the lowlands of the Pacific Coast and north to northern California, Oregon, and, exceptionally, the Okanagan Valley of British Columbia (eBird).

Spring migration appears to be protracted. While some return to the breeding grounds as early as late March, others may still be in the winter range in Colorado and Arizona in April and May. At Yellowstone, the first breeders "appear suddenly in March," with all territories occupied by early June.

Miller found no clearly demonstrable geographic variation within this subspecies, though he noted that birds might be smaller in Montana north of Yellowstone and larger in the Bighorn Mountains of Wyoming.

This abundant western junco is most readily identified by the shape of the dark hood, which is sharply set off from the back and the breast sides. *California, January. Brian E. Small*

OREGON JUNCO
Junco hyemalis [Oregon group]

Examples abound of the difficulty ornithology has faced in determining similarity and difference among the juncos. But there is no more eloquent illustration than that provided by the Oregon Junco subspecies once named *connectens*, "linking," a bird whose identity and affinities remained a subject of debate for more than 40 years after it was originally described—and for a full quarter of a century after it was proved not to exist.

Elliott Coues described and named *connectens* in the second edition of his authoritative *Key*, diagnosing it as "possessing in varying degree the characters of" both Slate-colored and Oregon Juncos, combining a colorful Oregon-like back with somber gray sides, or the gray back of a Slate-colored with pink flanks. Coues designated no type locality. He claimed instead that the new subspecies was found "wherever the breeding range of the two comes together, and elsewhere during the migration."

Bizarrely, Coues boasted, in an address delivered to the National Academy of Sciences in the spring of 1884, that he had described and named his new junco sight unseen, "before any specimens were received":

> A given set of sub-specific characters were hypothetically assigned to a bird (*Junco connecteus* [sic] Coues) from a particular region; and, upon receipt of specimens, the hypothetical characters of the presumed sub-species were confirmed.

That "confirmation" of Coues's "hypothesis" came in the form of a post-facto type specimen, a junco shot in Colorado by William Brewster from a flock of White-crowned Sparrows.

Though Coues was a member of the committees that compiled the first and second editions of the AOU *Check-list* in 1886 and 1895, respectively, his *connectens*, "a good subspecies," was "accidentally overlooked" both times. "In fact," he wrote in 1897,

> it also escaped my own memory, until it was brought to mind by the description of *J. h. shufeldti* by Mr. Coale, in . . . 1887; since which time I have been intending to bring up the case for final readjustment,

but have meanwhile been much preoccupied with other than ornithological affairs. Mr. Coale's *shufeldti* of 1887 is my *connectens* of 1884.

The AOU committee agreed, and the Eighth Supplement to the *Check*-list, published that same year of 1897, promptly revived Coues's *connectens*.

In 1901—more than a year after the death of Elliott Coues—Robert Ridgway "carefully examined and compared" the type specimen of *connectens*. He found it to be a "very nearly typical" Slate-colored Junco, with just a hint of the colorful back and flanks of an Oregon Junco; Ridgway asserted that the Brewster skin was actually of a young female Slate-colored or a hybrid between the Slate-colored and Oregon Juncos. He concluded, in language suggesting more than mild disapproval, that

> at any rate it has nothing to do with the form . . . to which the name *connectens* was unadvisedly applied by action of the A.O.U. Committee.

Unadvisedly or not, the AOU twice declined to judge the conflict between *connectens* and *shufeldti*. The resolution it finally offered in the 1910 edition of the *Check-list* was probably intended as a Solomonic compromise: Coues's scientific name *con-*

The male can be startlingly colorful, with an orange back and sides and a velvety black head. *California, January. Brian E. Small*

nectens was retained for a bird to be known by the English name of Shufeldt's Junco.

In his 1918 monograph of the juncos, Jonathan Dwight declared it appropriate to definitively

> dispose of the "*Junco hyemalis connectens*" of Coues . . . a curious mixture of fact and fancy . . . clearly a specimen of [the typical Slate-colored Junco with] the characters common to sex and season.

All juvenile juncos are finely streaked black above and below, but the overall colors and their distribution recall those of an adult. *California. Luke Tiller*

Four years later, however, Harry Swarth, after long study of the birds of northern British Columbia, mooted "a contrary opinion," namely, that the type specimen of Coues's *connectens*, which Brewster had shot 40 years before, was the representative of yet another junco taxon, the Cassiar Junco.

The AOU remained silent on the matter until 1931, when the new, fourth edition of the *Check-list* dispensed with the name *connectens* entirely. The committee justified its decision twice, in slightly different, and tentative, ways. In a footnote to the account of the Slate-colored Junco, the old *connectens* is described as "variously regarded" as a hybrid, a synonym, or the "distinct race" Swarth had asserted; the appendix to the *Check-list* offers the more definite statement that Coues's "name is considered to be based on a hybrid and not applicable" to the genuine Oregon Junco subspecies known, once again, as *shufeldti*.

The equivocation came to an end in 1933, when Alden Miller examined, "with much interest," the specimen once trumpeted as confirmation of the existence of Coues's "hypothetical" subspecies. Miller was able to refute Swarth's identification of the bird as a Cassiar Junco and, largely on the strength of its concave rather than convex lower hood, diagnosed it as "a fairly typical female" Slate-colored Junco. Miller published his findings in 1941, 57 years after Elliott Coues had first speculated on the identity of a bird that did not exist.

FIELD IDENTIFICATION

Among the most colorful and neatly marked of the passerellid sparrows, Oregon Juncos are often instantly recognizable in the field. A classic adult Oregon of either sex combines gray tail, rump, and wings with a brown or yellowish brown back, bright reddish or reddish brown flanks, and a deep gray or black hood.

It is not those colors, however, but rather their articulation that observers should evaluate in order to confirm an identification, especially where Oregon Juncos are uncommon. A goodly proportion of Slate-colored Juncos, especially females in their first winter, also show brown on the back and rust on the flanks, and the inexperienced or the ambitious can easily be misled by such birds. The presence of such color is less significant than its distribution, especially where the edges of the hood meet the back and the breast. A "good" Oregon Junco should show clear and rather abrupt contrast between the solid brown of the back and the black or medium to dark gray of the nape; in Slate-colored Juncos, that brown is more likely to spill over onto the nape and crown.

In a convincing Oregon Junco, the hood, whether gray or black, meets the white of the breast in a well-drawn line, and forms a sharp corner with the colorful breast side; the effect is somewhat like that of a Spotted or Eastern Towhee. Most importantly, the lower edge of the hood is convex, invading and truncating the white of the breast below it. In other words, the white of the underparts can be conceived of as a pointed oval "eclipsed" in front by the dark of the hood, or more imaginatively as an egg with its larger end carefully scooped out by a tarnished silver spoon.

The convex hood of the Oregon Junco is shared by the Pink-sided Junco. That latter bird is never black-headed, and always shows a strongly contrasting lore patch, satiny black in males and deep sooty in females; even the palest-headed Oregon Junco is uniformly gray on the face, lacking the squinting scowl so characteristic of the Pink-sided. The flanks and breast sides of most Oregon Juncos are yellowish pink or rust, unlike the Pink-sided's unusual cinnamon-pink, and the color is narrower and more evenly restricted, without the bulge toward the center of the breast visible in so many Pink-sided Juncos. The back of the Pink-sided Junco is a paler brown, usually without the rusty tones dominant in Oregons. Overall, the colors of the Pink-sided Junco are softer and more smoothly blended than the distinct blocks that make up the plumage patterns of the Oregon Junco.

The Cassiar Junco, and the often very similar hybrids and intergrades between the Oregon and Slate-colored Juncos, pose far greater problems, both intellectually and practically: it remains uncertain

The dark hood intrudes sharply onto the breast, creating a colorful corner between the breast and the front of the folded wing. Some paler-headed birds show decidedly blackish lores, but the breast pattern usually distinguishes them from Pink-sided Juncos. *New Mexico, January. Brian E. Small*

Some Oregon Juncos, most of them probably young females, can be nearly as dull as female Slate-colored Juncos, but differ consistently in the shape of the dark hood. These may recall the Pink-sided Junco, but note the Oregon's darker, less bluish head, deeper red back, more limited pink on the breast sides, and smaller size. *New Mexico, December. Brian E. Small*

what the Cassiar Junco is, whether a genetically discrete, geographically stabilized, self-replicating form or simply a scattered collection of randomly produced hybrids and backcrosses. In the field, male Cassiar and Cassiar-like juncos are essentially monochrome Oregon Juncos, reproducing the neatly defined patterns of the Oregon but in whites and shades of gray and black. Females and female-plumaged birds, on the other hand, are often completely indistinguishable from either Oregon or Slate-colored Juncos or, most signally, from hybrids and intergrades that blur the features of both. Exceptionally brown juncos with a hint of convexity in the hood and slightly brighter brown flanks and breast sides probably occur in each of those categories, and conscientious observers take note—and take notes—of such birds without surrendering to the temptation to assign them definite names.

Even individuals that seem to unequivocally combine features of the Slate-colored and Oregon Juncos may simply represent a point near the middle of the spectrum of intergradation rather than necessarily being the offspring of a mixed pair. Nowhere is a healthy agnosticism more appropriate than in junco identification.

Juvenile Oregon Juncos are seen only on the breeding grounds, where they undergo the preformative molt in late summer and very early autumn. Streaky and brown, they share the adult's white outer tail feathers, confiding manner, and sharp, cheerful calls. In the more northerly subspecies,

birds in their first winter often retain some juvenile greater coverts and tertials, making it possible to age many of them in the field, an important first step when confronted with an odd junco.

Oregon Juncos have the same varied array of call notes as the other northerly Dark-eyed Juncos, including a dry *dek* and sharp, faintly buzzy *tzit* flushing and flight notes. The robotlike *dew* call often runs into a short series when perched birds are alert or stressed. The long song, given only by males, is a short tremolo of often rather loud, musical notes, sweeter than the song of most Chipping Sparrows and less vague and rambling than the song of the Orange-crowned Warbler, which it can otherwise resemble quite closely. Both male and female Dark-eyed Juncos also have a longer, more varied "short-range" song, including "short whistles, trills, warbles, call notes . . . and other sounds."

RANGE AND GEOGRAPHIC VARIATION

The various Oregon Juncos breed south and west of the much more extensive breeding range of the Slate-colored Junco, from southern and southeastern Alaska across central British Columbia and southern Alberta to southwesternmost Saskatchewan, thence south to Montana, Nevada, southern California, and the mountains of northern Baja California.

British Columbia has some of the highest densities anywhere of breeding Oregon Juncos. Breeders are found on Haida Gwaii and Vancouver Island,

and as far north as 50 miles north of Prince George; *montanus* apparently comes into contact with breeding Slate-colored Juncos at Flood Glacier, 50 miles from the mouth of the Stikine River.

The Oregon Junco *montanus* breeds in the western third of Montana, with the rest of the state occupied by White-winged or Pink-sided Juncos. Oregon Juncos breed at scattered localities in northern Nevada, with notable concentrations in the Carson Range and other western mountains.

In California, *thurberi* is a common resident with a discontinuous breeding distribution in coastal and near-coastal areas of central California south to San Diego County; *pinosus* is resident south to San Benito and Monterey Counties. The two intergrade in San Luis Obispo County.

The two Oregon Junco subspecies breeding in Baja California are endemic residents of the Sierra Juárez and the Sierra de San Pedro Mártir.

Juncos of all forms are hardy birds, and some populations of Oregon Juncos are resident or exhibit only short-range or altitudinal movements in cold weather. Birds that do migrate begin the southward movement in September or October; early arrival dates at California's Salton Sea and along the lower Colorado River are in late September. Winterers arrive at the southern edge of the wintering range in Sonora in mid-October; there and in Chihuahua, recent winter "arrivals" have been recorded in late December and January (eBird), almost certainly reflecting the seasonal occurrence more of observers than of juncos. At the other end of their range, an adult male of the nominate race was collected in October on Wrangel Island.

Oregon Juncos winter regularly as far north as southern Saskatchewan, Alberta, central British Columbia, and coastal southern Alaska, with a scattering of records from the Aleutians as far west as Adak (eBird). They winter throughout California, including the offshore islands; unsurprisingly, they are least common in the desert habitats of the east and southeast. They also occur regularly in fall and winter in northern Baja California, south exceptionally to El Rosario (eBird).

They are common to uncommon winter residents in appropriate habitats in all of the United States' mountain states and on the western Great Plains, most abundant in the snow-free south. The subspecies *montanus* winters in western Texas east to the Edwards Plateau, but is rare or very rare to the east and south in the state. In Nebraska, birds of that same subspecies, possibly reinforced by the occasional *shufeldti*, are abundant in the west, but vastly outnumbered by Slate-colored Juncos in the east.

East of the Missouri River, Oregon Juncos are very low-density winterers in the states adjacent to the Mississippi River and south to Tennessee and northern Alabama (eBird). Birds identified as Oregons are seen every fall and winter, often at feeders, east to the mid-Atlantic coast, New England, Quebec, and the Canadian Maritime Provinces. The easternmost records are from Nova Scotia, with a date range from October into spring; the individuals collected or well photographed have proved to be of the expected subspecies *montanus*.

The spring migration of the Oregon Junco tends to be less conspicuous than that of the Slate-colored Junco, without the repeated waves that characterize the northbound journey of the eastern and northern bird. Most Oregons appear to leave the wintering grounds by April, with some lingering casually late into the month or even into May as far south as southern Arizona.

The Slate-colored Junco breeds over a much larger range than the Oregon, but is generally recognized as comprising only two or three subspecies. The Oregon Junco's more restricted breeding distribution, however, takes in a wider variety of habitats, a circumstance reflected in its geographic diversification into six or seven generally recognized races. The treatment below follows the taxonomy adopted in the two most recent authoritative world checklists, Howard and Moore and the IOC list.

John Townsend collected the type specimen of the nominate race *oreganus* on the wintering grounds. This subspecies breeds on the Pacific Coast from southeastern Alaska south to British Columbia's Haida Gwaii and Calvert Island. The plumage varies significantly between the sexes. The head and lore are medium-gray (black in males), the back and flanks bright pinkish brown (darker reddish brown in males); the division between the gray or black of the hood and the brown of the back is sharply defined. Both the back and flank are more reddish than in other Oregon Juncos, and the tail averages less white in this nominate race than in *shufeldti* and *montanus*.

Breeding south and east of *oreganus*, the two races *shufeldti* and *montanus* are often lumped by modern authorities. The former, named for the ornithologist Robert W. Shufeldt, occupies coastal forests from British Columbia south through western Oregon. The flanks are cinnamon-brown, the hood gray in females and black in males. This race is very slightly smaller and may be somewhat more reddish on the back than the similar *montanus*, the subspecies breeding from interior British Columbia and western Alberta south to eastern Washington and Oregon and northern Montana and Idaho. Most of the Oregon Juncos that occur as vagrants in eastern North America are thought to be *montanus*.

The commoner of the two races breeding in California is *thurberi*, found from southern Oregon through the coastal mountains to Marin, Napa, and

Sonoma Counties. This subspecies is also a common or abundant resident of both slopes of the Sierra Nevada, though it does not breed in the drier canyons from Inyo to Kern Counties; at the southern extreme of its range, this otherwise montane-breeding species has colonized the coastal lowlands of San Diego County. The principal distinction from *shufeldti* and from *montanus* is the "lighter, more pinkish back"; the head is darker than that of *montanus* and, at least in the Sierra, that of *shufeldti*.

The palest of the Oregon Juncos breeding north of Mexico is *pinosus*, named for the type locality, a grove at the Point Pinos lighthouse on the southern side of Monterey Bay. Its discoverer, Leverett M. Loomis, later the curator of ornithology at the California Academy of Sciences, noted that males of his new subspecies could be "distinguished at a glance from *thurberi*," the nearest breeding population, "by the decided slate-gray aspect of the fore breast, jugulum, and throat," different from the decidedly black plumage of those areas in *thurberi*. Miller, however, found overlap in the head color of the two subspecies, and emphasized instead the "ruddier" back and sides of *pinosus*; the back color is "russet" or cinnamon, the sides dull pinkish recalling the "less vivid types" of the Pink-sided Junco. It is also slightly smaller and longer-billed than *thurberi*. Entirely nonmigratory, *pinosus* is resident from the Golden Gate and Carquinez Straits south through San Benito and southern Monterey Counties.

Two Oregon Junco subspecies are resident in disjunct ranges in Baja California. The 1,800 square miles of the Sierra Juárcz, arising just south of the United States border and extending some 90 miles south, are home to the race *pontilis*, described by Harry Oberholser in 1919. The Townsend Junco, *Junco oreganus townsendi*, inhabits the next range to the south, the Sierra de San Pedro Mártir; it may show some very short-distance altitudinal movement in winter.

Oberholser's *pontilis*, so named ("bridging") for its range and appearance intermediate between *thurberi* and *townsendi*, is grayer-headed and duller-backed than the former, with flanks more extensively pinkish ("nearly vinaceous fawn").

The Townsend Junco was described in 1889 as a distinct species, a status it enjoyed until it was formally lumped by the AOU in 1908. The head is paler and grayer than in *pontilis*, the back usually duller. The tail averages slightly more white, and the coloring of the flanks is even more extensive, a feature that led Dwight to propose that the Townsend and Pink-sided (and Guadalupe) Juncos were conspecific.

In much of the West and Southwest, wintering sparrow flocks may contain representatives of two or more Oregon Junco subspecies. Observers should always be alert to difference, but they should resist the temptation to identify birds to subspecies in the field, especially given the likely abundance of intermediates.

Pale wing bars, produced by white or whitish covert tips, are frequent in this and in other juncos. *New Mexico, January. Brian E. Small*

This winter bird is the classic Cassiar-like junco, with a very dark hood sharply set off from the very slightly browner back and intruding well onto the breast; the flanks and breast sides are gray with a barely discernible overlay of pinkish. It closely resembles the lectotype now in the American Museum of Natural History. *Colorado. Cathy Sheeter*

CASSIAR JUNCO
Junco hyemalis cismontanus

A prominent politician once said, "It depends upon what the meaning of the word 'is' is," an observation applicable in all its profundity to the status and distribution—and even the existence—of the Cassiar Junco.

Jonathan Dwight provided the first account of this bird in his 1918 study of junco plumages, calling it "a *hyemalis* [Slate-colored Junco] darkened by the *oregonus* strain [Oregon Junco]." Four years later, Harry Swarth discovered juncos like those Dwight had described breeding along the Stikine River in British Columbia.

Dwight believed that his juncos were most likely mere intergrades or hybrids between Oregon and Slate-colored Juncos, for which he proposed the "convenient" label, in significant quotation marks, "*cismontanus*" ("from this side of the mountains"). Swarth, in contrast, argued that the 44 specimens he collected were representatives of

> a "good subspecies," a geographic race, in the sense that the birds over a certain area (of undetermined extent but undoubtedly a considerable stretch of country) exhibit a combination of characters distinguish-

ing them from other described forms, and they remain true to these peculiarities within as close limits as do most recognized subspecies.

In 1941, Alden H. Miller affirmed Swarth's assessment and applied Dwight's name *cismontanus*—without the hesitant punctuation—to what he described as a population "of hybrid origin, now stabilized" and self-perpetuating, which in the breeding season occupies a discrete geographic range "to the exclusion of other forms."

Not everyone accepts even such cautious formulations of the ontological status of the Cassiar Junco. But it remains a bird—whether a "good subspecies," a randomly produced hybrid, or some of both—that sparrow watchers should continue to look for and to document whenever it is encountered.

FIELD IDENTIFICATION

Dwight did not identify a type specimen for his *cismontanus,* a circumstance Miller corrected in 1941 when he also designated as the lectotype, a type selected from a series of the same taxon, the skin of a male collected February 13, 1905, in New Westminster, British Columbia, by Allan Brooks. Formerly in the collection of Jonathan Dwight and now no. 402559 in the American Museum of Natural History, that specimen provides the starting point for the identification of the Cassiar Junco.

If all Cassiar Juncos were as distinctive as the type specimen selected by Miller, a remarkably dark and neatly contrasting male, identification would be far less challenging than it so often is. Such well-marked Cassiar Juncos are distinctive, though most more subtly so than Miller's lectotype. In its overall slaty grayness, the Cassiar closely recalls a Slate-colored Junco, but the darkness of head, breast, and nape approaches that of a male Oregon, creating a more striking and clearly defined contrast between the "hood" and the rest of the upperparts than is visible in any but the very blackest of male Slate-coloreds.

As is generally the case in junco identification, more significant than the colors themselves is their distribution: the blackish nape is sharply set off from the grayer or brownish back, and the dark of the breast meets the paler gray of the flank and the white of the upper belly in a well-defined straight line, where the white underparts of a Slate-colored Junco curve up and into the breast. Thus, a male Cassiar Junco, like the Oregon Junco and unlike the Slate-colored, shows an abrupt corner where the breast and the flanks meet, and the white of the upper belly is concave. It is this character that allowed Miller to identify the type of Coues's *connectens* as in fact a Slate-colored Junco of the nominate subspecies.

Many male Cassiar or Cassiar-like Juncos are readily identifiable, but females are much more challenging. The female Cassiar shares the hooded appearance of the male and of the Oregon Junco, but the flanks are gray or strongly washed with gray over brown, thus recalling a Slate-colored Junco. Away from the breeding range, many are probably not distinguishable from female Oregon Juncos, while many others are likely identical in the field (or in the hand) to especially brownish, probably first-cycle, female Slate-colored Juncos. David Sibley's painting of the female of what he calls the "Slate-colored population nesting in the Canadian Rockies," which is based not on a single specimen but on a composite of the artist's observations in Alaska, Alberta, Missouri, New England, New Jersey, and California (David Sibley, pers. comm.), is most instructive, depicting as it does a bird that differs from the Oregon Junco shown on this book's facing page only in a nearly invisible tinge of gray on its brownish flank. A bird resembling the Sibley painting could certainly be overlooked in the field as an Oregon Junco, while an individual with an even slightly less abrupt transition between hood, flank, and breast would likely be indistinguishable from a female Slate-colored Junco.

The intermediate appearance of Cassiar Juncos reflects the historical source of this population. In designating it "of hybrid origin," Miller meant

This Cassiar-like individual is noticeably warmer on the back, and the flanks are more evenly mixed gray and pinkish. *Colorado. Cathy Sheeter*

Even in winter, apparent male Cassiar-like juncos such as this one are easily picked out in flocks of Oregon or Slate-colored Juncos. *Colorado. Cathy Sheeter*

that the founding birds, thousands of generations ago, were the offspring of mixed pairs, one parent an Oregon and the other a Slate-colored Junco; repeated pairings between these hybrid offspring and their descendants resulted in a population of birds consistently displaying the plumage characters observed today. The same scenario surely recurs even now, when the occasional pure Oregon or Slate-colored Junco strays into the breeding range of the other group and breeds with a mate of the "wrong" subspecies, producing young of intermediate appearance that can replicate the appearance of a true Cassiar Junco. The impossibility in any given case of eliminating hybridism makes it strictly accurate away from the breeding grounds to speak only of an "apparent" Cassiar Junco or a "Cassiar-like"

Junco rather than definitely assigning a winter bird to the subspecies *cismontanus*.

Cassiar Juncos are indistinguishable in their vocalizations from either the Oregon or the Slate-colored Junco.

RANGE AND GEOGRAPHIC VARIATION

Any pairing of Slate-colored and Oregon Juncos can presumably result in offspring that are visually identical to Cassiar Juncos. The subspecies *cismontanus*, the stabilized population of descendants of such crosses, breeds east of the coastal mountain ranges from the south-central Yukon into British Columbia, reaching the basin of the Stikine River. It also breeds east of the Continental Divide south to Jasper in the upper drainage of the Peace River in east-central British Columbia and west-central Alberta. In the east, it probably reaches the headwaters of the Dease and Liard Rivers.

Because the Cassiar Junco away from its breeding grounds cannot be definitively distinguished from hybrids between Slate-colored and Oregon Juncos, it is not possible to determine the migratory paths or winter range of this subspecies with any certainty. Where Cassiar-like birds occur in winter with regularity and in abundance, it seems likely that many or most are genuine *cismontanus*; such areas include southeast Alaska, the lower mainland of British Columbia, southern California, the southern Rocky Mountains, and west Texas. Improved awareness on the part of field observers has shown that juncos resembling the Cassiar are uncommon to fairly common in many parts of the Great Plains, in some places in apparent contradiction to the specimen record.

There are winter reports of Cassiar and Cassiar-like Juncos from much of the eastern United States and Canada. Many refer to brownish Slate-colored Juncos, while others show pronounced Oregon-like characters that suggest recent hybridization. Classic black-hooded, gray-flanked Cassiar-like males are reported in the East far less often, but birds of that type have been photographed as far east as Nova Scotia and New Jersey (Cathy Sheeter, in litt.). Even in those cases, there can be no certainty that the birds are not hybrids rather than genuine *cismontanus*, but it is important that they be recorded as "Cassiar-like" or "*cismontanus*-type" rather than simply being disregarded.

Variation in head color, back color, and flank color in the Cassiar Junco is confused by interbreeding with both Slate-colored and Oregon Juncos. Toward the northeast of the Cassiar Junco's breeding range, males become more and more Slate-colored-like, until their distinctive characters "to all purposes disappear in pure *hyemalis*."

Cassiar-like juncos in female plumage away from the breeding grounds are infrequently detected and can be extremely difficult to identify. The clear border between this bird's rather dark gray hood and brown back and the intrusion of the hood onto the breast are suggestive. *Colorado. Cathy Sheeter*

The male Slate-colored Junco is slaty gray, with a variably darker head and bright white underparts. *Minnesota, January. Brian E. Small*

SLATE-COLORED JUNCO
Junco hyemalis hyemalis and carolinensis

Abundant over much of the United States and Canada in winter, this easily recognized and confiding little bird has been part of American culture from the beginning. A "companion of every child" of European settlers, the junco was equally well known to the continent's native residents, whose names for the species were diligently recorded (and often imaginatively analyzed) by white explorers. Today, more than four centuries after John White first documented the species for European science, the Slate-colored Junco remains the sparrow species most likely to be recognized by non-birders.

It is also the only passerellid sparrow still in possession, if now increasingly tenuous, of a genuine folk name in American English. In mid-January 1749, the naturalist Pehr Kalm—one of the most prominent of Linnaeus's "disciples"—reported that when the temperature fell to minus 22 degrees Celsius in the Scandinavian immigrant settlement of Raccoon, New Jersey,

> a kind of small bird, called "snöfogel" by the Swedes and by the Englishmen "chickbird," came into the buildings.

English-speakers in the middle colonies also used the name "snowbird," and that is the common name, dating at least to Catesby's day and beyond, that can still be heard today over much of the species' winter range.

Familiarity is not always the same as understanding. Alexander Wilson found the snowbird in winter "by far the most numerous, as well as the most extensively disseminated, of all the feathered tribes that visit us from the frozen regions of the north," and considered them "almost half domesticated . . . about the barn, stables, and other outhouses . . . not only in the country and villages, but in the heart of our large cities." But no matter how abundant and well known, the snowbird was still an object of superstition. On a visit to New England, Wilson discovered that it was "pretty general[ly]" believed that the gray finches "crowding around the threshold early in the morning, gleaning up the crumbs, appearing very lively and familiar" in the cold season metamorphosed in the spring into brown birds with neat rusty caps.

> I had convinced a gentleman of New York of his mistake in this matter, by taking him to the house of a Mr. Gautier, there, who amuses himself by keeping a great number of native as well as foreign birds. This was in the month of July, and the Snow-bird

appeared there in the same colored plumage he usually has. Several individuals of the Chipping Sparrow were also in the same apartment. The evidence was therefore irresistible; but as I had not the same proofs to offer to the eye in New England, I had not the same success.

Some of the most sophisticated natural historians of their day entertained equally implausible ideas about the junco. Writing in Buffon's *Natural History of Birds*, the French ornithologist Philippe Guéneau de Montbeillard declared the "Jacobin bunting" a mere climatic variety of the Snow Bunting, a view unequivocally refuted 40 years later by Louis Pierre Vieillot, who had had field experience of both species during his stay in the young United States. Thomas Nuttall, in both editions of his *Manual*, essentially plagiarized Wilson's account of the Common Snow-bird, but added that the same species, so widespread in the lowlands of eastern North America, also occurred in southern Europe, where it was restricted to the highest mountains. Nuttall had perhaps been misled, understandably, by Charles Bonaparte's endorsement of Wilson's repeating of Catesby and Bartram's invalid name *nivalis,* but the lapse would expose him to the scorn of Maximilian zu Wied-Neuwied:

> Nuttall considered this bird to be the European Snow Finch, to which it has not the remotest similarity.

By Nuttall's day, the earliest formal book names for the American bird—including John Latham's "Black Bunting" and John Gould's "Winter Finch"—had been entirely displaced by the triumphant folk name Snow Bird, variously written as one word or two, with or without a hyphen, modified or not by a word such as "common," "blue," or "black." Through the year 1885, every published account, technical or popular, employed some version of "snowbird" for the gray and white sparrow's English name, while "*Junco*" was in use exclusively as the scientific name of the genus.

That changed abruptly in 1886. "Snowbird" still survives in the popular lexicon, but starting with the first edition of the AOU *Check-list*, the ancient and familiar was replaced by a new standard English name. "Junco," obviously, was imported directly from the scientific name, and "slate-colored" is likely borrowed from Alexander Wilson's description, in which he called "the head, neck, and upperparts of the breast, body, and wings . . . of a deep slate color." Ludlow Griscom would later claim that "amateur bird students . . . never objected to" the new name, an assertion that would seem to suggest

that birders have changed greatly over the past 130 years.

Though none of the other publications issued by the committee responsible for the first *Check-list* justifies, or even mentions, the change in the English name of the juncos, two motivations come to mind. By 1886, "Snowfinch," "Snowflake," and "Snow Bunting" had become well established in the literature, making "Snowbird" a potential source of confusion similar to that created by Guéneau or Nuttall. Perhaps more significantly, the decades immediately before the appearance of the *Check-list* had seen the discovery and formal description of a number of new members of the genus *Junco* from landscapes and latitudes not generally thought of as snowy. To call *Junco insularis* the "Guadalupe Snowbird" could easily have been found jarring by anyone familiar with the climate of that tropical island.

Such considerations meant and mean nothing to observers whose only interest in these birds is their cheering presence on a wintry doorstep or roadside. For them, the junco will always be the Snowbird.

FIELD IDENTIFICATION

Among the most neatly marked of the passerellid sparrows, Slate-colored Juncos are often instantly recognizable in the field. A classic adult male Slate-colored Junco combines an elegant dark gray tail, rump, and wings with a darker, often blackish hood covering the head and upper breast; the gray parts of the plumage contrast strikingly with the white outer tail feathers, belly, and breast.

Females and birds in their first winter are often less distinctive, though easily recognizable as juncos. A goodly proportion show brown on the back and rust on the flanks, and the inexperienced or the ambitious can easily be misled into identifying such birds as Oregon or Cassiar Juncos. The presence of such color is less significant than its distribution, especially where the upper breast meets the back and the breast. On the Slate-colored Junco, the color of the back—whether gray or brown—continues onto the nape without marked transition, and the dark of the front of the hood flows from the upper breast onto the breast sides and flanks. There is no pronounced corner where the hood meets the breast sides, and the border between the hood and the lower breast is concave, allowing the white of the underparts to curve up into the dark of the upper breast. Thus, the white of the underparts can be conceived of as a complete pointed oval, with the large end of the egg bulging into the breast.

The Cassiar Junco, and the often very similar hybrids and intergrades between the Slate-colored and Oregon Juncos, pose problems, both intellectually and practically: it remains uncertain what the Cassiar Junco is, whether a genetically discrete, geographically stabilized, self-replicating form or simply a scattered collection of randomly produced hybrids and backcrosses. In the field, male Cassiar and Cassiar-like Juncos reproduce the neatly defined patterns of the Oregon but in the Slate-colored's shades of gray and black. Females and female-plumaged birds, on the other hand, are often completely indistinguishable from Slate-colored Juncos or, most signally, from hybrids and intergrades that blur the features of both. Exceptionally brown juncos with a hint of convexity in the hood and slightly brighter brown flanks and breast sides probably occur in each of those categories, and conscientious observers take note—and take notes—of such birds without surrendering to the temptation to assign them definite names.

Even individuals that seem to unequivocally combine features of the Slate-colored and Oregon Juncos may simply represent a point near the middle of the spectrum of intergradation rather than necessarily being the offspring of a mixed pair. Nowhere is a healthy agnosticism more appropriate than in junco identification.

Juvenile Slate-colored Juncos are seen only on the breeding grounds, where they undergo the preformative molt in late summer and very early autumn. Streaky and brown, they share the adult's white outer tail feathers, confiding manner, and sharp, cheerful calls. In the northern, nominate subspecies, birds in their first winter often retain some juvenile greater coverts and tertials, making it possible to age many of them in the field, an important first step when confronted with an odd junco: if a suspiciously plumed bird is in its formative plumage, it may well be "just" a Slate-colored Junco, while an adult showing similar colors and patterns could prove more interesting, even if no more identifiable.

Slate-colored Juncos have the same varied array of call notes as the other northerly Dark-eyed Juncos, including a dry *dek* and sharp, faintly buzzy *tzit* flushing and flight notes. The robotlike *dew* call often runs into a short series when perched birds are alert or stressed. The long song, given only by males, is a short tremolo of often rather loud, musical notes, sweeter than the song of most Chipping Sparrows and louder, lower-pitched, and less vague and rambling than the songs of most Pine Warblers. Both male and female Dark-eyed Juncos also have a longer, more varied "short-range" song, including "short whistles, trills, warbles, call notes . . . and other sounds."

RANGE AND GEOGRAPHIC VARIATION

Between them, breeding Slate-colored Juncos of the nominate race *hyemalis* and southern *carolinensis* inhabit nearly all of the coniferous forest of northern and northeastern North America. Though the

The gray of the Slate-colored Junco's breast is continuous and concolorous with the gray of the flanks; the white belly intrudes in a curved line into the breast. *Alaska, June. Brian E. Small*

Slate-colored is often called the "eastern" form of the Dark-eyed Junco, it in fact breeds west to Alaska's Brooks Range and the Yukon Delta, far beyond the nesting range of any of the Oregon Juncos.

The Slate-colored Junco breeds throughout Alaska except for the extreme north and west; it is replaced on the state's southeastern coast by the Oregon Junco. In British Columbia, Slate-colored and Cassiar Juncos breed in the northern and northeastern parts of the province, where they meet with the Oregon Junco on the lower Stikine River. This is the breeding junco of boreal forest habitats in the Yukon and Northwestern Territories and southwestern Nunavut; it also breeds commonly across the northern two-thirds of Alberta and Saskatchewan and most of Manitoba, as well as all forested regions in the Canadian provinces to the east, including the Maritimes and Newfoundland and Labrador.

In the central United States, Slate-colored Juncos breed south to east-central Minnesota, central Wisconsin, and northern Michigan, including that state's upper peninsula. They are also found as nesters from Maine through western New England to New York, Pennsylvania, and western Ohio; they are rare or absent on the coastal plain south of southeastern Maine. The nominate subspecies intergrades with the Appalachian race *carolinensis* in south-central Pennsylvania.

"Pure" Carolina Juncos have been recorded breeding from northeastern West Virginia and western Maryland south through the mountains of easternmost Kentucky and western Virginia and North Carolina to eastern Tennessee, northern Georgia, and northwestern South Carolina.

The southern subspecies is largely resident, though some are altitudinal migrants, moving downslope into mountain valleys when the winter cold arrives. There are winter records of this race from central Maryland, central Virginia, central North Carolina, central Georgia, and the South Carolina coast.

The autumn movement of northern breeders begins in September and continues through early November, with most on the wintering grounds by early December. Females and adults tend to depart earlier than males and young birds; adults winter farther south than birds in their first winter.

The winter range of the Slate-colored Junco is even more vast than its breeding distribution, reaching from coastal Alaska and Newfoundland across southern Canada and south to Texas and, uncommonly, the Gulf Coast east to the Florida Panhandle. Most winter at middle latitudes east of the Rocky Mountains, but small numbers of Slate-colored Juncos (or of hybrid juncos visually indistinguishable from Slate-colored Juncos) occur every year in the northern Mexican states and in the western United States and Canada to the Pacific Coast, making this a bird that is entirely unexpected at very few places in North America. Wintering Slate-coloreds are uncommon in San Diego County, California; junco flocks there rarely contain more than three or four individuals, but there is an exceptional early winter record of a group of 21 Slate-colored Juncos in the year 2000. Scattered individuals also winter across southern Arizona, and there are regular reports from Sonora; the first specimen from Sonora, a male, was identified as of the nominate race, though "its pale gray sides contrast slightly with the head," suggesting perhaps that it was a hybrid, an intergrade, or a Cassiar Junco.

At the southern edge of their more easterly wintering range, Slate-colored Juncos winter widely across New Mexico south almost to the Mexico border (eBird). Slate-coloreds are rare to uncommon in Texas west of the Pecos and rare in south Texas; they are common to abundant in the eastern two-thirds of the state. They are uncommon but regular to the Gulf Coast at New Orleans and Mobile (eBird), and rare or very rare in Florida south of Alachua County, south to the Keys. As expected, all of Florida's specimens have been determined to be of the nominate race, with a single record of the Cassiar Junco.

Yellowish or pinkish on the flank may be a sign of Oregon Junco ancestry. This individual also shows a relatively clear contrast between head and nape and between breast and flank. *Alaska, June. Brian E. Small*

The spring migration begins in March, with males moving north first. In at least some areas in the East and Midwest, the first half of April sees repeated waves of northbound juncos, separated by days when the species can be virtually absent. Migrating flocks in both spring and fall can be very large, numbering in the hundreds or, exceptionally, the thousands.

An abundant bird with a very large breeding range that undertakes moderately long migrations, the Slate-colored Junco regularly appears outside of its expected range. It has been known for nearly 200 years as a stray to Iceland and Greenland, and there are now more than 50 European records from winter and, especially, spring in Great Britain and on the continent, including the Netherlands, Norway, Poland, and Gibraltar. Records from Ireland and Italy, originally suspected of pertaining to escaped captives, are now widely accepted as being wild birds. Other seagoing Slate-colored Juncos have reached the Bahamas, Jamaica, Puerto Rico, and the Virgin Islands.

In the Northwest, Slate-colored Juncos are casual on the coasts of northern and western Alaska and on the Bering Sea and Aleutian Islands west to Shemya. The first specimen from Asia was collected in Siberia sometime before 1848, when Alfred Malherbe offered it in trade to the Asiatic Museum of Calcutta, India. In June 1879, a freshly secured female was brought to J. A. Palmén on board the *Vega* at anchor off the Chukchi Peninsula; Palmén believed that the species "had never been encountered before within the borders of the Old World." There are now 16 records of Slate-colored Juncos from eastern Russia, including a pair that spent a summer month on Wrangel.

The Slate-colored Junco group comprises the very widespread nominate subspecies *hyemalis* and the southern breeding population *carolinensis*; some authorities also include the Cassiar Junco in this group, though its intermediate appearance and apparent hybrid origin could place it as easily among the Oregon Juncos.

Though the breeding juncos of the southeastern United States had been known for centuries, *carolinensis* was not recognized as a discrete subspecies until 1886, when William Brewster found that these sparrows—inevitably, he observed that "the mountain people called it 'Snowbird'"—were larger, paler, and more uniformly blue-gray above than their northern counterparts, with a darker, horn-colored bill. Later, in the same volume of the *Auk*, Brewster labeled the Carolina Junco a full species, a status affirmed without comment by the AOU in the first, 1889 Supplement to the *Check-list*.

Jonathan Dwight reacted by taking a collecting trip to Pennsylvania "expressly with a view to determining what sort of Juncos, if any, were found there." When he and Brewster examined the skins Dwight had secured, they found that the Pennsylvania specimens of *carolinensis* were "indisputably . . . the connecting links between the Junco that breeds in New England and his representative in Western North Carolina," and "clearly . . . only one end of a series . . . beginning with typical *J. hyemalis* to the north and extending southward along the Appalachian Mountain System." Dwight consequently re-lumped the Carolina Junco with the northern-breeding Slate-coloreds, and humorously styled the entire episode "an excellent text for a sermon upon the evils of nomenclature." Chastened, the AOU—again without comment—followed suit the next year.

A short-distance migrant breeding in the mountains from eastern West Virginia and western Maryland to northern Georgia and South Carolina, the Carolina Junco is slightly larger than the northern race of the Slate-colored; the two overlap entirely in winter. The brownish gray or, in adult males, gray of the upperparts is paler than in its northern relative, and the head, which is often noticeably more blackish than the back in the nominate race, shows no contrast with the breast or upperparts. The southern bird's bill is somewhat longer and heavier, and rather than pink or pinkish white, both mandibles are a light bluish horn. Juveniles are said to be grayer and more narrowly streaked above than those of the northern race.

Female and young Slate-colored Juncos can be very brown above and below. This individual was photographed on the breeding grounds of the Slate-colored, but the pinkish breast sides and flanks and the shape of the hood suggest some Oregon influence at some point in its family tree. *Alaska, June. Brian E. Small*

This is a classic pale gray White-winged Junco, with conspicuous jagged white wing bars formed by pale covert tips. *Colorado, December. Steven Mlodinow*

WHITE-WINGED JUNCO
Junco hyemalis aikeni

The 49 continental states of the third-largest country in the world are home to a scant dozen endemic birds, species that have never been recorded breeding outside the political boundaries of the United States. The number rises, of course, if the notion of endemism is extended to subspecies; even then there are few North American birds with ranges so small and so well defined as the White-winged Junco of the pine islands of the Great Plains.

This handsome junco was first collected by Charles H. Holden Jr., in the summer of 1869 in the Black Hills of Wyoming. Holden graciously named the new bird for his colleague Charles E. Aiken, "Colorado's pioneer ornithologist," who had an opportunity to see the junco himself in El Paso County, Colorado, in the winter of 1871–1872. The localities they studied were more than 200 miles apart, but Holden and Aiken nevertheless combined their results into a single publication, a species list prepared by Holden and extensively annotated with quotations from Aiken's Colorado experience. This mode of presentation had obvious disadvantages; worst of all, the formal citation of the new subspecies would make it seem as if the "always genial and courteous" Aiken had named it for himself, *Junco hyemalis* var. *aikenii* Aiken.

Robert Ridgway removed that potential source of embarrassment by naming and describing the new junco "correctly" in 1873, basing his account on half a dozen winter skins sent to him by Aiken. Ridgway endorsed Holden's original view that the White-winged was not a distinct species but rather a subspecies of the widespread Slate-colored Junco; he explained this race's large size as simply a climatic adaptation "to its alpine habitat," and pointed out that both the Slate-colored and Oregon Juncos could exhibit "quite a distinct tendency" to the same wing pattern that might otherwise characterize the White-winged as a distinct species.

Just how controversial Ridgway's approach was became obvious less than a year later. While the 1874 *History of North American Birds* reproduces verbatim Ridgway's argument for the merely subspecific status of the White-winged Junco, the plate in the very same volume labels its depiction of the type specimen as a full species, *Junco aikeni*. At almost exactly the same time, Elliott Coues published a series of notes from the railroading ornithologist T. Martin Trippe about the juncos wintering along the Colorado Front Range. Trippe found it "difficult to discuss, intelligently, the relations of these races or species," but he pointed out that the White-winged Junco's song was "higher and sweeter" than that of its congeners, that it was less gregarious, and that it regularly wintered at high elevations "rarely visited" by the other juncos. In the large series he collected, he discerned "obvious

Large, pale, and long- and thick-billed, this rather brownish White-winged Junco may be a young bird. Many White-wingeds are even plainer-winged than this individual. *Colorado, December. Steven Mlodinow*

approaches" to both the Slate-colored and the Oregon Juncos, but he did not observe the "intimate" and nearly complete intergradation linking those two junco forms; in fact, he wrote, "a specimen that cannot be decidedly referred to either [the White-winged] or [Oregon/Slate-colored intergrades] is unusual." Trippe concluded that the White-winged Junco was distinct from the others, though he, too, equivocated about its status as a species or subspecies.

The uncertainty persisted through the 1870s and early 1880s, with some authors treating the White-winged Junco at species rank and others not. The matter was settled with the appearance of the first edition of the AOU *Check-list* in 1886, in which *Junco aikeni* was accorded a place as specifically distinct from all other juncos, the status it would occupy for almost a century. Only in 1973 was the White-winged, along with the former Oregon and Guadalupe Juncos, folded into the AOU's newly minted Dark-eyed Junco.

FIELD IDENTIFICATION

From the very beginning, ornithologists, collectors, and field observers have been cautioned that the White-winged Junco's eponymous wing pattern is neither diagnostic nor invariably present. There are White-winged Juncos with entirely dark wing coverts, and white wing bars, in some cases impressively complete, occur regularly in individuals of the other Dark-eyed Junco forms as well. Nevertheless, the White-winged is a distinctive and usually easily identifiable bird, "even at gunshot range," and Alden H. Miller was willing to go even further than Trippe in writing that he had "yet to examine an equivocal specimen."

Leaving aside the presence or absence of wing bars, there are several important characters to note in the field. The White-winged is "much the largest of the juncos," a "giant species" whose superior size is always noticeable, and sometimes egregious, in the mixed sparrow flocks of winter. Males average three-quarters of an inch longer than Slate-colored Juncos, and they can outweigh Oregon Juncos by as much as 20 percent.

White-winged Juncos are often said to be a lighter gray than Slate-colored Juncos, a distinction that probably applies most clearly to females or older birds. Many, probably mostly males, appear darker in the field than the ashy gray sometimes said to be characteristic of all White-wingeds, but in the hand or at very close range in good light, even those individuals can be seen to be relatively pale; Miller speculated that some of the darkest individuals might be young birds with abnormally retained plumage.

Like many male Slate-coloreds, some White-winged Juncos show a decided darkening on the head, the deeper gray blending smoothly into the gray or slightly brown-tinged gray of the back and flowing uninterruptedly into the gray of the upper breast and flanks. Others, however, are essentially concolorous above; on these birds, the lore may be covered by a narrow, somewhat contrasting darker patch.

The bill color varies confusingly in White-winged Juncos. In a few birds it appears—at least in field conditions—to be nearly as pale pink as in a Slate-colored or Oregon Junco, while in most others it is an unusual shade of pinkish blue-gray, slightly darker on the upper than on the lower mandible. The pattern and color may recall the Red-backed and Yellow-eyed Juncos, but the White-winged's bill is never truly steely or blackish above.

More telling than the bill's color is its shape. The White-winged Junco's bill is deep and broad at the base, and usually notably long; it may have a slightly more curved culmen than the finer, straighter bill of the Slate-colored. In some individuals the bill is so long as to almost resemble that of a towhee.

Complete or nearly complete white wing bars are not unique to this form, but when they are present, they draw attention to the bird as surely as does its greater size. Miller found that slightly more than 3 percent of male and 1 percent of female Slate-colored Juncos had white covert tips creating wing bars; the respective figures in the White-winged Junco are greater than 90 percent and nearly 80 percent. Only 5 percent of Miller's male White-wingeds exhibited no white on the wing coverts, while 22 percent of females were plain-winged. In a few birds of both sexes, only the upperwing bar, formed by the white-tipped median coverts, is present, while the greater coverts show gray rather than white tips.

More informative is the presence of white in the secondaries and tertials. According to Miller, Slate-colored Juncos never show white on these feathers; though some have pale buffy edges that could be taken for white by the hopeful observer, those individuals are usually decidedly brownish overall, and thus very unlikely to be confused with the larger, grayer White-winged Junco, in which even the buffiest individuals show largely gray sides.

The species' English name notwithstanding, the most distinctive feature of the White-winged Junco in the field or in the hand is the tail pattern. Nearly one-quarter of male White-wingeds show some white on the second rectrix, and a few females show a trace of white even on the inner web of the first, central rectrix. Rectrix 3 shows at least some white in more than 90 percent of White-wingeds of both sexes, while that feather is entirely black in virtually all male and female Slate-colored Juncos. The remaining, outer three pairs of rectrices are also significantly more likely to be pure white than in the Slate-colored Junco.

The exact extent of white in each of a junco's tail feathers may be difficult to assess in the field, but

The bill of the White-winged Junco is long and sharply pointed. On many, the bill is also noticeably deep at the base. *Colorado, January. Cathy Sheeter*

In addition to its shape, the odd grayish pinkish color of many White-winged Juncos' bill can be a good identification clue. *Colorado, January. Cathy Sheeter*

White-winged Juncos with complete sets of rectrices are always strikingly white-tailed, a feature as likely as their large size or, at least in some individuals, pale aspect to draw the birder's attention. In many birds, the white of the rectrices is visible as a narrow but conspicuous band across the tip of even the folded tail from above, a circumstance unlikely to be observed in any other junco.

As in most other sparrows, the finely streaked juveniles are rarely seen without a parent in attendance. Individuals in their first plumage are more brownish than adults, and indeed moved even so cautious an observer as Elliott Coues to describe the birds at Sylvan Lake in the Black Hills as a new subspecies of the Slate-colored Junco, which he named *danbyi* for the principal of the nearby Custer High School. Coues had been misled by the absence of wing bars in the type specimen; in recanting, he noted that even these buffier young birds should

> not be mistaken for *hyemalis* at any age; the "aspect" in life, even at gunshot range, is distinctive; for one receives the impression of a large gray bird, more like *caniceps* [the Gray-headed Junco] than like *hyemalis*,

a comparison that remains evocative and useful.

Given the White-winged Junco's restricted range, interbreeding with other Dark-eyed Juncos appears to be relatively infrequent. In southeastern Montana, nesting White-winged and Pink-sided Juncos occur within 70 miles of each other; "conclusive" hybrids between the two have been collected on the Colorado wintering grounds and within the breeding range of the White-winged Junco in Powder River County, Montana. While white wing bars occur in a small number of "pure" Pink-sided Juncos (and in individuals of other populations of the Dark-eyed Junco), most hybrids and presumed backcrosses display intermediate back color and mixed flank colors; the hood can meet the lower breast either concavely, as in the Pink-sided, or convexly, as in the White-winged. The greater and median coverts can be white-tipped or plain, while the extent of white in the tail can be as great as in normal White-winged Juncos or only slightly greater than in normal Pink-sided Juncos. One hybrid described by Miller was "chiefly" White-winged in appearance, differing only in its smaller size, the convex lower border to the hood, and the extensively white tail; it had no white in the wing. Others are significantly closer visually to Pink-sided Juncos, but larger, with dull brown-gray backs, largely gray wing coverts and tertials, and gray-washed dull cinnamon flanks and sides bulging toward the center of the breast.

The White-winged Junco is a fairly tame but often inconspicuous bird on its breeding grounds, where it hops and scratches quietly on the ground in dry ponderosa pine forest and low woodlands of mixed oak and pine. Wintering birds often mix with other junco kinds, especially (as is expected given their range) with Oregon and Pink-sided Juncos. Dominance by race within such mixed flocks is difficult to determine and apparently varies regionally and from year to year, but some field observations have shown that—at least at a given site in a given year—White-winged Juncos are generally dominated by

The tertials of the White-winged Junco often show very broad gray or white edges. *Colorado, December. Cathy Sheeter*

Pink-sided and Oregon Juncos; the relationships were less clear in confrontations between White-winged and Slate-colored Juncos.

Like other juncos, feeding White-winged Juncos on the ground often flirt the tail, revealing the startling amount of white. Skilled or fortunate photographers can often obtain images showing the extent of white on at least the outer four pairs of rectrices.

The calls and songs of the White-winged appear to be indistinguishable from those of other Dark-eyed Juncos. As in the other races, fleeing birds issue a series of distinctive sharp, short notes on flushing. Males sing from a low- or midlevel perch in a pine or shrub, sometimes ascending into the treetop to sing their slow, uneven trills.

RANGE AND GEOGRAPHIC VARIATION

The White-winged Junco breeds along the length of the Pine Ridge in Nebraska; breeding has also been confirmed at the eastern end of that land feature in South Dakota, but it is not clear whether it also nests to the west in Wyoming.

The stronghold of breeding White-winged Juncos, holding by far the majority of the world population, is the ponderosa pine and aspen forests (and spruce at the highest elevations) of the Black Hills in southwestern South Dakota and extreme northeastern Wyoming, west to Bear Lodge Mountain (Warren Peak) and Devils Tower. There are outlier summer records as far north as Harding County in South Dakota and as far west as the Laramie Mountains in Wyoming, where one was discovered paired with a Gray-headed Junco in June 2014 (Christian Nunes, in litt.). There is also a late June specimen from Clear Creek, Colorado.

In southeastern Montana, White-winged Juncos breed in the forests of the Long Pine Hills, the Rosebud (Wolf) Mountains, and other suitable buttes and mountains in Carter, Powder River, southern Custer, Rosebud, and Big Horn Counties. The ranges of the White-winged and Pink-sided Juncos approach each other in the south-central portions of that last county.

Family groups begin to join into flocks of up to 30 birds on the breeding grounds in August and September. Many birds remain near or on the breeding range in winter, while others arrive on wintering grounds in the Nebraska Panhandle and Colorado's eastern foothills and mountains in mid- or late October. The winter range extends regularly south to northern New Mexico and extreme western Kansas, and more erratically and less frequently south to the Oklahoma and Texas Panhandles (eBird).

The easternmost documented records in Nebraska are from Lincoln County (eBird), some 200 miles southeast of the nearest breeders. There is an early winter record from Russell County, Kansas (eBird), but birds recently photographed slightly farther east in Reno County seem to be Slate-colored Juncos. White-winged Juncos are casual to very rare in west Texas, with a total of 14 accepted records as of 2014.

The general direction of the autumn migration takes most White-winged Juncos to the south and west, and vagrants may continue as far as western Colorado, Arizona, and California. White-winged Juncos have been documented in 6 of the past 80 winters in Arizona, with single birds in November 1971 in Yuma County, in December 2012 at the Grand Canyon, and in January 2008 (eBird) and February 1971 at Flagstaff; at least seven were collected in northern and eastern Arizona in the winter of 1936–1937, while four records were documented as far south as Pima County in 2000–2001.

In California, a White-winged Junco spent the winter of 1990–1991 at a feeder in Marin County, and another was observed in Riverside County in November 1996. The "extraordinary" winter of 2000–2001, in which out-of-range White-winged Juncos were discovered in western New Mexico and Arizona, brought single birds to Sacramento and Inyo Counties. The most far-flung record of the White-winged Junco in the West is of a bird photographed in Deschutes County, Oregon, from late February and mid-April 1987.

The northbound movement begins in March, when White-winged Juncos wintering at lower elevations in and near the Black Hills move upslope toward the breeding areas. Most birds have departed the more southerly wintering grounds by mid-April, and females are laying eggs in South Dakota by the end of May.

Spring vagrants are less frequent than autumn and winter wanderers, but it is that season that has produced some of the most remarkable records of this so very local sparrow. In early June 1911, a flock of 40 White-winged Juncos appeared in Lincoln County, Nebraska, 50 miles east of the normal wintering range and notably late in the season; one was examined in the hand. Even more extraordinary, a White-winged Junco was captured, measured, and photographed in April 1994 in Plymouth County, Massachusetts, almost 2,000 miles east of any locality where the bird was remotely expected; the published images confirm the identification, but it will never be known whether this individual had wintered nearby or been blown outlandishly off course on its migration north.

The Golden-crowned Sparrow in alternate plumage is a colorful bird, with a yellow, black, and white crown. *Alaska, June. Brian E. Small*

GOLDEN-CROWNED SPARROW
Zonotrichia atricapilla

In his *Conspectus generum avium* of 1850, Charles Lucien Bonaparte described a new species of sparrow from the south Pacific,

> gray and rust, rather well marked above with black spots; dull white below, with a grayish breast and brown flanks: the forehead yellow, bordered black on each side: the wings with two white bars: the tail slightly rounded.

Bonaparte named the bird *Zonotrichia galapagoensis* for the place of collection indicated on its label. Twenty years later, the type specimen in Paris was still unique, and the Galápagos sparrow remained a desideratum of collectors and curators elsewhere—until Osbert Salvin examined Bonaparte's bird and found it to be "only a specimen of the Californian *Z. coronata* [the name then current for the Golden-crowned Sparrow], to which a wrong locality has been assigned."

Imprecise, incorrect, and simply fantastic localities have been attached to the Golden-crowned Sparrow ever since its discovery by European science nearly 250 years ago. In May 1778, James Cook and the members of his third voyage were at anchor in Prince William Sound, then also known as Sandwich Sound; there they encountered "a small land bird, of the finch kind, about the size of a yellowhammer . . . of a dusky brown colour, with a reddish tail; and the supposed male had a large yellow spot on the crown of the head, with some varied black on the upper part of the neck." Both this bird and a female from Nootka Sound with her nest and eggs were painted by William W. Ellis, the artist on Cook's final voyage, "ad vivum"—not necessarily alive, but probably before they were prepared as specimens.

Ellis's paintings and the birds themselves returned to England with the surviving members of the expedition in 1780. The Prince William Sound specimen ended up in the famous museum of Ashton Lever, where it was described by John Latham and assigned the English name Black-crowned Bunting; the female from Nootka Sound was described under the same name by Latham's contemporary and colleague Thomas Pennant.

It was Latham, an authoritative figure in Anglo-American and Australian ornithology for more than half a century, who introduced what would be an equally long-lived uncertainty about the true nature and home of the Black-crowned Bunting. With a single question mark in the text, he raised a gentle doubt about the identity of his bird with Pennant's from Nootka Sound. Latham also committed a geographic error, one he would repeat in his account of the Boreal Owl: for Cook's Sandwich Sound, on the southern coast of Alaska, he substituted the Sandwich Islands, now known as Hawaii. Latham's inaccuracies created the specter of two similar but not identical species with widely separated ranges in the Pacific Ocean, and set the stage for close to two centuries of nomenclatural uncertainty.

For his 1789 edition of the *Systema naturae*, Johann Friedrich Gmelin mined every ornithological work he could in search of species still without scientific names. Among his most important sources were the catalogs of Pennant and Latham, neither of whom had assigned Linnaean binomials to any of the new species they described; in this way, the German systematist became the author of a great number of names of birds he had himself never handled.

In the case of the Black-crowned Bunting, Gmelin had only to translate the English epithet directly to generate a perfectly good scientific name, *Fringilla atricapilla*. As the authority for his Latin description, he cited Latham and, with the same question mark that that author had invoked, Pennant; combining the range descriptions in the two sources, Gmelin indicated that the species occupied "Natka" Sound and the Sandwich Islands.

In 1794, the young explorer Joseph Billings shipped several skins of the Golden-crowned Sparrow, collected on Kodiak Island, to Peter Simon Pallas at the St. Petersburg Academy of Sciences. Pallas recognized the birds as Latham's Black-crowned Bunting, but he either did not know or, more likely, chose to ignore the fact that Gmelin had assigned the species a scientific name half a decade earlier. Pallas would go on to name the bird anew, as *Emberiza coronata*, the epithet indicating the bright yellow oval on the center of the crown.

Gmelin's *atricapilla* did establish a tenuous tradition in the ornithological references of the early nineteenth century, including such important works as Bonaparte's *Geographical and Comparative List* and Audubon's *Birds of America* and *Synopsis* of 1839. Thomas Nuttall, who, with John K. Townsend, was the first European ornithologist since Cook's day to see the bird in life, introduced another alternative in 1840, when he renamed what he called the Yellow-crowned Finch *Fringilla aurocapilla*; that name, meaning literally "golden-crowned," was in occasional use for nearly two decades.

In 1858, Spencer Baird's account of the Golden-crowned Sparrow made explicit the ornithological establishment's discomfort with Gmelin's name *atricapilla*:

> Latham . . . describes a *black-crowned Bunting* from the Sandwich Islands, and incidentally mentions the present species as a variety from Nootka Sound. Gmelin bases an *Emberiza atricapilla* upon that name, and includes both original and variety. If his name can be retained for either one, however, it must be for the Sandwich Island species, which is very different from ours.

In other, plainer words, if Gmelin's *atricapilla* was conceived on two distinct taxa, then the name is entirely invalid or, at very best, must be restricted to refer to only one of the taxa it originally subsumed. Baird's assertion of the distinctness of the birds from Nootka Sound and from what he believed, thanks to Latham's geographical confusion, were the Hawaiian Islands was repeated even more anxiously by his younger colleague Robert Ridgway, who somewhat redundantly described Latham's Black-crowned Bunting as "essentially a totally different bird." Baird, Ridgway, and their contemporaries and successors sidestepped the confusion by adopting Pallas's *coronata*, the name under which this species entered the AOU *Checklist* in 1886.

It is plain, however, on reviewing Latham's comments that—whatever his laxity in matters geographic—his descriptions of both the Sandwich Sound specimen and the Nootka Sound specimen (and the published painting of the former) can apply only to the Golden-crowned Sparrow, and that the slight differences can easily be explained as the result of age or season. This was finally pointed out by Erwin Stresemann in 1949, who wrote in his study of the ornithological results of Cook's final voyage that the sparrows in question "obviously [belonged] to the same species" and that "there seems to be no reason for rejecting Gmelin's name [*atricapilla*] in favour of *Emberiza coronata* Pallas, 1811." Three years later, the American Ornithologists' Union changed the scientific species name from *coronata* to *atricapilla* and amended the type locality from Kodiak Island to Prince William Sound, Alaska.

FIELD IDENTIFICATION

This robust, full-bellied, long-tailed, and large-billed sparrow is nearly unmistakable for observers familiar with its less colorful plumages. Unfortunately, this is one of the species of western North American birds whose identification was made more difficult

In any plumage, the dark underparts and dark bill make this large sparrow look dignified, even somber. *Alaska, June. Brian E. Small*

by their inclusion in Roger Tory Peterson's first field guide: the description there and in subsequent editions of immature Golden-crowned Sparrows as "like large female House Sparrows" has left generations of birders more timorous than necessary when faced by a dull member of this species.

Size, shape, behavior, and any number of plumage features distinguish the Golden-crowned from the House Sparrow at any season. Winter birds, however, especially first-cycle individuals showing little or no yellow on the forehead, could be dismissed as particularly dull White-throated Sparrows. The two share a dark bill—quite unlike the bright bills of the White-crowned and Harris Sparrows—pale throat, and dingy gray-brown underparts, but the Golden-crowned is noticeably, if only slightly, larger and lacks the conspicuous broad white or tan supercilium and median crown stripe of the White-throated. The streaking of its upperparts usually appears coarser and scanter, and the wing pattern is brighter, with broader, more continuous white wing bars.

Each of those species is at best uncommon in the winter range of the other. The winter range of the White-crowned Sparrow, in contrast, entirely encompasses that of the Golden-crowned, and the two are often found feeding together on brushy edges. The Golden-crowned often makes a decidedly stockier impression, enhanced by its short neck, heavy bill, and horizontal posture. Like the

White-throated Sparrow, Golden-crowneds are very brown, while the White-crowned Sparrow is a strikingly gray bird. All White-crowned Sparrows, even of the pale, dull coastal races, have yellow or orange bills; the bill of the Golden-crowned is very different, blackish above with an odd purplish black lower mandible. Juvenile Golden-crowned Sparrows have rather plain heads with a pale, unstreaked throat; White-crowned Sparrows of the same age are densely marked with fine streaks from belly to chin, and lack the strong lateral throat stripe of the Golden-crowned.

Many passerellid sparrows molt using what is known as a complex alternate strategy, with distinct plumages worn by first-cycle birds (formative plumage, attained by the partial preformative molt), breeding adults (alternate plumage, attained by the partial prealternate molt), and nonbreeding adults (basic plumage, attained by the complete prebasic molt). The Golden-crowned, however, is one of only a relative handful in which each of these molts produces a visual aspect that is obviously different to human eyes.

Juveniles, heavily streaked on the breast and flanks, begin their preformative molt in July, on the breeding grounds; the molt of the head feathers and underparts is completed there before migration. The formative plumage that results is characterized by a complete absence of well-defined stripes on the crown, which darkens slightly toward the front and may show a faint tinge of yellow at the center. In extremely dark formative individuals, the area above the lore may be a saturated brown and the central crown noticeably yellowish, thus overlapping with the palest and dullest of basic-plumaged adults (Pyle).

Both formative-plumaged birds and basic-plumaged birds begin their prealternate molt in early spring. This molt involves chiefly the feathers of the head, such that by April the crown patterns of the two age groups—birds nearly one year old, and birds nearly two years old or older—are closely similar; the black crown stripes may be slightly less saturated in the younger birds, the yellow and white slightly less extensive, and the nape browner, but it is probably not safe to age Golden-crowned Sparrows in the field from about March to August, when that summer's juveniles become conspicuous (Pyle).

The adults' prebasic molt begins in July or August, on the breeding grounds; it can be completed on the breeding grounds or suspended for migration and completed afterward in the wintering range (Pyle). The head pattern of basic-plumaged birds is much duller than that of alternate-plumaged individuals, but averages brighter and better-defined than that in formative immatures, with some overlap as described above. The lateral crown stripes are broad, connected narrowly at the forehead, but the black is browner, and each feather is faintly tipped dull tan, creating a scaled effect at close range that

wears away by late winter; the forecrown is yellow or yellowish and usually unstreaked, the rear of the crown dull grayish or brownish.

In addition to plumage differences of age and season, adult Golden-crowned Sparrows also show average differences between the sexes. The crown stripes of females are duller than those of males of the same age, "but substantial overlap precludes reliable sexing of individuals" (Pyle) except perhaps in mated pairs.

Apparent hybrids between the Golden-crowned and other sparrow species are now regularly, if infrequently, detected in winter on the Pacific Coast. The first record of such a bird may have been a specimen, now apparently lost, collected in Sonoma County, California, before 1858, which Spencer Baird described as having

> an ashy streak above the eye bordering the black, similar to the pattern in [the White-crowned Sparrow], which, like the median stripe of the crown, is yellow anteriorly. There is a dusky line back of the eye. The dark stripes on the crown are more brownish than black, and considerably narrower . . . . The pattern of coloration . . . is precisely the same as that of [the White-crowned Sparrow], the median stripe on the head being yellow anteriorly and grayish posteriorly, instead of pure white . . . .

The lower mandible of winter birds can be paler yellowish; the head pattern is variable, with this individual among the most colorful at the season. *British Columbia, January. Brian E. Small*

Baird's description is insufficient to determine whether the "other" parent was a White-throated Sparrow or a White-crowned Sparrow; in any event, the "ashy streak" bordering the supercilium and the narrow crown stripes virtually rule out the possibility of a "pure" Golden-crowned Sparrow.

Hybrids within the genus *Zonotrichia* can exhibit plumage features intermediate between those of the parent species or a "mosaic" of characters unique to the parents. An apparent hybrid wintering in Michigan combined the yellow and gray crown patches, plain face, grayish scapulars, dark underparts, and white marginal wing coverts of a Golden-crowned with the whitish supercilium, yellow above the eye and lore, and white throat with dark lateral throat stripes of a White-throated Sparrow.

Hybridization between the Golden-crowned and White-crowned Sparrows appears to be much more common, though sometimes more difficult to detect; such birds typically differ from Golden-crowned Sparrows in showing "a broader supercilium with dark lower border, often meeting the lateral crown stripes at the back." Two spring specimens from California combined the golden and silvery gray crown of the Golden-crowned with the black crown stripes of a Gambel White-crowned Sparrow; one also showed pale underparts and a pink bill typical of that parent.

Some apparent hybrids of these two species show golden yellow in the supercilium above, behind, or, rarely, in front of the eye, a feature that can easily mislead observers and photographers into identifying them as the descendants of White-throated Sparrows. In fact, yellow or gold is present in those same areas in the Golden-crowned Sparrow, but it is "suppressed or obscured" by the melanin in the black areas of a typical Golden-crowned; hybrids with White-crowned Sparrows replace the black there with white or very pale gray, revealing the yellow otherwise masked. The narrowness and extent of the head stripes, especially whether the eye line curls onto the nape, are helpful characters in the assessment of parentage of a possible *Zonotrichia* hybrid.

The frequently heard call note of the Golden-crowned Sparrow year-round is a dull, unvoiced *tep*, sometimes recalling a giant Black-throated Gray Warbler and usually easily distinguished from the brighter, more metallic notes of White-crowned Sparrows; it is shorter, thinner, and drier than the husky *chep* of the Song Sparrow. This species also gives a sharp, slightly wavering *tzeeet* similar to that of the other *Zonotrichia* sparrows, both in flight and as a warning when perched in cover. Feeding birds call *chip, churr,* and *plear, plear, plear,* and there is, at least on the wintering grounds, a long, sweet, almost goldfinchlike descending whistle. Flocks occasionally engage in a sputtering jumble of calls, mixing *tep* notes with lower, fuller chips.

The well-known whistled song of the Golden-crowned Sparrow shares the wistful tone of the White-throated and Harris Sparrows. The three notes descending by whole steps are easily set to words, classically rendered "Mar-y had . . ." or "Three blind mice . . . ." "To the miners carrying their packs along the Alaska gold trails, the constantly repeated plaintive notes seemed to say 'I'm so weary'," while the truly discouraged heard the birds warning them "No gold here."

An especially appealing variant of the song begins with a typical breathy whistle, followed by two rapid whistles on the same pitch and a soft, loose trill: *peer peer-peer drdrdr*. In place of that terminal trill, some birds sing two slower whistles on nearly the same pitch, producing a song of five syncopated notes: *peer peer-peer peer peer*.

RANGE AND GEOGRAPHIC VARIATION

Even apart from the venerable Hawaiian canard and Bonaparte's *galapagoensis*, it was some time before this species was correctly assigned as a breeding bird to the extreme Pacific Northwest. Thomas Nuttall told Audubon that his first encounters with the Golden-crowned Sparrow had been with juveniles "running on the ground . . . on the central table-land of the Rocky Mountains, in the prairies," far from the species' actual breeding range and far out of the species' preferred habitat of damp, dark thickets. Audubon himself owned a juvenile identified as

Golden-crowned Sparrows in their first winter, wearing a formative plumage, may show almost no yellow on the crown. *British Columbia, January. Brian E. Small*

Golden-crowned Sparrows often look decidedly small-headed in the field, an appearance quite unlike the puffy head shape of the White-crowned Sparrow or the bull-necked, flat-headed appearance of the White-throated. *British Columbia, January. Brian E. Small*

this species collected by Nuttall's colleague and traveling companion John K. Townsend "on the Rocky Mountains" in July 1834; his published description, however, suggests that this bird was in fact a Green-tailed Towhee.

Adolphus Heermann claimed that the Golden-crowned Sparrow "occasionally breeds in California," and described in detail a nest with four eggs he had discovered near Sacramento. Two decades later, Thomas Mayo Brewer described another nest, again with a clutch of four eggs, said to have been collected by Thure Kumlien's son Ludovic in Shasta County; Brewer observed that "the nest and eggs of this species [had] hitherto escaped the notice of collectors," Heermann's report notwithstanding. Neither record, both expressly queried by Robert Ridgway as "probably erroneous," is accepted today, and the species is not now known to nest anywhere in the state of California (or in Hawaii).

Golden-crowned Sparrows breed in small numbers in the mountains of central Alaska, with populations known from the Talkeetna and Alaska Ranges; they breed throughout the Brooks Range, extending on the south slope into western Alaska. They are common to fairly common elsewhere in western Alaska, including the coast from Wales south and inland on the lower Kuskokwim River. Golden-crowneds breed commonly in southwestern Alaska and on the Aleutians as far west as Unimak Island; they are locally common to abundant on Alaska's south coast, but rare in the mountains of the southeast.

Yukon breeders occur on the Montana, White,

and Rancheria Mountains; territorial males have been found as far east as the Mackenzie Mountains and north to beyond the Arctic Circle. The species has also bred at tree line in the Mackenzie Mountains of the Northwest Territories, the same habitat it uses east to Banff and Jasper National Parks in Alberta.

In British Columbia, the Golden-crowned Sparrow is found in most of the province's mountainous regions in breeding season. The highest breeding concentrations are in the northern boreal mountains; most records in the province are from the coastal ranges and areas to the east, from Carmine Mountain northwest to Mount Garibaldi. This species is not known to breed in the mountains of the southern interior or in the Rocky Mountains south of Mount Robson; there also appear to be no breeders in mountainous habitats on Vancouver Island and Haida Gwaii, though there is a single nesting record from an anomalously low-elevation site on southern Vancouver Island.

South of the Canadian border, Golden-crowned Sparrows have bred in the northern Cascades of Washington, and three "possible" or "probable" instances were recorded during that state's 1987–1996 breeding bird atlas surveys. Given their scarcity in southern British Columbia, they are unlikely to be regular breeders in Washington.

Adults and young leave the breeding range starting in early August; most undergo the prebasic or preformative molt still in the north, but some suspend the molt and complete it on the wintering grounds, which are typically occupied by late

September or early October (Pyle). Southbound migrants are most common at higher elevations along the Pacific Coast, even appearing abundantly on California's Farralon Islands. As Pallas anticipated more than two centuries ago, apparent reverse migrants have been recorded in September and October on Wrangel Island and the Chukchi and Kamchatka Peninsulas in the Russian Far East some five times.

Inland, numbers of migrants are dramatically smaller, but there are scattered August and September records as far east as Montana, Wyoming, Colorado, and southeastern Arizona (eBird). A late September individual from Seal Island, Nova Scotia, was notably early, as most of the rare but regular strays to the Atlantic Coast do not arrive, or are not detected, before mid-October (eBird).

The heart of the Golden-crowned Sparrow's winter range stretches along the Pacific Coast from southern British Columbia into Baja California. In Canada, the largest numbers are found on southeastern Vancouver Island and in the Fraser River lowlands; the species also winters locally in small numbers in the low valleys of central and southern interior British Columbia. In Washington and Oregon, wintering birds are common west of the Cascades, particularly so around Puget Sound, but much scarcer east of the mountains.

Golden-crowned Sparrows are common winter residents of the western half of California, from sea level to about 3,000 feet in the foothills. They are uncommon in winter on the Farallon Islands and most of the Channel Islands, though usually much more abundant on Santa Cruz Island. East of the Sierra Nevada, Golden-crowneds are uncommon or fairly common at elevations up to 5,000 feet from northern Owens Valley and Mono County north. In San Diego County, winterers are concentrated in the higher mountains and on shady, north-facing slopes within 15 miles of the coast; they are scarce to fairly common elsewhere. As elsewhere in the arid Southwest, this species is a rare winter visitor to brushy oases in the state's eastern and southeastern deserts.

In Mexico, Golden-crowned Sparrows are uncommon winter residents of northern Baja California, with fewer than five records as of 2013 from Baja California Sur. The species has also occurred on Cedros and Guadalupe Islands off the Baja coast. Though there are only a few records from Sonora, the species' winter status in northern Sonora is presumably identical to that in adjacent Arizona, where Golden-crowned Sparrows are sparsely distributed but annual in occurrence.

A characteristic bird of the Pacific Coast of Canada, the United States, and Mexico, the Golden-crowned Sparrow is also among the species that can occur virtually anywhere in the nonbreeding season. Winter birds from Japan include three in January and February (eBird), and three of Nova Scotia's ten records are of individuals wintering at feeders. The extreme southeastern point of this sparrow's winter range as of 2016 appears to be Orange County, Florida, where one was observed in February 2005. The species remains scarce to rare anywhere east of British Columbia and the Pacific states, but it is entirely unexpected nowhere in North America. Indeed, given the unfamiliarity of most eastern birders with the winter plumages of this species, it is almost certainly underreported over most of the continent.

Wintering birds begin the northbound migration in April, arriving on the Alaskan breeding grounds as soon as late April and May. Apparently overshooting individuals regularly occur on St. Paul Island in May (eBird), and there are records for the same month in the Russian Far East.

Eastern winterers probably keep to a similar schedule. Three of Nova Scotia's records are of birds arriving or first discovered in April and May. Golden-crowned Sparrows have likewise "been detected surprisingly often in spring" in Massachusetts, and there is a late April record from New Hampshire, suggesting that there is a necessarily small but nevertheless perceptible northward movement among birds wintering along the East Coast of North America.

No subspecies are recognized.

WHITE-CROWNED SPARROW
Zonotrichia leucophrys

More than two and a half centuries after its first scientific description by Mathurin Brisson, the White-crowned Sparrow remains one of the most frequently misidentified of passerellid sparrows. The adult, with its more or less strikingly marked black and white head, is distinctive enough, but birds wearing the equally striking but quite different formative plumage easily mislead the inexperienced and the hopeful into finding in them any of a number of very superficially similar species, from Chipping Sparrows to Rufous-winged Sparrows to Golden-crowned Sparrows.

Brisson's type specimen for the bird he called *Passer canadensis* was a brown-crowned individual in formative plumage. The skin had been sent from the eponymous region by Jean-François Gaultier, who arrived in Quebec in 1743 and would soon take principal responsibility for Governor Roland-Michel Barrin de La Galissonière's scheme for a complete inventory of French Canada's scientific resources. Gaultier regularly shipped notable specimens to René Antoine Ferchault de Réaumur, who employed his cousin Brisson as curator of his extensive natural history collections, collections that served in large part as the basis for Brisson's six-volume *Ornithologie*.

On Réaumur's death in 1757, the collections and their curator were to have been transferred to the Académie des Sciences. The Comte de Buffon, however, Réaumur's rival and a powerful enemy of Brisson, eventually engineered the incorporation of most of the specimens into the natural history holdings of the Jardin du Roi; as the all-powerful *intendant* (director) of the Jardin, Buffon denied Brisson access to the specimens he needed, and Brisson abandoned the study of ornithology to spend the rest of his life teaching physics.

Buffon had Réaumur's Canada sparrow engraved for his *Histoire naturelle des oiseaux*. While the plate labels the bird with Brisson's name "moineau [sparrow] du Canada," the text, published in 1775, rather surprisingly reidentifies the Canadian as "a variety" of the Rock Sparrow *Petronia petronia*:

> We have named it the "soulciet," because it is a little smaller than the Rock Sparrow ["soulcie"], as all of the other animals of the New World are also smaller than those of the same species in the Old World.

Meanwhile, Johann Reinhold Forster had himself received an interesting specimen from Hudson

The adult White-crowned Sparrow is one of the most handsome and should be one of the most easily identified of sparrows. *Brian E. Small*

Bay. The English traders and functionaries there knew it as the Hedge Sparrow (the Dunnock, *Prunella modularis*), a misidentification even more far-fetched than Buffon's; Forster, however, recognized in it a new species, which he named *Emberiza leucophrys*, the White Crowned Bunting. The name makes clear, and Forster's detailed description confirms, that his "elegant little" bird was an adult, "with a black head and snow-white supercilium and crown stripe."

Given the striking differences between Forster's description on the one hand and Brisson and Buffon's descriptions and illustrations on the other, it is unsurprising that the earliest natural historians in the United States failed to connect the brown-crowned and the white-crowned birds. Neither did it help that the species was uncommon or even rare in the East, where, of course, most of the pioneering ornithologists were active: Alexander Wilson in his entire career saw only three individuals of "this beautifully marked species . . . one of the rarest of its tribe in the United States," and Audubon warned his "kind reader" that only in "the wild regions of Labrador" could one "form a personal acquaintance with . . . the handsomest bird of its kind," which in his experience occurred south of Canada only singly or in very small flocks,

Neither Wilson nor, as late as 1832, Thomas Nuttall knew, or at least recognized, the brown-crowned formative plumage of this species. Audubon did, though. Thanks to his extensive collecting, he knew that the difference was not one of sex, the males and females identical but for size; instead, he was

> convinced that these birds lose the white stripes on the head in the winter season, when they might be supposed to be of a different species.

His observations in Labrador led him to speculate that juveniles molted into a white-crowned plumage in the first weeks of their life, then joined the adults in acquiring their brown-striped heads, a molt regime complicated even by the standards of the *Zonotrichia* sparrows.

Thomas Nuttall encountered his first brown-crowned sparrow in Philadelphia, when John Kirk Townsend sent him a male specimen taken in late August in eastern Washington. Apparently, and oddly, unaware that Audubon had finally identified such birds as White-crowned Sparrows, Nuttall identified the bird with "crown deep chestnut, with a broad pale-brown medium band" as a new species, Gambel's Finch, *Zonotrichia Gambelii*, "somewhat allied to the" Golden-crowned Sparrow but smaller, with a brighter bill.

In a turn of events reminiscent of Robert Ridgway's rejection of the Ridgway Junco, the first doubt on the specific distinctness of the Gambel Finch was cast by none other than William Gambel, who in 1843 reported from California that these brown-crowned birds were "seen in almost every hedge in company with the [White-crowned Sparrow], to which it is closely allied, if not the same species." Four years later, Gambel had determined that the Gambel Finch described by "my friend Nuttall" was in fact the immature plumage of the white-crowned adult bird with which it consorted—and at the same time raised the possibility that these western sparrows might be "indeed different from the *leucophrys* of this [eastern] side of the continent," a sugges-

The formative plumage of this species is visibly different from that of the adult, but the gray ground colors, pale bill, and crown pattern identify it. This formative bird is a classic Gambel Sparrow. *Texas, November.* Brian E. Small

tion that would be borne out in the course of the unraveling of this widespread species' geographic variation.

By 1852, enough specimen material was available from the West that Adolphus Heermann could declare "conclusively that the *Fringilla Gambelli* [sic], Nuttall, is but the young" of the White-crowned Sparrow, a fact that entered the ornithological literature most authoritatively with the publication of Spencer Baird's great *Birds* of 1858 and has gone unquestioned since.

Oddly, one error repeatedly met with today—the notion that brown-striped birds are females, white-striped birds males—does not occur in the historical record of this species.

FIELD IDENTIFICATION

Few White-crowned Sparrows of any age or any subspecies are first noticed because of their head pattern. Instead, this large, long-tailed, full-bellied sparrow stands out in the field by its bold behavior and characteristically upright posture, unlike the more furtive habits and hunched attitudes of the Golden-crowned or White-throated Sparrows. White-crowned Sparrows feed confidently in short grass or on barren sand, necks stretched and small heads held high as they hop and scratch in search of seeds and small insects; flushed, they are as likely to settle, curious, atop a shrub or low tree as to disappear into a thicket.

Adults, once looked at, are unlikely to be confused with any other bird. In addition to shape and size, the clear gray or grayish tan breast, usually sparsely marked back, and tall, conspicuously marked crown are distinctive, even in the darker, less sharply patterned breeders of the West Coast. Formative-plumaged birds, with their brown and bright tan head stripes, can momentarily disconcert inexperienced observers, panicking them into a counterproductive concentration on superficial field marks; more than one immature White-crowned has been called a Rufous-crowned Sparrow or a Chipping Sparrow, species that would not even enter into consideration were size and structure assessed first.

In parts of eastern North America, where White-crowned Sparrows may be less common than in the West, birders were long taught to focus on details of the head and throat patterns to distinguish this species from the abundant White-throated Sparrow. Those differences are real and significant, of course, but the easiest way to identify these species is to ask whether the first impression in the field is of a gray bird—a White-crowned Sparrow—or of a brown bird—a White-throated Sparrow.

The vocalizations of various populations of White-crowned Sparrows may be better known to

Gambel Sparrows have a very different facial expression from the black-lored populations. *Alaska, June. Brian E. Small*

ornithology than those of any other North American species. Indeed, much of what is known about song acquisition, variation, and function in passerines has been learned by studying this sparrow in the field and in the laboratory.

One inventory of this species' call repertoire, conducted in the breeding range of the Mountain White-crowned Sparrow, found nine different call behaviors used by some or all sex and age classes. The most familiar call, heard from breeders and migrants, males and females, adults and juveniles, feeding birds and perched birds, is the well-known *pink*, a loud, sweet note closely recalling a Northern Cardinal's chip but fuller, less metallic, and with a very soft attack and very sharp, rapid decay. Like the *chink* of White-throated Sparrows or the *wink* of Harris Sparrows, this call is especially conspicuous on the wintering grounds in the early morning, when few other birds are vocal. Given in a wide variety of situations, this call may indicate "high arousal, or a state of indecision regarding subsequent behavioral acts." This call is said to be flatter in the populations breeding on the Pacific Coast.

Birds under stress and about to flee, in flight or on foot, may give a harsh whining call, sometimes in a rapid series. Each note is longer than the *pink* call, lower-pitched, and more clearly slurred, with a

This individual, from the zone of intergradation between pale-lored and black-lored birds, shows the bill color and lore pattern typical of a Gambel Sparrow—a reminder to use caution when identifying sparrows to subspecies or subspecies group. *Manitoba, June. Brian E. Small*

relatively long, ascending attack and long, descending decay. The common flight note, also given by nervous perched birds, is a very slightly buzzy *dsip*. Probably identical to the call labeled *SIP* by Hill and Lein, this note is surprisingly short, high, and thin for such a physically robust sparrow, vaguely recalling the slight and slender flight calls of a *Spizella* sparrow.

Other calls identified by Hill and Lein include a short trill, given by both sexes; a "flag" note indicating an intention to either attack or escape; a chatter and a rasping call, both given only by males; a broadly modulated *teez* given by nestlings and fledglings; and a mysterious "W-call," given only rarely and without an identifiable function. These calls are generally much more frequent on the breeding grounds than in migration or winter—obviously so in the case of the *teez* note of nestlings and juveniles.

Both males and, less frequently, females sing. Though it is usually instantly recognizable, the song of the White-crowned Sparrow varies individually and geographically, even within subspecies, creating a wide range of idiolects and dialects. What all songs share is the combination of rather clear, plaintive whistled notes with loose, broadly modulated buzzes in a relaxed series; many songs also include one or two complex slurred notes and end with a

low-volume trill or soft buzz. At a distance, only the buzzes may be audible, reminding some observers of the song of a Black-throated Blue Warbler or Clay-colored Sparrow.

Vocal differences among subspecies and local populations are most conspicuous in the sequence of song elements. The first note in the song is apparently always whistled, but the arrangement of the buzzes, note complexes, and trills that follow varies regionally. In the Sierra Nevada of California, for example, one study identified five local dialects within the breeding populations of *oriantha*; each male sang the same song type from year to year. As in all passerellids, song in White-crowned Sparrows is learned. The degree of variation and the number of dialects are thus lowest in areas with high population densities and large stretches of continuous habitat, where young males are likely to repeatedly hear songs of many adults; conversely, variation is most pronounced among small, isolated breeding units.

RANGE AND GEOGRAPHIC VARIATION

As a species, the White-crowned Sparrow is common to abundant over much of North America at one season or another. Apparently rare in the eastern United States 200 years ago, White-crowneds now feature prominently in mixed-species flocks there in migration, especially in October and early November.

White-crowned Sparrows occupy a vast northern breeding range, covering most of mainland Alaska, west to the Alaska Peninsula and north to the tundra, and east across the Yukon and Northwest Territories, mainland Nunavut, the northern prairie provinces, northern Ontario, much of Quebec, and Labrador. The highest numbers in British Columbia are in the northern mountains and along the Georgia Depression, including eastern Vancouver Island; breeders are absent from the mountains along the west coast of the province.

In Alberta, the breeding range includes the Rocky Mountains on the western edge of the province, extending east to Saskatchewan north of the Peace River. An isolated population breeds in the Cypress Hills on the border between those two provinces. White-crowned Sparrows breed across the northern third of Manitoba and Ontario, thence east across northern and central Quebec to southern Labrador and northern Newfoundland.

Of these northern populations, *gambelii* breeds west of Hudson Bay, *leucophrys* to the east, with intergrades where the two meet.

This species does not breed in the eastern United States, but is widespread in the West, represented by *oriantha* in the interior and the Pacific races *pugetensis* and *nuttalli* on the coast. Mountain White-crowneds are fairly common breeders in the western half of Montana, east to the Beartooth and Pryor Ranges, and breed throughout the mountains of Wyoming and western Colorado; their range extends south into the high elevations of north-central New Mexico.

The status of this species as a breeding bird in Arizona is not entirely clear. Reported as early as the nineteenth century in the San Francisco Mountains, breeding was not confirmed there until 1969. Singing birds have been encountered in June and July at Mount Baldy in the White Mountains, and fledglings have been observed on the north rim of the Grand Canyon. The White-crowned Sparrow is currently considered a "rare and local" nesting bird in the state.

White-crowned Sparrows of the *oriantha* race also breed throughout the mountains of Utah, Idaho, and northeastern Oregon. In Nevada, the species is concentrated as a nester in the Santa Rosa and Ruby Mountains, the Carson Range, and the Jarbidge region, with scattered records from mountains on the state's western and eastern borders. In California, Mountain White-crowned Sparrows are fairly common breeders at high elevations on both slopes of the Sierra, south on the west side to Tulare County and on the east side to Inyo County. There is also an outlier population on San Gorgonio Mountain in San Bernardino County, and the species has bred once in San Diego County.

The coastal *pugetensis* breeds from British Columbia's southwestern coast and the eastern shore of Vancouver Island south on the immediate coast to northwesternmost California, Del Norte and Humboldt Counties; there is said to be an inland population breeding in Oregon's Willamette Valley. White-crowneds breeding from Mendocino County south through central California to Point Conception are assigned to *nuttalli*. The contact zone between the two breeding populations is centered on Cape Mendocino, where there are "numerous phenotypic intermediates" and "substantial gene flow" between them.

The White-crowned Sparrows of the central California coast are more or less sedentary, but other populations are strongly and conspicuously migratory, leaving the breeding grounds from late August to September; the first arrivals appear on the wintering grounds from September to October, but in the East, *leucophrys* continues to migrate through the middle latitudes well into November. For the most part, the wintering range of each subspecies lies directly south of its breeding range, though some *leucophrys* from the eastern part of that race's range drift southwest as far as central Texas. There is an October record from as far south as Belize.

Western winterers occur as far north as southern British Columbia; small numbers now winter in Montana, where Christmas Bird Counts have

The overall silvery aspect and red, pink, or yellow bill immediately distinguish this species from the White-throated Sparrow. *Texas, November. Brian E. Small*

The pale rusty, yellow-billed populations of the Pacific Coast are a likely split from the other White-crowned Sparrows, differing not only in size and color but in breeding biology. *California. Luke Tiller*

recently recorded up to two dozen individuals a year. East of the Great Plains, the northern limit of regular wintering is about 40 degrees north latitude; this is a large, hardy sparrow, though, and well-stocked feeders lure individuals to occasionally linger much farther north.

White-crowned Sparrows are common winterers across northern Mexico, south to Baja California Sur and from Nayarit across northern Michoacán to Tamaulipas; they are erratic winterers on the Yucatán Peninsula. The species is uncommon in the United States as far south as the Gulf Coast states, but at least in Florida, winter numbers appear to be increasing. Records from the Caribbean extend to the Florida Keys, the Dry Tortugas, the Bahamas, Cuba, and Jamaica. One spent the winter of 1981–82 as far afield as the Netherlands, and it seems likely that most of the other five spring and summer records of the species from Europe are of birds that wintered in the Old World and went undetected until they moved north.

To the west of the White-crowned Sparrow's North American range, there are records of eleven strays encountered between May and October in the Russian Far East, and winter vagrants have also appeared in South Korea and Japan.

In spring, northbound birds have begun to move by late March, with some still passing through the middle latitudes in May. Rarely, individuals linger very late on the wintering grounds, in the West as far south as Arizona. Most, however, have arrived in the breeding areas by mid- to late May; the short-distance coastal migrant *pugetensis* returns to its nesting areas in early April.

The history of the "Intermediate Sparrow" can serve as an emblem of the difficulty ornithology has had in unraveling the variation exhibited by this extremely widespread species. Described by Ridgway in 1873 as a westerly and northerly variety of the White-crowned Sparrow, the Intermediate Sparrow had long been promoted to full species status by the time it entered the first edition of the AOU *Check-list* in 1886. At the end of the decade, however, Ridgway had the opportunity to examine more specimens, which proved to him the occurrence of "extensive intergradation" between the Intermediate Sparrow and the Gambel Sparrow—both then considered full species—and between the Intermediate Sparrow and the White-crowned Sparrow; Ridgway's recommendation that all three be considered "as merely geographical races of one species" was ratified by the AOU Committee in 1890.

When even more specimen material arrived at the Smithsonian, however, Ridgway determined that he had made one error and perpetuated another: his "Intermediate Sparrow" was in fact the bird Nuttall had described as the Gambel Sparrow in 1840, and what American ornithologists had been calling *gambelii* was in fact the still unnamed resident bird of the southern Pacific Coast. Thus, *intermedia* was simply a synonym of the true Gambel Sparrow, and Ridgway found himself obliged to assign the California coastal breeders a new name, *nuttalli*, a gracious and appropriate gesture but not one calculated to simplify the historical synonymies of the species. The renaming of the two western species was carried out, and the old *intermedia* removed, by the AOU and in Ridgway's *Birds of North and Middle America* in 1901.

More than a quarter century later, Joseph Grinnell discovered that the breeding *nuttalli* of central coastal California differed consistently from the birds of Oregon, Washington, and southwestern British Columbia, which Grinnell separated as a distinct subspecies *pugetensis*. Grinnell also took the opportunity to refute renewed grumblings that the various White-crowned Sparrows might represent different species:

> There is that approximate degree of uniformity of characters in the . . . forms to make of them excellent *sub*species; but the likenesses between them are so outstanding . . . that an indication of the really close mutual inter-relations among them would be lost by according the [now four] forms of *leucophrys* full specific rank.

The AOU adopted Grinnell's view in 1931. Almost immediately thereafter, Harry C. Oberholser split the nominate subspecies *leucophrys*, recognizing the birds breeding in the mountains of the interior West as a new race, *Zonotrichia leucophrys oriantha* ("flower of the hills"!). Black-lored like nominate *leucophrys*, Oberholser's new subspecies was described as "much paler" on the upperparts, breast, and flanks, with a grayish rather than brownish tone to the rump, back, and flanks. It was recognized as a valid race in 1944, and was retained in the fifth, 1957 edition of the *Check-list,* the last to list subspecies.

As if there had not been enough already, Clyde Todd warned of further "unfortunate nomenclatural complications" in 1948, when he argued that Forster's type, described in 1772, was actually a Gambel Sparrow, making it necessary to apply the subspecies name *leucophrys* to that race and to assign a new one, *nigrilora*, to the breeding birds of the eastern boreal forest. Perhaps less in conviction than exhaustion, the AOU Committee in 1953 rejected Todd's proposal and affirmed the alignment of the subspecies names as they would appear in 1957.

For the field observer, White-crowned Sparrows can be sorted into dark-lored and pale-lored "kinds." In dark-lored adults, the black of the lateral crown stripe meets the black of the broad eye line in front of the eye, creating the impression of a black helmet with white stripes; the pattern is similar in the formative plumage, though the contrast between dark brown and light brown is less and the pattern thus often less easily assessed.

In pale-lored adults, the black of the lateral crown

This very large-billed, pale-backed spring migrant is presumably a Mountain White-crowned Sparrow, *oriantha. California, April. Brian E. Small*

stripe does not invade the front of the white supercilium, and the black eye line starts at the back of the eye, leaving the area between the eye and the base of the lower mandible gray or whitish; the impression, especially in the poorly marked Pacific Coast races, is of a gray head with a series of alternating parallel stripes. The pattern is similar in the formative plumage, though the contrast between dark brown and light brown is less and the pattern thus often less easily assessed.

The nominate race of the White-crowned Sparrow, breeding across Canada east of Hudson Bay, is a dark-lored bird with a pink bill. It is large and large-billed, with relatively dark grayish brown upperparts sparsely and regularly streaked with brighter brown; the breast is brown and fairly dark. Formative-plumaged birds have a "moderately distinct" brown ear patch (Pyle).

Not all authorities still recognize *oriantha*, the Mountain White-crowned Sparrow, as distinct from the Eastern White-crowned; *Birds of North America*, for example, treats them as synonyms, presumably in affirmation of Richard C. Banks's lump of the two. The Mountain White-crowned appears to differ consistently, however, in the much paler ground color of its back and the decidedly paler gray breast and duller buffy flanks—characters difficult to assess confidently in the field, but worth noting especially in areas on the western Great Plains where either subspecies is plausibly as likely or unlikely as the other.

Perhaps the most promising field character for distinguishing Mountain and Eastern White-crowned Sparrows is the short tail of the latter. According to measurements published by Peter Pyle, the wing of the two subspecies is essentially identical in length, averaging very slightly shorter in the Mountain White-crowned than in the eastern race; the tail of the Eastern White-crowned, however, is as much as 0.4 cm shorter than that of the Mountain (Pyle), creating a proportional difference, at least in extreme individuals, that can be noticed in the field by observers familiar with one or the other. The slightly larger bill on average of *oriantha*, in combination with the longer tail, contributes to the more attenuated appearance of some birds of this race as well.

Classic Gambel Sparrows, especially adults, are quite different from either of these subspecies. They are roughly the same length and mass as both Eastern and Mountain White-crowneds, with a short-tailed look like that of *leucophrys*, but the Gambel Sparrow's bill is small, pale, and bright, usually with a yellowish or orange "candy corn" tone unlike the purer dark reddish pink of the others. The upperparts are dark and fairly densely streaked, as in the nominate, eastern race, but the breast is pale gray to

Birds like this winterer from the lower Rio Grande Valley are presumed to be representatives of the nominate, eastern subspecies, but could also be Mountain White-crowned Sparrows; the two are very difficult or impossible to distinguish in the field. *Texas, November. Brian E. Small*

grayish brown, as in *oriantha*. The marginal coverts of the underwing, white in both the Eastern and Mountain White-crowneds, are gray, sometimes with a yellowish overtone creating a "flash" at the bend of the wing.

The precise details of the head pattern can be difficult to see well in the field, but the "open" face pattern with pale lore and forward portion of the supercilium creates an expression unlike the much sterner look of the dark-lored subspecies. Rather than concentrating exclusively on the color of the lore, it is often more helpful to ask simply whether the lateral crown stripe meets the supercilium in front or not; this is ultimately simply a different way of posing the same question, but birders, like all thoughtful people, profit from having more than one approach to a problem.

The pale bill and greater expanse of white on the head can make Gambel Sparrows appear conspicuously bright in mixed flocks, while *oriantha* and *leucophrys* are more somber. This is true particularly of adults, but birds in formative plumage also show the distinctive pattern of their respective subspecies, though in rich brown and rufous rather than clear black and white.

Gambel and Eastern White-crowned Sparrows intergrade commonly in Manitoba, Gambel and Mountain White-crowned Sparrows in southeastern British Columbia. The descendants of mixed pairs may show indeterminate bill color and intermediate head patterns, with smudgy dark lores.

The coastal subspecies *pugetensis* and *nuttalli*, breeding north and south of Sonoma County, California, respectively, are sometimes thought of as simply yellow-billed Gambel-like sparrows. In fact, the coastal races are "biologically . . . dissimilar," both "morphologically, vocally, in breeding habitat, and in migration distance . . . and it has been suggested that they would be better treated as two separate species" from a more narrowly defined, interior-breeding White-crowned Sparrow. The more significant differences include the timing of reproductive readiness, molt, and daily activity.

Most immediately distinctive in the field is the upperparts color of these Pacific Coast birds. The shaft streaking of the back feathers is dark brown to nearly black, creating a dark impression above in spite of the lighter brown or buffy feather edgings. The back of birds of the remaining, interior races is paler and grayer, with light reddish shaft streaks and bright gray or whitish feather edges. The nape is also browner in *pugetensis* and *nuttalli*, grayer in *gambeli*, *oriantha*, and the nominate race.

Distinguishing the two Pacific Coast forms from each other is more difficult, even impossible. While (by definition) they occupy different breeding ranges, *pugetensis* White-crowneds winter throughout the resident range of their southern relative. California's *nuttalli* is larger than *pugetensis*, separable in the hand by the longer tarsus and the greater mass of breeding males. "Northern birds tend to be lighter and more reddish on both the rump and the upper back"; *nuttalli* also undergoes a less extensive prealternate molt, leaving it more worn and that much darker-backed in the spring and early summer than *pugetensis*.

An adult Harris Sparrow in alternate plumage is among the most dramatically patterned of American sparrows, with silver cheeks, black face, and bright bill. *Manitoba, June. Brian E. Small*

HARRIS SPARROW
Zonotrichia querula

Usually quick to pronounce an unknown bird "new," John James Audubon hesitated uncharacteristically long when his companions Edward Harris and John G. Bell collected three unfamiliar finches in May 1843. Audubon had clear suspicions from the very start, but it was a full 13 days before he finally declared himself

> truly proud to name it *Fringilla Harrisii*, in honor of one of the best friends I have in this world.

Audubon, however, had no right to give the bird a name at all. Not only was he not the discoverer of the species, but he somehow overlooked the fact that it had been described and formally named nearly four years earlier, by Audubon's colleague, friend, and correspondent Thomas Nuttall. In late April and early May of 1834, he, John K. Townsend, and their party "observed," and presumably collected, this sparrow along the Missouri River in Missouri and Kansas; in the 1840 edition of his *Manual of the*

Ornithology of the United States and of Canada, Nuttall named it the Mourning Finch, *Fringilla querula*, on the basis of its distinctive "long, drawling, faint, monotonous and solemn note."

Just how Audubon, who quotes Nuttall continually and at length elsewhere in the *Ornithological Biography* and the octave edition of *The Birds of America*, could have overlooked the *Manual*'s concise but unequivocal description has long been a mystery. In 1919, the artist and librarian Harry Harris sketched an implausible conspiracy between Nuttall and Townsend to keep the new sparrow a secret. Harris insinuated that Nuttall withheld the specimen in violation of his promise to supply Audubon with complete material from western North America, thus "reserv[ing] this discovery for his own book":

> Not only was posterity thereby deprived of a Havell engraving of the largest and handsomest of our Sparrows, but Audubon, being kept in the dark, was himself to later publish the bird as the discovery of his friend Edward Harris.

Nuttall's consistent good nature and generosity in dealing with Audubon make such a plot seem very unlikely—and in any event, Audubon must have read the description of the new species in 1840 with the rest of the ornithological world.

The true solution to just "why Audubon and his coworkers were in ignorance of their lack of claim to Nuttall's Mourning Finch" is actually to be sought in Edward Harris's contemporary comments. Audubon, Harris, Bell, and Isaac Sprague had almost certainly seen the description before embarking on their expedition; but another, far more exciting prospect appears to have driven Nuttall's finch far out of mind.

All three of the sparrows Bell and Harris shot on May 4 proved to be males, as did the birds taken on May 6, 7, and 8. On May 19, Harris wrote to his brother-in-law in New Jersey to explain why Audubon, even with so many specimens before him, still hesitated:

> We feared that it might have proved the male of Townsend's Finch, with which it agrees in measurements exactly.

Had that fear—or hope, more likely—been realized, it would have solved one of the greatest puzzles in American ornithology, the identity of the Townsend Bunting. On May 16, however, Harris shot a female "of the large new Finches," and the next day Sprague took another, from a flock of more than a dozen seen by the party. Finding that the female "resemble[d] the male almost entirely, . . . only a little paler in its markings," Audubon was satisfied that these finches represented after all a distinct species, and in the heat of the taxonomic moment, seems never to have thought to consult Nuttall and the *Manual* before announcing his "new" species—neither on the banks of the Missouri in 1843 nor the next year, when he published the name *harrisii* in the second edition of *The Birds of America*.

Rather like the Baird Sparrow, another—this time genuine—discovery from Audubon's Missouri River expedition, the Harris largely languished beneath the ornithological horizon for the next 30 years, with only occasional notices of local records from its winter range on the eastern Great Plains. Elliott Coues brought the bird back to observers' attention with a virtually affectionate account in his *Birds of the Northwest*, in which he confessed himself "struck with [the] size and beauty" of one of "the most characteristic birds of the Missouri region;" ever conscious of his scientific duties as a collector, though, he noted too that this "bird of imposing appearance—for a sparrow" was fond of taking a high perch, "and in this conspicuous position they may of course be readily destroyed."

Coues's stay on the Souris River in the autumn of 1873 was an opportunity to "destroy" a "fine series of specimens of this handsome and interesting Finch" and to learn a bit about its habits on its fall migration. What neither he nor anyone else could discover, however, was where those southbound sparrows originated.

> I presume the bird has some special, restricted breeding localities, of which, in due time, we shall learn.

He was right, but the Harris Sparrow would keep the secret of its nest and eggs until well after Coues's death in 1899, making it one of the last North American birds whose breeding grounds were unknown.

The problem was not immediately recognized. In May 1834, just upstream from Thomas Nuttall and just two weeks after the official discovery of the Mourning Finch, Prince Maximilian zu Wied-Neuwied had his first and only encounter with the bird later to be known as the Harris Sparrow. Apparently assuming that the birds were by then already on their breeding grounds, Maximilian asserted, with surprising confidence, that the species bred in the thickets along the Missouri River and near the mouth of the Platte—a claim that, of course, was never substantiated.

More than half a century later, Charles Bendire, the great nestor of American oology, reviewed the negative data from "the vast interior of the former Hudson's Bay Territory" and concluded "necessarily" that

> the summer home of Harris's Sparrow, if properly looked for, will be found along the foothills of the Bearpaw and Chief Mountains in Montana, along the Turtle Mountains in Dakota, and their centre of abundance probably near Duck Mountain, Manitoba, as well as in suitable localities in the Territories of Alberta and Assiniboia, south of Lat. 54°.

Indeed, Bendire reported that in June 1885, one of his collectors at Fort Custer, Montana, had brought him a nest and four eggs—"without, unfortunately, securing the parent"—that would almost "certainly prove to be those of Harris's Sparrow." He, too, was wrong, and he, too, would die—in 1897—before a genuine nest was discovered.

Real progress was first made in 1900, when the brothers Edward and Alfred Preble undertook a natural history survey of the Hudson Bay region. At Churchill, in the last week of July, they found Harris Sparrows "rather common" in scattered stands of dwarf spruce; they collected and described two "young just from the nest." Seven years later, Ernest

Thompson Seton set off with Edward Preble on an exploration of the Barren Grounds northeast of Great Slave Lake, in the course of which Seton, on August 5, discovered a grass nest "on the ground under a dwarf birch . . . [that] contained three young nearly ready to fly." Preble's remarks on the tour make it clear that he was not present for the momentous discovery, but there is no reason to assume that this first occupied nest of the Harris Sparrow was among the "lots of things" Preble would later gently suggest Seton had imagined during their travels.

Not until 1931 did ornithologists finally set eyes on what had become a minor holy grail, a Harris Sparrow nest with eggs. The searchers were no more skilled, the birds no less furtive—but

> the completion of the Hudson Bay Canadian Government Railway in 1929 opened up a vast territory . . . of interest for the ornithologist.

The end of the new line was in Churchill, where the Prebles had found Harris Sparrows to be so common at the turn of the twentieth century, and that was where a new expedition, composed of John Bonner Semple, George Miksch Sutton, Bert C. Lloyd, and Olin Sewall Pettingill Jr., would begin their search.

For three weeks the young ornithologists trudged five to eight miles each way, in a mix of "high enthusiasm and frank exasperation," from their lodgings in Churchill to the wooded areas where Harris Sparrows were most abundant. Finally, on June 16, Sutton flushed a bird from "the top of a mossy, shrub-covered, water-girt mound in a cool, shadowy spot" beneath a stand of tall spruce trees. The nest, four slightly incubated eggs, and the female were "collected at once." Over the next ten days, the Semple party found another nine occupied nests, and a nearly century-old quest came to a successful end.

FIELD IDENTIFICATION

Big and bold, the Harris Sparrow can hardly be confused with any other North American bird. New birders and birders from regions where this species is scarce sometimes expect it to resemble a male House Sparrow or even a Lapland Longspur, but neither shares the Harris Sparrow's long tail, long pink tarsus, finely streaked back, dotted white wing bars, glaringly white belly, and bright pink bill. Both the longspur and the Old World sparrow are considerably smaller, as well: half an inch longer than an Eastern Phoebe, a stout Harris Sparrow can outweigh a Stilt Sandpiper by several grams.

That large size is the first and dominant impres-

The sides of the head are buffy in a basic-plumaged adult, and the black crown and throat are scaled and spotted with white. *New Mexico, December. Brian E. Small*

sion this bird makes in the field, especially in winter, when it joins flocks of juncos, American Tree Sparrows, and White-crowned Sparrows—all sturdy birds, all immediately puny when a Harris bounds in cheerful assertiveness to the middle of the seed pile. The second impression, surprising to observers taught by the field guides to concentrate on the adult's black hood, is of a predominant whiteness. Very few passerellids can be said to be truly white below, but the Harris Sparrow gleams from undertail coverts to breast, the bright ground color contrasting strikingly with the chestnut streaks of the flanks and the black spots and blotches of the breast. Even from a fast-moving vehicle, birders quickly learn to identify the white blobs perched atop plum thickets and Siberian elms in rural winter landscapes.

Harris Sparrows appear to be curious and engaged, usually turning to face the human viewer. Seen from the rear, though, they are nearly as easily recognized. In addition to the superior size and long, square tail, the pale tan rump and lower back with a delicate silver wash are distinctive, as is the fine, clear chestnut, black, and whitish streaking of the upper back and scapulars. At any season, the chestnut or black "checkmark" bordering the rear of the ear coverts, visible from behind or from the side, is also characteristic.

The Harris Sparrow exhibits considerable variation in head pattern, depending on age, season, and to some extent sex. All Harris Sparrows replace the head feathers beginning in late summer or early autumn. In birds hatched earlier that summer, this replacement occurs in the course of the partial

Birds in their first winter typically have a largely white throat with variable black lateral throat stripes. *California, January. Brian E. Small*

preformative molt; in adults, it occurs in the course of the complete prebasic molt (Pyle).

The blackish crown feathers of formative-plumaged birds are edged and tipped with gray or pale brown; as those edges and tips wear over the winter, the crown, at first scaled or scalloped, becomes more solid black and thus more like the crown of the adult, which is mostly or entirely black at all seasons, with at most a few paler streaks in winter.

The pattern of the throat and breast reliably distinguishes birds in their first year of life from adults. In the 1970s, the ornithological ethologist Sievert Rohwer developed a convenient and helpful "studliness scale" for assessing the underparts of Harris Sparrows; the accompanying photographs, now readily available online, can be of use not only to banders but to field observers. Rohwer's study, based on the examination of museum specimens and monitoring of living birds over several consecutive years, determined that fully 95 percent of winter birds could be aged by the extent of black on the throat. Individuals whose throats are entirely white or show only four or five small black spots and a small black chin patch, and whose lateral throat stripes are narrow and pale, fall into studliness classes 1 to 5; the palest and most lightly marked, belonging to classes 1 to 3, can safely be aged as in

their first calendar year. The white throat probably functions as a "badge of subordination," permitting the peaceable integration of young birds into wintering flocks dominated by more experienced adults.

The lateral throat stripes are darker and wider, and the throat and chin more extensively black, with spotting and splotching connected to the breast band, in birds of classes 6 to 14; the darkest, showing little or no white on the throat, belong to classes 9 to 14, and can reliably be aged as adults. Birds in classes 3, 4, and 5 are most likely immatures, those in classes 6, 7, and 8 most likely adults; Rohwer recommends that aging in those cases be verified against the shape and wear of the central tail feathers, a character essentially impossible to see in the field.

There is no apparent difference in plumage between the sexes, whether immatures or adults, in winter. Adult females in alternate plumage average somewhat more lightly marked on the throat than males; birds that have been confidently aged as adults and that show a throat pattern assigned to classes 3, 4, or 5 are almost certainly females, while only males exhibit the extremely extensive hoods of classes 13 and 14. In the hand, many adult males and adult females can be sexed by wing chord (Pyle).

Alternate plumage is attained as the result of a prealternate molt beginning in late winter. This molt is partial in both adults and immatures, but adult birds molt most or all of the head feathers, replacing the buffy and white of winter with an extensively black crown, forehead, lore, supercilium, throat and breast, setting off the bright silvery sides of the head and the orange or pink bill. A few adult females, as noted above, may retain considerable white on the throat even in alternate plumage.

Perhaps because of its relatively small population and remote breeding area, the Harris Sparrow has less frequently been recorded interbreeding with other species than have its close relatives in the genus *Zonotrichia*. An apparent hybrid between this species and the White-crowned Sparrow collected in Ontario in May 1969 is largely intermediate in plumage between the presumed parental species. The streaked upper breast loosely recalls a Harris Sparrow, as do the buff-tipped gray and black feathers of the central crown, while the broad black lateral crown stripes, whitish supercilium, and absence of a "checkmark" at the rear of the ear coverts are White-crowned characters.

The Harris has also been reported to hybridize with the White-throated Sparrow (Pyle).

Harris Sparrows are vocal at all seasons. The best-known call is a loud, cheerful *cheenk*, fuller and richer than the similar notes of the White-crowned and White-throated Sparrows and less sharply metallic than the chip of a Northern Cardinal; there is also a drier, shorter *tink*. The thin *seep* call is longer and more wavering than the short, decisive *tzip* of the White-crowned.

The song, heard on the breeding grounds, from migrants spring and fall, and on warm winter days alike, is a loud, measured *pee pee pee*, the notes delivered slowly and well separated by long pauses; the mournful tone resembles that of the songs of the Golden-crowned and White-throated Sparrows, but all of the notes are on the same pitch.

Early observers repeatedly noted a further vocalization, described rather helplessly by Coues as "a queer chuckling note" and by Seton as "a warble somewhat like that of a bluebird." It is impossible to be certain at this remove, but these accounts probably refer to the "roosting call," a loose rolling chant occasionally heard as winter flocks assemble for the night in dense red cedar stands on the edge of the prairie.

RANGE AND GEOGRAPHIC VARIATION

In summer, the Harris is the Canadian sparrow par excellence, breeding nowhere in the world but at the forest-tundra edge from the westernmost Northwest Territories to the southern shores of Hudson Bay in Ontario, east at least to Fort Severn. The species' status in the Yukon is uncertain; one was singing on Herschel Island in June 2007 (eBird), but previous summer reports from the territory are inadequately documented. There are June and July records from as far north as Barrow and as far south as Juneau in Alaska (eBird), but the species breeds only in arctic Canada, making it that country's only endemic breeding passerine.

Harris Sparrows leave their breeding range in late August and early September, moving south at a leisurely pace. Early flocks "swarm across the prairie provinces," then, once they have reached the eastern Great Plains in southern Nebraska, Kansas, and northwestern Missouri, "the birds linger in abundance" through October and early November; they are vocal and conspicuous during this long layover, "sing[ing] as sweetly, if not as fully and volubly, in October as in May."

The route taken by these southbound flocks is nearly as well defined as the arctic breeding range. The continuously updated maps provided by eBird are especially eloquent: from September to November, Harris Sparrows are reported most frequently and in the largest numbers in a narrow north-south band from Winnipeg to Dallas; both numbers and frequency drop off dramatically west of the central plains states and east of the Missouri River. This picture is remarkably similar to that sketched in the 1880s by Wells Cooke, who described the Harris Sparrow's

> area of greatest abundance [as] the country for seventy-five to one hundred miles on each side of a due north and south line connecting Pembina, Dak., with San Antonio, Tex.

In Cooke's day, this species was still considered strictly accidental east of the Mississippi River; the western edge of its normal migratory route was unknown. In the nearly century and a half since, Harris Sparrows have been determined to be extremely low-density autumn migrants across much of the United States and Canada, from Alaska to Florida and California's Channel Islands to the Maritime Provinces. Even so, the appearance of one of these handsome sparrows anywhere away from the normal Great Plains route remains noteworthy—even west just to the short-grass prairies of Colorado and Wyoming or east just to Illinois and Wisconsin.

The great bulk of the world's Harris Sparrows winter on the southern Great Plains, with by far the largest numbers occupying the narrow band from eastern Nebraska and western Iowa to east-central Texas. At this season, too, the species can be encountered almost anywhere in North America north of Mexico. West of the Great Plains, Harris Sparrows are expected annually in small numbers, often at bird feeders, in every mainland western state and across the southern tier of Canadian provinces; one was also observed in extreme northern Chihuahua in January 2007 (eBird).

Harris Sparrows are less frequent and less common to the east of their core winter range. In Iowa, for example, the species is a common migrant and regular wintering bird along the Missouri River, but notably rare at any season in the Mississippi River Valley a scant 300 miles away. East of the Mississippi, the Harris Sparrow is sufficiently rare that (as of 2015) state and provincial records committees from Mississippi, Tennessee, Kentucky, Ohio, and Ontario east request written documentation of sightings. At the time of writing, the easternmost records of the species are from Newfoundland's Avalon Peninsula (eBird), the southernmost from Hendry County, Florida—but given its tendency to wander, this sparrow with the so closely circumscribed breeding range, abundant in winter only on such a narrow strip of wooded prairie edge, is possible anywhere.

The spring migration of this species, from the southern Great Plains to the edge of the Canadian arctic, is complex. Birds wintering in the southern portions of the range, in Oklahoma and Texas, begin to move north in early March, bringing them into the northern parts of the wintering range by early April. These early migrants appear to "stall" for a full month or more, at which point the entire population—southerly winterers and northerly winterers alike—make a dramatic push across eastern Kansas and Nebraska into the Dakotas and prairie provinces. Harris Sparrows can seem to be everywhere on the northern Great Plains on a bright day in early or mid-May, feeding in loose, restless flocks on grassy edges and singing from the tops of newly leafed trees.

This second, late-season push takes the sparrows to their breeding grounds in the arctic, where most arrive in late May and early June. Harris Sparrows are notorious, though, for dawdling on the northern Great Plains into mid-June or beyond, and some birds found to the east or west of the normal wintering range appear to be in no hurry, either, lingering late in spring, often at feeders, until the arrival of truly warm weather drives them on.

No subspecies are recognized.

Northbound migrants in spring are often discovered high in fruiting elms, where they alternate chewing and singing. *Kansas, April. Brian E. Small*

WHITE-THROATED SPARROW
Zonotrichia albicollis

One of the most abundant and familiar winter birds in much of eastern North America, the White-throated Sparrow somehow managed to evade formal scientific attention for some two and a half centuries after European settlement. It was William Bartram who first alerted ornithologists to the existence of the species, when the teenaged naturalist sent "a neat drawing in colours" to his father's correspondent George Edwards. Edwards engraved the drawing for his *Gleanings of Natural History* of 1760, adopting the English name "White-throated Sparrow" for the brown bird from "Pensilvania." Bartram himself would publish nothing about the sparrow for more than 30 years, finally listing in his *Travels* "the large brown white throat" among the birds that "arrive in Pennsylvania in the autumn, from the North, where they continue during winter, and return again the spring following, I supposed to breed and rear their young." Bartram assigned it the straightforward scientific label *Fringilla fusca*— but he was too late, and by 1791, European writers had already given the bird not one but three Latin names.

George Edwards, as was his custom, did not assign the new sparrow a formal Latin name. His account served as the source for Mathurin Brisson, however, who in the appendix to his *Ornithologie* called the bird "le moineau de Pensilvanie," *Passer pensilvanicus*. Johann Friedrich Gmelin knew and cited Brisson's account, but he declined to adopt the French ornithologist's Latin name. Instead, he translated Edwards's (and Bartram's) English name as *Fringilla albicollis*, publishing that name in his edition of the *Systema naturae*. Based on a description in Pennant of a bird from New York, Gmelin's *Fringilla striata*—"streaked finch"—also refers almost certainly to this species, which is strongly marked below in both juvenile and, often, formative plumages.

That last name was quickly subsumed, but the first two—the one older by nearly 30 years, the other bearing the posthumous stamp of Linnaeus—remained current in ornithological works through the first half of the nineteenth century. While Alexander Wilson had chosen to use Gmelin's binomial in the *American Ornithology*, his *albicollis* was "corrected back" to *pennsylvanica* in a subsequent edition of Wilson's work by William Jardine and Charles Bonaparte. Bonaparte himself used both, arguing first for the priority of *pensylvanica* but then definitively adopting *albicollis* in his *Conspectus generum*. Audubon asserted the validity of *pennsyl-*

The large size, rich brown upperparts, and bright head pattern readily identify white-striped adults. *Maine, June. Brian E. Small*

vanica in every one of his publications, and Prince Maximilian used it as late as 1858 for the birds he had seen on his North American tour a quarter of a century earlier.

It was in that same year of 1858 that Spencer Baird pronounced Brisson's species name invalid:

> As Brisson's nomenclature is not binomial, and his names merely literal translation into Latin from the French vernacular . . . I have followed Cabanis, Bonaparte, and most modern authors in rejecting them altogether.

Today, Brisson's generic names have been readmitted to use in ornithological nomenclature, but his species epithets are universally accounted invalid, making Gmelin's *albicollis* the only name in proper use for this species.

FIELD IDENTIFICATION

The White-throated is a large, neckless, rather sluggish sparrow, superficially similar as an adult to the White-crowned and Golden-crowned Sparrows and as a juvenile to the Song Sparrow. Some birds in their first winter remain heavily and extensively marked below, often with a large central breast spot; they, too, can be confused with Song Sparrows or other streaked sparrows by observers first concentrating on plumage features rather than size, shape, and behavior.

White-throated Sparrows share with White-crowned and Golden-crowned Sparrows a rather long, straight-edged tail; flat back and full belly; and large head. They differ most strikingly from White-crowned Sparrows in their overall brownness, without the bright pearly gray of that species' neck and breast; the White-throated's back is cryptically marked with relatively coarse black streaks on a dead-leaf ground, reflecting its fondness for shady duff, while White-crowneds are brighter gray above with sparser blackish or reddish streaking. The White-throated Sparrow's rather dumpy, uniformly brown aspect carries over to the bill, which is dull blackish gray, with none of the bright contrast provided by the pink, yellow, or reddish bill of the White-crowned. White-throated Sparrows are also very different in posture from White-crowneds, usually keeping their heads low and the axis of the body horizontal, unlike the stretch-necked vertical tendency so typical of the White-crowned.

White-striped birds have dramatic black and white stripes on the head and a well-set-off white throat; the yellow patch above the lore is particularly bright. *Maine, June. Brian E. Small*

Eastern birders are sometimes surprised to find that it is in fact the Golden-crowned Sparrow that reminds them more of the familiar White-throated. Formative-plumaged individuals of both are brown and dark-billed, with tan-washed flanks and a bit of yellow at the lore. The slightly larger Golden-crowned is somewhat grayer on the rump, breast, and sides of the neck, but especially dull White-throated Sparrows can be surprisingly similar at first glance. The White-throated's white belly, often more heavily streaked breast, and heavier lateral throat stripe are all obvious on a closer look, but it is almost certain that the odd stray of one species has been overlooked at times as the other.

Familiar as it is as a migrant and winterer over so much of North America, the White-throated Sparrow is far less well known to most observers on its breeding grounds, and the juvenile of this species is a bird few see regularly. Long-tailed, heavily streaked, and dark-billed, with a plain crown and wide black lateral throat stripe, juvenile White-throated Sparrows are strikingly similar to Song Sparrows. The White-throated's wingtip is longer and more pointed, and both the secondaries and the greater coverts are more lavishly edged rusty or chestnut than in the Song Sparrow. The head pattern is decidedly muted, especially on the crown, but good clues are provided by the broad white jaw stripe and the whitish or pale brown supercilium, sometimes with just a hint of yellow above the lore. Aging the bird first—by its lax plumage, sharp begging calls, or soft yellow gape—can lead to the correct identification, as can, of course, simply waiting for the parents to return.

The identification and appreciation of *Zonotrichia* sparrows is made more interesting by the relatively complex plumage variation these species show. In the White-crowned Sparrows, plumages differ conspicuously with age; in the Golden-crowned, with age and season; in the Harris Sparrow, with age, season, and sex; and in the White-throated, with age, season, sex, and morph.

The plumage variation in adults of the White-throated Sparrow has been known for nearly as long as the species itself. In 1785, Thomas Pennant reported a summer specimen from Newfoundland with "the yellow spot at the base of the bill very obscure, nor had it the white spot on the chin"; he "supposed" that this was the female. Alexander Wilson assured his reader that "all the parts that are white in the male are in the female of a light drab color," a diagnosis confidently repeated by Audubon 20 years on:

> In the female, the colours are similarly
> arranged, but much duller, the bright bay
> of the male being changed into reddish-

Tan-striped adults have a similar pattern to that of white-striped birds, but the crown stripes are dull creamy brown and brownish black, and the yellow at the bill base is dull; they are more likely to have a thin black streak separating the white throat from the white jaw stripe. *Manitoba, June. Brian E. Small*

brown, the black into dark brown, and the white into greyish-white. The white streak above the eye is narrower, shorter, and anteriorly less yellow, the greyish-blue of the breast paler, and the white spot on the throat less defined.

Joel Asaph Allen, "observing many birds singing in the garb of the female," investigated the matter more closely in the early 1860s. After collecting a number of these singing "females," Allen discovered "them invariably to be males" and concluded that

> the males do not attain their mature colors till the second spring. The young males sing equally well with the adults, and probably breed in this plumage . . . . This accounts for the great proportion of birds in the livery of the female, both in spring and fall, often observed.

Elliott Coues offered an alternative explanation: females might well sing, too, and it is "probable" that males molt in late summer into a distinct, and duller, nonbreeding plumage.

The truth is significantly more intricate and significantly more interesting. White-throated Sparrows occur in two distinct color morphs; the morphs have been thought to be determined rather simply by the presence or absence of a gene inversion on the second chromosome, but more recently, the difference has been attributed to a "supergene," a complex of some 1,000 genes affecting plumage color. Both morphs occur in both sexes, and are dis-

tinguished principally by the color—white or tan—of the median crown stripe.

> Regardless of sex, the white-striped birds are generally brighter colored and less streaked; they have more black on the lateral crown areas, less streaking on a wider and grayer chest band, less intense black on the [lateral throat stripe], and brighter yellow on the superciliary stripe . . . . A central breast spot occurs more frequently in the tan-striped birds.

It was long believed that the morphs differed only in alternate plumage, which is attained in a prealternate molt beginning in late winter and usually completed on the wintering grounds, though it is still underway "occasionally . . . on northbound migration or breeding grounds." That molt, which takes place in birds of both sexes and all ages, is largely restricted to the feathers of the head, and thus White-throated Sparrows are at their brightest and most contrasting from early spring to early summer. The head plumage then grows increasingly worn, the result of mechanical abrasion and exposure to sunlight, through the summer, to be replaced with the fresh, less colorful and sharply marked basic feathers.

While the differences in basic plumage are more subtle than in spring, one year-to-year study of banded wild birds determined that even in basic plumage 90 percent of individual White-throated Sparrows could be correctly assigned to morph. All basic-plumaged birds have some tan and brown in the crown stripes, but there are noticeable differences between the morphs in the amount of black and white also present. Age for age and sex for sex, white-striped birds in basic plumage averaged brighter, with a "significantly whiter" median crown stripe and supercilium and "significantly blacker" lateral crown stripes. The brightness of the basic-plumaged head can also be used in white-striped birds to distinguish birds in formative plumage from adults, all of the head stripes "significantly" brighter in the latter age class. The differences are less useful in tan-striped birds: in basic plumage, only males can be aged reliably by head pattern, namely, by the blacker lateral crown stripes of adults. The head stripes are brighter on average in basic-plumaged males of the white-striped morph than in females; again, the differences are not as helpful in sexing tan-striped birds, in which only the darker lateral crown stripe of adult males distinguishes them from females.

This complex "system" of head markings in all ages, all plumages, and both sexes is linked to an equally complex reproductive regime, in which the brightness of an individual's head pattern provides signals to other White-throated Sparrows about its age, sex, and potential behavior as a mate. There is no doubt that the sparrows are extremely adept at reading these signals: males and females alike adjust their reactions to an intruder based on the intruder's morph, and, remarkably, mate almost exclusively disassortatively by morph, tan-striped birds seeking white-striped birds for breeding and vice versa.

The advantage of such heterogamy lies in the different behavioral strengths of birds of each morph. White-morph birds of both sexes are more aggressive and thus more zealous in maintaining territories and resources, while tan-striped birds of both sexes are more solicitous in caring for eggs, nestlings, and juveniles. This Jack Sprat arrangement ensures that both members of a pair bring complementary attributes to the nesting effort—and genetically maintains the demographic balance of the two morphs.

White-striped White-throated Sparrows are more persistently vocal than their tan-striped conspecifics, but there is no difference in their repertoires. Both males and females sing, though song is much rarer in tan-striped females than in white-striped females.

The most commonly heard calls in this species include a long, slightly wavering creeper- or kingletlike *see*, often given by birds feeding invisibly in dense brush or understory; and a loud, somewhat liquid *chlink*, fuller and rounder than the corresponding note of the White-crowned Sparrow, often given by alert birds perched atop a thicket. Both the *chlink* note and a repeated two-syllable *ch-chink* in dotted rhythm are often heard at dusk as flocks assemble to roost.

The song of the White-throated Sparrow is one of the most easily learned of passerellid songs, a simple phrase of clear whistles. The first note or, more rarely, the first two notes are long and tentative; this is followed by a series of shorter, faster whistles, usually grouped into triplets or dotted triplets but sometimes into groups of two. The long introductory note is usually on more or less a single pitch, but it may be slurred up or down in some birds. The tone is slightly breathy, usually more so than in the Harris Sparrow or the Golden-crowned Sparrow, and buzzy notes like those of a White-crowned Sparrow are very infrequent. There is considerable individual variation, and some geographic variation, in song, with at least 15 different patterns recognized by one study; most individuals sing only one pattern, though, and the tone and combination of a long note or notes with hurried triplets are distinctive no matter what pattern a given bird prefers.

Over the course of a 12-year study of this species, Borror and Gunn observed significant changes in the frequency and distribution of song patterns, some of which declined nearly to extinction in some areas. They suggested that the song types behind classic transcriptions such as "hard times, Canada, Canada, Canada" and "Old Sam Peabody, Peabody, Peabody," clearly comprising two long introductory notes and a series of three triplet groups, were relatively common in the nineteenth and earlier twentieth centuries but had died out over much of the species' range by the mid-1960s, when only three of the 711 White-throated Sparrows studied still sang that venerable melody.

RANGE AND GEOGRAPHIC VARIATION

Traditionally thought of as an eastern bird, the White-throated Sparrow breeds across the boreal forests of Canada nearly to Alaska, its western extension similar to that of several "eastern" wood warblers. Such geographic patterns were the result of early deglaciation in what is now northern British Columbia, permitting the westward spread of northern and eastern organisms while their southern and western counterparts were still blocked in by ice. To a great extent, however, the White-throated Sparrow and similarly distributed songbirds still follow ancestral migration paths and occupy wintering areas established before the melting glaciers allowed their breeding areas to expand to the west.

The White-throated Sparrow is a fairly commonly to locally abundant breeder in forested valleys of the southeastern Yukon; it is rare north of Watson Lake, with scattered records from elsewhere

Winter birds with heavy streaking beneath are probably in their first year of life. *Kentucky, November. Brian E. Small*

in the territory. It breeds in the southern Northwest Territories and on the southern James and Hudson Bay Islands of Nanavut, thence across Ontario and the southern two-thirds of Quebec to the Maritime Provinces. White-throated Sparrows are common breeders in the boreal forests of northeastern and north-central British Columbia; they are absent as nesting birds from the province's coast. They also breed over much of northern and central Saskatchewan and in Alberta from the Bow River north, with a clear concentration of confirmed nestings across the south-central portion of that province; they are found as breeders across virtually all of Manitoba's boreal forest.

In the United States, the White-throated Sparrow breeds across the boreal forests of the northern tier of states from north-central North Dakota to Maine. The southern limits of the breeding range are in central Minnesota, northern Wisconsin, central Michigan, northern Pennsylvania, and the mountains of northern and central New York. The species is an erratic or former breeder in West Virginia (AOU) and in northwestern New Jersey. In Ohio, breeding was recorded several times in the first third of the twentieth century, but nesting has been confirmed only once since, in Ashtabula County in 1997; non-breeding summering birds are annual in the state in very small but possibly increasing numbers.

White-throated Sparrows leave their breeding grounds beginning in late August, arriving in eastern winter areas from as early as mid-September to as late as November. There are scattered and increasing winter records from within the Canadian breeding range, but most move south to cover most of the United States from the eastern Great Plains; winterers are most abundant in the Southeast, from east Texas north and east to Maryland (eBird). They are uncommon to rare in the Rio Grande Valley of Texas and in northeast Mexico. Common in winter

in the Florida Panhandle and northern portions of the peninsula, this species is rare in south Florida and only casual in the Florida Keys, though there are several winter records as far east as Bermuda (eBird) and one as far south as Aruba.

West of the Great Plains, small numbers of wintering White-throated Sparrows occur, often at feeders, as far north as southern British Columbia and the prairie provinces, as far south as northern Sonora and Chihuahua, and along the Pacific Coast from southeast Alaska to northern Baja California. Throughout this vast region, it is rare to see more than a tiny handful of White-throated Sparrows at a time, usually in company with White-crowned or Golden-crowned Sparrows or juncos; in San Diego County, California, for example, the maximum count is three, a number that would be equally notable at most sites in the West and Southwest.

In winter and migration alike, the White-throated Sparrow is given to appearing far to the east of its usual North American range. One of the most frequent of Nearctic passerines seen in Britain and Ireland, the species was first recorded in the Old World in 1867, when a female was shot in mid-August in Aberdeenshire, Scotland. The unusual date, from a period when the southbound migration from the breeding areas has scarcely begun, raised eyebrows, and the record was not accepted. The first English specimen was taken at a more seasonal date, in late March in Brighton, but was likewise rejected. The species was finally added to the official British and Irish list in 1960, when the record of a male shot in May 1909 on the Outer Hebrides was deemed acceptable.

There are now more than 50 accepted records of the species in the British isles, and it has also been recorded in Iceland, the Netherlands, Denmark, Finland, and Gibraltar. Unusually for an American passerine, the European records of the White-throated Sparrow are concentrated in spring, and many of them fall at just about the time that birds wintering in their normal range are on the way north, a circumstance suggesting that at least some of the Old World records are of birds that have wintered in Europe but gone undetected until moving into more heavily birded regions.

In eastern North America, that movement begins with males in mid-March; females, which winter on average farther south, leave the wintering grounds one or two weeks later, giving the males a head start to stake out attractive breeding territories. Peak spring migration is evident in the northern tier of states and the southern tier of Canadian provinces by late April, and the northern breeding range is occupied by early June, even as a few, probably non-breeding birds tarry to the south into early summer.

No subspecies are recognized.

Rufous-collared Sparrows in our area, at the northern edge of the species' range, differ from southern birds in their conspicuous head stripes. *David Krueper.*

RUFOUS-COLLARED SPARROW
Zonotrichia capensis

In this age of internal combustion, birders with time and money can travel nearly anywhere at a moment's notice to see nearly any bird on earth. Not long ago, it was instead the birds that came to the birders—as skins and skeletons, in wooden chests and glass jars, by sea and by land. Most arrived intact and properly labeled, but many were unpacked by European scientists disappointed to find that the specimens they had so looked forward to were damaged or, worse, unaccompanied by information about their origin and habits.

Over the centuries, that is the fate that has befallen even some of the most familiar and most widespread birds of the non-European world, among them a "pretty little bird" that John James Audubon would describe as new in 1839 and name for the physician, ethnographer, and secretary of the Philadelphia Academy of Natural Sciences, Samuel George Morton. Audubon reports that the type specimen of the Morton Finch, an adult male, had been "procured . . . in Upper California" by John K. Townsend. Fatally, the young collector sent his bird skins to Philadelphia nearly two years before his own return, with the result that Audubon, "frantic" to publish the novelties in the appendix to his *Ornithological Biography,* described several species before he could confer with Townsend— and supplied the missing details with convenient fantasies spun from whole cloth.

In the case of the "infamous Morton's Finch," the most important of the blanks so creatively filled in by Audubon was the locality at which the bird had been collected. Townsend never visited California. On examining the type specimen in Philadelphia 20 years later, Philip Lutley Sclater definitively pronounced it "nothing more than a Chilian [sic] specimen of" the Rufous-collared Sparrow, obviously collected not by Townsend—who was in Hawaii when the ship bearing his skins called in

Valparaiso—but by a well-meaning sailor who generously added it to the natural history treasures on their way to Philadelphia. It would be more than a century and a half before a genuine Rufous-collared Sparrow was seen in the wild north of southern Mexico.

Audubon was not the first to be confused about the homeland of this handsome sparrow. The Rufous-collared Sparrow seems to have made its earliest European appearance in one of the engravings prepared for the Comte de Buffon's *Histoire naturelle*. Sharing a plate with a strikingly yellowish House Finch, the well-drawn and instantly recognizable sparrow is labeled "Bruant, du Cap de Bonne-Espérance," the Cape of Good Hope Bunting.

It took the better part of a decade for Buffon and his collaborators to finish their text account of this species, which finally appeared in print in 1778. Meanwhile, the Netherlandish naturalist Pieter Boddaert had been carefully mining the plates from the *Histoire naturelle* for his own updating of the Linnaean *Systema*. In 1772, he published the first verbal description of this

> finch, which is rusty brown with a black head and a white stripe over the eyes; the belly is white, the tail feathers bluish from below.

Noting that the bird inhabits Africa's Cape of Good Hope, Boddaert called it the "Kaapsche Vink," the Cape Finch—but did not give it a latinizing binomial. That oversight was remedied four years later with the posthumous publication of the Supplement to Philipp Ludwig Statius Müller's edition of Linnaeus. Müller simply translated Boddaert's Dutch description into German, but added, crucially, a Linnaean name, *Fringilla Capensis*, indicating the species' putative range in southernmost Africa.

The first doubts were introduced two years later, by none other than Buffon and the coauthors of the *Histoire naturelle*. Introducing the "Bonjour-commandeur," the "good-morning bird," they observed that this Guyanan native

> so completely resembles the Bunting of the Cape of Good Hope . . . that Sonnini considers them one single species with two names . . . . Given that he confirms that this bunting is a resident of Guyana, it is more than likely that it is found at the Cape of Good Hope only when it has been carried there on a ship.

Boddaert eventually agreed, and in his 1783 concordance to the plates of the *Histoire naturelle*,

he rejected any name, latinizing or vernacular, suggesting an African origin for this species. Instead, he adopted the Buffonian "Bonjour Commandeur" and coined a brand-new scientific name, *Emberiza pileata*, the "crowned sparrow," not knowing or not caring that Müller had got there first with an authoritative, if "geographically erroneous name."

FIELD IDENTIFICATION

Large, ornately plumaged, and often quite tame, adult Rufous-collared Sparrows are virtually unmistakable when seen at all well. The sturdy body with full belly, rather long and square-tipped tail, and slightly crested head are characteristic, as are the black-striped soft gray head, rusty nape, and broken black breast band. The understated off-white of the belly contrasts neatly with the collar and the faint wing panel created by the rusty secondary edges.

Juveniles are less distinctly marked, but share the adult's robust proportions, stubby crest, and neatly dotted white wing bars. The head pattern is essentially identical, but the gray of the adult plumage is replaced by dull buffy brown, the black head stripes by slightly less well-defined dusky markings. The buff of the underparts, on adults restricted to the flanks and lower breast sides, extends across the breast as a broad diffuse band; sparse blackish streaking continues onto the throat and jaw stripe, which are whiter than the breast.

The patterns of the juvenile's underparts can bring to mind a Lincoln Sparrow, but that skulking species of shady thickets is darker and more finely streaked above, with inconspicuous wing markings and a buffy jaw stripe clearly separated from the white throat by a neat blackish lateral throat stripe. The head of the Lincoln Sparrow is decidedly more grayish than the yellowish buff of the juvenile Rufous-collared, and the white eye-ring is usually very noticeable. The Lincoln Sparrow's wingtip is longer and its bill much finer than the clunky bill of the Rufous-collared and other *Zonotrichia* sparrows.

Across its vast range, the Rufous-collared Sparrow, similarly to the White-crowned Sparrow, exhibits wide variation in song, dialects associated not only with geographic distribution but with specific habitat types within a region.

> The song typically comprises 2–4 rising or falling whistled introductory notes . . . almost always followed by a terminal trill of more or less identical rapidly falling notes . . . . The most striking variation is in the rate of delivery of the elements of the terminal trill . . . . Diversity runs from buzz-like trills of ~5-millisecond intervals to languid whistled trills in which the trill

elements are separated by 300–400 milli-seconds. In some regions, birds produce more complex songs, with double trills, either fast-plus-slow or slow-plus fast . . . .

The frequently heard calls include a squeaky, explosive *chink*, and a "melancholy" descending *tseeu*.

RANGE AND GEOGRAPHIC VARIATION

Stretching from Mexico and the Caribbean to Tierra del Fuego, the vast geographic range of the Rufous-collared Sparrow nearly matches that of the Song Sparrow—as does its geographic variability. "This protean species" is thought to have originated most likely in Central America, its range and physical diversity expanding during the Pleistocene. Today, most authorities recognize at least two dozen subspecies, of which one, the aptly named *septentrionalis* ("northerly"), approaches the region covered in this guide and may also have occurred much farther north as a wild stray. For citations to original descriptions of the subspecies not considered here, the interested reader can consult the world list published online by the International Ornithological Congress, which recognizes 29 races.

Whatever the number of subspecies recognized by a given authority, they can easily be divided into stripe-headed and gray-crowned races. The southernmost are gray-headed, without the strikingly broad lateral crown stripes shown by tropical breeders; in the Venezuelan subspecies *inaccessibilis, macconnelli,* and *roraimae,* the crown is almost entirely black, with only a hint of the gray median stripe. The extent of the black patch at the side of the upper breast also varies geographically: these patches are broadest, nearly forming a breast band, in northerly races such as *septentrionalis, costaricensis, antillarum,* and *orestera,* and most restricted in *matutina* of northern Brazil and Bolivia. The degree of contrast between the hindneck collar and the remainder of the upperparts depends on the richness of the brown back streaking; *carabayae* and *antofagastae,* from Peru and Chile, respectively, are rufous above, the colorful back blending into the collar.

Zonotrichia capensis septentrionalis—a large Rufous-collared Sparrow, with an only moderately defined rufous neck collar blending with the rich brown, rather narrowly black-striped back—is resident at elevations of up to 10,000 feet in Chiapas, northeast to San Cristóbal and Tumbalá. To the south, it ranges through the highlands of Guatemala to southwestern Honduras, north to Santa Barbara National Park and east to El Paraiso. In El Salvador, *septentrionalis* meets *costaricensis,* which shows a better-defined back pattern, with broader and blacker streaks and sharper boundary between the back and the nape collar.

In May 2010, a singing male Rufous-collared Sparrow was discovered in Clear Creek County, Colorado, 2,000 miles north of the species' usual range. The bird remained until at least August, and was thoroughly photographed and its song well recorded during its four-month stay. The sparrow was not collected or otherwise examined in hand; photographs were said to suggest that it was the representative of a northern Middle American race, while the sound recordings were thought more typical of an equatorial population. In any event, the possibility appears to be virtually ruled out that this individual originated from a southern, gray-crowned and long-winged subspecies; while those birds are austral migrants, moving north in the non-breeding season, the tropical populations of Mexico, Central America, and northern South America are sedentary within their ranges. No subspecies of the Rufous-collared Sparrow has an established pattern of vagrancy, but it will never be known whether the Colorado bird—or another that arrived in Montana in autumn 2017—was an escaped captive, a stowaway, or, just possibly, a wild vagrant that had made the journey on its own without human intervention.

Colorful but often secretive, with a bright white throat and bright olive upperparts, this species lacks the white supercilium and broad breast band of the much larger Collared Towhee. *Oaxaca, February. Amy McAndrews*

CHESTNUT-CAPPED BRUSH FINCH
Arremon brunneinucha

In an age when science is conducted in public museums and university laboratories, it can be startling to recall the extent to which early natural history and ornithology were obliged to rely on the generosity of private collectors who opened their cabinets to scholars and investigators. In the first half of the nineteenth century, one of those benefactors of science was the Bordeaux businessman Charles Brelay, who made his "beautiful ornithological holdings" available to his colleagues, and even gave away some of his most valuable specimens in the interest of advancing science.

Brelay made his fortune processing bristles for brush manufacturers, and he spent it on bird skins, for the most part, it seems, from the Americas. His wife, Aglaé Delmestre Brelay, was afflicted with the same passion for collecting, and their contemporaries admired

> the very special enthusiasm with which she herself has engaged in ornithology and collaborated with M. Brelay in forming his collection, which includes many thousand individual birds.

Frédéric de La Fresnaye was so impressed that in 1839 he overcame his resistance to the common practice of "giving new birds the names of women to whom a taste for ornithology is so often entirely foreign" and christened a new and "very beautiful species" of becard *Pachyrhynchus Aglaiae* in her honor.

La Fresnaye resumed his normally sober nomenclatural behavior when it came to naming another bird from the Brelays' collection. This one, characterized by a chestnut cap tending to brighter reddish at its edges, he named straightforwardly (if slightly imprecisely) *Embernagra brunnei-nucha*, the brown-naped "tanager-bunting." The Brelay specimen, said to have been taken in Mexico, would remain unique in European collections for several years, until others began to arrive in the 1840s and 1850s from such widely separated localities as Colombia, Costa Rica, Venezuela, and Peru, providing a first glimmering of how extremely widespread this common tropical sparrow truly is, with a range reaching from Mexico to Peru.

FIELD IDENTIFICATION

This is a large, long-tailed, short-winged, and colorful sparrow, unlikely to be confused with any species other than the remarkably similar—and only rather distantly related—Collared Towhee. Both are fond of dark, densely vegetated habitats where they can be difficult to see well, but the brush finch's thin metallic vocalizations are obviously different from the robust whistles and trills given by the larger and heavier towhee.

Once glimpsed, the brush finch's darker, duller upperparts and grayer underparts are distinctive. The towhee's black breast band is broader and messier than the fine line separating the brush finch's throat and upper breast. In addition, the towhee's black mask—created by the dark ear coverts and lore—is separated from its rusty crown and extensive black forehead by a conspicuously continuous white supercilium; in the brush finch, white in the face is restricted to a small spot between the black lore and the small black forehead patch, and the ear coverts and crown are set off only by a yellowish stripe (almost or entirely absent in the subspecies *apertus*, Plain-breasted Brush Finch).

Juveniles of both species are, happily, often accompanied by a parent. They are darker and duller than the respective adults, both irregularly streaked above and below. The head pattern of the juvenile Collared Towhee is a muted version of the adult's, the conspicuous supercilium off-white and vaguely streaked; even so, it provides a useful distinction from the juvenile brush finch, which shows no supercilium or only a hint of a paler yellowish stripe between crown and ear coverts.

The calls of the Chestnut-capped Brush Finch are surprisingly high, weak, and insectlike for a bird of its bulk. Single thin *peek* and *seep* notes sometimes run into a rapid excited series, especially when birds are agitated.

The rather variable song is no more impressive to human ears. Like the calls, the notes of the song are fine and high-pitched, most with a pronounced fingernails-on-the-chalkboard screeching tone, others with the hint of a liquid squeal behind them. Some songs comprise only a strictly rhythmic sequence of notes, *spee spee spee-ur-seep*; others incorporate soft popping sputters too irregular to be called a tremolo or trill.

RANGE AND GEOGRAPHIC VARIATION

Fairly common overall in the understory of mid-elevation forests, this species is resident in the mountains of South America from Venezuela to southern Peru and in foothills and highlands from Panama to southern and western Mexico. The Sierra de Tuxtla on the southern Veracruz coast is occupied by *apertus*, with *suttoni* resident in the mountains of Guerrero and central Oaxaca; the widespread nominate race occurs in eastern San Luis Potosí, Hidalgo, Puebla, and Veracruz, south to the vicinity of Presidio.

Once described as nearly invariable over a range extending from Mexico to Peru, the Chestnut-capped Brush Finch is now generally separated into nine or ten subspecies, three of which occur in the portions of Mexico covered in this guide. Those three northerly races are distinguished by the color and pattern of the underparts and by the precise details of the head pattern, including the extent of rusty color onto the nape and back and the prominence of the yellowish supercilium separating the black ear coverts from the rufous crown.

Originally named as a separate species, the race *apertus* differs from nominate *brunneinucha* most signally in the absence of the narrow black band at the lower edge of the white throat; the center of the breast and belly are whitish, while those areas are more uniformly and extensively gray in *brunneinucha*. The chestnut of the crown meets the black of the ear coverts directly, with no or only the hint of a yellowish supercilium dividing them. Found only in the Sierra de Tuxtla of southern Veracruz, this Plain-breasted Brush Finch, which "has no characters uniquely its own, but merely exhibits a different combination of the characters found in other races," was reduced to subspecies status by Kenneth C. Parkes a dozen years later.

In the same paper, Parkes named another distinctive Mexican subspecies for George Miksch Sutton. This race, *suttoni*, known from Guerrero and Oaxaca, is larger and longer-tailed than the nominate subspecies, with more extensively white underparts. The rusty color of the crown, neatly set off from the greenish back in other races, continues as two irregular stripes onto the nape and upper back; there is a yellow patch on the lower nape between these stripes, a character seen only rarely in the other Mexican races.

The forehead pattern of *suttoni* also differs from that of *brunneinucha*. In all Chestnut-capped Brush Finches, a short white line divides the black of the forehead below the rusty crown; in this race, that line continues a short distance into the crown, sometimes becoming a very faint yellowish median crown stripe reaching all the way to the nape patch. The yellowish supercilium is conspicuous and extensive, sometimes meeting the small white patch above the black lore. A further Oaxacan race, *nigrilatera*, was described in 1968, but is now generally considered indistinguishable from *suttoni*.

Like the "black" towhees, the brush finches belie the common notion of sparrows as small, brown, and streaked. The Green-striped Brush Finch occurs only at the southernmost edge of the area covered here. *Manuel Grosselet*

GREEN-STRIPED BRUSH FINCH
Arremon virenticeps

This somberly elegant sparrow entered ornithology in 1855, in a single sentence in Charles Bonaparte's report of an extended visit to the museums and private collections of England and Scotland.

> To the modest number of species in the genus *Buarremon* we must add *B. virenticeps*, Bp., from Mexico, similar to the Gray-browned Brush Finch *B. assimilis* but colored green, not gray, in the areas between the black stripes of the crown.

Almost simultaneously, Philip Lutley Sclater noted the existence of another specimen in Europe, this one in the royal museum of Berlin, where the curator, Martin Hinrich Lichtenstein, had labeled it *Fringilla quadrivittata*, the "Four-striped Finch." Scientific names handwritten in pencil on grease-stained specimen tags have no nomenclatural standing, leaving this among the several species for which Lichtenstein could have been the naming authority had he thought to publish his work.

FIELD IDENTIFICATION

Large, long-tailed, and retiring, this modestly plumed green and gray sparrow is unlikely to be mistaken for any other in its very restricted range. While the Green-striped shares the conspicuous bright white throat of the Chestnut-capped, the two do not overlap in range. The smaller, finer-billed, plainer-faced, dull-throated *Arremonops* sparrows are residents of damp lowland forests rather than the high-elevation conifers inhabited by the Green-striped Brush Finch.

With their dark striped heads and greenish upperparts, several saltator species can superficially recall the Green-striped Brush Finch—a similarity responsible for the assignment of this species to the tanager family in the nineteenth century. The Grayish Saltator nearly overlaps in range with the brush finch, but it is a large, noisy bird, far more brutish in appearance and in habit: no saltator could ever have been named *verecundus* ("retiring"), as is one race of this brush finch.

The full belly, moderately long and wide tail, and contrasting white throat, often puffed out, are generally sufficient to identify the Green-striped Brush Finch. In the shady thickets this species prefers, the greenish upperparts may look simply dark grayish brown, but the contrast between white throat and dark mask remains conspicuous.

The juvenile is duller than the adult, with broad, blurry streaks beneath and a yellow bill base.
Manuel Grosselet

The vocalizations of this species are similar to those of the Chestnut-capped Brush Finch. The most frequent call is a high-pitched, insectlike *seeei*, the song a long series of similar notes in a "slightly jerky, irregular rhythm."

RANGE AND GEOGRAPHIC VARIATION

Found only in western and central Mexico, the Green-striped Brush Finch occurs at elevations between 6,000 and 10,000 feet from the Pacific Slope in central Sinaloa to northern Nayarit and from Jalisco east to western Puebla. There is also a report from Guanajuato.

In mid-November 1934, Chester Converse Lamb extended the known range of this species—and of the genus—to southeastern Sinaloa, when he shot an apparent pair of Green-striped Brush Finches in the high mountains near Rancho Batel, two of more than 40,000 Mexican specimens Lamb collected over more than 20 years for Robert T. Moore. Moore, a wealthy and devoted dilettante and philanthropist, described the brush finch in 1938 as a new subspecies, *verecundus* ("retiring," a remarkable understatement given the withdrawn habits of this species). Moore found that Lamb's specimens differed from the nominate race in their darker and more extensively gray underparts, smaller overall size, shorter bill, and proportionally shorter tail.

Reexamining the type and three other specimens of *verecundus* a generation later, John William Hardy and Thomas Webber found that dark underparts were not unique to the Sinaloa population, and that the bills of three of the four Sinaloa birds were in fact longer than the average of those from central Mexico; they also discovered "wide overlap" between the nominate race and *verecundus*.

In the same year that Moore named *verecundus*, van Rossem described nine specimens from the Volcán de Colima as a distinct subspecies *colimae*, characterized by its brighter upperparts and nape; he also noted that the median crown stripe was more conspicuous and the supercilium brighter and more yellowish than in Bonaparte's *virenticeps*. In the decades that followed, some authorities recognized, if tentatively, this race as valid, but today the Green-striped Brush Finch is almost unanimously treated as a monotypic species.

BLACK-THROATED SPARROW
Amphispiza bilineata

By the late 1840s, John James Audubon was sinking rapidly and irrecoverably into dementia, and without his energy and guiding force, the Audubon family was obliged to supplement its always tenuous income. In 1849, John Woodhouse Audubon, the younger son of John James and Lucy Audubon, convinced his brother, mother, and wife that he should join a company of some 80 prospectors bound for the gold fields of California; caught up in the enthusiasm of the day, he assured his father's old friend Edward Harris that his participation would bring "at least $20,000" into the family's treasury.

The expedition was a complete failure. The party's guide deserted them once they reached the Rio Grande in Texas. Cholera sickened many of those who continued; Audubon's cousin John Howard Bakewell died from the disease. When Audubon finally returned to New York, it was without the promised thousands of dollars—but with substantial debt that the family was forced to make good.

It was likely small consolation, but Audubon's long journey did make some contribution to the scientific knowledge of the avifauna of the Southwest. Among the bird skins that made their way to the Academy of Natural Sciences in Philadelphia were representatives of the Bridled and Black-crested Titmice and the first known specimen of "one of the most remarkable finches yet discovered in America . . . quite unlike, even in general appearance, any other species of this country."

Had Audubon discovered the bird even just half a decade earlier, it would certainly have been named, described, and illustrated by his father. In 1850, however, the octavo edition of the *Birds of America* was complete, and the elder Audubon, just a few short months from death, was likely unaware even that his son had departed and returned. It was left instead to John Cassin to introduce *Emberiza bilineata* to science; four years after that first description, when the bird was known to be "numerous" and "extensively diffused" in Mexico, Texas, and New Mexico, Cassin devoted a longer account to "this curious little Finch," illustrated with a stiff and unsophisticated lithographed portrait by William E. Hitchcock of what he named the Black-throated Finch.

FIELD IDENTIFICATION

The Black-throated Sparrow is a bulky, deep-bellied, full-breasted sparrow with a rather long tail and large, square head. Both adults and juveniles are smoothly, evenly colored brownish above and whit-

Sturdy and elegant, the adult Black-thoated Sparrow can hardly be confused with any other species. The amount of white in the tail varies geographically. *Texas, March. Brian E. Small*

ish gray below, with a variable amount of brown tinging the sides and undertail coverts; the only strong patterning visible in birds of either age class is the bold striping of the slaty gray head.

Adult Black-throated Sparrows are virtually unmistakable. They differ from other sparrows and sparrowlike birds with black throats most notably in their unstreaked plumage, without the heavy coarse markings of a House Sparrow or the boldly striped flanks of a Harris Sparrow.

The juvenile Black-throated Sparrow is finely, usually inconspicuously streaked above and below. Lacking the eponymous black throat and breast, juveniles are often compared to adult Sagebrush and Bell Sparrows, from which they differ most signally in the Black-throated's unstreaked crown and very broad, long white supercilium on a deep blue-gray head; at any age, the Black-throated Sparrow's blackish tail is more strongly graduated and shows more white (especially in the nominate subspecies) than that of either of the two "sage" sparrows. While

The contrast between the plain brown back and blackish tail is often visible even in flight. *Texas, March. Brian E. Small*

the youngest juveniles are usually accompanied by an adult, some Black-throated Sparrows may retain juvenile characters such as streaking on the breast and back and a brown wash to the head into midautumn, well after having attained independence of their parents. Some birds do not complete their molt out of juvenile plumage until after arriving on the wintering grounds, as late as November (Pyle).

Perched atop a bush or cactus, this species shares with the "sage" sparrows the habit of nervously twitching its tail upward in a stiff staccato series. That motion is sometimes accompanied by a very short, often barely audible call note, a high-pitched, silvery *tsit* with a sharp attack and virtually no decay. The same call, repeated at irregular intervals, is given by anxious birds on the ground, but in sustained flight this species is usually silent.

The song of the Black-throated Sparrow is remarkably variable among individuals, and some males have as many as nine distinct songs. Singing begins in some areas as early as February, and is most frequent within an hour of sunrise. Most songs are simple, comprising a ticking introduction followed by a trill or buzz. In the most emphatic examples, loud and rhythmically strict, the concluding trill recalls the tremolo with which many Rufous-winged Sparrow songs end; more typically, the notes shimmer and blur like tiny cymbals touched with a brush. More complex songs insert a soft, loose buzz between the introductory chips and the concluding trill; the trill itself may be extended or repeated, either in identical form or as a series of variations.

> Black-throated Sparrows may also sing in flight; during song flight, [the] wings [are] fluttered and quivered dramatically and [the] head [is] thrown back, unlike normal direct flight.

RANGE AND GEOGRAPHIC VARIATION

This very widespread and generally common species is found from southernmost Baja California, Jalisco, Guanajuato, Hidalgo, and southern Tamaulipas north and west nearly to Canada. The eastern edge of the breeding range runs along the eastern Edwards Plateau of Texas, north to Shackelford and Palo Pinto Counties. The species breeds in the extreme west of the Oklahoma Panhandle (eBird) and is rare and local on the plains of southeastern Colorado from Baca to Fremont Counties; it is an uncommon summer resident along that state's western border south of Mesa County. There are June and July records from southwestern Wyoming (eBird), though breeding there is deemed only possible.

The juvenile is more subtly marked than the adult, but they share the black tail and broad supercilium. *Amy McAndrews*

Like the Sagebrush and Bell Sparrows, the Black-throated often twitches its tail upward when it is perched. *New Mexico, May. Brian E. Small*

Black-throated Sparrows breed throughout the deserts of New Mexico, Arizona, Nevada, Utah, and southern Idaho. In California, breeding occurs almost exclusively east of the axis of the Cascades and Sierra Nevada, from northeastern Siskiyou County to the Nevada border and south through the mountains of the eastern deserts; there are disjunct populations, not all of them present every year, in Kern, Ventura, Santa Barbara, Butte, and El Dorado Counties. The species is uncommon and local in the high deserts of southeast Oregon, west to Klamath and north to Wheeler Counties. The northernmost breeders are in Washington, especially in Kittitas, Yakima, and Benton Counties in the south-central portion of the state, north to Vantage in Kittitas County; the species is rare anywhere in Washington.

Black-throated Sparrows are not known to breed in British Columbia, but the species has occurred with increasing frequency since the first was collected there in 1959. Almost all of the 50 or so records for the province are from May and June, and it is virtually expected that two of these spring overshoots will eventually find each other and breed.

Only the three northerly subspecies are known to be migratory; some individuals of *deserticola* move more than 1,000 miles from the nesting grounds in Oregon and Washington. Autumn migrants leave the northern breeding range in late August and September, arriving in mixed-age flocks in the southwestern United States and northern Mexico beginning in late September. Those migrants winter from southeast California, southern Nevada, southwestern Utah, Arizona, and southern New Mexico into north-central Mexico; their status is uncertain in Texas, where *deserticola* presumably winters within the permanent ranges of both the nominate race and *opuntia*.

Spring migrants arrive on the northerly breeding grounds in April and May, while southern California and Arizona breeders are in place beginning in March. Overshooting birds are now nearly annual in British Columbia.

Unlike the Sagebrush Sparrow or other species with a similar United States breeding range, the Black-throated Sparrow is a regular, if always scarce, vagrant east of its breeding range virtually across the continent, most frequently in winter. Individuals have been recorded, often at feeders, from Quebec and, once, New Brunswick (eBird) to Florida, where a specimen was collected in February 1976. Out-of-range spring records in the West may represent overshooting northbound migrants, while May and June records from the East are probably of birds undetected over the winter.

Nine of 13 described subspecies are generally recognized today. The most widespread and to many birders the most familiar race is *deserticola*, described by Robert Ridgway from a May specimen taken in Tucson, Arizona, by Edward Nelson. Breeding in desert habitats from south-central Washington to central Baja California and east to northwestern Chihuahua, this Desert Sparrow is larger than the gray-backed nominate race, with a paler and browner back and much smaller white tail spots. The nominate race, discovered by John Woodhouse Audubon, is small, with the back medium to dark gray-brown; the white on the outer rectrix averages nearly twice as large as in *deserticola*. This is the resident race from north-central Texas east of the Pecos River south to eastern Coahuila, Nuevo León, and Tamaulipas.

The remainder of the Black-throated Sparrow's United States range is home to the subspecies *opuntia*, named for the cane cholla it is so fond of. Found from southeast Colorado and western Oklahoma south through Texas's Trans-Pecos to northeastern Chihuahua and northwestern Coahuila, *opuntia* is large, most falling between *bilineata* and *deserticola* in size but some even larger than the latter; its medium to pale gray back is tinged brown, and there is only slightly more white on the outermost rectrix than in *deserticola*.

The most widespread Black-throated Sparrow in Mexico is the race *grisea*, a large, quite dark gray bird with proportionally short bill and feet and restricted white on the outer rectrix. This subspecies is found from central Chihuahua to Jalisco, Hidalgo, and southwestern Tamaulipas, by way of Zacatecas, Aguascalientes, Guanajuato, Querétaro, southern Coahuila, and San Luis Potosí. It is a permanent resident throughout its range.

Another Mexican resident, *bangsi*, is common on the Baja California Peninsula south of about 26 degrees and on Santa Margarita, Magdalena, Santa Catalina, Santa Cruz, Carmen, San José, San Francisco, and Espíritu Santo Islands and on the Coronados south of that latitude. This smallest of all the Black-throated Sparrows resembles *deserticola*, which breeds as far south as 27 degrees on the peninsula and on Cedros, Natividad, and Ángel de la Guarda Islands, but is slightly paler above and slightly larger-billed, with a shorter wing and tail. Farther north on the east coast of the Sea of Cortez, the rather poorly named *pacifica* is a resident of the deserts of southern Sonora and northern Sinaloa, along with Tiburón and San Pedro Nolasco Islands;

it meets *deserticola* in Sonora at 29 or 30 degrees, but differs from that subspecies in its darker upperparts and smaller size.

Three subspecies are restricted each to a single island off western Mexico. Endemic to San Esteban, in the Sea of Cortez just southwest of Tiburón, *cana* is, as its name suggests, the palest and grayest of the Black-throated Sparrows. It is smaller and shorter-tailed than *deserticola*, with pale ashy gray back and flanks; it is similar in size to *pacifica*, the resident form of adjacent Tiburón Island, but pale and much grayer above. Less than 100 miles to the south, on Tortuga Island in the Sea of Cortez, *tortugae* is the darkest race, dark slaty above and extensively gray below, with only a narrow white median stripe stretching from the lower breast to the undertail coverts. The race *belvederei*, of Cerralvo (Jacques Cousteau) Island in the southwestern Sea of Cortez, is darker and grayer, with a somewhat more curved bill, than the nearby mainland's *bangsi*.

This species is closely associated over its entire range with *Opuntia* cactus, including chollas and prickly pears. *California, January. Brian E. Small*

The Five-striped is a large and long-billed sparrow, with strikingly patterned brown, gray, and black-and-white plumage. *Arizona, July. Brian E. Small*

FIVE-STRIPED SPARROW
Amphispiza quinquestriata

The discovery of the Five-striped Sparrow by European science remains, a century and a half after the bird's formal description, a striking instance of the gradual fossilization of supposition and surmise as "knowledge." "This well-marked species" was first published by Philip Lutley Sclater and Osbert Salvin in 1868, on the basis of

> a single indifferent skin, which has long remained without a name, in Sclater's collection. It was obtained by him some years ago from Mr. [John] Gould, who received it along with a collection of Mexican Hummingbirds.

A dozen years on, Sclater's was still the only known specimen. In the intervening years, Salvin and Frederick DuCane Godman had devised a theory about its provenance:

> Judging from the preparation of the skin, we believe that it was made by Floresi,

who resided for some time in the mining districts of Central Mexico, and who corresponded with Gould.

Damiano Floresi d'Arcais was an Italian mining engineer, supervisor in the 1840s of an English-owned mine in Jalisco; he was later sent to Panama as a member of a zoological expedition, where he died. Floresi's attainments as a miner included the introduction to Mexico of a method for retrieving silver from silver ore without the use of mercury. As an ornithological collector, he is best known as the likely source of the specimens from which his friend John Gould first described the Imperial Woodpecker.

Salvin and Godman's theory about the origins of the Gould/Sclater specimen edged closer to received fact in 1934, when A. J. van Rossem simply and uncritically asserted that "the type was received by Gould from Floresi"—and then deduced from that that "though the type locality is not known with certainty, it is most probably Bolaños, Jalisco," a conclusion drawn from an uncertain premise and the baseless assumption that Floresi had shot the bird himself and during his residence in Jalisco. Van Rossem's residual caution was thrown largely to the wind in 1957: at the same time as they affirmed that the type was only "probably" from Bolaños, Alden H. Miller and his coauthors pronounced without any qualification the nominate subspecies' range to be "Jalisco (Bolaños)." The American Ornithologists' Union *Check-list* now identifies the type locality as Bolaños, without comment but with a degree of specificity that can be described only as fictional.

FIELD IDENTIFICATION

Distinctive and distinctively beautiful, the Five-striped is a rather large, long-tailed, small-headed sparrow; adults are notably dark above and below. The long tail and slender bill give the bird an attenuated look unlike the blocky, stocky aspect of the somewhat similar Black-throated Sparrow; the two species prefer very different habitats, the Black-throated relatively sparse cactus-strewn grassland and the Five-striped steep thorny slopes. Habitat also neatly separates the Five-striped and the more social Sagebrush Sparrow, the latter a denizen of wide-open sagebrush flats with scattered bushes and trees; Sagebrush and Bell Sparrows are much less dramatically marked on the head, with brighter whitish breasts and brown-streaked flanks and breast sides. Hopeful birders visiting southeast Arizona for the first time have been known to misidentify even Lark Sparrows as this highly sought-after species. Both Lark Sparrows and Five-striped Sparrows have conspicuously marked heads and prominent central breast smudges, but they differ otherwise in size, structure, plumage, vocalizations, habits, and habitats: no Lark Sparrow lurks on the sheer cliffs of heavily vegetated canyons, and Five-striped Sparrows never gather in busy, confiding flocks to feed on urban ball fields and roadsides.

Such potential stumbling blocks aside, adult Five-striped Sparrows are readily identified even at a distance by their somber upperparts, breast, and flanks and contrastingly white belly and throat. At closer range, the blackish tail, rich brown back and wing coverts, very wide black lateral throat stripe, and short, narrow white jaw stripe and supercilium are diagnostic. The sexes are virtually identical, though females are said to average smaller, paler, and overall browner than the male.

Juvenile Five-striped Sparrows are rarely seen, skulking silently in dense, usually thorny vegetation. The blackish tail and unstreaked back and rump resemble those of adults, though the wing coverts are less colorful and the tertials duller slaty with rusty edges; the back may show obscure darker spotting. The underparts are tawny brownish to yellowish, grayer on the flanks, with a sparsely streaked brown band across the breast; the head pattern is subdued, with hints of a whitish jaw stripe and supercilium but without the adult's broad lateral throat stripes. Wolf discovered similarities between this plum-

The white belly is set off by the gray flanks and breast; there is a large, irregularly shaped black breast spot. *Arizona, July. Brian E. Small*

age and the juvenile Rusty-crowned Ground Sparrow, but given the only distant genetic relationship between the species, the shared characters of yellowish underparts, prominent breast band, and reduced head pattern are "fortuitous," as Wolf suggested might prove to be the case.

The song of the Five-striped Sparrow is structurally simple, "an introductory note followed by one to six, usually two" sequences of grouped notes. The introductory *tuck* or *tzip* appears to be quite stereotypical, but each male has at his command a startlingly large variety of note groups to follow it; one study identified 99 different such phrases, of which the most versatile male sang no fewer than 55 in his various songs. This wide range of available phrases and the birds' ability to combine and recombine them within the simple standard framework mean that any given male may sing more than 200 identifiably different songs.

Fortunately for the human observer, that daunting variety of vocalizations is noticeably consistent not only in its simple structure but in the general tone of the notes and note groups. Both the introductory note and those in the groups that follow usually have a distinctive loose, sloppy, lisped quality, with a lazy attack and trailing decay; the note groups can be separated by pauses of up to a full second: *tzip, slip-slip; slurp-slurp; tslit-tslit-tslit.* Birders from eastern North America may be distantly reminded of a Henslow Sparrow.

A note very similar to the song's introductory *tuck* or *tzip* is also used as a call by feeding or perched birds; the usual contact call is a single high-pitched *seet*. A "slurred chatter" may be given in disputes, and there is also a complex series of rushed notes at different pitches used in territorial confrontations and when the members of a pair are reunited.

RANGE AND GEOGRAPHIC VARIATION

In the north, *septentrionalis* is found on the steep cactus- and ocotillo-clad walls of brushy canyons and, locally, in grassland hills from the northern flank of Arizona's Santa Rita and Baboquivari Mountains south through the mountains of central Sonora, westernmost Chihuahua, Sinaloa, and western Durango; it may also breed in Nayarit (eBird). Apparently uncommon—or perhaps merely furtive —throughout its range, it is rare and irregular in its occurrence north of the United States border, where it was first collected in June 1957. Since at least 1969, the species has been known to be present in southern Pima and, especially, southern Santa Cruz Counties in Arizona; a systematic search conducted in 1977 discovered 57 individuals, most of them the more conspicuous males, in the Atascosa, Santa Rita, and Patagonia Mountains. With the apparent excep-

tion of the long-persistent populations in Sycamore Canyon and California Gulch, birds at this northern edge of the species' range appear to occupy a site for a few years and then, for reasons inscrutable, disappear. In any event, the Five-striped Sparrow's status rangewide is extremely difficult to assess given the remoteness of many preferred localities and the species' secretiveness over most of the year.

The southern, nominate race is even less perfectly known than the northern. It occurs in northern Jalisco, southern Zacatecas, and western Aguascalientes, in a narrow range that appears to be entirely disjunct from that of *septentrionalis.*

The nominate race is smaller and darker, with a larger central breast smudge, than the northerly subspecies *septentrionalis,* described in 1934 from specimens from Sonora and Guirocoba. Nominate *quinquestriata,* known from western Zacatecas, western Aguascalientes, and northern Jalisco, is presumably resident, while the northern race, breeding from extreme southeastern Arizona south through Sonora and Sinaloa, apparently undertakes southward movements in November, but it is unlikely that birds of the two subspecies ever come into contact, their ranges separated by some hundreds of miles in western Mexico. Northbound migrants appear to leave the wintering grounds by the end of March. While breeding takes place at elevations between about 750 and 6,000 feet, migrants and winterers have been found at sea level in southwestern Sinaloa.

Usually shy inhabitants of desert canyons, in summer male Five-striped Sparrows sing from conspicuous perches on ocotillo whips, mesquite branches, or even tall rocks. *Arizona, July. Brian E. Small*

The most overdetermined of sparrows, the Lark Sparrow should be unmistakable in any reasonable view. *Arizona, April. Brian E. Small*

LARK SPARROW
Chondestes grammacus

None of the early natural historians and ornithologists on the North American continent—not Mark Catesby, not William Bartram, not Ashton Blackburne, not Louis Pierre Vieillot, not Alexander Wilson—so much as suspected the existence of this lavishly marked sparrow; Audubon himself never saw it alive, but relied for his skeletal account (and notably poor engraving) of "this beautiful species" on the rudiments provided by Thomas Say, Charles Bonaparte, and Thomas Nuttall.

It was Say who discovered and first described the sparrow, just above the mouth of the Missouri River. In June 1819, he and his companions on the Long expedition to the Rockies shot "many specimens" of this species at Bellefontaine, Missouri Territory, and they would see the bird again the next spring at Engineer Cantonment, slightly north of what is now Omaha. Say observed that "they run upon the ground like a lark, seldom fly into a tree, and sing sweetly." He named the new species, however, not for its behavior but for its plumage. The curious epithet of Say's *Fringilla grammaca*—"badly selected . . . and badly spelled," grumped Elliott Coues—refers to the "lineated" head pattern, as if the stripes had been inscribed with a pen.

Say did not immediately give his new sparrow an English name. Two years after publishing the original description, he helped Charles Bonaparte prepare the account for the species in Bonaparte's *American Ornithology*; this time, the bird bore both Say's scientific name and a new English label, the Lark Finch. Relying on Say's experiences, Bonaparte wrote that

> like the Larks, they frequent the prairies, and very seldom, if ever, alight on trees; they sing sweetly, and often continue their notes on the wing.

Bonaparte also pointed out that "this very interesting new species" had gone unnoticed for so long because its range was restricted for the most part to the still wild country west of the Mississippi River, an explanation echoed by another European prince, Maximilian zu Wied-Neuwied, who on his 1832 tour of the West found the "stripe-headed bunting" only on the remote rolling hills of the upper Missouri.

By mid-century, though, it was evident that Lark Sparrows were on the move. The first record of the species from eastern North America was obtained "about 1845," when one was reported, and apparently collected, by the taxidermist Samuel Jillson in coastal Massachusetts. They first reached Ohio in 1860, and in the following decade, a young Robert Ridgway—long familiar with the Lark Sparrow in his home state of Illinois—observed that

> this species seems to be gradually extending its range to the eastward, probably in consequence of the general and widespread denudation of the forests, the country thus undergoing a physical change favorable to the habits of the species, having already become a regular summer resident in many sections of the country north of the Ohio [River].

The push east was not without its hazards, even in the nineteenth century. Writing from Illinois in 1887, H. K. Coale explained the Lark Sparrow's scarcity in the eastern portions of its range as in part due to the notorious agricultural pesticide Paris green, by which "a large number of the birds are annually destroyed."

Incidental poisonings notwithstanding, Roger Tory Peterson would include the Eastern Lark Sparrow in his 1934 *Field Guide* without comment or qualification, in the second, 1947 edition noting that its regular breeding range extended to "weedy fields and poor pastures" in Ohio and West Virginia.

Beginning in the mid-twentieth century, the Lark Sparrow has retreated noticeably toward its apparently original range in the Midwest and West. The last recorded breeding in Pennsylvania took place in 1930. Still nesting in more than 40 Ohio counties in the 1930s, the species is now formally listed as endangered in that state. In the 1970s, Lark Sparrows were still "locally uncommon" in southern Indiana, though already "very rare" elsewhere in the state; by the end of the century, breeding was noted "in only 2 percent of the state's 647 priority blocks . . . scattered along Indiana's western margin." The species remains locally fairly common in appropriate habitats in Illinois, but its breeding range overall today more closely resembles that occupied in the days of Wilson and Audubon than the wider distribution encouraged by the extensive agricultural practices of the nineteenth and early twentieth centuries.

FIELD IDENTIFICATION

"If you are going to introduce a beginner to sparrows, this is where you should start," with this large, conspicuous, tame bird that is "arguably the handsomest of sparrows." Those twin opinions—that the Lark Sparrow is beautiful and that it is readily identifiable—have been part of the common birderly wisdom for the entire two centuries since the species' discovery.

This species shows almost every head marking possible, in colorful contrast to the rest of its more discreet plumage. *Arizona, April. Brian E. Small*

Worn adults in summer are noticeably plain-backed and dull-winged, but the small white spot at the base of the primaries remains conspicuous.
California, July. Brian E. Small

The adult Lark Sparrow, with its long wing, fan-like, lavishly white-tipped tail, and strikingly heavy maquillage, is indeed virtually unmistakable. Large and heavy, with a long tail and strikingly small head, Lark Sparrows sometimes seem to be all breast, an impression heightened by the uniformly bright whitish color of the underparts. On the ground at any distance, adults appear smoothly light brown above and white below, a combination rare in passerellid sparrows. The central breast blotch can be conspicuous, and the strongly marked head is obvious even when the precise pattern cannot be discerned. The small whitish "pocket" at the base of the folded primaries varies in perched birds from glaringly evident to barely visible.

Seen from below, perched Lark Sparrows are creamy white below, with buffy-tinged flanks and a bright undertail created by the white outer rectrices. The breast splotch is invariably noticeable, but most striking is the throat pattern, a snowy patch bordered by bold black lateral stripes.

Those stripes are just the beginning of the Lark Sparrow's riotously overdetermined head pattern, the inspiration for such genuine folk names as "quail head" and "snake-bird, from the supposed resemblance of its striped head to that of a snake." Most sparrows have visible streaks or stripes on the face or crown, but the Lark Sparrow exhibits nearly every possible head marking, lacking only a median throat stripe: its combination of lateral throat stripes, jaw stripes, whisker stripes, eye lines, supercilia, lateral crown stripes, and median crown stripe makes the species the darling of artists charged with illustrating avian topography.

Juvenile Lark Sparrows are quite similar to adults, and some especially brightly colored and lightly marked individuals can be puzzling in the field. Most, though, are readily distinguished from adults by the heavier, coarser streaking on the upperparts and the irregular blackish streaks and flecks on the white vent, flank, and breast; the adult's head markings are present in juveniles, but typically narrower and browner, without the rich chestnut characteristic of the adult pattern.

Lark Sparrows undergo two molts in their first calendar year, in addition to the molt leading from the nestling's down to the juvenile's first plumage. The juvenile's body plumage is replaced in the summer range at some point between June and August, producing a supplemental plumage; the preformative molt follows between August and November (Pyle). That second molt is variable in extent, but can be nearly complete in some individuals; nearly half replace all of their primaries, secondaries, and tail feathers, but retain most of the primary coverts (Pyle), a feature unlikely to be of much use in aging birds in the field.

Say and Bonaparte explained the English name "Lark Sparrow" in part by reference to the species' loud and varied song. Low-pitched and rich, the song recalls to many observers' ears less any of the familiar larks than a lazy crossbill or a canary at half speed, alternating tenor trills and baritone gulps with soprano chips and chirps. In the nineteenth century, the Lark Sparrow was a greatly desired cage bird in some regions, a pair of young birds fetching as much as four dollars in Sacramento in the 1860s. Robert Ridgway, who knew the sparrow both from his childhood in Illinois and his western travels as a young man, claimed

> that the delightful song of this bird has no parallel among the North American *Fringillidae* . . . it is pre-eminently superior to that of all the other members of this family . . . in vigor and continuity unsurpassed, if not unequalled, by any other North American species.

Ridgway's enthusiasm is justified, if his conclusion not inarguable. The song usually begins with a short flatulent introductory note, followed by a series of loose trills, buzzing grunts, and high-pitched squeaks in a variable sequence; the slow tempo, with long pauses between phrases, gives it a distinctively conversational, even disjointed quality, almost mimidlike at times. It is not clear whether differences in the sequence of phrases are geographic or individual.

The most familiar call note, given by perched birds and in flight, is a very short, sharp, high-pitched *tst*, with an abrupt attack and virtually no audible decay; often heard from high overhead, this call can bring to mind a wood warbler or an insect until the distinctive whitish underparts and white tail of the Lark Sparrow are seen.

RANGE AND GEOGRAPHIC VARIATION

The Lark Sparrow now breeds across most of the United States between the Mississippi River and the Pacific Coast, and on the grasslands of western Canada and northern Mexico.

East of the Mississippi, the species breeds, or territorial males are regularly present, in extremely small numbers in natural and human-maintained sandy grasslands in western North Carolina and northern South Carolina (eBird); localities there are abandoned when pines and other vegetation become too dense. A low-density breeder in Kentucky, the species there has "a small and scattered, but persistent, nesting population," its numbers "probably not limited by habitat availability" but by the fact that the state is at the extreme eastern edge of the Lark Sparrow's range. The breeding range continues south through central Tennessee, where nesting birds are rare and local, to north-central Alabama and Mississippi, at least formerly.

In the North, Lark Sparrows breed across most of Illinois into western Indiana, and along the Mississippi River in southern and central Wisconsin and Minnesota, where it is uncommon and local. In Canada, Lark Sparrows breed in the southwestern corner of Manitoba and across southern Saskatchewan and Alberta to the Rocky Mountains; the species is almost entirely restricted in British Columbia to the Okanagan and Similkameen Valleys in the south-central portion of the province.

Lark Sparrows breed in appropriate habitat in most of Texas, absent only from the driest areas and uncommon in the Piney Woods of the east, where the nominate race is found. They breed across New Mexico and the eastern two-thirds of Arizona, and in Mexico from northernmost Sonora east through most of Chihuahua and Coahuila and south to Zacatecas.

In the far West, breeding occurs east of the Cascades in Washington and Oregon, north into south-central British Columbia. Breeders are largely absent from the Mojave Desert of Nevada, Arizona, and California, though a few nest in irrigated citrus plantings on the lower Colorado River in those last two states and the species is a fairly common breeder in California's Coachella and Imperial Valleys. The species is widespread in the remainder of California west of the Cascades and Sierra; they are less common as breeders on the east slope of the Sierra Nevada. A common breeder in southern California's San Diego County, the Lark Sparrow is found there from the highest elevations down to the deserts, but typically avoids areas within five miles of the ocean coast. The species also breeds in northern Baja California.

The southbound migration begins early, in northerly parts of the range by mid-July, and proceeds in a staggered rhythm, such that in some areas migrants appear while local adults are still engaged in breeding. Migration on the northern Great Plains is noticeably earlier than that of the Vesper Sparrow. Wintering areas in northern Mexico are reached as early as August, and obvious movement has generally concluded across the species' range by November, though "nomadic tendencies" lead Lark Sparrows to show up virtually anywhere in North and Middle America in autumn and winter, from central Alaska and the Canadian Maritimes to the Caribbean, Bermuda, and Panama.

The regular wintering grounds extend from southwestern Oregon south through California to Baja California Sur and from southern Arizona, New Mexico, and north-central Texas south to Oaxaca; the species is also a regular but rare to uncommon winterer in the Florida peninsula. In the northern interior, there is a concentration of wintering Lark Sparrows in the northern Great Basin of Utah and southern Idaho. Though the species is an early autumn migrant, it appears to be relatively hardy, and can occur virtually anywhere in North America in the colder season.

As in many short- and mid-distance passerines, the spring migration is more concentrated and more predictable. The northbound movement begins in March, and peak numbers of migrants are abroad through much of the West in late April and May; birds are on the breeding grounds by the end of May. The two records from southern England are from late May and early June, corresponding to arrivals at the northernmost edge of the usual breeding range, but it is possible that these individuals reached Europe by ship.

The sparrows originally described in the 1820s under two different names by Thomas Say and William Swainson were considered simply identical until 1880, when H. K. Coale and Robert Ridgway determined that there were consistent differences between birds collected in the eastern portion of the range and those taken in the West and Mexico. Those differences are not particularly pronounced, and the two currently recognized subspecies are not distinguishable in the field.

The western race *strigatus*, which breeds from the Canadian prairies west to British Columbia and south to California and western Texas, is paler above than the nominate *grammacus*, with finer, sparser streaking on the upperparts and less vividly rusty head markings (Pyle).

Autumn and winter birds on the Atlantic Coast can apparently represent either subspecies.

LARK BUNTING
Calamospiza melanocorys

A characteristic summer bird of the western Great Plains, the Lark Bunting is the only sparrow to have been designated an official "state bird." *Colorado, June. Brian E. Small*

On June 3, 1876, General George Crook—supposed to be in hot pursuit of the Cheyenne and Sioux still stubbornly insisting on their right to live on lands legally theirs—left the site of Fort Reno on the Powder River of central Wyoming. As was their wont, Crook and some of his officers broke the tedium of the march by indulging in their favorite natural history pastime:

> The singing of meadow-larks and the chirping of grass-hoppers enlivened the morning air, and save these, no sounds broke the stillness except the rumbling of our wagons slowly creeping along the road . . . . Captain [Azor H.] Nickerson shot five Missouri skylarks, which Gen. Crook preserved as specimens. Gen. Crook found under a small sage-brush a nest of the white-ringed [sic] black-bird, with six small turquois-blue [sic] eggs; the ordinary complement is five.

Nickerson's "skylarks" were Sprague Pipits, then as now an uncommon and sought-after Great Plains specialty. The "black-bird" responsible for the general's prize set of turquoise eggs was one of the most abundant passerines of the northern prairies, the strange and striking Lark Bunting.

The first European scientist to see this large and in many ways aberrant sparrow had been Prince Maximilian, who in mid-July 1833 collected several on the upper Missouri River. As Maximilian would later rather stridently point out, this was nearly a full year before Thomas Nuttall and John Kirk Townsend took their specimens on the Platte—specimens that would serve Townsend as the basis for the description of a new species he called *Fringilla bicolor*, the Prairie Finch.

Whatever their struggles for nomenclatural priority, Townsend, Nuttall, and the prince were of one voice in their praise of the bird's exuberant song, and in their conviction that "in its habits and behaviors, it is extremely similar to the Bobolink." Nuttall, indeed, would go so far as to join the two black and white grassland birds in a single genus, *Dolichonyx*, and to call the Lark Bunting the "Western, or Prairie Reed Bird."

Though Townsend in his original description found that the species "presents many of the habits as well as the song of the Bob-o-link," he also recognized that it was "somewhat allied" to the Vesper Sparrow, a fact that led him to classify it formally in the bunting genus *Fringilla*. Charles Bonaparte found the bird more closely reminiscent of some of the Old World buntings, and in 1838 coined the genus name *Calamospiza*—"reed bunting"—to include only this species. Presumably unaware of Bonaparte's new name, Audubon erected a genus of his own for the bunting, which he called *Corydalina*, "larklike."

The priority of *Calamospiza* was never in serious doubt, however, and for decades to come—with the signal exception of Nuttall and his "Prairie Reed Bird"—this species was known as *Calamospiza bicolor*. Then, in 1885, the Norwegian-American zoologist Leonhard Stejneger determined that Linnaeus had applied the name *Fringilla bicolor* to a tanager, the Black-faced Grassquit, in 1766, making Townsend's original species name for the Lark Bunting invalid and requiring a new epithet for the bunting. Stejneger's new coinage was *melanocorys*, "black lark," its second element echoing Audubon's abortive genus name and alluding to this species' larklike behavior.

FIELD IDENTIFICATION

Male Lark Buntings in the black and white plumage they wear from late winter to midsummer are among the most distinctive of native North American passerines. Many non-birders still know the bird as the Bobolink, echoing the impressions of so many early natural historians, but that error is less a misidentification than the application of a genuine folk name in a way that does not accord with the formal practice of nomenclature: those non-birders are not so much confusing two species as simply asserting the irrelevance of the species as category.

Observers to whom such things matter, though, will have no difficulty distinguishing male Lark Buntings from male Bobolinks. Quite apart from the significant structural differences, the conspicuous white patches shown by both are in entirely different places, on the back and rump in the Bobolink and on the wing coverts in the Lark Bunting, a distinction easily seen whether the birds are in flight or on a perch.

A less obvious source of identification uncertainty is the Pin-tailed Whydah (*Vidua macroura*), an African widowbird introduced and now locally established in the Caribbean; this attractive bird is common in captivity across North America, and escaped individuals with imperfect tails have been mistaken for male Lark Buntings. Even without their extravagant trains, whydas differ abundantly from the bunting in their much smaller size, white collar and underparts, and colorful soft parts—distinctions that may not come immediately to mind when the observer is startled by a small black and white passerine on the ground.

Lark Buntings of all ages and both sexes share a distinctive silhouette, making of them a particularly good illustration of the principle that most birds are best identified by beginning at the rear. The Lark Bunting's white-edged, white-tipped tail is short and broad. The wing is notably pointed, with long outer primaries, but the tertials—the broad, innermost secondaries—are nearly as long, on perched birds covering all but the very tips of the folded primaries. As a practical matter, the long tertials protect the rest of this ground-dwelling species' wing feathers from dirt and abrasion, but the impression created is simultaneously comical and sinister, as if the bird had draped itself in a bedspread to frighten its fellows. Sprague and American Pipits have similarly elongated tertials with a similar protective function, but their proportionally longer tails and more slender bellies are quite unlike the stubby rear and rotund body of the bunting.

Brown Lark Buntings—females of all ages and males other than adults in breeding plumage—are chunky, fist-shaped birds with short necks, square heads, and large, thick bills; the size of the bluish bill makes the dark eye look small and beady. The overall color of these birds when encountered in the field is brown—dusty, muddy, dirty brown. The underparts are paler, with a creamy white ground color, but heavily, coarsely marked brown or blackish brown; the streaks often coalesce into a large ragged spot on the center of the breast. The white of the wing coverts, this species' signature plumage character, appears as a pale slash or, especially on young birds, is almost concealed by the overlying scapulars. The pale markings on the outer rectrices are often visible at the side and the tip of the short, broad tail, though they can be obscure or even absent in worn plumage.

Brown birds have the thick-necked, bulky-headed aspect of all Lark Buntings. Unlike the uniformly plumed breeding-plumaged male, their head pattern is heavy and striking, with broad creamy jaw stripe and supercilium joining across the pale lore.

The black back, round body, and very heavy conical bill neatly distinguish alternate-plumaged males from the superficially similar Bobolink. *Colorado, June. Brian E. Small*

Adult males in winter retain a large black blotch on the chin and throat; otherwise the darkest parts of the head are the dark brown, black-streaked crown and the mottled blackish whisker separating the brown ear coverts from the jaw stripe. The black lateral throat stripe varies considerably in width, but is always present and usually conspicuous.

Lark Buntings are at least as distinctive in flight as they are on the ground. The outer web of the outermost rectrix is broadly edged with buffy white or, in males, white; all but the central pair of tail feathers are tipped with the same color, creating a conspicuous lacy pattern somewhat like that of the Lark Sparrow. There is a pale patch on the upperwing, buffy in females and covering the outer four or five greater coverts; in males, the patch is white and extends across all of the greater coverts, only the innermost being black on its inner web. Square in males, the wing patch is more oblong or crescent-shaped in females.

Nearly as characteristic is the underwing. Rarely much of a field mark in passerellid sparrows, the dark of the Lark Bunting's wing from beneath is striking in both sexes, especially in males. The flight feathers are dark dusky brown or, in males, dusky blackish, with very narrow paler tips, while the coverts in both sexes are a shade darker. As flying birds flap, they alternate pale flashes from above with dark surfaces from below, adding to the

Sexual plumage dimorphism is unusual in American sparrows, but it is taken to an extreme in this species. *Colorado, June. Brian E. Small*

"twinkling" impression made by this species' undulating flight habit.

Juvenile Lark Buntings molt into their formative plumage beginning in July or August, completing the molt on the southward migration or on the wintering grounds (Pyle). Birds in their first winter differ from adults in having retained the outer primary coverts of juvenile plumage; those feathers in adults

The female and male resemble each other much more in basic plumage; the female usually has a white throat. *Arizona, January. Brian E. Small*

are wide and blunt-tipped, with noticeably paler edges in females, while in formative-plumaged birds of both sexes, most or all of the outer primary coverts are narrow and tapered at the tip, with paler edges faint or absent (Pyle). Males in formative plumage have the outermost primary covert blacker than females (a character extremely difficult to see in the field) and generally show some black feathers mixed with the brown of the body, but, like females of all ages, lack black on the chin and throat.

Adults of both sexes are brownish in winter, after undergoing a complete prebasic molt in late summer (Pyle). The general tone of males at this season is grayish, while females are more straightforwardly brown; males have heavy black mottling and spotting on the chin and throat, a character lacking in females before the prealternate molt in March (Pyle). Males' outer primary coverts are black, females' dark brown (Pyle), a difference difficult to see in the field. Because the adults' prebasic molt includes all of the flight feathers, while the juvenile's preformative molt typically includes only the outermost four primaries, a visually obvious contrast between the outer and the inner primaries allows the observer to age such a bird as in its first winter of life.

All birds undergo prealternate molt in late winter or early spring; in some birds in the first year of life, this molt may simply be a continuation of the preformative molt commenced in the preceding late summer (Pyle). The resulting plumage is worn from March to July. Males are black, the adults more or less entirely so and first-cycle birds with a brown admixture; females at this season can be aged consistently only by the presence or absence of contrast in the primaries and primary coverts (Pyle).

The distinctive flight call, heard from flocks and individuals at all times of the year, is a hollow, breathy hoot, trailing off at the end; some observers on the southwestern wintering grounds find this note remarkably like one of the calls of the Phainopepla. Adult Lark Buntings are generally quiet when perched, giving an occasional hoot or faint buzz.

The Lark Bunting is justly celebrated for its flight song, first described by John Kirk Townsend:

> While the flock is engaged in feeding, the males are frequently observed to rise suddenly to a considerable height, and poising themselves over their companions, with their wings in constant and rapid motion, they become nearly stationary. In this situation, they pour forth a number of very lively and sweetly modulated notes, and at the expiration of about a minute, descend to the ground, and course about as before.

The male in basic plumage is usually darker on the wing, face, and throat than the female. *Texas, March. Brian E. Small*

Townsend's field companion Thomas Nuttall called the bird "one of the sweetest songsters of the prairie," while Joel Asaph Allen compared the Lark Bunting's performance to that of another gifted singer, the Yellow-breasted Chat; Allen and his collectors "naturally applied to it the cognomen of the 'Black Chat.'"

As Allen's comparison aptly suggests, the Lark Bunting's song comprises three to eight phrases at different pitches and of different tonal qualities, most of them slow but following one on the other without significant pauses. The phrase types, not all of which are sung by every male, include slurred, cardinal-like *sweet, cher-wheat,* and *weeta* notes; lower-pitched, chatlike *chug, chut, toot, churt,* and *chew* notes; and trills ranging from low buzzes to high, insectlike tremolos. Songs given by perched birds are usually shorter than those delivered in flight; singing birds often strike a distinctive pose with lowered head, as if scanning the ground beneath their perch.

RANGE AND GEOGRAPHIC VARIATION

As is true of many other prairie-breeding species, the Lark Bunting's summer range expands and contracts with weather conditions, extending both east and west in wet years and retreating to the core range in dry. The geographic center of the usual distribution is in the eastern Panhandle of Nebraska, coincidentally more or less exactly where Nuttall shot the type specimens in 1833. Breeding regularly occurs north to southeastern Alberta and extreme

All ages and both sexes share the stubby tail, very long tertials, sturdy feet, and grosbeak-like bill. *Texas, March. Brian E. Small*

the autumn migration of Lark Buntings from their breeding grounds begins early. The first arriving adults are seen in Texas, southeastern Arizona, and northern Sonora in July, with peak southbound migration in August and early September in Colorado and Kansas. Males are the first to leave the northern range, followed two to four weeks later by the females and the summer's young.

The nomadic tendency of the Lark Bunting is most pronounced of all on the wintering grounds, where birds may be locally abundant one year and scarce or absent the next. From August to May, flocks can be found anywhere on the dry plains and deserts of the American Southwest from the Four Corners south to Hidalgo and Baja California Sur, and from south-central Texas and Tamaulipas west to the Colorado River. Wintering birds are scarce in interior southern California, but sometimes occur in the hundreds; farther west, autumn migrants have occurred on the Farallones and the Channel Islands, and there are winter records of single birds or very small flocks from nearly the entire California coast, including at least ten records of one or two wintering individuals in San Diego County. Wintering occurs regularly as far east as central Tamaulipas and the lower Rio Grande Valley of Texas.

The spring migration begins by the first of March, but is notably protracted, with some birds arriving on the northern breeding grounds while others are still on the winter range in late May. The "heaviest migration appears to occur along [the] 102nd meridian, with peak migration at 38 degrees north by [the] second week of May." Arrival time varies from year to year, but in general this is the latest of the short-grass breeders to make its springtime appearance on the northern Great Plains.

Given its abundance, its fairly large range, and the irregularity of its seasonal movements, it is no surprise that the Lark Bunting is a frequent, if always unpredictable, stray to North American sites far from its "normal" distribution. Vagrants can occur virtually anywhere on the continent at any season, but are most frequent—or at least most often detected—in late spring and fall in coastal regions, where dunes and beaches mimic the sparsely vegetated plains and desert flats the species prefers. Males in their black and white plumage are conspicuous and hard to miss, but brown birds—females, juveniles, and males in winter dress—are probably overlooked at times by observers unfamiliar with the species.

No subspecies are recognized.

southwestern Manitoba, and south to the panhandles of Oklahoma and Texas and immediately adjacent New Mexico. Both the western and the eastern edges of the breeding range stretch and contract. In most years, birds are common to abundant east of the Continental Divide in Montana and Wyoming but rare to the west; in Colorado the species is abundant on the eastern plains but rare to uncommon in sagebrush shrub in mountain parks and on the northwestern mesas. The eastern edge of the normal breeding range cuts through the center of the Dakotas, Nebraska, and Kansas before curving noticeably to the west in Kansas and Oklahoma.

In moist years on the eastern Great Plains, Lark Buntings have bred in small numbers as far east as western Minnesota, Iowa, and Missouri, but they are never expected, far less common, in these areas. The breeding range may also be extended irregularly as far west as western New Mexico, northern Utah, and southern Idaho. The species has bred at least once in northeastern Arizona, and in 1878, several pairs nested successfully in San Bernardino County, California, after a wet winter.

Even by the standards of prairie-nesting birds,

The very small size, fine bill, long notched tail, and bright head pattern readily identify an alternate-plumaged adult Chipping Sparrow. *Michigan, May. Brian E. Small*

CHIPPING SPARROW
Spizella passerina

Early observers were of one mind in describing the Chipping Sparrow as abundant and familiar in eastern North America, a confiding little bird fond of human company and human landscapes. And yet the first scientific accounts devoted to the species are oddly brief, even perfunctory. William Bartram, the first great American-born naturalist, simply listed "the little house sparrow or chipping bird" with no information about its habits, and Alexander Wilson, the Father of American Ornithology, gave a scant page to this "sociable" species, with no mention of its distinctive song. Even John James Audubon presents little more than a compilation—not unfailingly accurate—of his predecessors' observations.

The one exception to this odd reticence is provided by the most important American ornithologist most birders have never heard of, Louis Pierre Vieillot. Scion of a Rouen merchant family, Vieillot represented their business interests in the French colony on Hispaniola before fleeing to the young United States shortly after revolution broke out in Saint-Domingue in 1791. He passed much of the decade traveling and working in the eastern United States, finally returning to France in 1798—four years after Wilson arrived in Philadelphia, and four years before the Scots immigrant turned from school teaching to ornithology.

Vieillot spent much of his time in America afield, watching and collecting birds. Encouraged by the Comte de Buffon to publish his observations, he wrote a four-volume, illustrated natural history of North American birds; the first two volumes appeared in fascicles beginning in 1807, but the others—including the volume treating the American sparrows—were never printed, though they survive in manuscript. Happily, Vieillot appears to have reused some of the unpublished material in articles he prepared a decade later for the *Nouveau diction-naire d'histoire naturelle*, including an entry dedicated to the bird he named "titit":

> This is one of the first species to return from the south and appear in spring in the central United States, where it frequents gardens, open woods, and orchards; it is in those same localities, which it occupies through the entire warm season, that its nest should be sought, at the tip of a branch of a fruit tree. The nest is made of

spindly grass stems, carelessly arranged and situated in such a way that this tiny cradle would seem to be clearly visible from virtually any angle; the clutch is of four or five bluish green eggs with weak rusty-gray spotting at the large end. The male's song seemed to me to pronounce the syllables *ti, ri, ri, ri, ri, ti,* repeated several times with varying strength and vivacity; it is heard only in the spring. After breeding, all the families in an area gather, and in the autumn form rather large flocks which sometimes join with other species on the way south.

This account of the Chipping Sparrow's habits and the plumage descriptions that follow it were based on observations made more than ten years before the publication of Wilson's *American Ornithology*. Vieillot cites Wilson as the authority for the epithet *socialis*, but his entry in the *Nouveau dictionnaire* is based on his own observations—and, brief as it is, more detailed than anything that would be published on the species for decades to come, one more illustration of why this still virtually unknown Franco-American ornithologist's work deserves far more attention than has ever been accorded it.

FIELD IDENTIFICATION

Distinctive in their bright-headed alternate plumage, Chipping Sparrows in winter and juveniles can be a source of confusion for inexperienced observers.

All Chipping Sparrows are notably small, long-tailed, flat-backed, slim-bellied birds with a small head and bill. Birders who know only the neatly patterned adults of summer may be puzzled by the same birds in their winter plumage, which is attained by the complete prebasic molt on the breeding grounds in late summer (Pyle). A correct assessment of the structural features suffices to eliminate all other sparrows except those sharing the genus *Spizella*. The Chipping Sparrow's tail tends to be more blackish than those of the smaller, smaller-billed Brewer and Timberline Sparrows, and the conspicuous silvery rump is unlike the decided tan of the Field and Clay-colored Sparrows. In the eastern United States and Canada, the Chipping Sparrow's back is brighter chestnut and more boldly streaked black than any of its congeners', but *arizonae* birds in the West and Southwest can appear washed out even in the field.

Winter Chipping Sparrows have a broad gray collar on the neck sides and nape, similar in tone to the color of the underparts but usually slightly darker

The forehead in alternate plumage is black with one or two very short white lines. Western birds average paler and duller than eastern birds. *British Columbia, June. Brian E. Small*

than the more bluish or silvery gray of the rump. Two narrow dark brown stripes cross the nape to connect the brown of the crown to the brown of the upper back; in many birds, the feathers making those stripes have very broad buffy to rusty edges, blurring the stripes and letting the brown wash into the gray between them. Thus, seen from the rear, the nape is gray at the sides, separated by blurry brown lines from the rather poorly defined grayish to gray-brown center. Others may show a more neatly "zoned" pattern with clear rusty lines dividing the soft gray collar into three discrete areas, as in the Clay-colored Sparrow; such individuals are still identifiable using the characters of rump color, back color, and head pattern. It is not entirely clear whether one of these two nape patterns is more frequent in basic-plumaged adults and the other in formative birds.

The head pattern of basic- and formative-plumaged Chipping Sparrows comprises several markings, none of them large or conspicuous on its own but all contributing to what is often a distinctive look in the field. The crown, which may show a broad buffy central stripe, usually shows a mixture of rufous and black on the sides, brighter than any of the other streak-crowned *Spizella* species. A buffy supercilium, broader behind the eye, where it stops abruptly on meeting the gray collar, and significantly narrower between the eye and the bill base, is emphasized by a fine narrow black eye line; that line varies in darkness and definition among the subspecies, but is always present and always darker and neater than any lateral throat stripe. In most birds, it continues onto the lore, creating a subtly more sinister facial expression than the more curious and welcoming aspect of the Clay-colored Sparrow. More importantly, more consistently, and usually more visibly, in the Chipping Sparrow that line splits a narrow but conspicuous white eye-ring both before and behind, a feature much easier to assess in the field than the precise pattern of the lore.

Juvenile Chipping Sparrows are quite different in plumage from adults, though they share the long dark tail, small head, and slender bill. They are readily identified when attended by their parents; in this species, however, juvenile plumage, especially in western populations, can be held through the southbound migration in autumn, with some birds still showing streaked underparts as late as January (Pyle). The very presence of streaks at so late a date is strongly suggestive of the Chipping Sparrow, the other members of the genus usually showing an unstreaked breast and belly by October. Juvenile Field Sparrows, which share the blackish tail of Chipping Sparrows in the same plumage, show very little streaking on either the crown or the breast, thus more closely resembling adults. Juvenile Chip-

Some winter birds show considerable rust in the finely streaked crown; the supercilium is less white than in summer and the bill and tarsus paler. The silvery rump is conspicuous in flight. *New Mexico, December. Brian E. Small*

ping Sparrows are also more heavily streaked below than Brewer and Timberline Sparrows of the same age, and are proportionally shorter-tailed, with larger, darker bills.

Juvenile Clay-colored Sparrows are more difficult to distinguish. The rump is brown; unfortunately, the rump color of some juvenile Chipping Sparrows is an indeterminate gray-brown, closely approaching the tan shown by Clay-coloreds. The Clay-colored is less heavily and less darkly streaked beneath (Pyle), and the underparts are buffier. Its eye-ring is rather an eye crescent, strongly defined below the eye but inconspicuous or even absent above, while most juvenile Chipping Sparrows show a face pattern like that of the basic-plumaged adult, with a dark eye line breaking the eye-ring at front and rear and continuing across the lore to the base of the bill.

One sometimes useful field character separating Chipping and Clay-colored Sparrows of any age is the difference in their relative wing and tail lengths. In birds with completely grown remiges and rectrices, the Chipping Sparrow is decidedly longer-winged, the Clay-colored shorter-winged. A correct assessment of proportions requires excellent close views, typically through a spotting scope, and extensive field experience with one or the other of the two species, usually the more widespread Chipping

Sparrow; but the careful observer may pick out a stray individual of one in a flock of the other simply on this slight difference in shape.

The usual call note of the Chipping Sparrow is a short and clear *tseep*, with a relaxed attack and a wispy decay; it is given both by perched birds and in flight, rapidly repeated when birds are startled or in the course of squabbles between feeding individuals. There is also "a unique *zeeeeeee* alarm" call given in the "presence of hawks."

The Chipping Sparrow's simple trilling song is uniform in structure across populations, but each individual's version comprises a different type of syllable, repeated at a different rate. At their extremes, these songs can be distinguished even by human ears, making it possible to track the movements and behavior of individual males in a park or backyard. While most birds sing a fairly dry, slow tremolo of chips, the syllables just too fast to be counted, some utter a higher-pitched, faster buzz, quite similar to the song of a Worm-eating Warbler; others sing more slowly, the individual notes sweet and audibly slurred, closely recalling some Pine Warbler songs. Unlike the songs of the two parulids, the Chipping Sparrow's trill is not noticeably softer at beginning or end, and there is no discernible variation in the tone or pitch of the repeated syllables, which are only rarely doubled. A behavioral difference can also provide a clue to help distinguish the sparrow from the Pine Warbler: both commonly sing from large pines, but the Chipping Sparrow sings repeatedly from a favored perch, while the warbler is heard now from this side, now from the other side of the tree as it sings and feeds, feeds and sings.

RANGE AND GEOGRAPHIC VARIATION

As late as 1839, it was claimed that the Chipping Sparrow was restricted in the breeding season to areas east of the Missouri River and north of Texas. In fact, this is one of the most widespread passerines in the Americas, nesting from Alaska to central America and from Newfoundland to Baja California.

In Canada, Chipping Sparrows breed from the north-central Yukon and Northwest Territories across British Columbia, the prairie provinces, most of Ontario, and the southern half of Quebec through the Maritimes and southern Newfoundland; numbers have declined "considerably" in Nova Scotia in the past half century.

This species is an uncommon breeder in portions of Alaska, including along the rivers in the southeast and the Tanana River Valley of east-central Alaska; there are scattered breeding records from elsewhere in the state, north and west to Fairbanks and Fort Yukon. In Washington and Oregon, most Chipping Sparrows breed east of the Cascades; the species also occurs fairly commonly in western Washington in the northern Olympics, on Mount Rainier, and on the Fort Lewis prairies. It is a widespread breeder in California, including the mountains of the eastern deserts and most of the Channel Islands, but absent from the desert lowlands, the southern coastal slopes, and the Central Valley. Chipping Sparrows are uncommon breeders in coniferous mountain forests of northern Baja California.

On the Great Plains, breeding is limited by the availability of conifers suitable for nesting; as a result, the species does not occur as a breeder over large stretches of Nebraska south of the Platte Valley, eastern Colorado, Kansas, and Oklahoma. As might be expected, Chipping Sparrows breed, though uncommonly, in eastern Texas's Piney Woods region; they are common residents on the Edwards Plateau. In the southeastern United States, breeders are absent from coastal Louisiana and from most of Florida, where there have been no confirmed breeding records since the 1940s.

At the southwestern edge of its breeding range in the United States, this species is restricted to open mountain forests in northern Arizona and New Mexico. It also breeds, less commonly, in the southern "sky island" ranges of those two states and, abundantly, in the Guadalupe and Davis Mountains of west Texas. In Mexico, Chipping Sparrows breed in open montane pine-oak forest in the states of Sonora and Chihuahua south through the Sierra Madre to Chiapas and east to Nuevo León and San Luis Potosí. They are fairly common residents in Belize and Honduras, uncommon and local in El Salvador, and common on both slopes and in the Atlantic highlands of Guatemala.

The northern races *passerina, arizonae,* and *atremaea* (and other described subspecies now synonymized with them) are migratory. At least some western populations appear to undertake a "molt migration," in which adults leave the breeding grounds soon after nesting to undergo the prebasic molt, only then continuing to the wintering range. July sees significant numbers of adult Chipping Sparrows on the plains of eastern Colorado (T. Floyd, in litt.) and in the lowland deserts of southeastern Arizona, where they do not breed. Migration begins later in the East, with peaks between mid-September and mid-October in the Mississippi Valley and on the eastern Great Plains; the first winterers arrive in Florida and Texas, and numbers increase into November. This species is a very rare visitor to Bering Sea islands; one was photographed on Wrangel Island in October 1981.

Increasingly, Chipping Sparrows are found in winter north into the breeding range, where individuals frequent feeders and brambly field edges. Regular wintering occurs across the southern tier of the United States, north to Oklahoma, southern Tennessee, the Delmarva Peninsula, and southern New Jersey; those limits seem likely to shift farther north as time passes.

The western subspecies *arizonae* winters well south into Mexico, overlapping at that season with *atremaea* and possibly *mexicana* as far south as Guerrero and Veracruz. At times, birds of the eastern, nominate race penetrate Tamaulipas and Nuevo León.

The movement north is discernible in southern regions by the beginning of March, with arrivals on the breeding grounds ranging from late March in the Southeast to mid-May in the Northwest. Northbound migrants proceed more rapidly along the coasts than inland.

Study of the distribution of unique segments of mitochondrial DNA in the Chipping Sparrow has not revealed any consistent pattern matching the geographic variation in size and plumage exhibited by the species. Nevertheless, five subspecies are generally recognized today; individuals in migration and winter are probably not distinguishable visually in the field.

The widespread nominate race *passerina* is a dark-backed, proportionally short-tailed, medium-small Chipping Sparrow (Pyle). The type locality of "Canada" was arbitrarily and gratuitously restricted in 1955 to the vicinity of Quebec City; the subspecies breeds from Minnesota and Newfoundland south to east Texas and northern Florida and winters in the southern portion of the breeding range and into eastern Mexico.

Western *arizonae* is larger and paler, with a proportionally smaller bill and less contrasting head pattern. In describing his "curious" new subspecies, Elliott Coues noted the similarity of breeding-plumaged adults to immature individuals of the eastern, nominate race, a canny view if an exaggerated one. The upperparts of *arizonae* are colder brown and less rusty than those of *passerina*, and the rump paler gray; the ear coverts of the western bird do not contrast as strikingly with the supercilium.

Exactly where *passerina* gives way to *arizonae* in the breeding season is subject to debate, but the two most likely meet on the eastern Great Plains, whence *arizonae* breeds west to the Pacific Coast, south to Baja California, and north to eastern Alaska. Birds of the Pacific Coast from southern British Columbia to southern California may be slightly darker and brighter brown above, and have been described as a distinct subspecies *stridula*.

Breeders on the western Great Plains, east to western Ontario and west to Alaska, assigned the name *boreophila*, have been said to be larger and drabber than the nominate *passerina* and darker, with a grayer and more distinctly marked head, than *arizonae*; like *stridula*, these birds are now generally allocated to *arizonae*, but if *boreophila* is recognized, its breeding distribution is a large wedge dividing the eastern and western races on the western plains.

To the south, the inscrutably named subspecies *atremaea*, first collected in southernmost Chihuahua, is "very much blacker" above than either *arizonae*, which breeds immediately north, or *mexicana*, which breeds farther south. The upperparts are broadly and contrastingly streaked black and buffy chestnut, and the dark gray breast is distinctly different from the white of the throat. This race is resident from the type locality and Durango east to Nuevo León and Jalisco. Chipping Sparrows from the northern Transvolcanic Belt of Mexico, said to average darker above and deeper-billed than other *atremaea* and described as a subspecies *comparanda*, are now generally considered to belong to *atremaea*.

The race *mexicana*, resident across the Mexican highlands from Nayarit to Veracruz and south into Chiapas and northern Guatemala, is larger than the nominate race, with more rufous feathering in the dark back. Birds from Oaxaca and Guerrero, now generally ascribed to *mexicana*, have been described as a separate subspecies *repetens*, recalling ("repetens") "the geographically remote *S. p. arizonae.*"

The southernmost of all Chipping Sparrows belong to the subspecies *pinetorum*, a resident of eastern Guatemala, Honduras, and Nicaragua. This dark sparrow was originally described as a separate species, and compared, oddly, not to more northerly Chipping Sparrows but to the Field Sparrow, from which it was said to differ in its brighter colors and heavier bill. Indeed, the first lumping to which the Guatemalan sparrow was subjected was not with the Chipping Sparrow but with the Field. Not until 1884 did Robert Ridgway suggest that this taxon was in fact a Chipping Sparrow; Ridgway would eventually hint that the earlier confusion had its roots in the reddish bill of the type specimen, collected in March: "the color of the bill would undoubtedly be black in a summer specimen." These birds have a rich dark chestnut crown and gray underparts, with dark gray neck and ear coverts. The *pinetorum* birds of northern El Salvador have been described as a separate subspecies, *cicada*, but the distinction is now not generally thought valid.

Juvenile plumage, which can be held into very late autumn, is finely streaked below, but shares the adult's richly colored upperparts and long black tail. *Oregon, August. Brian E. Small*

CLAY-COLORED SPARROW
Spizella pallida

The young Ernest E. Thompson—later famous under the name Ernest Thompson Seton—spent much of 1882 with his brother in southern Manitoba, watching the birds of the prairie and carefully recording their habits and behavior. All that spring, he was

> puzzled by a singular song that is uttered by a small sparrow which frequents scrubby localities. The song, if it may be so called, may be represented by the syllables "*scree, scree*," sometimes repeated two or three times . . . a sound like a fly in a newspaper.

Finally, on June 29, the mystery was solved, when Thompson shot one of the singers "in the very act," and discovered that it was a Clay-colored Sparrow, or "ashy-nape."

Two years later, as Thompson sat writing at his window, he spied six Clay-coloreds "rambling and foraging" in the yard outside, "picking up a hundred things which [he] could not see at all" before the "family of shattucks" continued on their way and out of sight.

"Ashy-nape" may well have been a genuine folk name, referring to one of this species' most conspicuous plumage characters. Thompson's use of the name "shattuck," however, leads back to the earliest scientific history of the Clay-colored Sparrow, when not even John James Audubon knew exactly what the bird looked like.

The "Clay-colored Buntling" was first described to science by John Richardson and William Swainson in 1831, on the basis of two specimens sent from Carlton House, Saskatchewan, where it was said to be "as familiar and confident as the common House Sparrow of England." In the middle of that decade, John Kirk Townsend supplied Audubon with two skins collected in June 1834, and thus Audubon was able to depict the species on one of the last plates in his *Birds of America* and to provide a painstaking description in the final volume of the *Ornithological Biography*. Never having seen the bird in life, Audubon relied for his brief account of the Clay-colored's habits on the *Fauna boreali-americana* and skeletal notes provided by Thomas

Nuttall, who had observed the species on the Great Plains. The difficulty is that neither Audubon's description nor, especially, his painting of the Clay-colored Sparrow at all matches the bird originally named and described from Saskatchewan—a striking example of how difficult it could and still can be to determine identity and difference using only what Audubon himself called "a mere synopsis" of characters.

The confusion might have been cleared up on Audubon's own journey up the Missouri River in 1843, but it was not. Audubon reported Clay-colored Sparrows seen at Bellevue, Nebraska, on May 9, and in northeastern Nebraska on May 17, when Isaac Sprague shot two. A week later, Sprague collected another; but Audubon expressed some uncertainty about those identifications, noting that the only authority available to him was still Richardson and Swainson's unillustrated description.

On his return from the Northwest, Audubon determined that the birds seen and collected on the prairies were in fact not identical to what he had painted and described in the 1830s as the Clay-colored Sparrow. Rather than critically reviewing that earlier identification, however, he decided that the sparrows from his recent expedition were a new and "handsome little species . . . quite abundant throughout the country bordering on the Upper Missouri." He took the opportunity to paint the new bird and to name it for "one of the amiable gentlemen who accompanied me on my voyage to the coast of Labrador" in 1833, young George C. Shattuck.

What Audubon never discovered was that his Shattuck Bunting was actually the true Clay-colored Sparrow, as described by Swainson in 1831. The plate showing the "Shattuck's" in the smaller, second edition of *The Birds of America* is not one of his best, but the bright silver collar and strong face pattern leave no doubt that this, and not Audubon's earlier plate, is in fact the first published illustration of the Clay-colored Sparrow.

By the time 40 years later that Ernest Thompson was watching Manitoba's "ashy-napes," the confusion had been unraveled and Audubon's original, ersatz "Clay-colored Sparrow" correctly identified as an example of the then undescribed Brewer Sparrow. It is a measure of just how persistent Audubonian authority was in the nineteenth century that Thompson, fully aware that his birds were formally, officially Clay-colored Sparrows, could still reach into the past to call them simply "shattucks."

FIELD IDENTIFICATION

The Clay-colored is a small, long-tailed, but often rather plump-seeming sparrow, unlikely to be confused with any other birds than the Chipping, Brewer, and Timberline Sparrows.

Like Clay-colored Sparrows, winter Chipping Sparrows have a broad gray collar on the neck sides and nape. On the Clay-colored, two narrow, finely black-streaked stripes cross the nape to connect the brown of the crown to the brown of the upper back; the gray square or rectangle between the stripes is obvious and well defined. Thus, seen from the rear, the nape is extensively clear gray at its sides, separated by neat, narrow lines from the prominent gray center. Some Chipping Sparrows may show a similar pattern, but such individuals should still be distinguishable by rump color, back color, and head pattern.

The head pattern of the Clay-colored Sparrow comprises nearly as many markings as the Lark Sparrow's. The crown is dominated by a broad buffy or whitish central stripe, averaging more distinct than in any other *Spizella* species. The lateral crown stripes are brown to dull rusty, with prominent black streaking. The buffy or whitish supercilium is broad over its entire length, with less clouding or flecking than in Chipping and Brewer Sparrows.

John James Audubon's "Shattuck Bunting" was in fact a Clay-colored Sparrow. *Birds of America* 1844. *Image from the Biodiversity Heritage Library. Digitized by Smithsonian Libraries*

N° 99.   Pl.493.

Shattucks Bunting

At close range, the head pattern, combining well-marked whisker and eye line with a buffy lore, is distinctive; the central crown stripe is broad, pale, and sparsely marked. The eye-ring is inconspicuous at any distance. *Texas, April. Brian E. Small*

The narrow eye line is mixed black and brown, and either just reaches the back of the eye or stops where it meets the upper rear point of the white crescent beneath the eye; famously, that line does not continue in front of the eye, leaving the lore plain buffy; that detail is difficult to see in the field, but the contrast between the open, gentle expression it creates and the more sinister expression of the Chipping Sparrow is clear. Neat dark stripes define the remainder of the face: the dark eye line nearly meets an equally dark, equally distinct whisker at the back of the ear coverts, which are themselves distinctly flecked dark brown.

Juvenile Clay-colored Sparrows are not significantly different in plumage from adults. The sparse, irregular breast streaks are usually lost in the preformative molt in late summer, but a few individuals may still show such streaking on arrival on the winter grounds; other characters should also be confirmed before ruling out a lightly marked juvenile Chipping Sparrow. The Clay-colored's rump at all ages is brown, though some birds may show a grayish admixture, especially when the feathers are ruffled to reveal the grayer bases. The eye-ring is faint or absent above the eye, but there is a clear white crescent below; most juvenile Chipping Sparrows show a face pattern like that of the basic-plumaged adult, with a dark eye line breaking the eye-ring at

front and rear and continuing across the lore to the base of the bill.

One sometimes useful field character separating Chipping and Clay-colored Sparrows of any age is the difference in their relative wing and tail lengths. In birds with completely grown remiges and rectrices, the Clay-colored Sparrow is decidedly shorter-winged, the Chipping longer-winged. A correct assessment of proportions requires excellent close views, typically through a spotting scope, and extensive field experience with one or the other of the two species, usually the more widespread Chipping Sparrow; but the careful observer may pick out a stray individual of one in a flock of the other simply on this slight difference in shape.

As the species' early taxonomic history suggests, distinguishing some Clay-colored Sparrows from some Brewer or Timberline Sparrows can be more difficult. Even in the hand, "dull" Clay-colored Sparrows and especially heavily marked Brewer or Timberline Sparrows can be puzzlingly similar, and a few will be impossible to identify with certainty in the field. The Clay-colored Sparrow's eye crescent is usually apparent only below the eye, while the other two species are characterized by a complete narrow eye-ring, only occasionally with a small break at the rear; the ring is as bright above the eye as below, though on average perhaps slightly duller overall in the Timberline than in the Brewer Sparrow. The eye-ring in the latter two species is made even more conspicuous by their dull supercilium and less noticeably marked ear coverts; in the Clay-colored, the supercilium and the jaw stripe are essentially equivalent in brightness. The clearly "zoned" nape of the Clay-colored Sparrow, with bright gray sides and center, is replaced in the Brewer and Timberline Sparrows with a narrower, pale gray-brown collar that is finely streaked black across its entire width.

Two characters that can be useful in the hand or in good photographs were first pointed out by Peter Pyle and Steve Howell in 1996. The whisker—the blackish line bordering the ear coverts below—in the Brewer and Timberline Sparrows is evenly narrow throughout its length, while in the Clay-colored Sparrow, that line widens toward the rear, such that it nearly meets the rearmost point of the eye line. More subtly still, the wing bars of the Clay-colored Sparrow are broader and more continuous, while in the other species, the dark centers of the wing coverts extend onto the pale tips as narrow "points," lending the wing bars a more jagged appearance.

Except for the mysterious and little-known Timberline Sparrow, classic individuals of any of the *Spizella* species are readily identifiable using only the characters adduced in any of the usual field guides. Observers confronted by an atypical bird will have a better chance at narrowing the possibilities of its identity if they have taken time to learn well the Chipping Sparrow and whichever of the other species is more common in their region; even then, certainty can prove elusive, especially given the range of geographic and individual variation shown by these species—and the possibility that they may rarely hybridize to produce young of intermediate appearance.

The usual call of the Clay-colored Sparrow, given perched or in flight, is a short, hesitant *dzip*, with strong attack and very little audible decay. The song is traditionally described as insectlike, a series of three or four, and occasionally as many as eight, widely separated buzzes, each buzz slightly faster and less coarse than the final note of a Blue-winged Warbler's song.

RANGE AND GEOGRAPHIC VARIATION

A classic species of brushy prairies on the northern Great Plains, the Clay-colored Sparrow has extended its breeding range in recent decades to Christmas tree farms and reclaimed strip mines as far east as Quebec, New York, and Pennsylvania; at the same time, the nesting distribution in the core of its historic range appears to be shifting north. As a migrant, the Clay-colored Sparrow occurs in low densities virtually across the United States and southern Canada, most abundantly in the center of the continent but also regularly, and probably increasingly, on the Atlantic and Pacific Coasts. The winter population is concentrated over much of Mexico, including Baja California Sur, with smaller numbers in southern and western Texas; small numbers are also found at that season in the southwestern United States and on the East and West Coasts.

Regular breeding takes place as far north as the southern Yukon and northeastern British Columbia; in the latter province, Clay-colored Sparrows are common in the central interior, north to the Sinkut River, and in the northeast from Tupper Creek to Fort Nelson, with the highest numbers in the Peace Lowland. The breeding range continues across most of Alberta, southern Saskatchewan, and southern Manitoba into southernmost Ontario and extreme southwestern Quebec; it has also bred in northern Ontario since at least 1931, including the southern shores of James and Hudson Bays. The

Breeding birds usually show very little "clay" color above or below, instead giving the impression of a smart gray and brown bird with a heavily marked head. *Minnesota, June. Brian E. Small.*

species has occurred annually in southern Quebec since the 1960s, but nesting is only infrequently confirmed. As of 2015, there were still no confirmed breeding records from New Brunswick or Nova Scotia, though a dozen territorial males were discovered in field work for the provinces' second breeding atlas.

There are scattered summering and breeding records through much of eastern Washington, and several of the three dozen documented records for the species in Idaho involve nesting or possible nesting. Common breeders on the Montana prairies, Clay-colored Sparrows nest across North Dakota, northern South Dakota, Minnesota, and northern Wisconsin and Michigan into New York, where nesting localities tripled between 1980 and 2005. In New England, the first decade of this century provided confirmed breeding records for Vermont, Massachusetts, and Maine. Breeding is

The rump varies from gray-brown to brown, but apparently never shows the bright bluish silver tone typical of the Chipping Sparrow. *Minnesota, June. Brian E. Small*

still rare but probably increasing in Pennsylvania and eastern Ohio. There are summer records from Indiana and Illinois, and the species appears to be returning to northern Iowa after an absence of some 40 years in the mid-twentieth century.

The southern limits of the western breeding range are uncertain and possibly changing. Persistent rumor notwithstanding, none of the breeding reports, and few of the summer reports, for Nebraska are documented, and there are said to be no breeding records for Wyoming or Colorado.

The autumn migration begins in late August, when northern nesters leave the breeding grounds, and is complete by the end of November, when virtually all birds have reached the winter range. The major route for southbound birds is the western Great Plains, where hundreds or at times even thousands can be seen in a day on roadsides, feeding on sunflower seeds, grass seeds, and insects; migrants there commonly flock with Vesper, Lark, and Chipping Sparrows. The species occurs in much lower numbers across the rest of the continent in the autumn, with multiple individuals each year, especially in October and November, from New England to, rarely, the Florida Keys and even the Dry Tortugas. On the West Coast, numbers are highest on the California coast and in that state's northern desert oases; most records are from the Farallones.

Scattered individuals winter on both coasts of the United States and, rarely, inland, but from November to March, the Clay-colored Sparrow is a decidedly Mexican bird. North of the border, wintering is regular only in Texas, where Clay-coloreds are uncommon to rare in southmost Texas, rare and irregular north to the Edwards Plateau and west to the Davis Mountains, and rare to very rare along the Gulf Coast and in the western Trans-Pecos. The species is very widespread in Mexico, ranging from northern Sonora and southern Baja California east to Nuevo León, thence south to Veracruz, Oaxaca, and Chiapas. It is listed as a vagrant in Guatemala, and has also been recorded in Belize; winter birds have strayed to Cuba and the Bahamas.

Spring migrants hew to a path slightly farther east than in the fall, and are noticeably more common east of the Missouri River in April and May than in September. The wintering grounds are abandoned beginning in March, and migration proceeds by "moving in waves northwards, in flocks of 25 to 100" or more. Clay-colored Sparrows can be abundant on mid-May days on the eastern Great Plains, joining warblers in the greening trees, where the sparrows dine on new elm seeds between fits of flatulent singing. Arrival on the northern breeding grounds is in May, again, sometimes in flocks numbering in the thousands.

No subspecies are recognized.

BLACK-CHINNED SPARROW
Spizella atrogularis

It is easy to determine who deserves the credit—or should shoulder the blame—for the scientific names assigned to any of the world's 10,000 and more species of birds. The matter of just which ornithologist, explorer, or lighthouse keeper's cat actually "discovered" the bird can be far more difficult to settle. Not only are the precise historical circumstances of these first encounters often elusive, but credit is frequently obscured by the way in which those circumstances are recounted, with the participants' roles sometimes defined less by their actions than by their place in a scientific hierarchy.

The most recent source to explicitly name the discoverer of the Black-chinned Sparrow was published nearly half a century ago, and has gone apparently unquestioned since.

> In 1851, while on an expedition in Mexico, J. Cabanis discovered this species and named it *Spinites atrogularis*.

Unfortunately, in 1851, Jean Cabanis was in fact busy half a world away, serving as principal curator of the University Museum in Berlin and, in his leisure moments, writing the great catalog of his friend and colleague Ferdinand Heine's private ornithological collections. He was not in Mexico, and neither was the Black-chinned Sparrow that he would describe and name: the bird, a male "adult in abraded plumage," reposed in a cabinet in the Prussian capital, placed there by Cabanis's mentor and predecessor, Hinrich Lichtenstein.

Eighty-two years later, A. J. van Rossem examined the type, then "in very fair condition," and discovered on the specimen's stand the name of the German explorer Lichtenstein credited with having collected the bird in Mexico, Heinrich Alwin Aschenborn. Aschenborn, a month younger than Cabanis, was a gifted student, and at the age of 20 he was awarded a stipend to travel through Germany, Denmark, and Sweden, on the condition that on his return he take a law degree and assume an administrative post in the German government.

> Thus, in the year 1838 I proceeded to Wrocław, where Professor Otto was rector, and registered in the law faculty of the university of that city. Here I attended courses for one year. When that year was over, I was seized by a great desire to see America, surely a more exciting prospect than law school.

Male Black-chinned Sparrows typically show more black on the throat and lore than females in breeding plumage. *California, April. Brian E. Small*

The voyage from Hamburg to Veracruz was 83 days, and the overland trip from Veracruz to Mexico City, Aschenborn's base for his American visit, took four more. The young explorer had justified his journey as an expedition to collect botanical specimens, but in a letter to a friend in Germany written not long after his arrival, Aschenborn confessed that his interests had already broadened to include all natural curiosities:

> In my mind I hear you asking: "How are you actually spending your time in that forsaken country?" I can tell you in just a few words. I am looking around, amusing myself as well as I can, even perhaps learning something new from time to time, and collecting every notable natural object that I can come up with. If you could only see me on my scientific wanderings!

Yours truly is out in front, with my dagger and vasculum hanging from straps; and behind me a Mexican soldier, with the face of a devil . . . loaded down with my hunting pouch and a shotgun; and in the very rear, a half-naked native with a roll of blotting paper . . . . While I gather plants, my companions shoot birds and look around for other products of nature, such as hummingbird nests, stones, snakes, lizards, etc. Then, once I am back home, I am busy for several days in drying the plants, stuffing the birds, and preserving the other animals in spirits.

The first Black-chinned Sparrow known to science was, apparently, among them. Jean Cabanis published its description and the name he found on

This species' very long, narrow tail is especially prominent in a side view. *California, April. Brian E. Small*

the specimen label. That label had been attached by Hinrich Lichtenstein, the curator who received the Mexican shipment in Berlin. The sparrow was stuffed and mounted by Alwin Aschenborn in Mexico City—and shot, perhaps, by Aschenborn, but more likely, if we take his words seriously, by a low-ranking soldier or a Native guide, the true, and forever anonymous, discoverer of the Black-chinned Sparrow.

FIELD IDENTIFICATION

Black-chinned Sparrows can scarcely be confused with any other passerellid species: the small size, long tail, rusty back, and overall gray aspect are virtually unique in all age and sex classes. The Black-chinned is sometimes described as juncolike, an almost certainly unknowing reminiscence of Couch's allocation of the species to *Struthus* a century and a half ago, but juncos are big, gregarious, conspicuous birds with broad tails and big, blocky heads.

Nevertheless, observers in eastern North America with no experience of the Black-chinned Sparrow and an archaic faith in the "field marks" system of bird identification continually misidentify juncos with brown-tinged backs as this southwestern species. More perplexing at first glance are certain hybrids (and backcrosses) between the Slate-colored Junco and the White-throated Sparrow, which can share with the genuine Black-chinned a neatly streaked brown back, solid gray head, and pink bill. Size, structure, and habit should nevertheless readily distinguish these birds; it is also worth recalling just how extremely rare this species is anywhere east of its breeding range.

The most usually heard call is very short and rather faint, with a sharp attack and fast, higher-pitched decay; inconspicuous and insectlike, it can easily be overlooked as the creak of a boot or the tinny jingle of a zipper pull.

The Black-chinned Sparrow's song is subject to considerable geographic variation. In Arizona, Nevada, and points farther east, the song is a simple, pleasant series of four or five slightly slurred introductory notes followed by a thinner, drier ascending trill or buzz; it often recalls a dry Field Sparrow's song. In California and Baja California, the introduction is reduced to a single note or absent entirely, leaving only the trill, while some birds in New Mexico and Texas add one or two notes at the end of the song.

Another song, given only early in the morning, also omits the introductory notes, but conjoins two trills in rapid succession, the first usually descending and the second ascending. Birds singing this song repeat it at a faster rate than when giving the "typical" song.

Superficially juncolike in plumage, the Black-chinned Sparrow is structurally very different: tiny, with a very long tail, small head, and small bill. *California, April. Brian E. Small*

RANGE AND GEOGRAPHIC VARIATION

Casual in spring north to southwestern Oregon, where it has nested (eBird), the Black-chinned Sparrow breeds locally in California's chaparral-covered coastal foothills and mountains and, in the interior, in mixed brush habitats. In the inner Northern Coast Ranges, this species is found north to Lake, Trinity, Glenn, and irregularly Marin Counties; it is scarce in the southern hills between Alameda and Contra Costa Counties south to southern San Benito County, where it is "reasonably common" in the Diablo Mountains of San Benito and Fremont Counties. Black-chinned Sparrows are fairly common to common in the Southern Coast Ranges between Monterey and San Luis Obispo and Santa Barbara Counties, and east to the Mount Pinos area, Transverse Ranges, San Bernardino Mountains, and Eagle Mountains of Riverside County; the breeding range extends south in the Peninsular Ranges to southwestern Imperial County. In San Diego County, the heart of the range of the subspecies *cana*, this is one of the commonest breeding birds of rugged chaparral-clad slopes in the foothills and mountains above 1,500 feet in elevation. On the west flank of the Sierra Nevada, Black-chinned Sparrows are uncommon and very local, rare and erratic in occurrence north of Yosemite; on the east side, they occur in a narrow band west of the

Owens Valley, with occasional records from farther north. The species also breeds in the mountains of the southeastern deserts, including the White, Inyo, Panamint, Coso, Argus, Kingston, Granite, New York, and Providence Mountains of Inyo and San Benito Counties. The race *cana* breeds into northern Baja California as well, south to about 30 degrees north; a nesting record for Isla Cerralvo is considered hypothetical.

This species has "a sporadic, southerly distribution" in Nevada; it breeds on dry slopes in the southernmost Mojave region, and should be looked for in the northwestern part of that region as well. There are May records as far north as Lincoln, Lyon, and Washoe Counties (eBird). The Utah occurrences of this species are heavily concentrated in the southwestern portion of the state, with erratic springtime reports as far north as Millard County, on the Nevada border, and as far east as San Juan County, nearly to Colorado; a cluster of April records from Canyonlands National Park and Natural Bridges National Monument suggest that breeding may occur there at times (eBird).

Successful breeding of the Black-chinned Sparrow in Colorado was recorded in 2012, after the species first appeared in Mesa County a year earlier. In New Mexico, Black-chinned Sparrows breed in suitably steep, brushy habitats across the southern

The back and scapulars are chestnut with long, neat black shaft streaks, creating a saddled appearance between the silvery gray nape and lower back. *California, April. Brian E. Small*

and central portions of the state, north to Taos and San Miguel Counties; as expected, they avoid the eastern plains at any season.

In Arizona, Black-chinned Sparrows are most common as breeders in central and northwestern regions below the Mogollon Rim, at elevations between 3,800 and 7,700 feet. They also breed on the western slope of the Kaibab Plateau and locally in northern Navajo County, and the Arizona Breeding Bird Atlas project discovered hitherto unknown populations in the Harcuvar, Harquahala, and Kofa Mountains, southwest of what had been thought to be the limits of the species' range. None of those areas are especially well birded, unlike southeast Arizona, where this species is relatively scarce, frustratingly so for many of the annual ornithopilgrims

hoping to see it there; the Black-chinned Sparrow breeds in the Chiricahua, Mule, Huachuca, Whetstone, Baboquivari, Rincon, Galiuro, Dragoon, and Santa Catalina Mountains, with the most accessible birds usually found in the latter two ranges.

The easternmost breeders in the United States are found in the mountains of Texas's central Trans-Pecos region. On the Mexican mainland, there are spring and summer records from Sonora of birds in appropriate breeding habitat. The nominate race is a common resident on the Mexican Plateau from Durango, southern Coahuila, central Nuevo León, and southwestern Tamaulipas south through Zacatecas, San Luis Potosí, Jalisco, Guanajuato, Querétaro, Hidalgo, Tlaxcala, Puebla, Michoacán, and Guerrero; it is very uncommon in interior Oaxaca, where the species reaches the southeasternmost point in its range.

The nominate race is resident within its range, as are some populations in southeastern Arizona, southern New Mexico, and Texas; there is some altitudinal movement from the steep cliffs used by breeders down to the mouths of brushy canyons for the winter. More northerly Black-chinned Sparrows are strongly migratory, leaving the nesting areas in August and early September. Migrants of this shy, not especially gregarious bird are only rarely detected, but a very few have been found slightly east of the expected corridors in autumn, including Texas records from Garza, Howard, Lubbock, Midland, Randall, and San Patricio Counties, and a September individual from Prowers County, Colorado (eBird). Autumn is also when Black-chinned Sparrows have been recorded on rare occasion from the California coast and islands.

Birds breeding in Baja California and in California west of the Sierras winter in Baja California Sur, while more easterly populations move into Arizona, Sonora, and Chihuahua. Arizona's winters are concentrated in the southeast, though some spend the season "in most years" in the mountains of Yuma County as well. The species is only casual in the lowlands of the lower Colorado River Valley.

The northbound migration probably begins early, with southern breeding localities occupied in March and April. In the extreme north of the range, in northern California and southern Oregon, arrivals commence in mid-May, "with peak numbers probably in late May."

The range of the Black-chinned Sparrow is not large by sparrow standards, but probably because it is only a short-distance migrant, the species has evolved a modest degree of geographic differentiation. Three (Pyle) or four subspecies are generally recognized.

The nominate subspecies is large and dark, with a distinct reddish tone to the back and a brown wash on the wing and underparts in fresh plumage; in individuals with black on the face, that color extends onto the lores and narrowly onto the forehead. As many as two-thirds of females show a black chin, sometimes as extensive as that in the male.

Breeding in California's coastal foothills, *caurina* ("northwestern") is also dark, with deep cinnamon tones to the back but little or no brown elsewhere in the plumage; the white of the belly is less extensive than in other Black-chinned Sparrows. The head is darker than that of the adjoining subspecies *cana*, and the ear coverts and sides of the chin are darker slaty than in the nominate race. The overall darkness of birds from the San Francisco Bay area had been noted even before Miller described *caurina*, but was often attributed to the "considerably soiled" state of many specimens. Examination under the microscope showed no significant difference in the amount of dirt sullying the feathers of specimens of *cana* and *caurina*, and washing the birds in gasoline and in xylol made the difference in tone "even more apparent than formerly," proving that *caurina*'s darkness was not merely artifactual.

The other Pacific race, *cana*, breeds south to about 30 degrees north in Baja California. It is smaller and slightly darker than the race breeding to its east, *evura*, and smaller, shorter-tailed, and paler than *caurina*, with which it intergrades from Ventura County, California, north. The gray underparts of *cana* are washed with brownish.

The breeding distribution of the subspecies *cana* is sometimes described as extending to the species' entire United States range east of that occupied by *caurina* (Pyle). Most authorities, however, recognize a fourth race, *evura* ("well-tailed"), paler than nominate *atrogularis* and sandier on the back; the male's lores are duller and its chin patch less extensive. Easterly individuals of *evura* are larger and longer-tailed than *cana*, while the two approach each other more closely in measurements in the western portions of *evura*'s range.

By definition, the breeding ranges of the four subspecies are discrete. In winter, though, *evura* and *caurina* may overlap in western Arizona; they are unlikely to be confidently distinguishable in the field.

Winter birds of both sexes show little black on the head. *Texas, March. Brian E. Small*

WORTHEN SPARROW
Spizella wortheni

Thirty years after this rare scrubland bird was formally described to science, Frank M. Chapman of the American Museum of Natural History could write that "few of our birds have a briefer history than this Sparrow." Ten years later, the bird was still so poorly represented in museums that John E. Thayer—eponym of the Arctic gull—twice sent professional collectors into the field to secure specimens. The second effort was wildly successful: over three months in Tamaulipas, beginning in May 1924, Thayer's collector Wilmot W. Brown took more than 20 skins and a similar number of nests and egg sets. Brown's specimens even today make up some of the most important material for any study of this species.

The first attempt, conducted by Frank B. Armstrong, had been far less productive. Sometime before 1915, Thayer commissioned Armstrong, "the most successful collector in Texas" and the owner of "a natural history business that was unrivaled during his time," to spend the summer searching for Worthen Sparrows in the vicinity of Silver City, New Mexico. Armstrong was a canny choice: as a young man, he had secured the type specimen of the western race *arenacea* of the Field Sparrow, and could be expected to have the distinctions between that bird and Thayer's desideratum firmly in mind.

In hindsight, the decision to send Armstrong to New Mexico was less felicitous. But of the eight specimens known at the time, the first had in fact been taken "on the flat near the town of" Silver City on June 16, 1884, by Charles H. Marsh, "Territorial Taxidermist." It was logical to assume that the type locality would still harbor the species, but Armstrong and a disappointed Thayer were obliged to report failure to see "any sign of the Worthen Sparrow."

Nearly a century and a half after Marsh shot the first specimen, the Worthen Sparrow has yet to reappear north of the Mexico border, leaving the Marsh skin the only evidence of the species' occurrence so far north and west. Though there is essential unanimity today that the type was in fact collected by Marsh and near Silver City, that record has occasionally been the subject of uncertainty and even doubt—circumstances most likely created not by Marsh or by Robert Ridgway, the author of the formal species description, but by Charles K. Worthen, the man who played perhaps the smallest role in the discovery of the sparrow that bears his name.

Like Marsh, and like Armstrong and Brown

The plain wing of this adult Worthen Sparrow is a significant distinction from the Field Sparrow. *Nuevo León. Amy McAndrews*

after him, Worthen's interest in natural history was largely commercial. Beginning his career as an illustrator,

> his tastes led him to engage in the collection and sale of natural history specimens, in this way becoming well known to the naturalists of the country, and especially to museum curators.

Among those curators was Worthen's fellow Illinoisan Robert Ridgway. In 1884, Worthen exchanged 118 bird specimens with the Smithsonian, among them the type of the new sparrow Ridgway would soon name for him.

Worthen's eulogist would describe him as "always intelligent and trustworthy," but like too many of his contemporaries dealing in naturalia, Worthen was less than scrupulous in labeling the specimens he sold. He regularly removed the original tags of objects he received from field collectors, replacing them with labels bearing his own name and his own address in Warsaw, Illinois. In the case of the new *Spizella*, this regrettable practice created a flurry of confusions and hard feelings.

Ridgway, believing that Worthen's name on the tag identified him as the original collector of the sparrow, graciously named the bird for him in September 1884, with thanks for his having "kindly presented [!] the type specimen to the National Museum." There was no mention of Marsh, an oversight—innocent on Ridgway's part, less so on Worthen's—that Marsh would correct in January 1885, when he published in the *Ornithologist and Oologist* his own notice of the new species, the type of which, he clarified, he himself had "on the 16th of June . . . shot on the flat near the town."

Marsh ended his brief note with a paraphrase of Ridgway's published diagnosis of the specimen, a not unusual and not inappropriate gesture; nevertheless, his failure to provide a full citation for Ridgway's original description drew a rebuke from the editor of the *Ornithologist and Oologist* and the greatly trumped-up accusation of "plagiarism" that persists even today. Ridgway himself delicately put the blame where it belonged: on writing to the magazine to correct a typographical error in the bird's scientific name, he discreetly pointed out that Worthen had in fact twice failed to provide the correct identity of the bird's collector—and professed himself now "very glad to give Mr. Marsh due credit for its discovery." In April 1885, Marsh wrote to Ridgway to declare that he was satisfied, and assured him that he harbored

> no bad feelings in the matter of the *Spizella wortheni* . . . . If any one is at fault in the matter, it's Mr. Worthen, and I hardly think he did as he would like to be done by.

Even at this remove, it is hard not to agree.

FIELD IDENTIFICATION

The only species likely to be confused with the Worthen Sparrow, the similarly rusty-capped Field Sparrow barely invades northeastern Mexico in winter. Elliott Coues's typically confident dictum that the two species are "quite distinct" remains puzzling more than a century after it was issued; indeed, others have found them so similar as to be indistinguishable even in the hand. The truth lies in between, and a silent Worthen Sparrow should be identifiable in the field with good views and caution.

Slightly smaller than the Field Sparrow, the

This rare Mexican sparrow is blanker-faced than the otherwise similar Field Sparrow; also note the black feet, unlike the pink tarsus and toes of the Field. *Nuevo León. Amy McAndrews*

Worthen Sparrow is also shorter-tailed, especially in comparison with the western Field Sparrow race *arenacea*. It is grayer than the Field Sparrow in general aspect, with a gray forehead and gray-washed breast; in the Field Sparrow, the rusty crown extends to the base of the bill, and, especially in the eastern, nominate race, the breast is tinged with brown (Pyle), often concentrated into diffuse peach-colored patches on the breast side. The rump is grayer than in the Field Sparrow, and the auriculars are also gray, usually without the brown or rusty wash of that species; most Worthen Sparrows appear to have only a faint rusty line behind the eye or none at all. The white eye-ring is often said to be brighter than that of the Field Sparrow, a distinction not consistently apparent in photographs.

The Worthen Sparrow shows only weak wing bars, while the white tips of the greater and median coverts are conspicuous in Field Sparrows.

A traditional field mark for the Worthen Sparrow, the color of tarsus and toes varies seasonally. While winter birds show a dark brown to black tarsus, the feet are pink in at least some birds during nesting season, rendering them useless in identification.

The clearest distinction between the Field Sparrow and the Worthen Sparrow is not visual but vocal. The song of the Worthen Sparrow

> consists of a thin, introductory *seep* . . . followed by 10 [to 17] *chip* notes that sound flat and harsh . . . .

Some males in Zacatecas, presumably *browni*, were found to slur the introductory note upward, others down.

The dryness of the concluding notes—more a rattle than a trill—recalls many Chipping Sparrow songs; some observers hear a similarity to juncos. The trill follows directly on the introductory *seep* without transition; it can be higher or lower in pitch than the introduction, and is delivered in a steady, even rhythm, without the "bouncing ball" acceleration typical of the Field Sparrow's song.

The call is "a high, thin, fairly dry *tssip* or *tsip*, at times repeated rapidly."

RANGE AND GEOGRAPHIC VARIATION

With no records since 1884, it is virtually certain that no Worthen Sparrows persist north of the Mexico border, and the apparent population once present in Chihuahua seems to have been extirpated since 1959. As of 1993, Wege, Howell, and Sada were able to adduce additional records from Coahuila, Zacatecas, Nuevo León, San Luis Potosí, Tamaulipas, Veracruz, and Puebla; all records from the previous decade had come from the border areas of Coahuila and Nuevo León. Since then, inspired largely by those same authors' work, birders and ornithologists have systematically sought this species in its Mexican range, finding it most recently near Juan Aldama, Majoma, and Pozo de San Juan in Zacatecas; near Charcas and Vanegas in San Luis Potosí; and near Catarino Rodríguez in Nuevo León (eBird). By far the greatest number of recent sightings have been obtained along the border between Coahuila and Nuevo León, across an area stretching from San Rafael (Galeana) in the latter state west to El Fraile in the former (eBird).

There is "no adequate evidence" of long-distance migration in this species, though local movements in Coahuila and Nuevo León can bring flocks to areas where the species does not normally occur. Recent counts of winter flocks, tallied between mid-August and early March, have been as high as 150 individuals (eBird).

Two subspecies of the Worthen Sparrow are generally recognized, a circumstance perhaps surprising given the species' very small global range. Known from western Zacatecas, the race *browni*, named for Wilmot W. Brown, was described as significantly darker above, including the tail, with the crown less cinnamon and the back and scapulars buffier than in the nominate race, which occupies the rest of the species' range. As of early 2016, *browni* apparently survived in the vicinity of Torreón-Río Grande in Zacatecas (eBird).

One of the most challenging North American sparrows to identify, the Timberline differs visually from the Brewer and Clay-colored Sparrows in only such small details as the head and flank markings and bill size. *Alberta, July. Chris Wood*

TIMBERLINE SPARROW
Spizella taverneri

This species offers sparrow watchers everything they might want: rarity, a limited breeding range, an almost unknown winter range, and identification challenges so great that it is often deemed "almost impossible" to distinguish reliably in the field. Away from the restricted nesting areas in Alaska, the Yukon, British Columbia, Alberta, and Montana, most observers simply yield to despair, certain that the Timberline Sparrow is out there, but doubting their capacity, or anyone's capacity, to actually find and identify it.

The historical record gives hope. Many birds were discovered by science more or less incidentally, recognized as different and new only once the bleeding little corpse was in hand. The Timberline Sparrow, among the brownest and most subtly marked of all our brown, subtly marked passerellids, is, at least on occasion, sufficiently distinctive to draw the attention of even those who have never knowingly seen it before.

In 1924, Allan Brooks joined Harry Swarth on a collecting trip in the Atlin region of northwestern British Columbia . . . . While together once, they noticed a bird whose "appearance did not accord with anything we knew in the region, and Brooks started at once in pursuit. With some difficulty, for the birds were wary, he secured one of them."

The next year, Swarth and Brooks would affirm the novelty of their prize when they described the Timberline Sparrow, naming it *Spizella taverneri* in honor of the Canadian biologist Percy Taverner.

The first migrant specimen from the United States, a female taken in New Mexico in October 1931, was also recognized as different in the field. Seth B. Benson, curator of mammalogy at the University of California's Museum of Vertebrate Zoology, noted the bird's small bill, broad streaking, and gray ground color, and identified it as this species before pulling the trigger. Thirty years later, in February 1962, Amadeo Rea detected "a previously unheard sparrow" singing along a willow-lined creek near San Diego, California; it was the first Timberline Sparrow ever recorded on the West Coast "or indeed anywhere west of extreme eastern Arizona." It is not easy, and few are the birders today who confirm their identifications with birdshot, but these examples should inspire today's observers to look and listen carefully whenever and wherever the Timberline Sparrow is a possibility.

FIELD IDENTIFICATION

Like its close relative the Brewer Sparrow, this is a slight-bodied, long-tailed, gray-brown sparrow with little contrast between the sandy upperparts and dull off-whitish underparts. In the field (and even in the hand), the Timberline Sparrow should be carefully distinguished from both the Brewer and the Clay-colored Sparrows. Even then, there may be puzzling individuals that appear to fall closer to one or the other, or even to be intermediate between those two congeners.

The Timberline Sparrow is slightly larger than the Brewer Sparrow, with a slightly longer tail on average and, importantly, a proportionally shorter, more slender, and darker bill; in absolute terms, the bill of most Timberline Sparrows is no longer than that of the smaller Brewer. The flank of the Timberline Sparrow may average slightly darker, and the streaking of the back and crown are stronger and darker; the ground color of the upperparts is grayer than the sandy tone shown by Brewer Sparrows. Unlike the adult Brewer Sparrow, Timberline Sparrows of all ages exhibit a "tendency" to scattered narrow dark brownish streaks on the flanks and breast.

The head markings of the Timberline Sparrow approach those of the Clay-colored Sparrow: there is a variably conspicuous gray collar, and the streaked crown shows a variably broad median stripe; the supercilium is notably distinct, as are the narrow dark eye line and whisker. The eye-ring is complete, as in the Brewer Sparrow, but is not as bright white as in that species. Most Brewer Sparrows show the barest hint of a diffuse lateral throat stripe, while that mark can be dark and conspicuous in Timberline Sparrows. All the same, lightly marked individuals no doubt go unnoticed among wintering flocks of Brewer Sparrows in the Southwest.

In examining hundreds of skins of Brewer, Timberline, and Clay-colored Sparrows, Swarth and Brooks claimed to have "found no specimen of equivocal character," but heavily marked Timberline Sparrows are among the greatest identification challenges faced by the field birder. The distinctions between those birds and Clay-colored Sparrows are for the most part differences of degree, requiring close and extended views and some familiarity with the more common species.

On average, the Clay-colored Sparrow's gray collar is clearer and brighter than that of the Tim-berline Sparrow; the latter species is more likely to show extensive fine streaking at the collar's ends. The median crown stripe is also broader and less marked in the Clay-colored; in both species, it is bordered by notably dark lateral areas, which tend to be more solidly colored in the Clay-colored and more streaked in the Timberline. The complete eye-ring of the Timberline Sparrow—shared by the Brewer Sparrow—is unlike the whitish crescent below the eye of the Clay-colored. Both species have the ear coverts outlined above by a fairly dark eye line and below by a darker whisker; the Clay-colored Sparrow's whisker is said to widen toward the rear of the ear coverts, that of the Timberline to remain evenly narrow over its entire length. While the Timberline's bill is slender and brownish, the Clay-colored Sparrow's pinkish bill is deeper at the base.

Nonetheless, "the possibility of overlap between Clay-colors at the 'soft' end of the spectrum vs. Timberlines at the 'sharp' end" means that some birds will inevitably be unidentifiable in the field, especially in migration and winter, when adults are in their slightly duller-headed basic plumage.

Juvenile Timberline Sparrows differ from juvenile Brewer Sparrows in essentially the same ways as adults. Birds still entirely in juvenile plumage should be seen only on the breeding grounds, where range, habitat, and ideally the presence of adults in attendance on their young should suggest an identification.

If visual identification is challenging, the vocal distinctions between singing males are somewhat more pronounced. Both the Brewer and the Timberline Sparrows have two song "types," a short song comprising one to three different trills and a long song made up of five to ten different trills and syllables. The short song of the Timberline Sparrow appears to be relatively little known, but its long song differs from that of the Brewer in lacking that species' distinctly buzzy tone, thus making a "more musical and tinkling" impression. In Alaska, and probably throughout their range, Timberline Sparrow males frequently repeat segments within the long song; the only component of the Brewer Sparrow's long song to be repeated is "a series of descending sweet notes" without counterpart in the songs of Timberline Sparrows.

The short song of the Timberline Sparrow probably also averages "less broadband and more musical," but there may be overlap—and it is more difficult for the human observer to gain an accurate impression of a song of such brief duration.

RANGE AND GEOGRAPHIC VARIATION

Identification uncertainties and the remoteness of many presumed breeding areas leave this species' summer range imperfectly known nearly a century after its discovery. In Alaska, where it was first found in 1992, the Timberline Sparrow breeds very locally in the southeastern portion of the state along the Yukon border, with records on the upper Cheslina and Chisana Rivers; the breeding range extends east to the Yukon's Kluane Range and probably northwest into the Alaskan Mentasta Mountains. Alaska's Timberline Sparrows breed at a higher altitude than those in adjacent Canada, using willows rather than birches for nest sites.

The first Timberline Sparrows were collected in the vicinity of Atlin, British Columbia, in the extreme northwestern corner of the province. This species probably breeds over most of the northwestern third of British Columbia, where it occupies open areas dotted with shrubs and small trees, for the most part above timberline. In a small area of southeastern British Columbia, centered on Elko and Sparwood, Timberline Sparrows have been found breeding below treeline, on rocky slopes with scattered willows and other low woody vegetation; the identity of Brewer-like sparrows breeding at high elevations in the southern Okanagan is uncertain. In Alberta, the Timberline Sparrow breeds in high-elevation meadows in the Rocky Mountains from Waterton north to Jasper National Park.

The southern limits of this species' summer range remain unclear. In Montana, Timberline Sparrows are known to breed in Glacier National Park and the area of Lewis and Clark National Forest; nesting birds are found in stunted conifers at elevations between 5,500 and 7,500 feet and in willow thickets as low as 5,000 feet. Though there are no breeding records for Wyoming, it is thought that the species may be present in alpine areas in the northwestern portions of that state. Still mysterious is the summer occurrence of Timberline-like sparrows in suitable breeding habitat at high elevations in the Colorado Rockies; it is not certain whether these birds are in fact Timberline Sparrows—or the altitudinal range of the Brewer Sparrow (in the narrow sense) is wider than believed. The Timberline-like appearance of some Brewer Sparrows breeding in northern California is equally perplexing.

Timberline Sparrows are only rarely detected and identified in migration. Birds leave British Columbia in August and early September, and seem unlikely to linger much beyond that in any part of their high-elevation breeding grounds. The autumn migration probably takes most individuals straight south through the Mountain West; there is a New Mexico specimen from mid-October, but migrants of this species are otherwise very poorly represented in the ornithological record, whether as specimens or sight records. The easternmost plausible report is of one in northeastern Quebec in October 2014; photographs show a fairly extensive and lightly marked gray collar, distinct fine blackish shaft streaks on the flank and breast sides, and a well-defined narrow lateral throat stripe and whisker.

The winter range of the Timberline Sparrow is virtually unknown. It probably lies for the most part in the deserts of northwestern Mexico, but may extend into portions of west Texas and perhaps elsewhere in the southwestern United States. There is also a February specimen from San Diego County, still the only specimen known from California more than 50 years after it was collected. A bird photographed in eastern Virginia in the winter of 2011 and 2012 and accepted by that state's records committee appears to be more likely a Timberline than a Brewer Sparrow (eBird).

The movement north in spring appears to be leisurely and late. The only Arizona specimens attributed to this species are both from mid- to late May, though a few have been seen as far north as Washington in April and early May. In southern Alberta, Timberline Sparrows arrive as much as three weeks after Brewer Sparrows. Breeding adults do not arrive on the Alaskan breeding grounds until early or mid-June, about the time that Brewer Sparrows are fledging young elsewhere.

No subspecies are recognized.

BREWER SPARROW
Spizella breweri

Tiny, delicately slender, and finely marked, this long-tailed, small-billed sparrow is the voice of flat desert brushlands. *California, June. Brian E. Small*

The Boston physician and ornithologist Thomas Mayo Brewer had been dead nearly 20 years when the ever irascible Elliott Coues directed his pen at him in a final virulent postscript to the "Sparrow Wars" of the late 1870s:

> Everybody knows that Dr. Brewer made a fool of himself about the Sparrows for years, and the fact that he then died does not alter the other fact of what he did when he was alive. Many other persons . . . did the same; but Dr. Brewer's foolishness was more conspicuous, because he pretended to be an ornithologist. The harm he did is incalculable, and his name deserves to be stigmatized . . . . Dying makes a great difference to the person chiefly concerned, but has no retroactive effect upon the events of his life, and only sentimentalists allow it to influence their estimate of personal character.

The "sparrows" at issue here were introduced House Sparrows, of course. Coues would continue to prosecute his case against the non-native pests—and, less righteously, against Brewer for daring to defend them—for the rest of his life, even well after Brewer's death in 1880.

At times, both before and after the acrimonious climax of the Sparrow Wars in 1878, Coues spoke of Brewer with some of the respect owed a colleague 30 years his elder. In the brief history of ornithology first published in the second edition of the *Key*, Coues listed Brewer among the nobility of American science, and lavishly praised his work in writing far the greater part of the prose in the great *History of North American Birds* prepared in collaboration with Baird and Ridgway. In the revised edition of his *Check List*, Coues singled Brewer out as "long the leading oölogist of North America." In the final edition of the *Key*, though, published after Coues's own death on Christmas Day 1899, Coues could bring himself to describe the eponym of the Brewer Sparrow—a man once praised by John Cassin for "the highest abilities and social qualities . . . [and] an ardor and devotion to Ornithological science rarely paralleled"—as only "Dr. T. M. Brewer, of Boston."

FIELD IDENTIFICATION

In good views, the Brewer Sparrow can be confused only with another of the brown *Spizella* species. The plumage differences are real but subtle, at times excruciatingly subtle, and not all individuals can be definitively identified in the field or even in the hand: Audubon notoriously mistook what could have been the type specimens of the Brewer Sparrow for the Clay-colored, and even Coues fell into the same error in Montana in 1874. Modern birders who find this genus challenging are in excellent company.

The Brewer Sparrow is a notably tiny, notably long-tailed bird of open scrubby flats; the observer's first impression is usually one of uniform grayness, with little contrast between the bird's upperparts and its lower parts. Both the neck and the bill of this slender bird are small, and the relatively plain face gives it a bland expression. The white eye-ring, bright in most birds but duller in others, isolates the small eye on the face, creating an aspect simultaneously feckless and surly.

Some Brewer Sparrows show fairly conspicuous gray on the neck sides and nape, though never as much as on a typical Clay-colored Sparrow. On the Clay-colored, two narrow, finely black-streaked stripes cross the nape to connect the brown of the crown to the brown of the upper back; the gray square or rectangle between the stripes is obvious and well defined. In Brewer Sparrows, even those showing the most extensive gray collar, the fine black and brown streaking of the crown continues across the nape to meet the somewhat coarser streaks of the upper back; unmarked gray, if there is any, is restricted to the very ends of the collar on the sides of the neck. This impression of continuous streaking is also unlike that shown by Chipping Sparrows.

The head pattern of the Brewer Sparrow is the most weakly marked of brown western *Spizella* species. The crown may show a poorly defined median stripe, buffy brown-gray with darker streaks; the dark streaking of the sides of the crown is narrow and sparse, and there is no conspicuous rust as in most Chipping and Clay-colored Sparrows. The faintness of the dark line extending back from the eye and the pale lore mean that there is no brightly

The fine black crown streaking may show a very little rust, and some birds show the hint of a grayer central stripe. *California, June. Brian E. Small*

Many are very plain-faced, while others have darker whiskers and eye lines; the lore is always plain, and the eye-ring usually conspicuous. *California, June. Brian E. Small*

contrasting supercilium; instead, that area fades into the slightly grayer neck sides and nape and is often essentially continuous and concolorous with the brownish ear coverts. Timberline Sparrows have a darker, better-defined eye line, which more clearly divides the relatively strong supercilium from the gray-brown ear coverts; both species have a complete eye-ring, averaging slightly brighter white in the Timberline.

Clay-colored and Timberline Sparrows also show a darker and more contrasting whisker below the ear coverts; in the Clay-colored Sparrow, that whisker broadens toward the rear, while it is of even width across its length in the Brewer and Timberline Sparrows. Clay-colored and Timberline Sparrows and basic- and formative-plumaged Chipping Sparrows also have relatively heavy lateral throat stripes. In direct comparison with Clay-colored or Chipping Sparrows, Brewer Sparrows are strikingly small-billed—but the Timberline Sparrow's darker bill averages even smaller and thinner.

Juvenile Brewer Sparrows are not strikingly different in plumage from adults. They are buffier, with a warmer brown crown and rump, and an even less well-defined face pattern, though the eye-ring is conspicuous. The sparse, fine dark streaking on the breast and flanks is lost in the preformative molt, which is completed on the breeding grounds before

The underparts of the adult are plain whitish gray, with unstreaked flanks. *California, April. Brian E. Small*

the southbound migration; Timberline Sparrows show similar but less extensive streaking even as adults.

The usual call of the Brewer Sparrow, given perched or in flight, is a short, hesitant *dsip*, with strong attack and very little audible decay. A "soft twittering" is said to be given by females.

The song of this species is one of the great natural phenomena of the American West, heard on the breeding grounds and in migration and winter. Robert Ridgway first heard it on the sage flats of California:

> The song of Brewer's Sparrow . . . for sprightliness and vivacity is not excelled by any other of the North American Fringillidae, being inferior only to that of the [Lark Sparrow] in power and richness, and even excelling it in variety and compass. Its song, while possessing all the plaintiveness of tone so characteristic of the eastern Field Sparrow, unites to this quality a vivacity and variety fully equalling that of the finest Canary.

Like the Timberline, the Brewer Sparrow has two song "types," a short song comprising one to three different trills and a long song made up of five to ten different trills and syllables. The long song of the Brewer Sparrow is a long series of sustained tremolos of different tonal quality and at different pitches; unlike the Timberline, many of the phrases are distinctly buzzy and low-pitched, some of them recalling the slow flatulent buzzes of the Clay-colored Sparrow, others similar to the bubbling trills of the Chipping Sparrow. Each tremolo in a single song is different; at least some Timberline Sparrows repeat segments within the long song. The only component of the Brewer Sparrow's long song to be repeated is "a series of descending sweet notes."

The short song of the Brewer Sparrow draws on a lexicon of more than 50 discrete trill types. This song typically combines a fast, high-pitched trill with a following slow, lower-pitched trill.

RANGE AND GEOGRAPHIC VARIATION

One of the most widespread birds of western plains and sagebrush flats, the Brewer Sparrow breeds from the Sierra Nevada of California and the Great Basin deserts of Washington and Oregon east to the sage and short-grass prairies of extreme western Nebraska and the Dakotas; the breeding range extends north to the Okanagan Valley of southern British Columbia and the southernmost border areas of Alberta and Saskatchewan, and south to the desert scrub and sagebrush of southern Nevada, northern Arizona, and northern New Mexico.

British Columbia's breeding Brewer Sparrows are restricted to the shrub steppe of the southern interior west of the Okanagan River and from the Marron River south; nesting occurs at elevations from 1,000 to 2,500 feet. Numbers are highest in the areas of White Lake and Richter Pass. Brewer Sparrows are local and uncommon in extreme southern and southeastern Alberta, west to Taber and Milk River and north to the region of Bindloss and Empress. In Saskatchewan, summer reports are concentrated in the southwestern portion of the province, east to Grasslands National Park (eBird). A July record from near Brandon, Manitoba, may represent the easternmost breeding for this species (eBird).

In much of its United States range, the Brewer Sparrow, Sage Thrasher, Sagebrush Sparrow, and Greater Sage Grouse form a quartet of classic Great Basin sagebrush specialists. In eastern Washington, the Brewer Sparrow is a common breeder in big sagebrush habitats with a well-developed understory of native grasses. Oregon's Department of Fish and Wildlife considers the Brewer Sparrow an abundant breeder east of the Cascades, especially so in the southeastern quarter of the state. The species is common in suitable sagebrush habitat across the bulk of southern Idaho into Montana. Brewer Sparrows breed virtually throughout Wyoming, Utah, and Nevada. The summer range extends south through much of Arizona north and east of the Kaibab Plateau and in the Arizona Strip, north of the Colorado River; breeding occurs at elevations between 4,300 and 7,500 feet. In New Mexico, the southern limit of regular breeding lies just south of Highway 60 (eBird).

The western edge of the breeding range falls along the ridge of the Cascades in Washington and Oregon, and in California, the crest of the Sierra Nevada south to southern Kern County. Significant breeding areas in California include the Klamath Basin, the Modoc Plateau, and the basin and range landscapes south to the Mono Basin and the White and Inyo Mountains; Brewer Sparrows also breed in the desert mountains of Inyo and San Bernardino Counties. Smaller, disjunct populations are found in the Lanfair Valley, the eastern San Bernardino Mountains, the Mt. Pinos area, and the Carrizo Plain in San Luis Obispo County. West of the Sierra crest, this species is a common breeder in southeastern Tulare County and, more locally, in brushy habitats in Butte County. Long suspected of breeding in San Diego County, Brewer Sparrows were first confirmed to nest there in 2001 in Montezuma Valley and nearby areas.

The eastern limits of the Brewer Sparrow's range are somewhat less certain. In Colorado, where the species occupies essentially all suitable habitat west of the Rockies, breeding has been confirmed as far east as Weld, Washington, and Las Animas Counties; observations of possible or probable breeding

Streaked juveniles can be difficult to distinguish from Clay-colored and especially Timberline Sparrows, but the faint facial markings and fine crown streaking continuing down the nape are good clues. *Oregon, August.* *Brian E. Small*

have been made in Phillips and Yuma Counties as well. The breeding range in North Dakota comprises only the southwestern corner of the state in Golden Valley, Billings, Slope, and Bowman Counties; it is listed there as a Species of Conservation Priority. In South Dakota, breeding records are confined to the western edge of the state, in Harding, Butte, and Fall River Counties, where the species is said to be fairly common.

Nebraska lists this sparrow as an At-risk Species. Once thought to be restricted entirely to areas immediately bordering on Wyoming, the breeding range has been found this century to reach as far east as southwestern Box Butte and central Morrill Counties; possible breeding has also been recorded in extreme southwestern Dundy County, on the borders with Colorado and Kansas. An old report of breeding far to the east in central Nebraska is not credible. There are breeding records for Kansas in Morton, Finney, and Stevens Counties; the species is considered endangered in that state. The first breeding in the Texas Panhandle since the 1950s was recorded in 2008.

Autumn migration can begin as early as late July, but for the most part Brewer Sparrows start their southbound movements later than the western Chipping Sparrow, leaving the breeding grounds in September or early October. The usual migration routes are oriented north-south, but fall sees birds spread west to the southern California coast and east to western Kansas, Oklahoma, and the western third of Texas. There is a late September record from southeasternmost Oklahoma (eBird), and the species has occurred at least twice in late November on the Illinois shore of Lake Michigan (eBird). One of the three October reports for Nova Scotia is well documented.

The first winterers arrive in the southwestern United States in early and mid-September. Sociable and noisy, Brewer Sparrows are common and conspicuous from September through March in southeastern California, much of Arizona, New Mexico, and the Trans-Pecos in Texas; the species is rare and irregular elsewhere in Texas, as far south as the lower Rio Grande Valley. In Mexico, where arrival dates extend from mid-September to mid-October, Brewer Sparrows are fairly common in winter in Sonora, all of Baja California, northern and eastern Chihuahua, and central Mexico south to Jalisco, Guanajuato, San Luis Potosí, and Nuevo León.

Winter vagrants are detected slightly more frequently than autumn strays. One was collected in eastern Massachusetts in December 1873, and there are winter records from northwestern Arkansas, Louisiana, and northern Illinois as well (eBird). A bird that spent several weeks in eastern Virginia in the winter of 2011 and 2012 was photographed, identified, and accepted by that state's records committee as a Brewer Sparrow, but photographs show it to have been quite heavily marked on the head and distinctly streaked below, suggesting that it may in fact have been a Timberline Sparrow.

Unlike the Timberline Sparrow, whose northward movements are deferred by the lingering harshness of its high-elevation nesting habitat, the Brewer Sparrow is an early spring migrant, with transient flocks conspicuous in the southwestern United States by early March. The first birds arrive on the Great Basin breeding grounds later that month, and the northernmost breeders reach Alberta by the end of May. Springtime strays, some of which may be returning winterers, have been reported in Maine, Illinois, Michigan, and Minnesota (eBird); most such records are from mid- or late May, when most Brewer Sparrows have reached the breeding grounds but Timberline Sparrows are still on the move.

When the Timberline Sparrow is accorded full species status, the Brewer Sparrow is generally treated as monotypic. Brewer Sparrows breeding in northern California are said to average more heavily marked—and thus more Timberline-like—than birds in the remainder of the range, a claim said to be "contentious." Amadeo Rea suggested that eastern Washington, Oregon, and California might be occupied by an undescribed race differing in fresh plumage from nominate *breweri* in its more reddish, heavily streaked back; gray median crown stripe; darker ear coverts; and cinnamon-cream wing bars and tertial edgings.

FIELD SPARROW
Spizella pusilla

Bright upperparts, pink-washed breast and head, and a beautiful silvery song identify the Field Sparrow throughout its wide range. *Michigan, May. Brian E. Small*

The plaintively beautiful song of this common and familiar eastern sparrow has inspired more than one human rhapsody. Just before the turn of the twentieth century, Simeon Pease Cheney, "an old lover of the birds," called the Field Sparrow

> surely a fine singer . . . . When we consider the genius displayed in combining so beautifully the three grand principles of sound—length, pitch, and power—its brevity and limited compass make it all the more wonderful. Scarcely anything in rhythmics and dynamics is more difficult than to give a perfect *accelerando* and *crescendo*; and the use of the chromatic scale by which the field sparrow rises in his lyric flight involves the very pith of melodic ability. This little musician has explored the whole realm of sound, and condensed its beauties in perfection into one short song.

Oddly, the Field Sparrow's vocal perfection went entirely unnoticed by the first European scientists to encounter the bird. Mark Catesby, who painted what appears to have been this species on his travels through the American Southeast, does not mention any vocalizations at all in the brief prose account he dedicated to it. Louis Pierre Vieillot, usually one of the best observers of the birds he saw in his years in the United States in the 1790s, compares the species' song to that of the Chipping Sparrow,

> but its phrases are longer, while at the same time no more pleasing, and the voice resembles that of a cricket.

Alexander Wilson, to whom we owe the species' current scientific name, went so far as to claim that the Field Sparrow "has no song; but a kind of chirrupping not much different from the chirpings of a cricket." Wilson's corrector and posthumous champion Charles Lucien Bonaparte was at first "disposed to call it *E[mberiza] locustella*, a name taken from its voice, which is similar to the chirpings of a cricket." Audubon was only slightly more flattering in describing the song as "remarkable, although not fine," and confessed himself unable to provide any more detailed description.

It took a full 100 years from Catesby's publication of his account of this "Little Sparrow" for the first full description of the Field Sparrow's song to appear in print. Having traveled through much of what is now the eastern United States, the English botanist and naturalist Thomas Nuttall wrote that

> Our little bird has a pretty loud and shrill note, which may be heard at a considerable distance, and possesses some variety of tone and expression . . . a short trill, something like the song of the Canary, but less varied, and usually in a querulous or somewhat plaintive tone . . . . These tones are also somewhat similar to the reverberations of the Chipping Bird, but quite loud and sonorous, and without the changeless monotony of that species. In fact, our bird would be worthy a place in a cage as a songster of some merit.

The very long tail, white eye-ring, and small, pink bill are often visible at distances where other characters are obscured. *Michigan, May. Brian E. Small*

Nuttall's suggestion was never taken up on any large scale. Perhaps as a result, the Field Sparrow's song remained "little known or appreciated" for another full generation, until at last Thomas Brewer declared the species "a very varied and fine singer . . . very pleasing," a verdict concurred in by any modern birder fortunate enough to hear its silvery trills.

FIELD IDENTIFICATION

Small, rotund, and long-tailed, with a small round head and petite bill, the discreetly colorful Field Sparrow is one of the most easily learned members of the family. At any age and season, it is unlikely to be confused with any other sparrow but the very rare Worthen's; the two may on rare occasion overlap in winter in Nuevo León and Coahuila (eBird).

Slightly larger than the Worthen Sparrow, the Field Sparrow is also longer-tailed, especially the western race *arenacea*. It is rustier in general aspect, with a rusty forehead and variably pinkish- or brown-washed breast; in the Worthen Sparrow, the rusty crown stops short of the base of the bill, and the breast is tinged with gray (Pyle). The rump is also browner than in the Worthen Sparrow, and the Field Sparrow's ear coverts are brown or rusty, bordered above by a short rusty line behind the eye. The white tips of the greater and median coverts are also more conspicuous in Field Sparrows, forming neat paired wing bars.

A traditional field mark for the Worthen Sparrow, the color of tarsus and toes varies seasonally. While winter birds show a dark brown to black tarsus, the feet are pink in at least some birds during nesting season, rendering them useless to identification. The Field Sparrow's tarsus and toes are pink, even bright pink, at all seasons.

The clearest distinction between the Field Sparrow and the Worthen Sparrow is not visual but vocal. Male Field Sparrows sing two songs, the familiar sweet "simple" song and a less frequently heard "complex" song. The simple song, heard as soon as males arrive on the breeding grounds in spring, is a series of clear, bright slurred whistles accelerating into a fast but still musical tremolo; some birds pause briefly in the middle of the tremolo, thus ending their song with two or even three discrete trilled phrases. Most songs remain on or near a single pitch, but in some the tremolo is higher or, less frequently, lower than the introductory whistles; occasionally, the notes in the concluding portion of the song are slowed and rise in a chromatic scale, recalling the similarly rising, but buzzy, song of the Prairie Warbler with which this species shares overgrown fields in much of its range.

The complex song, commonly given when males eject an intruder and "signal[ing] heightened aggressive tendencies," begins with a tremolo and ends with more widely separated slurred notes. While all the notes of the simple song are typically clear and "pure" in tone and evenly slurred, the notes of the tremolo in this complex song are harsher and more modulated.

Field Sparrows call frequently at all times of the year. The *tsee* call is short and sharp, with a clear attack and rapid decay; the chip note, perhaps the most frequently heard call, is high and bright, often surprisingly loud for such a small bird and sometimes calling to mind a distant Northern Cardinal or White-throated Sparrow.

RANGE AND GEOGRAPHIC VARIATION

Most North American sparrows of limited distribution are western birds; only a few are restricted to an eastern breeding range, chief among them the Field Sparrow. The northwesternmost summer record is of a June bird just west of Calgary, Alberta; breeding was not confirmed there or in the case of a July record from Saskatoon County, Saskatchewan (eBird). In Montana, Field Sparrows are fairly common breeders east of the Continental Divide, especially south of the Missouri River, though there are records of singing birds from farther west in the 1920s and 1930s.

Elsewhere on the western Great Plains, Field Sparrows breed locally and uncommonly west to eastern Wyoming and Colorado. Texas breeding populations are oddly disjunct, occupying the northeastern corner of the state, the eastern Panhandle, and the northern Edwards Plateau; nesting also takes place in Brooks and Kenedy Counties, and perhaps along the coast from Victoria to Nueces Counties.

The northern edge of this species' breeding range runs across eastern Montana, North Dakota, and southern Minnesota and Wisconsin. In Ontario, Field Sparrows are locally common breeders in the southeast, while the northernmost regular nesting probably takes place in southernmost Quebec. They are rare breeders in New Brunswick.

At their southern limit, Field Sparrows breed from central Texas east across northern Louisiana, Mississippi, Alabama, and Georgia; in Florida, the species is uncommon to rare as a breeder in the Panhandle and extreme northern peninsula, irregularly as far east as Nassau County.

The southbound migration appears to begin in northern populations in early September, with arrivals on the wintering grounds complete by early December. Scattered autumn vagrants have occurred to the west of the breeding range, west to the lower mainland of British Columbia; Vancouver Island; and Neah Bay, Washington (eBird); there are a dozen records from California (eBird), some of them of birds lingering through the winter.

The upper wing bar, formed by extensive white tips to the median coverts, may be more conspicuous than the lower. *Michigan, May. Brian E. Small*

Western populations, belonging to the subspecies *arenacea,* tend to be grayer and duller. *Texas, April. Brian E. Small*

Wintering Field Sparrows are most abundant in the southeastern United States, from southern Virginia to central Texas. Wintering now regularly occurs north to Kansas and southern New England, and stray individuals can be found virtually anywhere in the breeding range. The southern limit is reached in Coahuila and northern Tamaulipas and in Nuevo León, where specimens representing both traditionally recognized subspecies have been collected.

Spring arrivals in the northern portions of the breeding range are usually in April, with males arriving up to two weeks earlier than females. At least on the eastern Great Plains, arrival dates have moved forward significantly in recent decades, such that March birds are no longer remarkable.

There are two currently recognized subspecies of the Field Sparrow, differing in relative tail length, overall size, and plumage. The eastern, nominate race *pusilla* and the western *arenacea* ("sandy brown") meet in the breeding season on the eastern Great Plains, and largely share the southeastern winter range.

The type specimen of *arenacea,* a female, was secured near Laredo, Texas, in mid-November 1885 by Frank B. Armstrong; the skin is now in the Harvard Museum of Natural History. The Boston collector Arthur P. Chadbourne formally described the new subspecies a year later. As the name he gave it suggests, this race is less rusty and more "brownish-ash," with a paler, less prominently streaked back than the eastern bird. The brown tinge on the throat, ear coverts, and underparts is paler, grayer, and less extensive. Overall, the western Field Sparrow is larger, longer-tailed, and heavier-billed than its eastern relative.

Intergradation between the two in southeastern South Dakota and eastern Nebraska produces birds that are darker and redder than genuine *arenacea* and paler and grayer than genuine *pusilla* (Pyle). Indeed, geographic variation in this species may be entirely clinal in size and in plumage tone, making it impossible to draw clear boundaries between populations, and some authorities consider the species monotypic.

Oddly shaped, with a tail too short, feet too long, and bill too large, the Grasshopper Sparrow is handsomely and complexly plumaged at close range. *Ohio, May. Brian E. Small*

GRASSHOPPER SPARROW
Ammodramus savannarum

In the autumn of 1808, the first volume of *American Ornithology* fresh from the press, Alexander Wilson set off for New York and New England in search of new subscribers to his work. One of the last stops on the tour, in late October, was Staten Island, where, as he admitted to an acquaintance, the "lukewarm" nature of "the slender countenance he received to aid him in his vast undertaking, was somewhat depressing to his feelings."

Those sour impressions obviously kept Wilson from taking the pleasure he might otherwise have found in the discovery of a new sparrow, but he found the bird just as uninspiring as the human inhabitants of Staten Island:

> This small species is now for the first time introduced to the notice of the public. I can, however, say little toward illustrating its history, which, like that of many individuals of the human race, would be but a dull detail of humble obscurity.

Wilson revealed no greater enthusiasm in naming his discovery, which he called simply the Yellow-winged Sparrow, *Fringilla passerina* ("little sparrowy finch").

Wilson's claim of priority was not only half-hearted but wrong. The inconspicuous but hardly dull, humble, or obscure little bird we now know as the Grasshopper Sparrow had been introduced to European science a century and a quarter earlier, when a young Hans Sloane discovered it during his 1687–88 stay in Jamaica. Sloane's description of what he called the Savanna Bird was first published 25 years later in John Ray's *Synopsis methodica*, then, accompanied by a remarkably poor illustration, in the account of his own issued in 1725. Sloane's notes on the bird's behavior were apt:

> They sit on the Ground in the Plains, and run thereon in the manner of Sky Larks, as low as they can, to avoid being discovered, and when rais'd, fly not far nor high, but light again very near.

The first observations of the species on the North American mainland were carried out in the 1790s by Louis Pierre Vieillot, who encountered it on open fields and hillsides in New York, where he also found its nests. He noted, too, that the bird's calls resembled those of a pipit, and pointed out for the first time that the species is migratory, wintering in the southern United States.

A better bibliographer than Wilson, Vieillot only hesitantly claimed his "Meadow Sparrow" as a new bird, remarking that Sloane's Savanna Bird

> exhibits a very great similarity to the male of this species in its shape, color, habits, and pointed tail feathers; it [the continental bird] hardly differs but for the fact that its plumage is brighter and the colors more distinct.

This account, unfortunately, did not appear in print until 1817, though it was presumably prepared a decade earlier for one of the unpublished volumes of Vieillot's *Histoire naturelle des oiseaux*. Nevertheless, Charles Bonaparte took Vieillot sternly to task.

> He is . . . censurable for not quoting Wilson, whose work he had, doubtless, constantly before him.

Bonaparte was himself uncertain about the identity of Wilson's sparrow with that introduced into the scientific literature by Sloane, just one of a long series of uncertainties about the bird. Sloane had in fact not only compared his new species to a lark, but labeled the plate in his *Voyage* "*Alauda pratorum minor rostro breviore*," "a rather small lark of the meadows, with a rather short bill." It was up to Jacob Theodor Klein in 1750 to point out that

> it has a short, thick, pointed bill . . . . Therefore it is not a lark.

Klein classified it instead, for the first time, as a sparrow, calling it *Passer pratorum*, the Meadow Ground Sparrow. Ten years later, Mathurin Brisson retained the bird in the catch-all genus *Passer*, but commemorated the site of Sloane's original discovery in a new species epithet, *jamaicensis*.

As the first scientific name assigned the species after 1758, Brisson's has a cogent claim to validity. The American Ornithologists' Union, however, has used the specific epithet *savannarum* for this species from the start, the name coined by Johann Friedrich Gmelin in his edition of the Linnaean *Systema* and based, obviously, on Sloane's "Savanna Bird"; Thomas Nuttall appears to have been the first scientist to positively assert the identity of the continental bird with the Jamaican species, a view definitively adopted by the AOU just before the publication of its first *Check-list* in 1886.

The short, notched tail and heavy bill give the intricately marked Grasshopper Sparrow a decidedly misshapen appearance. *Ohio, May. Brian E. Small*

FIELD IDENTIFICATION

Small and stubby-tailed, with a flat head and outsized bill, the Grasshopper Sparrow is distinctive and easily identified when seen well, a circumstance that obtains mostly in the spring and summer and mostly for territorial males. Otherwise, this secretive species can be maddeningly hard to see in the dense grass it inhabits, and especially in winter and migration, brief views and high hopes can mislead field observers into mistaking this bird for another, rarer species.

Inexperienced birders, and birders whose exposure to the Grasshopper Sparrow has been limited to the pale, dull birds of the western United States and Canada, are often startled by how bright and buffy adults representing other populations can be. Nelson Sparrows of any race are extensively deeper yellowish or orange on the head and breast, with gray back and nape and a dark crown; the only similar color on a Grasshopper Sparrow's head is restricted to a chrome-yellow or dull rusty patch above the lore. The slender, fine-billed LeConte Sparrow is likewise much more colorful. Both the Nelson and the LeConte Sparrows are typically found in heavily vegetated wet areas, even on migration, while Grasshopper Sparrows prefer drier, upland grasslands at all seasons.

Perhaps the classic "confusion species" is the Henslow Sparrow, which breeds in the same habitats as the Grasshopper Sparrow throughout its range. Both are large-billed, flat-headed, short-tailed sparrows of dense tall grass; both can be remarkably difficult to see well, or at all, except when the males sing from low perches. Henslow Sparrows of any age are rather dark, with noticeably chestnut secondaries. Though some Grasshopper Sparrow subspecies are rather dark above, the impression created by most is of an overall sandy grayness; on perched adults, the gray greater coverts can offer a striking contrast with the brownish primaries and secondaries.

Adult Henslow Sparrows are finely streaked across the buffy upper breast, forming a neat necklace that may recall the underparts pattern of a juvenile Grasshopper Sparrow, which shows scattered, irregular blackish streaks on its dull yellow-washed breast and flanks; the Grasshopper Sparrow lacks the chestnut wing, richly colored white-striped back, dark head, and complex head pattern of the Henslow. At the same time, juvenile Henslow Sparrows, largely without streaking below, have been mistaken for adult Grasshopper Sparrows. Like the adults, Henslow Sparrows of that age are dark birds, with rusty wings and somberly colored backs; their dark head and bill are quite unlike the pale face and striped crown of the Grasshopper Sparrow. At any age, the Grasshopper Sparrow also differs in its slightly longer wing and somewhat squarer tail.

On the Arizona wintering grounds, Grasshopper Sparrows are continually misidentified as the much scarcer, and much more "desirable," Baird Sparrow. The two species both have a complex and colorful back pattern, distinguishing them from the Savannah Sparrow, but they differ markedly in breast and head pattern; the Baird Sparrow's blackish tertials are quite different from the colorful rusty, gray, and black of the Grasshopper Sparrow.

A new and still little recognized source of confusion is the increasing abundance of various escaped or introduced finchlike birds with plain plumage. The Northern Red Bishop (*Euplectes franciscanus*) is established as a resident in southern California, and occasional individuals can be seen anywhere in North America. Males can hardly be mistaken for any passerellid sparrow, but female bishops—short-tailed, streaked above and buffy below, with large bills and striped crowns—have been misidentified as Grasshopper Sparrows, even in the hand. Bish-

ops are even shorter-tailed, with finely streaked uppertail coverts and rump and a coarsely black-streaked medium-brown back; the underparts are buffy with muted streaks, the crown finely streaked blackish. The bill is even more outsized than that of the Grasshopper Sparrow, and both mandibles are pinkish or yellowish, while the Grasshopper Sparrow's bill is dark horn-gray above and dull pinkish gray below. The bishop's sharp chipping notes and jumbled rattling song are unlike the finer calls and hissing song of the sparrow.

That song, to which the Grasshopper Sparrow owes its English name, was not cogently described for centuries after the bird itself had first been collected on Jamaica. Vieillot seems to have known only the high call notes, which he compared, not inaptly, to those of a Meadow Pipit (*Anthus pratensis*). Wilson, already unimpressed, found the bird's vocalizations no more inspiring:

> It has a short, weak, interrupted chirrup, which it occasionally utters from the fences and tops of low bushes.

That description suggests that Wilson never knowingly heard the species (though it is reminiscent of the bird's Jamaican folk name "tichicro"). Thomas Nuttall's account of the bird's singing behavior is less bland and less vague, but can hardly apply to the Grasshopper Sparrow:

> They perch in sheltered trees in pairs, and sing in an agreeable voice somewhat like that of the Purple Finch, though less vigorously.

Not even Audubon was able to offer a convincing description of a song he must have heard hundreds of times, styling it unhelpfully "an unmusical ditty, composed of a few notes weakly enunciated at intervals."

The first recognizable descriptions of this species' familiar song were provided almost simultaneously by two different ornithologists. Spencer Trotter, writing in 1909, credited Elliott Coues with the first comparison to the stridulations of an orthopteran:

> a humble effort, rather weak and wheezy, but quite curious, more resembling the noise made by some grasshoppers than the voice of a bird.

But Robert Ridgway's description of the song as bearing "a close resemblance to the note of a grasshopper" was published in the same year as Coues's. It is probably impossible to determine which of the two hit upon the similarity before the other, though

Coues appears to have been the first to give currency to the English name "Grasshopper Sparrow" for this species.

That name commemorates the male Grasshopper Sparrow's primary song, a thin, high-pitched, but distinctive phrase beginning with two or three abrupt ticking notes and ending with a thin, hissing rapid tremolo; its high pitch and low volume make it hard for many human ears to hear at any distance. Often heard at night on the breeding grounds, the song can be confused with that of a Savannah Sparrow, but the Grasshopper Sparrow's song is briefer, with more sharply separated introductory notes and a single, much dryer and less broadly modulated concluding trill; in pattern, but by no means in volume or tone, the Grasshopper's song resembles some Eastern Towhee vocalizations, while the Savannah Sparrow's can be thought of as sharing the Seaside Sparrow's descending fricatives: the Grasshopper sings a simple, dry, thin *tik tip tzeee*, the Savannah a more complex, more resonant, broader *tip tip tip dzdhzdhzh zhzh*.

Male Grasshopper Sparrows also have a second, less frequently heard song, uttered both perched and in a faintly ridiculous, stiff-winged song flight. This "sustained song," which can be given separately or immediately following a performance of the simple, primary song, is a sustained reeling series of squeaky whirring notes; some notes have the pizzicato quality of an American Woodcock's sky song, while others may recall the musical whines of a Common Yellowthroat or the buzzes of a Clay-colored Sparrow. That eclectic combination and the length of the song—up to 15 seconds—are distinctive.

Both members of a mated pair also sing a short trill; the male's trill is longer and descending, while the female's weaker, shorter song lacks that descending conclusion.

The most frequently heard calls of this species are high, fine, and insectlike, usually faintly stuttering: *d'd'dit*. There are also a bright, clear, evenly pitched *tsee*, and a sotto voce *tip*.

RANGE AND GEOGRAPHIC VARIATION

Widespread but very local—and generally uncommon to rare—over much of the United States and extreme southern Canada east of the Rocky Mountains, the Grasshopper Sparrow is also found as a breeder in disjunct populations in California, eastern Washington and Oregon, the desert Southwest, central Mexico, the Caribbean, Belize, Honduras, Panama, and Colombia and Ecuador. Southern populations are at least partly sedentary, but the northerly subspecies *pratensis* and *perpallidus* move as far south as southern Central America and the Caribbean to winter.

In the Northwest, Grasshopper Sparrows are rare to uncommon breeders in the Okanagan and southern Similkameen Valleys of British Columbia, where loose colonies of up to six pairs occupy sagebrush steppe from 1,000 to 2,500 feet in elevation. The British Columbian distribution is continuous with the species' range in the grasslands of eastern Washington and northeastern Oregon, where the Grasshopper Sparrow is uncommon, local, and erratic in occurrence from one year to the next.

In California, where the species has dramatically declined in recent decades, the Grasshopper Sparrow has a "patchy" breeding distribution, including the Shasta Valley of Siskiyou County, the west slope of the Sierra Nevada, and low hillsides in the Central Valley and foothills. On the northern California coast, the species breeds from Humboldt County south where suitably open grassland habitats occur. On the central coast and in the San Francisco region, the species is widespread and uncommon to fairly common along the coast and in the Diablo, Gabilan, and Temblor Ranges. It has "retreated greatly" in southern California, and now occurs in only small numbers from Santa Barbara County south to San Diego County, with many of the remaining birds confined to military installations. Grasshopper Sparrows also breed in extreme northern Baja California.

The Grasshopper Sparrow may no longer nest in Nevada, where it has traditionally been described as a summer resident in the extreme north; if it does persist, it is rare even there. It is also uncommon to scarce across southern Nevada, northeastern Utah, western Wyoming, and northwestern Colorado and west of the Continental Divide in Montana. A small breeding population of *perpallidus* Grasshopper Sparrows, discovered in the early 1970s in Yavapai County, Arizona, has persisted into this century.

Though declining and widely threatened by habitat loss, the Grasshopper Sparrow remains a locally common breeder on the Great Plains, from southern Alberta, Saskatchewan, and Manitoba south in favorable years to much of Texas; the species is a rare to uncommon breeder in the Trans-Pecos region, and rare or absent in the Piney Woods.

East of the Great Plains, Grasshopper Sparrows breed—often locally and in very small numbers—across the United States to the Atlantic Coast, south to central South Carolina and north to southeastern Ontario, southern Quebec, and southernmost Maine and New Hampshire; in 2006, there were 19 known territories in that last state, with at least nine breeding pairs that produced "very few documented fledglings," a typical assessment for this species at the eastern and northern edges of its range.

The bright, pale southwestern subspecies *ammolegus* breeds only in Pima, Santa Cruz, and Cochise

The extensive silvery bases of the greater coverts produce a gray panel on the folded wing. *Ohio, May. Brian E. Small*

Counties, Arizona, and in adjacent New Mexico and Sonora. The range of the Florida Grasshopper Sparrow, whose population has probably sunk to below 80, as of 2017, comprises only Kissimmee Prairie Preserve in Okeechobee County and Three Lakes Wildlife Management Area in Osceola County.

Only *pratensis* and *perpallidus*, the widespread breeding subspecies of the United States and Canada, are strongly migratory, though the southerly races probably also undergo some short-distance seasonal movement. Autumn departures from the northern breeding grounds appear to take place between late August and early October, with the earliest arrivals on the wintering grounds in September. Fall vagrants occur both north and south of the expected breeding and wintering ranges; approximately one bird a year is detected in Nova Scotia, and *pratensis* can wander as far east as Bermuda and as far south as Panama.

Winter sees Grasshopper Sparrows widespread across the southern tier of the United States from Texas east; in the West, winterers regularly occur in southern New Mexico, Arizona, and California. The southern limits of the winter range are uncertain, thanks largely to the species' secretive habits, but they are known to be uncommon winter visitors as far south as Costa Rica and western Panama. Most Mexican winterers, especially on the Pacific Slope, are representatives of the subspecies *perpallidus*; *ammolegus* has been recorded away from the Sonoran breeding grounds, south to Morelos, in autumn and spring, and *pratensis* is a scarce winter resident of eastern and southern Mexico.

The spring migration is late by sparrow standards. Typical arrivals on the northerly breeding grounds are not until May in British Columbia, western Nebraska, and New Hampshire.

Currently, as many as 12 subspecies of the Grasshopper Sparrow are recognized, ranging—disjunctly in some southern populations—from the Canadian prairies to the Caribbean and northern South America. The number of races considered valid and their nomenclature have been extremely unstable over time; the treatment here relies on Pyle and on the account in *Handbook of Birds of the World*.

The nominate race, discovered by Hans Sloane and ultimately named by Gmelin, is endemic to Jamaica. This is a short-winged, short-tailed Grasshopper Sparrow, with plumage variously described as "slightly darker" than that of continental birds or "paler" on the back and more darkly streaked on the nape. The resident Grasshopper Sparrow of Puerto Rico, *borinquensis*, is smaller on average, with warmer buff on the crown stripe, back, tertials, and uppertail coverts; the ear coverts, breast, flanks, and undertail coverts are "distinctly buffy" cinnamon to clay-brown. Resident on Hispaniola, *intricatus* is larger-billed and darker above than the nominate subspecies, with darker brown ear coverts, breast, and flanks; it differs from *borinquensis* in its blackish upperparts and duller buff. The palest of the Caribbean subspecies is *caribaeus*, resident on Bonaire and Curaçao. It is notably smaller than nominate *savannarum*, with a considerably smaller bill and more brownish lateral crown stripes; on Bonaire, it is known popularly as *raton de cerro*, the "cliff mouse."

The southernmost Grasshopper Sparrows in the world are representatives of the subspecies *caucae*, resident in the upper Cauca Valley of southern Colombia and northwestern Ecuador. Discovered in 1911 on the American Museum of Natural History's great Colombian expedition, *caucae* is dark and richly patterned above, with black and dark chestnut feathers edged with gray; the underparts are strikingly pale, with less buff below than in any other subspecies.

The race *cracens* of Guatemala is dark above and bright below. Panama's *beatriceae,* named for Beatrice Wetmore, the wife of the ornithologist Alexander Wetmore, is the palest of all Grasshopper Sparrows, with light gray upperparts and pinkish buff undertail coverts, flanks, breast, and throat. It is grayer above, with a much paler median crown stripe, than the widespread Middle American *bimaculatus,* whose underparts are rich yellow. The nape streaking of *beatriceae* is distinctly reddish, unlike the blackish brown streaks seen in *caucae*; the Panama race is also larger-billed and less brownish above and less yellowish below than the pale *caribaeus.*

While *beatriceae* is restricted to the Pacific Slope of southern Coclé Province, the provinces of Chiriquí and Panamá have breeding populations of the widespread *bimaculatus,* first described from Temascaltepec, Mexico, by William Swainson. Breeding and probably mostly or entirely resident from central Panama north to the states of Veracruz and México, and possibly to Zacatecas, *bimaculatus* is relatively long-winged and long-tailed, with a small, slender bill and pale rusty upperparts. This subspecies and the widespread western *perpallidus* have at times been considered identical, a circumstance van Rossem found "difficult to understand," as *perpallidus* is decidedly larger and paler than the Middle American bird named by Swainson.

In comparison with the eastern subspecies *pratensis, perpallidus* is longer-winged and thinner-billed, with paler upperparts dominated by tawny and dull gray rather than bright gray and blackish. The breast band is dull buffy in *perpallidus,* brighter and tending to orange in *pratensis.* Where they meet on the eastern Great Plains, these two subspecies intergrade.

The thickest-billed of the United States races is the very rare and endangered Florida Grasshopper Sparrow, discovered and described in late 1901 by Edgar A. Mearns. Overall, *floridanus* is smaller, longer-legged, and much darker above and paler buffy or whiter below than *pratensis,* with

> chestnut upper surfaces much reduced in amount and replaced by black; lateral dark areas of crown almost solid black; spotting of nape and scapulars almost black; interscapular region [back] much blacker than in

pratensis, which winters more or less commonly in the extremely limited south-central Florida range of *floridanus.*

The most colorful of the Grasshopper Sparrows is the very locally distributed *ammolegus,* restricted as a breeder to the high grasslands of northeasternmost Sonora, northwesternmost Chihuahua, and adjacent Arizona and New Mexico. "Easily distinguished," *ammolegus* differs from *perpallidus* in its slightly larger size and long bill; the pale upperparts are dominated by chestnut or rufous, and there is "much less, sometimes almost no, black on the back." The underparts are also pale, the light buffy breast with faint, sparse streaks on the side. The two subspecies overlap in the winter, and can often be at least tentatively distinguished given good views in the field.

Western birds are duller above, with less chestnut in the feathers of the back. *Kansas, April. Brian E. Small*

This drab and secretive sparrow is unlike any other bird found north of Mexico. *Texas, April. Brian E. Small*

OLIVE SPARROW
Arremonops rufivirgatus

Destruction is usually easy to see. Change, on the other hand, can be harder to detect and its effects harder to assess, especially when the result is what seems to human eyes, at least, the replacement of one more or less "natural" habitat by another.

The "Brown Striped Olive Finch" was first described in 1851 by George Newbold Lawrence, probably on the basis of a specimen collected at Fort Ringgold, on the lower Rio Grande near Brownsville, Texas. The species' known range was promptly extended south of the border to at least Nuevo León, but it would be almost half a century before Olive Sparrows were found north of the Rio Grande flood plain in Texas.

Frank Chapman of the American Museum of Natural History spent the first month of spring 1891 camped some 150 miles north of Brownsville, in search of specimens for the museum's collection. He found

> birds abundant both in species and individuals, and of the 190 odd species observed, at least eighty-five were common.

Olive Sparrows were "not uncommon" in the chaparral, though "much more frequently heard than seen." Chapman notes that his were the first specimens to be taken this far north, near Corpus Christi, an area that, as he is at pains to point out, was even then "by no means ornithologically new ground."

Less than a month after Chapman's departure for New York, Samuel Nicholson Rhoads arrived in Corpus Christi, where he spent three weeks collecting for the Academy of Natural Sciences of Philadelphia. Rhoads, too, found the Olive Sparrow "thoroughly at home in the Corpus Christi and San Patricio chaparral, and secured their nests and fully fledged young," a notable change from just a dozen years earlier, when "it was not a member of the fauna of this region." Rhoads explains the change over the course of the 1880s in the species' status:

> . . . the growth of dense mesquite chaparral, which now forms such a marked feature in the landscape of southeastern Texas, and is steadily encroaching upon the tithe of open prairie yet remaining, is directly due to the grazing of immense herds of cattle which have . . . robbed the prairie fire of its fuel and at the same time distributed the seeds of mesquite,

creating the dense thickets required for the north-ward spread of the Olive Sparrow even as such grassland obligates as the Burrowing Owl disappeared in the region with "the usurpation of their prairie domain by the now ubiquitous chaparral," an early and striking lesson in what would only much later be known as ecology.

FIELD IDENTIFICATION

In the area covered by this guide, the Olive Sparrow can hardly be confused with any other passerellid. Green-tailed Towhees wintering in this species' permanent range are similarly olive-green above, but the towhee is markedly larger and thicker-billed, with darker, grayer underparts and a far more strikingly patterned head; the tarsus and toes of the Olive Sparrow are pinkish, while those of the towhee, especially in winter, are dark gray to blackish, a difference sometimes surprisingly useful in the dark thickets both species frequent.

The Green-backed Sparrow, *Arremonops chloronotus*, overlaps with the Olive Sparrow on the Yucatán Peninsula; the best distinction is the former species' decidedly greenish flanks and bright yellow undertail coverts.

The Olive Sparrow is a rather heavy, dumpy sparrow with very short wings and a long, broad, round-tipped tail. The eponymous greenish upperparts usually appear simply brown in the dense, dimly lit vegetation this bird dwells in, and the lateral crown stripes are often less prominent than illustrations might suggest, particularly in the eastern parts of the species' range. That very lack of contrast can be one of the most important field characters for identification, visible long before the head pattern can be discerned.

Juvenile Olive Sparrows are not often seen unaccompanied by their parents. They are browner overall than adults, with sparse, blurry dark brown spotting and streaking on the upperparts and crown; the crown lacks the adult's well-defined lateral stripes. The greater and median coverts of the wing are tipped yellowish brown, creating two vague wing bars. The underparts are buffier than those of the adult, with variably extensive dull chestnut streaking on the flanks and breast. Birds in this plumage could conceivably be misidentified as juvenile cowbirds or *Passerina* buntings, but are smaller than cowbirds, with slighter bills and narrower brown tails, and larger than *Passerina* buntings, with heavier bills and longer tails.

The calls of the Olive Sparrow include dry, nervous *tit-tt-tit* notes and a long, rather towheelike *bzee*. The monotonous song, soon familiar to even the most casual observer in areas where this species is common, is a series of faintly downslurred *tdip* notes, gradually accelerating into a slow, loose tremolo; some songs are vaguely cardinal-like in their sweet tone, while others are dryer and more staccato in articulation.

RANGE AND GEOGRAPHIC VARIATION

Much of the historical difficulty in determining the relationships among and within populations of *Arremonops* sparrows lies in the fact that many occupy decidedly disjunct ranges, making it a matter of speculation how they might behave if their breeding ranges overlapped.

The three subspecies groups of the Olive Sparrow are no exception. The Yucatán birds, of the subspecies *verticalis* and *rhyptothorax*, are restricted to the northern and eastern portions of that peninsula, south to Belize and probably northern Guatemala; they are separated by some 300 miles from the nearest birds of the Atlantic group, in Veracruz, and by nearly that much from the isolated population in central Chiapas.

Nominate *rufivirgatus* is the most northerly of the Olive Sparrows, its resident range extending from southern Tamaulipas north into Texas, where this species is common in the extreme south and reaches its northern limits from the southern edge of the Edwards Plateau east to the western edge of Calhoun County. They have been recorded as far west along the Rio Grande as northern Coahuila and Terrell County, Texas (eBird). The range of the Olive Sparrow in western Mexico is less precisely known, but there are records of *sinaloae* north to the El Salto Reservoir in southern Sinaloa (eBird).

The squat shape and dull greenish upperparts are useful characters when the bird is seen well—as is this species' reluctance to be seen well. *Texas, April. Brian E. Small*

The gray head is marked by four rusty brownish stripes, which can be difficult to see well in the field. *Texas, February. Brian E. Small*

Apparently resident throughout its range, the Olive Sparrow may nevertheless be subject to short-distance movements inspired by local conditions.

Of the eight generally recognized subspecies of the Olive Sparrow, the nominate race *rufivirgatus* ranges from northern Veracruz north into southern Texas. This is a relatively long-tailed race, with buffy grayish flanks, bright brown lateral crown stripes (whence the scientific name), and an olive median crown stripe. Resident from central Veracruz south to Oaxaca, *crassirostris* has a slightly shorter wing and tail and much heavier bill; its back is dark olive, the flanks deep buffy olive, and the undertail coverts dark orangish buff. This southerly subspecies has occasionally been accorded full specific rank, though it intergrades with its northern relative in Veracruz.

The four proportionally shorter-tailed subspecies that occupy western Mexico have also at times been considered distinct from the two northeastern races. Described from specimens collected near Mazatlan, *sinaloae* is dominated by gray tones in the plumage of its crown, ear coverts, flanks, and upperparts. This subspecies is resident south through Nayarit; resident to its south, from Jalisco to western Oaxaca, the slender-billed race *sumichrasti* has darker chestnut lateral crown stripes, a brighter back, and buffy to grayish throat and ear coverts. The Chiapas River Valley of southern Mexico is occupied by *chiapensis*, which has darker green upperparts, deeper buff underparts, and a darker gray median crown stripe than *sumichrasti*.

The first race of this Pacific group to be described was *superciliosus*, making that the name to be applied if they are treated as a distinct species. The race *superciliosus* (whether considered one of the races of the Olive Sparrow or the nominate subspecies of the "Pacific Sparrow") is medium olive above, with buffy gray flanks, grayish buff ear coverts and median crown stripe, and dark chestnut lateral crown stripes; it is a fairly common resident from Nicaragua to central Costa Rica.

The Yucatán Peninsula is home to two Olive Sparrow subspecies; *verticalis* is brownish olive above, with olive-gray flanks and median crown stripe and black-streaked chestnut lateral crown stripes, while *rhyptothorax* is paler, with whitish belly and gray-brown flanks.

Startlingly colorful and well marked, this tropical sparrow can hardly be mistaken for any other. *David Krueper*

BRIDLED SPARROW
Peucaea mystacalis

Nearly a century after the Harris Sparrow was first named, one of the most famously long searches in North American ornithology ended in 1931, with "man's first recorded glimpse of the eggs of the Harris Sparrow in the wilderness summer home of the bird." The joyous explorers' pride in their discovery—even if expressed in such mock-heroic tones—was justified. The Harris Sparrow was not the last holdout among the North American passerellids, however; that honor probably goes to the Bridled Sparrow, whose eggs were not seen by ornithologists until 2001, almost exactly 150 years from the time that Gustav Hartlaub described the species from specimens in Bremen and Hamburg.

J. Stuart Rowley encountered a pair of Bridled

Sparrows with a large chick in July 1963, and in 1967 found a nest with three young. The nest with eggs continued to elude science, however, and the search was set back by Rowley's early death in an accident in the mountains of Oaxaca in 1968. Finally, 33 years later, in August 2001, a nest and two eggs were discovered, and eventually collected—some 300 yards from a popular botanical garden in Puebla. The nest was surrounded by three cactus pads and concealed by an overhanging vine, a placement as unlike the stunted conifers of the tundra edge as possible, but obviously even more effective in avoiding detection than the long-sought nest of the Harris Sparrow.

FIELD IDENTIFICATION

A handsome and distinctively patterned bird, the Bridled Sparrow—named in English and in scientific nomenclature for its white jaw stripe—may occur together with its close relative the Black-chested Sparrow, from which adults are easily distinguished by their dark throat and pale breast, the opposite of the pattern in the adult Black-chested.

Juveniles of the two species are more challenging. Bridled Sparrows in this plumage appear to be more coarsely streaked and mottled above and on the breast and head than juvenile Black-chested Sparrows. Like the adult's, the tone of the juvenile Black-chested Sparrow's upperparts is overall rusty; the rusty tinge is more clearly restricted to the scapulars in the Bridled Sparrow, while the remainder of the back is brownish gray. The juvenile Bridled Sparrow's white or pale yellowish throat is separated from the whitish jaw stripe by a relatively heavy dark lateral throat stripe, apparently often lacking in the Black-chested.

The juvenile plumage in both species is held only briefly; late-hatched birds may begin their preformative molt even before the juvenile feathers of wing and tail are completely grown.

The primary song of the male Bridled Sparrow begins with a short, sharp note, followed by a squeaky *pitchew* and a liquid tremolo. There is also a "chatter duet" given by both members of a pair when they are reunited after a separation; this twittering vocalization comprises ten or more scratchy notes at slightly different pitches and in an irregular rhythm. Other calls include a loud, rich *dzip* and a short, sharp alarm note.

RANGE AND GEOGRAPHIC VARIATION

The Bridled Sparrow is a resident of dry slopes from Puebla to northwestern Veracruz and in the mountains of northwestern Oaxaca. It is local in extreme eastern México State and eastern Morelos. The northernmost recent records are from just west of Xalapa, the southernmost from near Puerto Escondido and Huatulco (eBird). This species occurs in treed habitats, including arid hillsides, open thorn forest with large cactus, and successional patches inside stands of low oaks.

No subspecies are recognized.

The Black-chested is one of the most dramatically patterned of all New World sparrows. *Amy McAndrews*

The Black-chested is one of the most dramatically patterned of all New World sparrows. *Amy McAndrews*

BLACK-CHESTED SPARROW
Peucaea humeralis

Photography has become one of the most important tools in the birder's field kit, and the inexpensive digital camera seems poised to replace the time-tested binocular for many younger observers. But the role of the photograph in bird identification is not entirely new, and even almost 150 years ago, when photography was an arcane, expensive, and time-consuming art, there were signs that a photographic image could replace the traditional study skin in some matters.

In 1877, the Mexican congress created the Geographical and Exploring Commission of the Republic of Mexico. Charged at first with the publication of a new map of the country, the Commission soon found its duties expanded to include "bringing together an extensive collection of specimens of the different branches of natural history," an ambitious enterprise placed under the direction of Fernando Ferrari-Pérez.

The Commission's budget did not increase in full proportion to its workload, to Ferrari-Pérez's great regret and his collectors' great disadvantage:

> The considerable expense necessary for the acquisition of the very costly instruments to be used by the astronomical and topographical division of the Commission has not yet permitted in the natural history division the formation of a library sufficiently complete to insure the success of its work.

The solution, apparently proposed by the Mexican president, Porfirio Díaz, was to submit some of the specimens north, where they could be "exactly identified by comparison with specimens in the rich collections of the United States." The risk of that plan came clear in August 1884, when a steamer on its way to Washington burned in Havana, destroying 123 boxes of skins and "all of the manuscripts pertaining to the collections."

Some of the specimens could be replaced, and Ferrari-Pérez accompanied this new lot to the United States National Museum, where he consulted with Robert Ridgway and Leonhard Stejneger in identifying the birds—among which the Smithsonian scientists found no fewer than five new species, including a strikingly "elegant" sparrow from the state of Puebla, which Ridgway formally described in 1886 and named *Amphispiza ferarriperezi*.

When Ridgway's description reached London, Frederick DuCane Godman and Osbert Salvin wondered whether Ferrari-Pérez's sparrow might not already be known. Traditionally, such a suspicion would have been confirmed by comparing specimens, or at least drawings; in this case, however, perhaps with an eye to the unfortunate incident in Havana's harbor, the skins returned to Mexico City while it was a photograph of the male and female cotypes that made the perilous voyage across the Atlantic. As the British ornithologists suspected, they had "no difficulty in recognizing them as representing" a species actually described 35 years earlier by Jean Cabanis, a conclusion made possible by a technology that would come to affect birding in ways then still inconceivable.

FIELD IDENTIFICATION

The adult Black-chested Sparrow is "unmistakable and striking," with a long, dark, noticeably graduated tail and bright reddish back, scapulars, and lesser wing coverts. The broad black breast band and white throat, jaw stripe, and spot at the base of the bill make it difficult to confuse this bird with any other in its range.

Juveniles are uniformly brown above with sparse and faint darker streaking. The flanks are brownish or buffy, the remainder of the underparts dull whitish, with variable dusky streaks. The greater coverts are tipped and bordered with brown-buff rather than, as in the adult, with white.

Juvenile Black-chested Sparrows undergo a preformative molt, with replacement of the body feathers complete about two months after hatching. The formative plumage that results is worn until the following spring; it is duller than the adult plumage, with the head and breast dull gray where the adult is blackish and the crown with brown streaking. It is possible, however, that some formative-plumaged birds are not distinguishable in the field from adults.

The calls of this species include a decisive ascending *tzeep* and a soft, prolonged liquid chatter. There is also a stuttering, rising *pititit* or *pitza*, given by perched birds and in flight.

Wolf distinguished two song types in the Black-chested Sparrow, one probably given only by the male and the other a "chatter duet." The first song type is given beginning at dawn, but gradually replaced after sunrise by chatter duets; it comprises "an introductory figure or phrase followed by a trill of varying length." The introduction varies from a simple downslurred bisyllabic squeak to a series of two or three cardinal-like notes, while the trill tends to be slow and loose, with the bright, cheerful tone of an Eastern Towhee's song.

In the chatter duet, sung by mated pairs, the male and the female each produce different syllables or phrases; Wolf suspected that the male's contributions were more variable than the female's.

RANGE AND GEOGRAPHIC VARIATION

The Black-chested Sparrow is a true Mexican endemic, occurring in thorn forest between 700 and 5,000 feet in Jalisco, Colima, Michoacán, Guerrero, Morelos, and southern Puebla. It is also common in dry scrub in northern and extreme southwestern Oaxaca. The species has also been reported from the state of México (eBird). It is presumed to be resident throughout this range.

Ludlow Griscom described a new subspecies of the Black-chested Sparrow, *asticta*, on the basis of a specimen from Colima, collected in 1889 by W. B. Richardson. Griscom found that this bird from the very northern edge of the species range differed from the nominate population in its lack of black streaking on the upperparts and from the lack of contrast between the rump and back. This species is now considered monotypic, and Griscom's race "but doubtfully valid."

Bulky and stout-legged, this handsome sparrow can hardly be confused with any other in its range. *Manuel Grosselet*

STRIPE-HEADED SPARROW
Peucaea ruficauda

At the age of 26, Adrien Louis Jean François Sumichrast had wearied of studying the fauna of his native Europe, and in the autumn of 1854 set off with the young naturalist Henri de Saussure for a year's exploration in the Caribbean and Mexico. After just four months in Mexico, Saussure wrote that he had "had [his] fill, ten times [his fill] . . . Mexico is a horrible country." He and two others of the party took the first opportunity to return to Europe, but

> for all the "horrors" of Mexico, Sumichrast was so taken by the country that he chose to stay, giving the rest of his life to the scientific exploration of Mexico. He discovered vast numbers of species of mammals, birds, insects, and his favourite creatures, reptiles.

Sumichrast supported himself for the next quarter century as a commercial collector, supplying specimens to museums throughout Europe and North America. Around 1870, the Smithsonian Institution commissioned Sumichrast to undertake "an extended exploration of the Pacific side of the Isthmus of Tehuantepec, Southwestern Mexico," and between 1872 and 1876, more than 1,700 bird skins, representing more than 300 species, arrived in Washington. George Lawrence, asked by Secretary Joseph Henry to work up the collections, registered his pleasure with Sumichrast's work:

> The specimens sent (which are of a remarkably fine character) bear testimony to Professor Sumichrast's efficiency as an industrious and energetic collector, and the many valuable notes manifest his accuracy and intelligence as an observer.

But Lawrence was at the same time disappointed that so many of the skins had come to him unaccompanied by "biographies," accounts of the species' life history. Sumichrast responded to Lawrence's query with one of the most cogent excuses in the history of ornithology:

> I regret to be unable to tell you certainly which are the biographies and notes that I forwarded to the [Smithsonian] Institu-

tion. Almost all of my books and papers were carried off in 1871 during the pillage of my house in Juchitan,

in the course of Porfirio Díaz's revolt against the regime of Benito Juárez.

One of the specimens Sumichrast was able to send from Juchitán, in southeastern Oaxaca, was a bird that Lawrence identified as a Stripe-headed Sparrow, a species originally described in 1853 by Charles Bonaparte. In 1886, Osbert Salvin and Frederick DuCane Godman reexamined a specimen in their collection, obtained from Sumichrast by way of the dealer Adolphe Boucard and also labeled as a Stripe-headed Sparrow. The English ornithologists discovered that the head pattern did not after all fit that species, and described the Sumichrast specimens as a new species, *Haemophila lawrencii*, the Tehuantepec Ground Sparrow.

Sumichrast did not live to see this further species credited to his efforts. He died, a victim not of revolution but of cholera, in Chiapas in the autumn of 1882, four years before Salvin and Godman described and named *lawrencii*—and almost 20 years before Robert Ridgway would revise the bird's taxonomic status to that of a subspecies of the Stripe-headed Sparrow.

FIELD IDENTIFICATION

This is a large, handsome, chubby sparrow with a long, broad tail and striking and distinctive head pattern. Even members of the relatively small northern races seem heavy, even bulky in the field, with coarsely streaked buffy upperparts and bright white breasts. Even juveniles, which are obscurely mottled on the breast and have paler, slightly less well-defined head stripes, are virtually unmistakable.

Stripe-headed Sparrows are fairly noisy, frequently giving soft *zsee* calls and, when alarmed, loud nasal *tchumk* notes, the latter sometimes in a long, slightly accelerating series.

Males sing two different songs. One, the *pechew* song, comprises a brief, abrupt one- or two-syllable introduction followed by a cheerfully frantic phrase of five or six squeaky doubled notes, *chew—che-pechew-pechew-pechew-pechew*. They also sing a chattering song, sometimes joined by the female, whose contribution to the duet may be less varied than the male's notes.

RANGE AND GEOGRAPHIC VARIATION

Two of the four subspecies are found in disjunct distributions in southwestern Mexico. Occupying the largest range of any, the northern *acuminata* is fairly common in brushy scrub habitats

The Stripe-headed Sparrow has a very long, broad, towhee-like tail. *Manuel Grosselet*

Even quite young juveniles have the distinctive broad head stripes. *Manuel Grosselet*

from sea level to 5,000 feet on the Pacific Slope from southern Durango and Nayarit south to southeastern Guerrero and east to Morelos and southern Puebla. In southern Oaxaca and extreme southwestern Chiapas, *lawrencii* is recorded from the Pacific Coast north to Santa Efigenia and west to Tehuantepec City, the type locality. The species is also uncommon in western Oaxaca from the border with Guerrero to the area northwest of Puerto Escondido, where the subspecies involved is thought to be *acuminata*.

The two southern races, *connectens* and the nominate *ruficauda*, occupy the Pacific Slope of Central America from southeastern Guatemala to northwestern Costa Rica. A 2002 record of five individuals of this sedentary species observed in extreme northwestern Guatemala is remarkable,

falling geo-graphically almost exactly halfway between the range limits of *lawrenceii* and *connectens*.

Overall size increases in this species from the north to the south, with the subspecies *lawrencii* largest and *acuminata* smallest, the heaviest representatives of the former weighing nearly twice as much as the lightest of the latter. The northern *acuminata*, often known as the Colima Sparrow, is whiter below and less rusty-tailed than the nominate race; the head markings are blacker, and the ear coverts lack the pale spot at the rear seen in other subspecies. Ludlow Griscom described the Guatemala race *connectens* as both "very distinct" and an "intermediate form"; it is grayer above than the nominate race, but with a brighter tail than *lawrencii*.

Within this species' narrow range, the reddish-streaked upperparts, yellow wing flash, and buffy flank and breast band are distinctive in the good views Bachman Sparrows only grudgingly give. *Florida, February.* *Brian E. Small*

BACHMAN SPARROW
Peucaea aestivalis

The stuttering history of this species' repeated near-discovery in the eighteenth and early nineteenth centuries finally came to an end in 1823, when Hinrich Carl Lichtenstein published a catalog of duplicate specimens to be put up for sale by the Zoological Museum in Berlin. Among them was a new finch, which Lichtenstein described rather unhelpfully as "very similar to the preceding species" in his list, the South American Grassland Sparrow, but with underparts entirely ashy gray and the tail graduated. Lichtenstein gave the duplicate specimen, which had come from the state of Georgia, the name *Fringilla aestivalis*.

That name was not original with Lichtenstein, but had already been assigned the North American sparrow by Karl Illiger, first director of the Berlin museum, who had received the bird in June 1812 from the German collector Friedrich Wilhelm Sieber. Sieber did not take the specimen himself, but

had acquired it from a London dealer or collector in exchange for skins he had collected in Brazil. The specimens Sieber brought from London to Berlin had almost certainly been collected in Georgia or South Carolina by John Abbot, who supplied specimens and drawings to British collectors and who at least twice painted a bird he called the "Summer Sparrow," plainly the origin of Illiger's name *aestivalis*. One of the three surviving specimens that passed from Abbot to an unknown London dealer to Sieber to Illiger to Lichtenstein is the type of the Bachman Sparrow, collected in Abbot's Georgia and labeled in Lichtenstein's hand *Fringilla aestivalis*.

In the same year in which Lichtenstein published his *aestivalis*, another specimen of the sparrow—also collected by John Abbot in Georgia and also delivered to a British collector or dealer—was described by another ornithologist, John Latham. While Lichtenstein provided merely a brief Latin description and name, Latham could offer not only an exhaustive plumage diagnosis but an authoritative comment on the seasonal distribution, habitat, nest, and eggs of the species "called the Summer Sparrow." As a curiosity, Latham adds that birds of this species are sometimes seen quite white . . . . Mr. Abbot informed me of such an [sic] one being taken from a nest, in which all the others were of the common colour.

Unfortunately, Latham failed to assign the sparrow a scientific name.

For more than a decade, John Abbot had single-handedly furnished European ornithology all it knew about this species. But Abbot was also well connected with his American colleagues and contemporaries, among them Alexander Wilson, whom he supplied with specimens, drawings, and notes after their first meeting in Savannah in 1809. Writing from Philadelphia in January 1812, Wilson thanked Abbot for a package of drawings and birds meant for himself, William Bartram, and Benjamin Smith Barton; he also urged Abbot to send on several apparently long-awaited specimens, among them a male Chuck-will's-widow, a Chuck-will's-widow egg, a "striped wren," and, tantalizingly, "the beautiful rare Sparrow you mention." It can never be known for sure, but it is possible that Abbot had promised Wilson yet another specimen or painting of the "Summer Sparrow."

Had Abbot kept that promise, Wilson would have had the opportunity to name the bird a decade before Lichtenstein, and ornithology might today know not the Bachman but the Abbot Sparrow.

A decided latecomer in the long series of introductions of this species to science, John James Audubon can nonetheless claim credit for both its current genus name and both of the English names that have been applied to the bird since Abbot's "summer sparrow" was abandoned. John Bachman, father of the Audubon sons' wives and Audubon's collaborator in the *Viviparous Quadrupeds*, in April 1832 encountered a sparrow in South Carolina that he did not recognize; shortly thereafter, he "obtained many specimens of it" in the open pine forests just north of Charleston. On examining the skins, Audubon named the bird as a new species, the Bachman's Finch, *Fringilla bachmanii*, "to testify the high regard in which I hold that learned and most estimable individual, to whose friendship I owe more than I can express on this occasion."

Just five years later, Audubon determined that Bachman's sparrow was not just a new species but the representative of an entirely new genus, which he named *Peucaea*, from a Greek word meaning "pine," reflecting the habitat to which Bachman had assigned it. He also changed the species' English name to the Bachman's Pinewood-Finch, unwittingly providing a neat solution to the problem of what to call the birds when, almost 50 years later, William Brewster and Robert Ridgway determined that Audubon's bird from Bachman and

The large silvery bill and graduated tail are often visible at greater distances than the intricate patterns of upperparts and wing. *Florida, February. Brian E. Small*

Lichtenstein's bird ultimately from Abbot represented two races of a single species. Thus, the two entered the AOU *Check-list* in 1886 as the Pinewoods Sparrow and the Bachman's Sparrow, *Peucaea aestivalis aestivalis* and *Peucaea aestivalis bachmanii*, respectively.

FIELD IDENTIFICATION

Apart from its furtive behavior over most of the year, this species presents few identification challenges in its usual range in the American Southeast. Bulky, long-tailed, and flat-headed, the Bachman Sparrow, especially the northern and western subspecies *bachmani* and *illinoensis*, at first glance might seem to share the attractive rusty and pinkish tones of the Field Sparrow, another sweet-voiced species of brushy areas. The Field Sparrow is more slender and slightly longer, with a long, narrow, blackish tail and dusky wingtip; its back is more densely striped black than the Bachman Sparrow's more open pattern of gray and rust, and the Field Sparrow's neat white wing bars are unlike the more uniformly patterned wing of the Bachman. The Bachman Sparrow shows an odd tawny buff breast and flanks, while the pale buffy breast of many Field Sparrows is more pinkish cinnamon. Overall, the dull-billed Bachman Sparrow strikes the observer as a somber reddish bird, the Field Sparrow as a crisply marked rusty or pink bird; that difference holds for the streak-breasted juveniles of the two species as well.

The more serious concern for most birders will be the prospect of distinguishing an apparent vagrant Bachman Sparrow from Cassin or Botteri Sparrows. All three are usually very secretive away from the breeding grounds, making it difficult to obtain the detailed views necessary to identify them with confidence.

The ranges of the Bachman and the Botteri Sparrows do not overlap, but it is conceivable that in Texas a Bachman might stray south or a Botteri north into the range of the other species. Visually, the two are extremely similar, so much so that they have at times been considered conspecific. Botteri Sparrows of the subspecies *texana* are grayer above, with duller, slatier back streaks and a less prominent crown pattern than in the Bachman Sparrow. The uppertail coverts of the Botteri are grayish brown with wide black shaft streaks, while the Bachman has brighter brown coverts with much finer, less conspicuous black markings. Where the secondaries and greater coverts of the Bachman Sparrow's wings are quite uniformly reddish, the Botteri Sparrow's secondaries are dark brown and its greater coverts broadly gray-edged, sometimes creating a vague paler panel on the folded wing. These are very subtle differences, and the two might not be reliably distinguishable by sight in the field; should a Bachman Sparrow ever appear in the range of the reddish *arizonae* race of the Botteri, or vice versa, the difficulties in identifying a silent stray would likely be insurmountable, requiring the bird to be examined closely in the hand.

Both the Bachman Sparrow and the Southwest's Cassin Sparrow have been recorded in areas of North America where neither species is expected, from the Great Lakes to the mid-Atlantic coast. The challenge is clearly illustrated by a May record of a *Peucaea* sparrow from Seal Island, Nova Scotia, identified in the field as a Bachman Sparrow and in photographs, correctly, as a Cassin. These two species, too, have been lumped by ornithologists in the past, but in the good views they only infrequently give, vagrants should be identifiable. Cassin Sparrows are decidedly gray, with buff restricted to the slightly streaked flanks. The long gray tail shows

Pudgy and short-winged, this species is a declining specialty of southeastern pine forests. *Florida, February. Brian E. Small*

The flanks sometimes show short dark streaks. *Florida, April. Brian E. Small*

dull whitish, rather than pale brownish, tips to the outer two feathers and more pronounced fine black barring on the central tail feathers. The oddly patterned uppertail coverts are brownish gray with black tips and white edges, creating a scalloped look that continues onto the back. The wings are gray, the greater coverts with black tips and neat white edges, and the crown is marked with fine blackish streaks. Cassin Sparrows are smaller-billed than the Bachman (and the Botteri) Sparrow, making them seem subtly smaller-headed and higher-crowned than the Bachman or the Botteri Sparrow.

Juvenile Bachman Sparrows share the richly colored wings and broadly streaked upperparts of the adult. Their more extensively buffy underparts are marked with short, narrow dark streaks on the flanks and across the breast, usually extending onto the throat; the crown is also marked with fine black streaks. Juveniles molt the body plumage on the breeding grounds in summer, then undergo a complete molt, including the feathers of wings and tail, in autumn (after the southbound migration in migratory populations) (Pyle).

Male Bachman Sparrows have a repertoire of three songs: the primary song, a whisper song, and an "excited" song. The variability of the primary song in this species is both geographic and individual. One Florida bird was noted as singing 39 different primary songs, while almost 250 distinct song types were recorded in Florida and Ohio. The most frequently heard is a simple, pleasing phrase comprising a sweet whistled note and a light, loose tremolo; John Bachman described it as deceptively similar to

the song of an Eastern Towhee. The whistled note is sometimes slurred up or down, and the tremolo can be higher or lower in pitch than the introduction; the effect is often rather Field Sparrow–like, though the tremolo is usually noticeably slower, with more space between the notes. Some songs reverse the sequence of these two elements, while others add a second tremolo or insert one or more additional slurred whistles at the beginning or end.

The whisper song resembles the primary song in pattern and variability, but is "barely audible to observers at any distance from [the singing] birds." Once thought to be largely restricted to the earliest stages of nesting, it can apparently in fact be given at any time during the nesting cycle, and is sometimes interspersed in bouts of primary singing.

The "excited" song, given by a perched bird or, occasionally, in flight, is a jumbled set of slurred whistles and tremolos, a continuous whirling series of primary songs uttered fast and without a pause, usually ending with a more typically enunciated primary song. This song can recall the warbled song of a Grasshopper Sparrow or even the song of a Botteri Sparrow.

The adult's calls include a sharp *tink* and a high, thin *sit* with rapid attack and decay. There is also "a series of chip calls run together with sufficiently short intervals between them that the result sounds like a trill of the Chipping Sparrow," though somewhat slower and less evenly pitched. That "chittering" trill functions both as a reunion duet between the members of a mated pair and as an aggressive call.

RANGE AND GEOGRAPHIC VARIATION

Once found with variable frequency north to approximately 40 degrees north, from Illinois to Pennsylvania, the Bachman Sparrow is now of regular occurrence only from southeastern Oklahoma to the North Carolina coast, south to east Texas and the Florida peninsula (eBird). Its greatest abundance as a breeder is in the open pine forests edging the Gulf of Mexico from Mississippi to Florida.

The Ozarks of southeastern Missouri are the northwestern limit of the Bachman Sparrow's current range, but even there it is rare as a nesting bird. Oklahoma records in this century range as far north as Osage County (eBird), but are concentrated in the extreme southeast corner of the state in Atoka, Pushmataha, LeFlore, and McCurtain Counties (eBird); this distribution is similar to that found over the course of a study conducted in the late 1990s. In Indiana, the species was known as a breeder from the 1880s until the early 1980s; Bachman Sparrows underwent a significant decrease in the state in the late 1950s, and by 1993 the remaining suitable habitat was unoccupied. The species' decline set in even earlier in Illinois; once a breeder perhaps statewide, the Bachman Sparrow has not been recorded there since the 1970s. In Ohio, where the species once nested north to the central part of the state, no Bachman Sparrows were detected during surveys conducted for the most recent breeding bird atlas. Pennsylvania's last breeding record is from 1937.

The Bachman Sparrow remains a locally common breeder in Arkansas, with records this century clearly concentrated in the Ozark and Ouachita Mountains (eBird). The species' only remaining breeding site in Kentucky is Fort Campbell, in Christian County (eBird); Bachman Sparrows nest on both the Tennessee and the Kentucky portions of that facility. Elsewhere in Tennessee, nesting apparently continues in Lewis, Hickman, Wayne, and Lawrence Counties in the west-central part of the state (eBird). There appear to be no breeding records from West Virginia this century, and eBird provides only a single Virginia sighting, in July 2002, that might represent twenty-first-century nesting (eBird).

In Texas, the Bachman Sparrow is rare to locally uncommon in the Piney Woods region of east Texas, west to Leon and Henderson Counties and south to Harris County. Louisiana's Bachman Sparrows are concentrated in the mature pine forests of the eastern Florida and west-central parishes, often where an open grassy understory is artificially maintained for the benefit of Red-cockaded Woodpeckers. The species is uncommon and local over much of Mississippi, but apparently absent

The juvenile is brighter and more extensively buffy below than the adult, with fine streaks forming an irregular breast band. *Georgia, August. Lois Stacey*

as a breeder west of that state's Loess Bluff, in the Delta region, and in areas adjacent to the Mississippi River. The Bachman Sparrow is an uncommon breeder on Alabama's inland coastal plain along the Gulf Coast, rare in the mountains and Tennessee Valley.

In Florida, where many observers see and hear their first Bachman Sparrows, the species is a locally uncommon to fairly common breeder in the Panhandle and south through the peninsula to Charlotte, Glades, and Palm Beach Counties; most breeders are of the nominate race, with *bachmani* occupying the Panhandle east to Calhoun and Franklin Counties. Bachman Sparrows breed in mature open pine forests across much of Georgia (eBird), most commonly on the coastal plain and more locally in the southern and northern Piedmont and mountains. In South Carolina, Bachman Sparrows breed widely across the state except on the coastal plain and in the mountains of the northwest. The species now finds its northeastern limit as a breeder in North Carolina, where it is now a fairly common breeder on the southern coastal plain locally from Croatan National Forest southward, almost exclusively in appropriately managed longleaf pine woodland; it is also fairly common at a few sites in the Carolina sandhills, especially Fort Bragg and the Sandhills Game Land.

The old alternative name Pine-woods Sparrow obscures the fact that this species breeds in two quite different habitats in different portions of its range. In the south and east, Bachman Sparrows are found almost exclusively in open pine forests with a grassy understory punctuated by shrubs and short trees. In the north and west, however, the preferred habitat is—or, in most such areas, was—abandoned pastureland,

> usually fallow for four years or more, which is well grown up with goldenrods and asters, various grasses, and the miscellaneous composites and weeds typical of dry, eroded slopes. The presence or absence of pine seedlings seems to have no bearing . . . . This species is practically confined to hill country, almost never appearing in the valleys or even on the lower slopes of the hills. A typical territory is near the top of a slope where eroded gullies have been healed and are covered with shrubs, particularly blackberry bushes.

While southern populations of this species appear to be resident, migration is documented in both *bachmani* and *illinoensis*. This species' extremely reclusive behavior away from the breeding grounds makes it difficult to detect migrants and wintering birds and thus to establish their seasonal timing. Arrival on the northern breeding grounds appears to take place, or to have taken place, in April and May; fall migration begins in August.

The present-day winter range of the Bachman Sparrow extends from east Texas across the Gulf states and peninsular Florida north to central North Carolina (eBird).

Given the obvious southward retreat of the Bachman Sparrow over the past half century, it is difficult to know whether older records north of the current distribution represent moribund relict populations, pioneers in search of new habitats, would-be recolonizers of a former range, or genuine vagrants. One of New Jersey's two accepted reports of the species is of a singing male in June in pine habitat, while the other, from May 1918, seems more likely to have been a spring overshoot. New York's three records also include a singing male, from May 1940 "on a hillside clearing overgrown with meadow grasses and scattered small aspens and conifers," a description clearly reminiscent of the habitat described for breeders in the north; the other two records are from Long Island in April and early June. There are Michigan specimens from April 1944 and May 1946, both males and both identified as of the subspecies *bachmani*. One was collected in April 1917 at Point Pelee and another in May 1928 on Ontario's Ryerson Island; there have been other sight reports, but these are the only accepted records for Canada of a species otherwise endemic to the United States at all seasons.

The Bachman Sparrow was long held to comprise only two subspecies: the nominate form (the old Pine-woods Sparrow) of the coastal plain in South Carolina, Georgia, and Florida, and the race *bachmani* over the rest of the species' range in the southeastern United States. The subspecies *illinoensis*, first described by Robert Ridgway from southeastern Illinois, was later lumped with *bachmani* by Ridgway and Brewster, but 50 years later, it was resurrected for application to the birds found at the western and northwestern edges of the species' range; birds representing this western race are much paler above, with rusty brown streaking replacing, entirely or nearly so, the black marks shown by the nominate subspecies.

Audubon's *bachmani* is intermediate between the other two races, brighter and rustier than *aestivalis* and darker brown and more heavily black-streaked above than *illinoensis*. The breast of *bachmani* has a buffy wash, sometimes with an orange tinge, similar to the underparts of *illinoensis*, while the nominate subspecies is grayer below than either.

Intergradation between *bachmani* and *illinoensis* has been recorded in Tennessee; *bachmani* and *aestivalis* are said to intergrade in Georgia.

CASSIN SPARROW
Peucaea cassinii

Always given to the oracular, Elliott Coues did little to lift the fog surrounding the identity and identification of the Cassin Sparrow when he wrote of the species that "a peculiar character of marking raises groundless suspicion of immaturity." Clearly pleased with it, Coues would repeat this obscure sentence, with no further explanation, in each of the last four editions of his *Key*.

The allusion is to the uncertainty and confusion that had stuck to the Cassin Sparrow since its discovery in April 1851 by the 30-year-old Samuel Washington Woodhouse. On collecting a male on the prairies near San Antonio, Texas, Woodhouse at first misidentified the specimen as a Savannah Sparrow, "but upon examination it proved to be totally distinct." Woodhouse failed to find a second individual of what we now know is a very reclusive species, and all of the natural history collectors assigned to the railroad surveys conducted between 1853 and 1856 were able to supply the Smithsonian with a grand total of three additional skins identified by Spencer Baird as Cassin Sparrows.

One of those skins, taken in June 1855 by Caleb Kennerly on the border between Sonora and Arizona, was not quite identical to the others, and Baird cautiously concluded that the bird was "probably considerably older" than the rest, which he "considered as in quite immature plumage." Not until 1874 did Baird and his colleagues Thomas Brewer and Robert Ridgway correct the error following "a more recent examination of additional material" by Ridgway. The earliest describers, Baird among them, had been misled by the first specimens' scaled and spotted upperparts, a pattern typical of the juveniles of many other sparrow species; Kennerly's bird from the Southwest was in fact not an "adult" Cassin Sparrow but a Botteri Sparrow.

Once again, it seems to have been Elliott Coues's offended vanity that kept him, even 25 years later, at such pains to point the ancient error out. Coues admits that he, too, had once believed the Cassin and Botteri Sparrows were a single species in different age-related plumages, but that "upon examination of ample material with Mr. Ridgway" at the Smithsonian, the two ornithologists had agreed that in fact two species were involved. Coues and Ridgway published that conclusion separately in 1873—Ridgway in the pages of the *American Naturalist* and Coues in the first edition of his *Check List*—but the following year, when Baird and his colleagues published the *History*, only Ridgway was cited as having uncovered the mistake. The suspi-

Generally grayer than the Botteri Sparrow, at close range and in fresh plumage this species reveals a beautifully complex pattern of crescent and fringes on the upperparts. *Texas, September. Brian E. Small*

cion is probably not groundless that Coues resented being overlooked, and that he continued to voice that resentment even after his death, in the fifth and final, posthumous edition of his *Key*.

FIELD IDENTIFICATION

In lingering close views, an adult Cassin Sparrow in fresh plumage is intricately patterned and distinctive, unlikely to be confused with any other North American bird. Unfortunately, birds of this species are often very shy, and views are usually brief and imperfect. More significantly, the complicated molt schedule of the Cassin Sparrow, like that of its close relative the Bachman Sparrow, means that "individuals may appear to be molting almost constantly," and even perched in the open, a Cassin Sparrow only infrequently "looks like its picture in the birdie books."

The intricate upperpart pattern nearly disappears as the feathers wear and bleach in the desert sun. *Arizona, July. Brian E. Small*

The long, extravagantly rounded tail and flat head and heavy bill of the Cassin Sparrow are enough to eliminate several sparrows frequently confused with this species. Both the Grasshopper and the Savannah Sparrows are found in similar habitat, and both, like some Cassin Sparrows, show a touch of yellow above the lore; the Grasshopper Sparrow also has a neat white eye-ring, a character often conspicuous on Cassin Sparrows. Both, however, are strikingly shorter-tailed and more heavily and neatly marked on the head. The Grasshopper Sparrow is brighter above and below, with a very large bill, while the Savannah Sparrow is neatly streaked on the breast and the back.

The overall bland, grayish aspect of the Brewer Sparrow, its neat eye-ring, and its finely black-streaked crown have misled hopeful observers on the wintering grounds to misidentify that species as the Cassin. Brewer Sparrows are tiny, very long- and narrow-tailed, with petite pinkish bills; they are both more social and more confiding than Cassin Sparrows, and frequently perch atop bushes and fences to be admired at the human observer's leisure. At all seasons, Cassin Sparrows prefer grassland to the bleaker desert flats and sagebrush steppe so favored by the Brewer Sparrow.

The much less common and much more locally distributed Botteri Sparrow shares its entire range in the United States and northern Mexico with Cassin Sparrows. In these northern portions of its range, the Botteri generally prefers somewhat more densely shrubby grassland than the Cassin, but it is not unusual to hear both singing from a single mesquite-dotted pasture. In Arizona, Sonora, and Chihuahua, Botteri Sparrows are decidedly more reddish above and brighter tawny beneath than the grayish Cassin. Botteri Sparrows in Texas and eastern Mexico are much duller and more similar to Cassin Sparrows, but all Botteri Sparrows are streaked above, with buff-edged blackish tertials and a large, noticeably down-curved bill that passes smoothly into the flat crown; Cassin Sparrows are spotted and scaled above, with narrowly white-edged gray-brown tertials and a relatively small, fine bill that meets the high forehead in a pronounced stop. Especially in worn plumage, the gray greater coverts of the Cassin Sparrow produce a subtle paler panel on the folded wing, which is more uniformly brown on the Botteri. At close range, the rump feathers of fresh-plumaged Cassin Sparrows have a unique complex pattern of pale bases, neat black subterminal spots, and narrow white tips and edges; the back feathers and scapulars have brown bases, neat black subterminal spots, and relatively broad gray tips and edges. The result, quite different from the more normal "sparrowlike" aspect of the Botteri Sparrow, is that "peculiar character of marking" that created so much confusion in the nineteenth century.

At any distance in the field, Cassin Sparrows are strikingly gray. The patterns of the wing and upperparts can be appreciated only at close range, and even then only when the bird is in fresh plumage; the paler areas of the feathers wear quickly, by physical abrasion and exposure to sunlight, and for much of the year, many Cassin Sparrows are distinctively plain, the Orange-crowned Warblers among the passerellid sparrows.

In fresh plumage, the long, rounded tail of the Cassin Sparrow is gray with faintly but decidedly barred central feathers; the outer two or three tail feathers have pale tips of variable extent and brightness, only rarely conspicuous and never pure bright white. The tips wear more rapidly than the rest of

The Cassin and other brown *Peucaea* sparrows may look rumpled at any season, even in fresh plumage. *Texas, September. Brian E. Small*

There are often narrow, sparse streaks on the flanks and breast sides. This bird also shows the typical barred tail feathers and finely streaked crown. *Texas, September. Brian E. Small*

the feather, producing the ragged, wedge-shaped tail typical of Cassin Sparrows for much of the year. Flank streaking, often invoked as a character distinguishing Cassin from Botteri Sparrows, is not especially useful in the field: faint even when fresh, the streaks wear away rapidly on Cassin Sparrows, while the slightly darker feather shafts on the Botteri Sparrow's flank sometimes line up to produce very weak, irregular short streaks.

Fresh juvenile Cassin Sparrows are rarely seen. Careful attention to the patterns of the tail, wings, and back is essential if they are to be identified with any confidence. Juveniles are more richly colored above and below than their parents, buffy brown rather than dull pale gray, and the faintly yellowish underparts are spotted and streaked with black. The tail is browner than the adult's, but shares the pale tips of the outer feathers and the barred pattern of the central rectrices created by the "serrated" extension of the shaft streaks onto the web. The uppertail coverts and rump are darker than in the adult, but the bases of the feathers are paler than the subterminal spot, and the adult's white fringe and tip are echoed in the juvenile by buffy brown. The back feathers are less complexly patterned, but the dark centers and paler (buffy rather than gray) edges still produce a clearly scaled and spotted aspect. The greater coverts are buffy to slightly reddish, only faintly contrasting with the rest of the folded wing, but as in the adult, they are tipped with pale buffy or whitish to create a single narrow pale wing bar.

Most or all Cassin Sparrows molt three times in their first year of life (Pyle). While still on the breeding grounds, juveniles undergo a first molt of the body feathers, producing a supplemental plumage; some birds also replace wing coverts, one or more of the tertials, and the central tail feathers (Pyle). This

The juvenile is heavily scaled above and finely streaked on the breast; the head is often tinged yellow. *Colorado, July. Steven Mlodinow*

molt can take up to two months, usually ending by September. The preformative molt begins almost immediately thereafter. This molt, unlike the pre-supplemental molt, is complete, involving the body feathers and the flight feathers of wings and tail; it is completed no later than November, either on the breeding grounds or on the wintering grounds (Pyle). The first prealternate molt is "extended and gradual," with body feathers slowly being replaced between late February and August; thus, birds that are between about seven months and one year old "have a mixture of older, abraded [formative] feathers and newer, less worn, less faded [alternate] feathers" before beginning their molt into definitive (adult) basic plumage in their second August or September (Pyle).

The definitive prebasic molt takes an average of about six weeks, beginning in some individuals as early as August, ending in others as late as the end of November. The adult's extremely prolonged pre-alternate molt "begins as early as mid-Feb[ruary], continues at low intensity to late June, and intensifies in July and Aug[ust]." While in many songbird species this molt involves little more than the head, in the Cassin Sparrow it results in replacement of all or nearly all the body feathers and occasionally one or more tertials and central tail feathers (Pyle).

It is obvious that "some replacement of body feathers occurs almost continuously in this species" (Pyle), and the combination of new, fresh feathers and retained, worn feathers is responsible for the disheveled appearance of many individuals of this species over most of the year.

Many birders see their first Cassin Sparrow during the birds' breeding season, when the male's sad, silvery song renders visual identification secondary. Adolphus Heermann was the first to note the distinctive song flight:

> Rising with a tremulous motion of its wings some twenty feet or more, it descends again in the same manner to within a few yards of the spot whence it started, accompanying its entire flight with a lengthened and pleasing song.

Oddly, Heermann had nothing more to say about the song, though that was what had alerted him to the presence of the bird in the first place. Others on encountering the species would also remark on "its sweetly modulated song," but an actual description of the notes and phrases was not published until 1875, when Henry Henshaw finally provided an account of the Cassin Sparrow's "very plaintive, but quite pretty and attractive" melody:

> It begins with a low tremulous trill, followed by slow and plaintive syllables, the last of which is softer and more prolonged, and in a lower key. Though little varied, and on this account somewhat monotonous, it yet possesses an indescribable sweetness and pathos, especially when heard, as is often the case, during the still hours of the night.

Few observers would not agree that this is one of the most evocatively beautiful and regularly melodious of all passerellid songs. It typically starts with one or two very quiet but round, rich notes followed by a broad, loose, silvery tremolo in an impassioned crescendo; the bird inserts the most vanishingly momentary of pauses after the tremolo, then sings two finer, faintly trilled notes, the last lower by a full step than the one before it. Heard from a great distance across the desert grassland, the descending conclusion can strike the ear as disconcertingly similar to the spring "phoebe" song of a Black-capped Chickadee.

Males may also sing a secondary song, similar to the familiar primary song but introduced by a "series of chips, trills, and buzzy notes"; this song is given at early stages of the nesting cycle, presumably at least in part to reinforce the pair bond. The "whisper song," performed in late summer, is "very soft [and] consists of a few preliminary notes, then [an] assortment of trills"; very variable, this song may go on for several minutes.

When not singing, Cassin Sparrows are relatively quiet. They may give a stuttering *tsitsi* when flushed, or a soft, dull *tip* in alarm at the approach of a human observer.

RANGE AND GEOGRAPHIC VARIATION

This most widespread and by far most common of the *Peucaea* sparrows north of Mexico is a nomadic breeder, its summer range expanding and contracting with patterns of rain and drought in the Southwest and on the western Great Plains. "This opportunistic bird migrates to areas where precipitation levels are . . . high enough to provide sufficient vegetative cover and sufficient food sources." In wet years, the range can be extended west as far as the Lower Colorado River Valley in Arizona and even to California's San Bernardino Mountains and north to the sand-sage steppes of southwestern Nebraska and eastern Wyoming (eBird); in Texas, exceptional "irruptions" have brought breeding birds east to Harrison and San Augustine Counties, while a singing bird was present in Bossier Parish, Louisiana, in May 2011 (eBird). It seems likely that the southern limits of the breeding distribution in Mexico are similarly flexible.

The "normal" breeding range of the Cassin Sparrow is limited to the southwestern United States and northern Mexico. In Arizona, where breeding was first confirmed only 50 years ago, the species is fairly common most years on grasslands with sparse bushes at elevations between 2,200 and 5,600 feet, from central Pima County through Santa Cruz and Cochise Counties and north into southern Pinal and Greenlee Counties; they have also been recorded as breeders or possible breeders in southern Yavapai and Apache Counties. The breeding range extends south into northern Sonora and Chihuahua east to Tamaulipas and south on the central Mexican plateau to Zacatecas and San Luis Potosí.

Taking into account the sporadic, weather-related appearance from year to year, Cassin Sparrows are widespread and variably scarce to fairly common breeders across northern and eastern New Mexico, eastern Colorado, southwestern Nebraska, and the western half of Kansas and Oklahoma. They are common to abundant in the Panhandle and western plains of Texas, and locally common to rare west of a line from Tarrant County to Matagorda Bay.

Cassin Sparrows are migratory at least in the northern portion of their breeding range, but the timing of autumn departures and the winter distribution as a whole are uncertain given the fact that this species is extremely difficult to find when males are not singing. Further complicating the picture is the possibility, still unproved, that some individuals may breed in more easterly regions of the species' range, then move west into New Mexico and southeast Arizona to breed again. Typically, breeding birds appear to leave Oklahoma in the second half of September, in line with late dates from Texas and western New Mexico; "distinct waves" of fall migrants have been observed in the Big Bend region of Texas in the first three weeks of September.

Wintering birds are regularly, if generally infrequently, detected in southeastern Arizona and in the Trans-Pecos and Edwards Plateau regions of Texas, south to the coastal prairies of that state. The Mexican range shifts southward to northern Nayarit and Guanajuato in the winter, with occasional records as far south as Colima and Jalisco.

The usual dates for spring arrivals in northern Texas and Oklahoma appear to fall in April, while typical arrival dates (if such there be for this species) in southwest Nebraska and eastern Colorado are in mid- to late May.

Vagrant Cassin Sparrows are almost certainly more frequent than the relatively few records of this furtive, plainly plumed bird suggest. The species has occurred some 50 times in California, with more than three-fourths of the state's reports from early May to early July; half of those spring and early summer records are accounted for by the presence in May and June 1978 of 15 singing males in San Bernardino County. Fall birds, encountered almost exclusively on Southeast Farallon Island, have been recorded from early September to early November. Elsewhere in the West, vagrants have been found in Curry County, Oregon; southwestern South Dakota; Utah; Nevada; and northernmost Baja California.

East of the species' usual range, May Cassin Sparrows have occurred on the Lake Michigan lakefront in Illinois, while both of Indiana's records are from June. Another June bird appeared in Shelby County, Ohio, in 2011 (eBird). The eight documented records from Ontario are evenly split between spring and autumn (eBird).

On the East Coast, there are September records from St. Mary's County, Maryland, and from Island Beach State Park, New Jersey; a bird reported from Mount Desert Rock, Maine, in September 1986 has apparently not yet been reviewed by that state's records committee. Another appeared on western Long Island in October 2000. The only spring vagrants thus far detected far to the east have been mid-May individuals on Cape Cod (eBird) and Nova Scotia. The wide geographic span of vagrancy in this species and the remarkable number of records from Ontario suggest that this inconspicuous bird is overlooked east of the Great Plains in migration.

No subspecies are recognized.

Bright Botteri Sparrows from the southwestern deserts are remarkably similar in plumage to the eastern Bachman Sparrow. *Arizona, July. Brian E. Small*

BOTTERI SPARROW
Peucaea botterii

In 1898, Robert Ridgway turned a fresh eye on a bird that he and his colleagues had identified a quarter of a century earlier. Where the 24-year-old museum assistant had seen simply a badly worn Botteri Sparrow, the 48-year-old "undisputed king of birds" at the Smithsonian was able to convince himself that the skin, collected in eastern Mexico, in fact represented a new species, "similar to [the Botteri Sparrow] but very much darker." Ridgway formally described the bird and assigned it the scientific name *Aimophila sartorii*, in honor of the collector, the German-Mexican agriculturalist and natural historian Florentin Sartorius.

This Huatusco Sparrow—the English name commemorating the locality where Sartorius had shot the type specimen—would prove short-lived as a distinct species. Less than three years after the publication of his description, Ridgway determined that the bird merited only subspecific rank, as a race of the Botteri Sparrow. Later authorities have not accorded it even that status; with the exception of a brief attempt at a revival in the 1960s, the Huatusco Sparrow has been treated as a Botteri Sparrow, whether of the nominate subspecies or of one of the Central American races, or, most frequently, as nothing more than an intermediate population linking the "true" subspecies to its north and to its south.

For the better part of a century, *sartorii* has thus been a label for something that does not, strictly speaking, exist. Valid or not, however, the name survives in the synonymies as a reminder of the complexity of science's prolonged confrontation with the Botteri Sparrow—and of the twisting series of historical circumstances that brought the type specimen to Ridgway's desk in Washington.

Born in Hessia in 1796, Carl Christian Sartorius fled his native land, a forged passport in his pocket, at the age of 23, having been arrested and sentenced to prison for his outspoken revolutionary leanings. With a small group of like-minded emigrés, Sartorius planned to found an agrarian utopia in the New World, to be governed in a way that would set a shining example for their disunified and oppressive homeland. The company settled near Huatusco, in west-central Veracruz, where the settlers quickly turned their attention from sugarcane and coffee to more immediately profitable commodities such as silver, copper, and lead; the steady income from his mining interests made it possible for Sartorius to acquire a large plantation, El Mirador, a mixture of cultivated fields and wild canyons and jungle.

Sartorius possessed the educated amateur's passion for natural history, and he is recorded as "constantly taking notes and writing up his impressions of the country, which seemed to him totally new, exotic, and picturesque." Soon he was submitting specimens to botanists and entomologists in London, Paris, and Berlin, and eventually became a regular collaborator with the Smithsonian in Washington, supplying plants and animals of all kinds from the area between Xalapa and Orizaba.

Sartorius's two sons were both born in Mexico. Otto, the first of the Sartorius children, was killed at an early age in a hunting or collecting accident. The younger son, Florentin, born in 1837, was sent to Germany to study agriculture; he returned to El Mirador in 1857, where he assumed responsibilities for the family's sugar production before buying a pharmacy in Huatusco. His elder brother's sad fate notwithstanding, Florentin, too, became an active field collector.

Among the specimens Florentin Sartorius sent north from El Mirador was a large sparrow, collected on July 12, 1860. This was the bird, still held today in the collections of the Smithsonian, to which Ridgway would eventually—if temporarily—assign the name *sartorii*, a name that now stands for a chain of historical events, leading from Hessia to Veracruz to Washington, that would likely be forgotten entirely if not for the skin of a single sparrow.

FIELD IDENTIFICATION

Botteri Sparrows are plain by passerellid standards, but they hardly deserve the scorn cast on their modest appearance over the years. In lingering close views, an adult Botteri Sparrow in fresh plumage is humbly handsome, and unlikely to be confused with any other bird in its range. Unfortunately, birds of this species—like the Cassin and Grasshopper Sparrows with which it shares its grassland habitat—are often shy, and except when males are singing, the views are usually brief and imperfect.

The long, rounded tail, flat crown, and thick, heavy bill of the Botteri Sparrow are enough to eliminate several "confusion species." Both the Grasshopper and the Savannah Sparrows are found in similar habitat. Both, like some Cassin Sparrows, show a touch of yellow above the lore; the Grasshopper Sparrow also has a neat white eye-ring, a character usually inconspicuous or even invisible in the field on Botteri Sparrows. Both the Grasshopper and the Savannah Sparrows are strikingly shorter-tailed and more heavily and neatly marked on the head. The Grasshopper Sparrow is brighter above and below, with a short, swollen bill, while the Savannah Sparrow is neatly and finely streaked on the breast and the back.

The more reddish populations of the Botteri Sparrow and the Bachman Sparrow are extremely similar in plumage. The Bachman Sparrow's bill is slightly finer, and its back and scapulars are darker and rustier and its tail spots grayer and more conspicuous than those of the northern Botteri races. The Botteri Sparrow has a less conspicuously marked face, with a duller, less well-defined supercilium, and the underparts show less contrast between the breast and belly. These subtle differences of degree can only rarely be assessed in the field. The Botteri Sparrow, unlike the Bachman, has no established pattern of vagrancy, but any *Peucaea* sparrow discovered out of its expected range must be identified with great caution.

The Botteri Sparrow shares its entire range in the United States and northern Mexico with Cassin Sparrows. In these northern portions of its range, the Botteri generally prefers somewhat more densely shrubby grassland than the Cassin, but it is not unusual to hear both singing from a single mesquite-dotted pasture. In Arizona, Sonora, and Chihuahua, Botteri Sparrows are decidedly more reddish above and brighter tawny beneath than the grayish Cassin. Botteri Sparrows in Texas and eastern Mexico are much duller and more similar to Cassin Sparrows, but all Botteri Sparrows are streaked above, with buff-edged blackish tertials and a large, noticeably down-curved bill that passes smoothly into the flat crown; Cassin Sparrows are spotted and scaled above, with narrowly white-edged gray-brown tertials and a relatively small, fine bill that meets the high forehead in a pronounced stop. Especially in worn plumage, the gray greater coverts of the Cassin Sparrow produce a subtle paler panel on the folded wing, which is more uniformly brown on the Botteri. At close range, the rump feathers of fresh-plumaged Cassin Sparrows have a unique complex pattern of pale bases, neat black subterminal spots, and narrow white tips and edges; the back feathers and scapulars have brown bases, neat black subterminal spots, and relatively broad gray tips and edges. The result is quite different from the streaked, more normal "sparrowlike" aspect of the Botteri Sparrow.

In fresh plumage, the long, rounded tail of the Botteri Sparrow is dusky gray; the outer two or three tail feathers have brownish tips of variable extent and brightness, only rarely conspicuous and never white. Flank streaking, often invoked as a character distinguishing Cassin from Botteri Sparrows, is not especially useful in the field: faint even when fresh, the streaks wear away rapidly on Cassin Sparrows,

Many individuals are grayer, a tendency especially pronounced in the eastern and southern portions of the Botteri Sparrow's large range. *Arizona, July. Brian E. Small*

while the slightly darker feather shafts on the Botteri Sparrow's flank sometimes line up to produce very weak, irregular short streaks.

Secretive and hard to see, juvenile Botteri Sparrows are buffier above and below than their parents, and the faintly yellowish underparts are spotted and streaked with black, most heavily on the breast and flank. The dark centers and pale buffy brown edges of the back feathers still produce a clearly streaked aspect. The greater coverts are buffy to slightly reddish, broadly tipped and edged with paler buff.

The molt schedule of the Botteri Sparrow is imperfectly known (Pyle), but as in other *Peucaea* sparrows, the juvenile's preformative molt can be extensive, and there is likely also a prolonged pre-supplemental molt. "Limited year-round replacement of body feathers likely occurs" (Pyle), and the combination of new, fresh feathers and retained, worn feathers is responsible for the disheveled appearance of many individuals of this species over most of the year.

Most birders see their first Botteri Sparrow during the birds' breeding season, when the male's homely but cheerful rattling song renders visual identification secondary. The familiar primary song has three parts: a soft introduction; a long, dry tremolo; and a brief coda, sometimes omitted.

The introduction comprises a series of two to six hesitant paired notes; each note pair is of a different quality, often two broad, smacking chirps followed by two short slurs: *trp trp teew teew*. The tremolo that follows is low-pitched and loose, with very little "musical" quality; it has been aptly compared to the end of a Black-chinned Sparrow song. The coda, given at a very low volume or not at all, is a short phrase of two or three downslurred whistles; unlike the tremolo, which is conspicuous from considerable distance, the introduction and coda are often inaudible to the human observer, who may see the bill open but hear nothing.

After the eggs have been laid, males may switch to a second song type, made up of "short, variable introductory notes, followed by trills and whistles . . . repeated . . . for up to several minutes without a gap between repetitions"; this performance may continue for more than ten seconds without pause, then be taken up again.

Botteri Sparrows often sing in flight, but they do not perform the stereotyped floating song flight of the Cassin Sparrow. Instead, Botteri Sparrows sing while flying between song perches or from a song perch to the ground.

When not singing, Botteri Sparrows are relatively quiet. They may give a stuttering *chip'chip'chip* when startled, or a soft, faintly metallic *tik* in alarm at the approach of a human observer. The usual contact call is a "piercing, high-pitched, ventriloquial *seep*," lower-pitched and heavier than the corresponding *see* calls of Cassin Sparrows.

RANGE AND GEOGRAPHIC VARIATION

Botteri Sparrows are uncommon to locally common breeders on the coastal prairies of Texas from Kleberg County south, with irregular records north to San Patricio County and west to Duval County. These birds are of the race *texana*, whose range extends south along the coast of eastern Mexico to northern Veracruz, where intergradation occurs with *petenica*. There are also summer records from Jeff Davis County (eBird) and Presidio County, presumably pertaining to the western *arizonae*. That subspecies is an uncommon breeder in Grant and Hidalgo Counties in southwestern New Mexico, and is likely responsible for records east to Doña Ana County in that state and in northeastern Chihuahua (eBird).

The Botteri Sparrow is fairly common but local in southeastern Arizona grasslands from the New Mexico border west through Cochise and Santa Cruz Counties to at least the upper Altar Valley; the western limits of the breeding range in Pima County are unclear because of a lack of observations from the Tohono O'Odham Nation. Breeding has been regular this century in Pinal County on the northwest flank of the Santa Catalina Mountains, and there is a recent summer record from Graham County (eBird), both localities perhaps undergoing recolonization after the species was apparently extirpated locally at the end of the nineteenth century.

Botteri Sparrows are common in summer in grasslands with scattered oaks and mesquites in northeastern Sonora, where they occur at elevations as low as 2,000 feet. On Mexico's northwest slope, *arizonae* is found south to northern Durango. The more southern subspecies breed, in many cases disjunctly, south through the Mexican plateau to Veracruz, southern Oaxaca, northern Chiapas, northern Guatemala, and Belize. The Botteri Sparrow is uncommon to fairly common in eastern Honduras, though apparently rare in Guatemala. In Costa Rica, the species is said to be rare and local on the Pacific Slope of the Guanacaste, between 1,300 and 1,400 feet. Throughout its southern range, the Botteri Sparrow is more likely than in the north to inhabit grassy clearings in open pine and oak woodland, a habitat reminiscent of that used by the Bachman Sparrow in the southeastern United States.

Southern populations of this species are sedentary, but *arizonae* and *texana* are at least in part migratory. Breeding birds arrive in south Texas in mid-April, and most have left the state by late September. In Arizona and northern Sonora, spring arrival is notably late, with most birds not appearing until mid-May. Most have left Arizona by October, though there are several recent records of apparent overwintering; it is not clear whether those reports represent a northward range expansion or simply better detection and identification.

All Botteri Sparrows are shy and rarely seen when the males are not singing, making it very difficult to evaluate their distribution in winter. In the northwest of the range, there are winter records from southern Sonora and Chihuahua; in the east, there are early spring specimens of migrants of the race *texana* from Guerrero and Morelos. This species should be carefully sought—and carefully identified—across northern Mexico between October and April.

The currently recognized subspecies fall into

The stuttering song is usually delivered from a perch, though it may begin or end in a short fluttering song flight. *Arizona, July. Brian E. Small*

two groups: a northern group comprising *botterii, mexicana, goldmani, texana,* and *arizonae*; and a southern group comprising *petenica, vantynei, spadiconigrescens,* and *vulcanica.* Not all authorities recognize *mexicana,* considering it instead a southwestern representative of the nominate race.

The southern races are small and dark, northern birds larger and paler above. The striking *petenica* is black above with sooty brown streaks, contrasting with a whitish throat, breast, and belly; the wing coverts have brown edges. Ridgway's *sartorii* and the particularly dark birds of central Tabasco, originally described as a separate subspecies *tabascensis,* are considered synonyms of *petenica.*

The highlands of central Guatemala are occupied by *vantynei,* which is browner and paler above than *petenica* but darker brown above than *botterii*; the underparts are darker, with more brown or dusky tinge, than either of those subspecies. The edges of the wing coverts and secondaries are rustier in this race than in *petenica.* Resident in northern Honduras and Nicaragua, *spadiconigrescens* ("chestnut-blackish") is browner above than *petenica,* with chestnut-brown rather than sooty streaking on the back and a grayish breast contrasting with the pale throat and belly; the wing coverts and secondaries have broad rusty edges. The southernmost race, the large *vulcanica,* resembles *petenica* in the darkness of its upperparts, but the rusty edgings of the wing and back are broader and more conspicuous, and the underparts are "more intensely smoke-gray and brown."

Resident in central and southern Mexico, the nominate *botterii* is pale and relatively uniformly colored. The upperparts and wings are gray-brown, the feathers with blackish shaft streaks and gray edges, and the underparts pale buffy; the scapulars have broad rusty centers. The streaking on the back of *mexicana,* a subspecies often synonymized with *botterii,* averages broader and more blackish. Resident from southernmost Sonora to Nayarit, *goldmani* is smaller and shorter-billed than the nominate race, with a darker flank and breast; at its geographic and phenotypic extreme, *goldmani* shows chestnut edges on the tertials, inner secondaries, greater coverts, and uppertail coverts.

Of the two subspecies that occur in northernmost Mexico and the United States, *arizonae* is the more abundant and the more familiar to most birders. It is paler than the nominate race, with a duller, paler but more reddish back and paler underparts; the upperparts show more gray on average in *texana,* a difference that can sometimes be striking even in the field.

The eponymous "shoulder" patch can be concealed, but the long, graduated tail and heavy, yellow-based bill distinguish this locally distributed species from other sparrows in our area. *Arizona, April. Brian E. Small*

RUFOUS-WINGED SPARROW
Peucaea carpalis

At old Fort Lowell, on the site of what is now downtown Tucson, Charles Emil Bendire shot what Elliott Coues in his formal description would call "a homely little bird, not particularly remarkable for anything." Bendire would later admit that he was himself in part responsible for that unfavorable impression:

> I took my first specimens on June 10, 1872, and after spending many hours in vain in trying to locate them [in] my ornithological library . . . nothing was left for me to do but to try and make a few skins to send East for identification. I believe this was one of my first attempts in this line, at least on so small a bird. I managed to strip the hide off in some way . . . and my skins after they were done looked as if a dog had chewed them for a short time.

Bendire eventually collected several more specimens, which Coues was pleased to find on inspection "brighter and purer" than the first he had received.

Writing from the southwestern frontier, Bendire told Coues that the Rufous-winged Sparrow was "very common" in the area around Fort Lowell, and the presence there of what was already recognized as a very local species made the area a site of ornithological pilgrimage in the years to come. Henry Henshaw, who had come west, with Spencer Baird's recommendation, as a naturalist to the Army's Wheeler Survey, visited Tucson to find himself "turned loose" among the Rufous-winged Sparrows; echoing Coues, he described the bird as unassuming and difficult to identify:

> In its appearance, the Rufous-winged Sparrow has little to attract attention, and, in its behavior, is so much like its commoner and less desirable associates, that I found difficulty in properly distinguishing between this and the other sparrows, many of which I killed by mistake.

The indefatigable collector was happy, though, to find the sparrow and his other target species at Fort Lowell as common as Bendire had promised: "to birds there is no limit," he wrote C. Hart Merriam. Henshaw secured "a large suite of specimens," and soon enough it was part of every visiting ornithologist's routine to collect his own series of this very local and very little-known bird.

The species remained common around Tucson through the 1870s, but Henshaw's hopeful assertion that there was "no limit" was wrong. The wildly varying population fortunes of the Rufous-winged Sparrow in the nearly 150 years since the first ones were shot are without parallel in the history of North American ornithology. Little is known about the species' historical abundance in its Mexican range, but the Arizona populations have been carefully tracked ever since Bendire first took aim at his type specimen. The population around Fort Lowell in fact declined notably over the next 15 years; Frank Stephens found it only "sparingly" in the traditional localities in 1881, and a specimen taken in 1886 was the last found in Arizona until 1915. By 1931, the species was widely thought to be extirpated from the United States.

Rufous-winged Sparrows were rediscovered in the Tucson area in the late 1930s, where they had last been encountered more than half a century before. Twenty years later, in 1956, "a great irruption" brought large numbers to southeast Arizona. On the historical United States breeding grounds, a peak in numbers in the early 1970s was followed by a noticeable crash when precipitation totals plummeted in 1973, and this species was still one that had to be intentionally sought out in southeast Arizona in the 1990s.

Since the end of that decade, Rufous-winged Sparrows have been common once again and increasing in their limited United States range. In 2003, the species reached the western end of Arizona's Guadalupe Canyon, and in January 2011, Rufous-wingeds were recorded for the first time in the eastern canyon, in Hidalgo County, New Mexico (eBird). It is impossible to know whether this species will continue to expand its range in the United States, or its current status is simply a high point in the Rufous-winged Sparrow's historically erratic pattern of occurrence and abundance.

FIELD IDENTIFICATION

The Rufous-crowned Sparrow is found throughout the range of this species; though their preferred habitats differ—rocky, brushy canyons in the case of the Rufous-crowned, and grassy mesquite bosques in the case of the Rufous-winged—the two can be found near each other at many sites where gulches

The finely streaked back shows less rusty than even the dullest western Chipping Sparrow. *Arizona, April.* Brian E. Small

The streaked rusty crown, rusty eye line, and pale-based bill can recall the geographically unlikely American Tree Sparrow. *Arizona, July. Brian E. Small*

dissect desert grasslands. The Rufous-crowned Sparrow is larger, heavier, and above all darker than this species, more nearly recalling a somber Canyon Towhee than the relatively slender, slight, and pale Rufous-winged. The two differ vocally as well, with all of the calls of the Rufous-crowned lower-pitched and fuller; the song of the Rufous-crowned Sparrow is a sustained jumble quite unlike any of the trilling songs of the Rufous-winged.

The entirely allopatric Cinnamon-tailed Sparrow is much larger and more richly and coarsely marked above, with a distinctly rufous tail.

The principal sources of identification confusion within the Rufous-winged Sparrow's range are not other species of *Peucaea* or *Aimophila* but rather the rusty-capped members of the genus *Spizella*. The Chipping Sparrow, which is abundant in winter in the same habitats frequented by Rufous-wingeds and often found in association with them, is usually adduced as the most likely confusion species; that bird is significantly smaller and more slender, with a longer and narrower tail particularly noticeable in flight. The small head and narrow bill of the Chipping Sparrow make the bird look much less "normally" proportioned than the Rufous-winged Sparrow, and the upperparts of the Chipping are considerably more varied, with a dark tail and silvery rump contrasting strongly with the richer

chestnut and black streaking of the back. Chipping Sparrows have a distinctive head pattern comprising a thin black eye line that continues onto the lore, unlike the broader, shorter rusty line behind the eye of the Rufous-winged Sparrow; the throat is at most vaguely bordered by a single lateral throat streak, while in Rufous-winged Sparrows two well-defined streaks set the jaw stripe off from the throat and the ear coverts. The contact notes of winter Chipping Sparrows are light, high-pitched, and silvery, very different from those of the Rufous-winged.

American Tree Sparrows share the rusty crown and eye line and bicolored bill of the adult Rufous-winged Sparrow, but are square-tailed and black-footed, with bright chestnut and black upperparts. The breast and flanks, entirely grayish white in Rufous-winged Sparrows, are strongly tinged buffy peach, and the tail is black with noticeable bright white feather edges. There are no records of the American Tree Sparrow from the range of the Rufous-winged, and the Rufous-winged Sparrow's apparent expansion this century has taken it east, not north, and thus no closer to the winter distribution of the American Tree Sparrow.

Usually encountered in family groups with their parents, juvenile Rufous-winged Sparrows closely resemble adults. They are buffier on the head and flanks, and the throat and breast are marked with

The "double whiskers" starting at the base of the bill, comprising a black lateral throat stripe and a black whisker, are unlike the plainer cheek of the Chipping or Tree Sparrow. *Arizona, July. Brian E. Small*

sparse dark streaks. The light rusty crown with brown streaking and generally whitish underparts in combination with the long, rounded tail, rather heavy bill, and rotund body are unlike any other breeding passerellid of the desert. The timing of nesting and of molt in this species varies from year to year with rainfall, and birds in juvenile plumage can be seen at any time in summer and fall into November (Pyle).

Male Rufous-winged Sparrows sing nearly throughout the year, with peaks discernible in wet springs and especially during the late summer monsoon period, when many nest. Nearly three dozen different song types have been described in this species. The basic pattern common to most songs is a series of chipping or lisping notes followed by a trill; the trill may be higher or lower than the introductory notes. Some songs lack the initial notes, and the trill can then be reminiscent of the song of the Canyon Towhee. In some songs, the trill is very fast and dry, resembling that of a Chipping Sparrow or Worm-eating Warbler; in others, it is slow and loose, like a Pine Warbler's or distant Northern Cardinal's song. A warbling vocalization, also given only by males and in some individuals reminiscent of a brief Common Yellowthroat song, may serve

as pair recognition or a summons to reunion when feeding groups have scattered.

The most distinctive call note of this species is a frequently heard low-pitched and rich *plipt*, fuller than the thin chips of many other sparrows; this note or one very like it is repeated to form the introduction of some male songs. There is also a high, silvery *teep*, recalling the calls of the Black-throated Sparrow, or even an insect, but with a decided and abrupt decay.

RANGE AND GEOGRAPHIC VARIATION

The Rufous-winged Sparrow occupies one of the most restricted ranges of any North American passerellid, limited to desert hackberry thickets and open mesquite woodlands with an understory of grasses and sparse forbs in the southern reaches of the Sonoran Desert. In the United States, it was long considered an Arizona "endemic," but the species has recently expanded to the east in the New Mexico portions of Guadalupe Canyon. The species' stronghold in the United States remains south-central Arizona, north to Pinal and extreme southeastern Maricopa Counties, west to western Pima County, and, increasingly, east to Cochise County. Southward in the Sonoran Desert, this species is common

Like the other desert-dwelling sparrows in its genus, the Rufous-winged Sparrow's plumage can be decidedly unkempt at any season. *Arizona, April. Brian E. Small*

to abundant in appropriate habitats in Sonora, and occurs in Sinaloa as far south as 23 degrees north latitude at Mazatlán (eBird).

Rufous-winged Sparrows are resident throughout their range, but some flocking and movement is apparent in the nonbreeding season. Normally found at elevations between 2,000 and 4,000 feet, stray individuals are occasionally found as high as 5,000 feet in the oak canyons of southeastern Arizona, where they may visit feeders.

There are three generally recognized subspecies of the Rufous-winged Sparrow; the distinctions among them are subtle, and essentially invisible in the field. Northern birds are larger and paler, southern birds smaller and browner, but variation is clinal, making it difficult to delineate regional populations.

The nominate *carpalis*, the most familiar subspecies to most birders, breeds from southeastern Arizona south to north-central Sonora. It is large and pale, with gray-toned upperparts and whitish underparts. Somewhat smaller and browner, the race *bangsi* is finer-billed than *carpalis;* it is resident in southeastern Sonora and northern Sinaloa, intergrading with the southernmost subspecies *cohaerens* at El Fuerte. That race is found in central and south-central Sinaloa, between about 23 and 26 degrees north; it is darker and browner above than *bangsi*, with the back more broadly and more conspicuously streaked, and darker gray on the breast and throat.

ACKNOWLEDGMENTS

NOTES

INDEX

ACKNOWLEDGMENTS

I long suspected that the topos of universal thanks to the "birding community" was in fact an attempt to proleptically disqualify any and all potential reviewers. Now, though, faced with the challenge of thanking all those with whom I have been fortunate enough to watch sparrows over the years, I too find my gratitude extending to every one of the friends, teachers, and field companions who have guided my thinking about not just the passerellids but all the birds of the air, the ground, and the brush pile. Every member of the American birding community with whom I have been in contact has helped form my approach to birding, and my singling out by name just a few should not be understood as excluding the many who have been kind enough to answer questions, to ask questions, even to argue intelligently and provocatively about what it means to know and to learn. Jon Dunn, Ted Floyd, Ruth Green, Alan Grenon, the late Betty Grenon, Christian Hagenlocher, Homer Hansen, Chris McCreedy, Wayne Molhoff, Christian Nunes, Ross Silcock, Darlene Smyth, Denis Wright, and Soheil Zendeh all taught me more than they may know, in person and in correspondence, and I am grateful to them, as I am to my Facebook friends, and to my friends on Facebook, particularly the moderators of several sparrow "groups" whose efforts in providing and policing a forum for discussion and debate have changed the nature of bird identification for better and forever. Brooke McDonald and David Griffin kindly read portions of the manuscript, and I hope that they see their critical eye in the final text.

The reference desk and interlibrary loan office at Firestone Library provided their usual stellar assistance; Judith K. Golden and Lennart Beringer were equally helpful in finding and passing on to me any number of obscure, even arcane literary sources. The New York Public Library and the libraries of the Grolier Club and the American Museum of Natural History offered access to many of the rare items in their care; particular thanks are due Mai Reitmeyer for her kindness in saving me more than one trip across the Hudson. The curators and collections managers at several museums were equally generous, including especially Peter Reinthal and Melanie Bucci at the University of Arizona Museum of Natural History, Nate Rice at the Academy of Natural Sciences in Philadelphia, and Paul Sweet at the American Museum of Natural History.

This book would be far less appealing if not for the kindness and collegiality of the photographers whose images I am privileged to reproduce here. As a verbal rather than a visual soul, I found organizing such a quantity of visual material even more daunting than expected, and I am grateful to each of these artists, most especially to Brian E. Small, for their help in keeping things straight. At Houghton Mifflin Harcourt, Lisa White brought to her role an editorial acriby matched only by her patience and persistence, and to admit that this book would never have been finished without her help is as understated as it is obvious. My wife, Alison Beringer, supported this project with her usual generosity and forbearance; dedicating this book to her is thanks far too modest for her constant help.

NOTES

INTRODUCTION

PAGE

3   Passerellidae as understood by Klicka et al.: Klicka, J., et al. 2014. "A Comprehensive Multilocus Assessment of Sparrow (Aves: Passerellidae) Relationships." *Molecular Phylogenetics and Evolution* 77: 177–182.

4   "any bird that looks sparrowy": Blanchan, N. 1922. *Birds Worth Knowing.* New York: Doubleday.

5   "altogether impracticable . . ." : Wilson, A. 1808. *American Ornithology.* Vol. 1. Philadelphia: Bradford and Inskeep.

5   "great perplexity to the student": Ibid.

6   "considers it the best": Wilson, A. 1828. *American Ornithology.* 3 vols. Philadelphia: Collins and Co. and Harrison Hall.

6   Ambiguous Sparrow: Nuttall, T. 1832. *A Manual of the Ornithology of the United States and of Canada.* 1st ed. Vol. 1. Cambridge: Hilliard and Brown.

6   extensively illustrated guide: Baird, S., et al. 1874. *A History of North American Birds.* 3 vols. Boston: Little, Brown.

6   Coues's *Key*: Coues, E. 1872. *Key to North American Birds.* 1st ed. Salem: Naturalists' Agency.

7   "a difficult problem": Baird, S., ed. 1870. *Ornithology, From the Manuscript and Notes of J.G. Cooper.* Cambridge: University Press.

7   the head as the site of the distinguishing marks: Peterson, R. 1934. *A Field Guide to the Birds.* 1st ed. Boston: Houghton Mifflin.

7   two full pages of sparrow "busts": Robbins, C., et al. 1966. *Birds of North America.* 1st ed. New York: Golden Press.

8   "How providential": Blanchan, *Birds Worth Knowing.*

9   "the generic approach": Kaufman, K. 1990. *A Field Guide to Advanced Birding.* Boston: Houghton Mifflin.

9   "To identify a sparrow": Iliff, M. 2014. "Emberizids." In J. Alderfer, ed. *National Geographic Complete Birds of North America.* 2nd ed. Washington: National Geographic Society.

12   Ambiguous Sparrow: Nuttall, *Manual,* 1st ed.

13   "traveling suit": Chapman, F. 1916. *The Travels of Our Birds.* London: D. Appleton and Co.

14   "form a well-supported clade": American Ornithologists' Union. 2010. "Fifty-first Supplement to the American Ornithologists' Union *Check-list of North American Birds*." *Auk* 127: 726–744.

16   "quite unusually long" wings: Baird, S., with J. Cassin and G. Lawrence. 1858. *Birds.* Explorations and Surveys for a Railroad Route vol. 9. Washington: Beverly Tucker.

16   "scarcely tenable": Coues, E. 1873. "Notes on Two Little-known Birds of the United States." *American Naturalist* 7: 695–697.

16   German ornithologist and bibliographer: Giebel, C. 1872. *Thesaurus ornithologiae: Repertorium der gesammten ornithologischen Literatur und Nomenclator sämmtlicher Gattungen und Arten der Vögel.* Vol. 1. Leipzig: Brockhaus.

16   "peculiarly exasperating": Coues, E. 1872. "Giebel's *Thesaurus*." *American Naturalist* 6: 549–551.

16   not in every case one another's closest relatives: Klicka et al., "Comprehensive Multilocus Assessment."

16   first applied to the Grasshopper Sparrow: Allen, H. 1908. "A List of the Genera and Subgenera of North American Birds, with Their Types, According to Article 30 of the International Code of Zoölogical Nomenclature." *Bulletin of the American Museum of Natural History* 24: 1–50.

16   designated by Maynard himself: Stone, W. 1907. "Some Changes in the Current Generic Names of North American Birds." *Auk* 24: 189–199.

16   recommendation offered by Klicka et al.: Klicka et al., "Comprehensive Multilocus Assessment."

17   Howard and Moore's *Complete Checklist*: Dickinson, E., and L. Christidis, eds. 2014. *The Howard and Moore Complete Checklist of the Birds of the World.* 4th ed. Vol. 2. Eastbourne: Aves Press.

18   characters Swainson adduces: Swainson, W. 1837. *On the Natural History and Classification of Birds.* Vol. 2. London: Longman, Rees, Orme, Brown, Green, and Longman.

18   moved into *Passerherbulus*: American Ornithologists' Union. 1909. "Fifteenth Supplement to the American Ornithologists' Union *Check-list of North American Birds*." *Auk* 26: 294–303.

18   coined in 1905 by Harry C. Oberholser: Oberholser, H. 1917. "Notes on the Fringilline Genus *Passerherbulus* and Its Nearest Allies." *Ohio Journal of Science* 17: 332–338.

18   joined there in 1973 by the LeConte Sparrow: American Ornithologists' Union. 1973. "Thirty-second Supplement to the *Check-list of North American Birds*." *Auk* 90: 411–419.

18   a once-again broadened *Ammodramus*: American Ornithologists' Union. 1982. "Thirty-fourth Supplement to the American Ornithologists' Union *Check-list of North American Birds*." *Auk* 99 (supplement): 1CC–16CC.

19   they proposed that Oberholser's *Ammospiza* be revived: Klicka et al., "Comprehensive Multilocus Assessment."

19   *Thryospiza*, coined by Oberholser: Oberholser, "Fringilline Genus *Passerherbulus*."

20   the great Linnaean genus *Fringilla*: Forster, J. 1772. "An Account of the Birds Sent from Hudson's Bay, With Observations Relative to Their Natural History, and Latin Descriptions of Some of the Most Uncommon." *Philosophical Transactions* 62: 382–433.

20   moved it into the genus *Passerina*: Vieillot, L. 1817. *Nouveau dictionnaire d'histoire naturelle.* Vol. 25. Paris: Déterville.

20 Indigo Bunting, the Bobolink, and the Snow Bunting: Vieillot, L. 1816. *Analyse d'une nouvelle ornithologie élémentaire*. Paris: Deterville.

20 William Swainson: Richardson, J., and W. Swainson. 1831. *Fauna boreali-americana: Or the Zoology of the Northern Parts of British America*. Part 2. London: John Murray.

20 John James Audubon: Audubon, J. 1839. *A Synopsis of the Birds of North America*. Edinburgh: Adam and Charles Black.

20 a new genus *Spizella*: Bonaparte, C. 1832. *Saggio di una distribuzione metodica degli animali vertebrati*. Rome: Antonio Boulzaler.

20 Clay-colored, Chipping, and American Tree Sparrows: Bonaparte, C. 1838. *A Geographic and Comparative List of the Birds of Europe and North America*. London: John van Voorst.

20 Cabanis objected: Cabanis, J. 1851. *Museum Heineanum: Verzeichniss der ornithologischen Sammlung des Oberamtmann Ferdinand Heine*. Vol. 1. Halberstadt: R. Frantz.

21 the single species *Spizelloides arborea*: Slager, D., and J. Klicka. 2014. "A New Genus for the American Tree Sparrow (Aves: Passeriformes: Passerellidae)." *Zootaxa* 3821: 398–400.

21 "leaden skies above and the snow below": Parkhurst, H. 1894. *The Birds' Calendar*. New York: C. Scribner's Sons.

22 "five white lines": Sclater, P., and O. Salvin. 1868. "Descriptions of New or Little-known American Birds of the Families Fringillidae, Oxyrhamphidae, Bucconidae, and Strigidae." *Proceedings of the Zoological Society of London* 1868: 321–329.

23 "exceedingly distinct Sparrow": Ridgway, R. 1883. "Notes Upon Some Rare Species of Neotropical Birds." *Ibis* ser 5. 4: 399–401.

23 "considerable doubt": Ridgway, R. 1901. *The Birds of North and Middle America*. Bulletin of the United States National Museum, Bulletin 50, pt. 1. Washington: Government Printing Office.

23 in spite of the similarity: Storer, R. 1955. "A Preliminary Survey of the Sparrows of the Genus *Aimophila*." *Condor* 57: 192–201.

23 rejected as premature: Phillips, A., and R. Phillips Farfan. 1993. "Distribution, Migration, Ecology, and Relationships of the Five-striped Sparrow, *Aimophila quinquestriata*." *Western Birds* 24: 65–72.

24 nearest relative of the Black-throated Sparrow: DaCosta, J., et al. 2009. "A Molecular Systematic Revision of Two Historically Problematic Songbird Clades: *Aimophila* and *Pipilo*." *Journal of Avian Biology* 40: 206–216.

24 finally reunited: American Ornithologists' Union, "Fifty-first Supplement."

24 subsequent molecular study: Klicka et al., "Comprehensive Multilocus Assessment."

25 *Passerina* by Vieillot: Vieillot, L. 1817. "La Passerine des prés, *Passerina pratensis*." *Nouveau dictionnaire d'histoire naturelle*. Vol. 25. Paris: Déterville.

25 *Emberiza* by William Jardine: Wilson, A. 1832. *American Ornithology, or The Natural History of the Birds of the United States, With a Continuation by Charles Lucian Bonaparte, Prince of Musignano; the Illustrative Notes, and Life of Wilson, by Sir William Jardine, bart*. Vol. 1. London: Whittaker, Treacher, and Arnot.

25 and by Audubon: Audubon, *Synopsis* .

25 path to the taxonomic future: Swainson, W. 1827. "A Synopsis of the Birds Discovered in Mexico by W.

Bullock, F.L.S. and H.S., and Mr. William Bullock, jun." *The Philosophical Magazine* 2nd ser., 1: 364–369 and 433–442.

25 to the genus *Coturniculus*: Baird, *Birds*.

25 the genus's type species: Bonaparte, *Geographic and Comparative List*.

25 the lead of Robert Ridgway: Ridgway, *Birds of North and Middle America*.

25 the Baird, Grasshopper, LeConte, Saltmarsh, and Nelson Sparrows: American Ornithologists' Union. 1903. "Twelfth Supplement to the American Ornithologists' Union *Check-list of North American Birds*." *Auk* 20: 331–368.

25 *Passerherbulus*: American Ornithologists' Union, "Fifteenth Supplement."

26 *Arremonops* sparrows of tropical forest: Klicka et al., "Comprehensive Multilocus Assessment."

26 "very great differences of coloration and considerable differences of form": Ridgway, *Birds of North and Middle America*.

27 declined to recommend that the genus be split: Storer, "Preliminary Survey."

27 "probably not as closely related to each other": Wolf, L. 1977. *Species Relationships in the Avian Genus Aimophila*. Ornithological Monographs 23. Lawrence: American Ornithologists' Union.

27 "probably not as closely related to each other": Ibid.

27 miscellaneous nature of the traditional *Aimophila*: Carson, R., and G. Spicer. 2003. "A Phylogenetic Analysis of the Emberizid Sparrows Based on Three Mitochondrial Genes." *Molecular Phylogenetics and Evolution* 29: 43–57.

27 DaCosta and his coauthors: DaCosta, et al., "Molecular Systematic Revision."

27 *Peucaea*, erected by Audubon: Audubon, *Synopsis*.

27 for the Bachman Sparrow: DaCosta et al., "Molecular Systematic Revision."

27 formally placing them in the resurrected *Peucaea*: American Ornithologists' Union, "Fifty-first Supplement."

STRIPED SPARROW

30 *Animals in Menageries*: Swainson, W. 1838. *Animals in Menageries*. London: Longman, Orme, Brown, Green & Longmans, and John Taylor.

30 the skins he was working with: Sclater, P. 1861. *Catalogue of a Collection of American Birds Belonging to Philip Lutley Sclater*. London: N. Trubner and Co.

30 "an exceedingly common species": Salvin, O., and F. Godman. 1889. "Notes on Mexican Birds." *Ibis* ser. 6, 1: 232–243.

30 "a bird of much character": Chapman, F. 1898. "Notes on Birds Observed at Jalapa and Las Vigas, Vera Cruz, Mexico." *Bulletin of the American Museum of Natural History* 10.2: 15–43.

30 "very easy to shoot": Stone, W. 1890. "On Birds Collected in Yucatan and Southern Mexico." *Proceedings of the Academy of Natural Sciences of Philadelphia* 42: 201–218.

30 still encountered today: Choate, E. 1985. *The Dictionary of American Bird Names*. Rev. ed. by R. Paynter. Boston: Harvard Commons Press.

30 miscellaneous genus *Embernagra*: Gray, G. 1870. *Hand-list of Genera and Species of Birds*. Part 2. London: Trustees of the British Museum.

30 the new genus *Plagiospiza*: Ridgway, R. 1898. "Descriptions of Supposed New Genera, Species, and Subspecies of American Birds. I: Fringillidae." *Auk* 15: 223–230.

31 as Bonaparte himself later recognized: Bonaparte, C. 1856. "Excursions dans les divers musées d'Allemagne, de Hollande et de Belgique, et Tableaux paralléliques de l'ordre des Échassiers." *Comptes rendus hébdomadaires des séances de l'Académie des sciences* 43: 410–421.

31 its new genus name: Van Rossem, A. 1942. "Bonaparte's Types of *Oriturus wrangeli* and *Oriturus mexicanus*." *Auk* 59: 449–450.

SONG SPARROW

32 the specimen he illustrated: Burtt, E., and W. Davis. 2013. *Alexander Wilson: The Scot Who Founded American Ornithology*. Cambridge, MA: Harvard University Press.

32 "the earliest, sweetest, and most lasting songster": Wilson, A. 1810. *American Ornithology, or the Natural History of the Birds of the United States*. Vol. 2. Philadelphia: Bradford and Inskeep.

32 five nesting pairs: Wilson, A. 1811. *American Ornithology, or the Natural History of the Birds of the United States*. Vol. 4. Philadelphia: Bradford and Inskeep.

32 "So nearly do many species of our Sparrows": Wilson, *American Ornithology*.

32 several suggestive details: Catesby, M. 1731. *The Natural History of Carolina, Florida and the Bahama Islands*. Vol. 1. London: privately published.

32 George Edwards: Edwards, G. 1764. *Gleanings of Natural History*. Pt. 3. London: privately published.

32 John Latham: Latham, J. 1783. *A General Synopsis of Birds*. Vol. 2. London: Benjamin White.

32 Thomas Pennant: Pennant, T. 1785. *Arctic Zoology*. Vol. 2. London: Henry Hughs.

32 "called in New York, the Shepherd": Ibid.

33 *Fringilla fasciata* and *Fringilla ferruginea*: Gmelin, J. 1788. *Caroli a Linné . . . Systema naturae per regna tria naturae*. Vol. 1. Leipzig: Georg Emanuel Beer.

33 Alexander Wilson knew Gmelin's list: Burtt and Davis, *Alexander Wilson*.

33 Bartram's 1791 *Travels*: William Bartram. 1791. *Travels through North and South Carolina, Georgia, East and West Florida, the Cherokee Country, the Extensive Territories of the Muscogulges or Creek Confederacy, and the Country of the Chactaws. Containing an Account of the Soil and Natural Productions of Those Regions; Together with Observations on the Manners of the Indians*. Philadelphia: James and Johnson.

33 Wilson named the sparrow: Wilson, *American Ornithology*.

34 he changed the scientific name: Nuttall, T. 1840. *A Manual of the Ornithology of the United States and Canada*. 2nd ed. Boston: Hilliard, Gray, and Co.

34 "the proper specific name of the Song Sparrow": Scott, D. 1876. "The Proper Specific Name of the Song Sparrow." *American Naturalist* 10: 17–18.

34 Robert Ridgway: Ridgway, R. 1880. "On Current Objectionable Names of North American Birds." *Bulletin of the Nuttall Ornithological Club* 5: 36–38.

34 Elliott Coues: Coues, E. 1880. "Notes and Queries Concerning the Nomenclature of North American Birds." *Bulletin of the Nuttall Ornithological Club* 5: 95–102.

34 coined for another bird: Oberholser, H. 1899. "The Names of the Song Sparrows." *Auk* 16: 182–183.

34 "necessary to revert to the long-familiar name": American Ornithologists' Union. 1901. "Tenth Supplement to the American Ornithologists' Union *Check-list of North American Birds*." *Auk* 18: 295–321.

34 *Fringilla cinerea*: Gmelin, *Systema naturae*.

34 the AOU concluded: American Ornithologists' Union. 1903. "Twelfth Supplement to the American Ornithologists' Union *Check-list of North American Birds*." *Auk* 20: 331–368.

34 "the Song Sparrows again become": American Ornithologists' Union. 1908. "Fourteenth Supplement to the American Ornithologists' Union *Check-list of North American Birds*." *Auk* 25: 343–399.

34 misapplied Gmelin's name: Wilson, A. 1811. *American Ornithology, or the Natural History of the Birds of the United States*. Vol. 3. Philadelphia: Bradford and Inskeep.

35 still classified as *Fringilla*: Audubon, J. 1839. *A Synopsis of the Birds of North America*. Edinburgh: Adam and Charles Black.

35 Baird's monumental *Birds*: Baird, S., with J. Cassin and G. Lawrence. 1858. *Birds*. Explorations and Surveys for a Railroad Route. Vol. 9. Washington: Beverly Tucker.

35 "all fail to separate the genera": Paynter, R. 1964. "Generic Limits of Zonotrichia." *Condor* 66: 277–281.

35 the world checklist begun by James L. Peters: Paynter, R., and R. Storer. 1970. *Check-List of Birds of the World: A Continuation of the Work of James L. Peters*. Vol. 13. Cambridge: Museum of Comparative Zoology.

40 The North Sea records: Lewington, I., et al. 1991. *A Field Guide to the Rare Birds of Britain and Europe*. London: HarperCollins.

40 netted on Fair Isle: Davis, P., and R. Dennis. 1959. "Song Sparrow at Fair Isle." *British Birds* 52: 419–421, pl. 70.

40 Russian Far East and Japan: Brazil, M. 2009. *Birds of East Asia*. London: Christopher Helm.

40 "Marshall, sling-shot": Phillips, A., et al. 1964. *The Birds of Arizona*. Tucson: University of Arizona Press.

40 "plasticity of organization": Ridgway, R. 1901. *The Birds of North and Middle America*. Bulletin of the United States National Museum, Bulletin 50, pt. 1. Washington: Government Printing Office.

40 Michael A. Patten and Christine L. Pruett: Patten, M., and C. Pruett. 2009. "The Song Sparrow, *Melospiza melodia*, as a Ring Species: Patterns of Geographic Variation, a Revision of Subspecies, and Implications for Speciation." *Systematics and Biodiversity* 7.1: 33–62.

42 the nonbreeding range of *caurina* and possibly of *rufina*: Johnson, O., P. Pyle, and J. Tietz. 2013. "The Subspecies of the Song Sparrow on Southeast Farallon Island and in Central California." *Western Birds* 44: 162–170.

42 the correlation between humidity and feather darkness: Burtt, E., and J. Ichida. 2004. "Gloger's Rule, Feather-degrading Bacteria, and Color Variation Among Song Sparrows." *Condor* 106: 681–686.

42 breeding on islands: Uy, J., and L. Vargas-Castro. 2015. "Island Size Predicts the Frequency of Melanic Birds in the Color-polymorphic Flycatcher *Monarcha castaneiventris* of the Solomon Islands." *Auk* 132: 787–794.

42 over much of California: Johnson et al., "Subspecies of the Song Sparrow."

43 *Zonotrichia fallax*: Baird, S. 1854. "Descriptions of New Birds Collected Between Albuquerque, N.M., and San Francisco, California, During the Winter of 1853-54, by Dr. C.B.R. Kennerly and H.B. Möllhausen, Naturalists Attached to the Survey of the Pacific R.R. Route, Under Lt. A.W. Whipple." *Proceedings of the Academy of Natural Sciences of Philadelphia* 7: 118–120.

43 uncharacteristic confusion: Coues, E. 1872. *Key to North American Birds*. 1st ed. Salem: Naturalists' Agency.

43 correctly treated as a single subspecies: Ridgway, R. 1896. *A Manual of North American Birds*. 2nd ed. Philadelphia: J.B. Lippincott.

44 type specimen was taken in July 1898: Nelson, E. 1899. "Descriptions of New Birds from Mexico." *Auk* 16: 25–31.

44 "along the Río Lerma drainage": Patten and Pruett, "Song Sparrow, *Melospiza melodia*."

SWAMP SPARROW

46 "history of this obscure and humble species": Wilson, A. 1811. *American Ornithology, or the Natural History of the Birds of the United States*. Vol. 3. Philadelphia: Bradford and Inskeep.

46 *Passer palustris* the Reed Sparrow: Bartram, W. 1791. *Travels Through North and South Carolina, Georgia, East and West Florida, the Cherokee Country, the Extensive Territories of the Muscogulges, or Creek Confederacy, and the Country of the Chactaws*. Philadelphia: James and Johnson.

46 "overlooked by the naturalists of Europe": Wilson, *American Ornithology*.

46 John Latham: Latham, J. 1790. *Index ornithologicus*. Vol. 1. London: privately published.

46 James Francis Stephens: Stephens, J. 1815. *Aves*. Vol. 9 in G. Shaw, ed. *General Zoology, or Systematic Natural History*. London: G. Wilkie et al.

46 Hinrich Lichtenstein: Lichtenstein, H. 1823. *Verzeichniss der Doubletten des Zoologischen Museums der Königlichen Universität zu Berlin*. Berlin: T. Trautwein.

46 Charles Bonaparte: Bonaparte, C. 1826. *Observations on the Nomenclature of Wilson's Ornithology*. Philadelphia: Anthony Finley.

47 Thomas Nuttall: Nuttall, T. 1832. *A Manual of the Ornithology of the United States and of Canada*. 1st ed. Vol. 2. Boston: Hilliard, Gray, and Company.

Nuttall, T. 1840. *A Manual of the Ornithology of the United States and Canada*. 2nd ed. Boston: Hilliard, Gray, and Co.

47 Audubon: Audubon, J. 1839. *A Synopsis of the Birds of North America*. Edinburgh: Adam and Charles Black.

47 "best to retain Wilson's name": Baird, S., with J. Cassin and G. Lawrence. 1858. *Birds*. Explorations and Surveys for a Railroad Route. Vol. 9. Washington: Beverly Tucker.

47 *georgiana* was correctly applied: Ridgway, R. 1885. "Some Emended Names of North American Birds." *Proceedings of the United States National Museum* 8: 354–356.

47 Elliott Coues: Coues, E. 1875. "Fasti ornithologiae redivivi: Bartram's 'Travels'." *Proceedings of the Academy of Natural Sciences of Philadelphia* 27: 338–358.

47 "The naming of our birds": Coues, E. 1887. *Key to North American Birds*. 3rd ed. Boston: Estes and Lauriat.

47 "there is no doubt": Coues, E. 1903. *Key to North American Birds*. 5th ed. Vol. 1. Boston: Dana Estes and Company.

47 on the instigation of Elliott Coues: Coues, *Key*. 3rd ed.

47 *Birdcraft*: Wright, M. 1895. *Birdcraft: A Field Book of 200 Song, Game, and Water Birds*. New York: Macmillan.

48 catch-all genus: Paynter, R., and R. Storer. 1970. *Check-List of Birds of the World: A Continuation of the Work of James L. Peters*. Vol. 13. Cambridge: Museum of Comparative Zoology.

48 coiner of the genus *Zonotrichia*: Richardson, J., and W. Swainson. 1831. *Fauna boreali-americana: Or the Zoology of the Northern Parts of British America*. Part 2. London: John Murray.

48 with the Saltmarsh and Seaside Sparrows: Swainson, W. 1837. *On the Natural History and Classification of Birds*.

Vol. 2. London: Longman, Rees, Orme, Brown, Green, and Longman.

48 *Passerculus* alongside the Savannah Sparrow: Bonaparte, C. 1838. *A Geographic and Comparative List of the Birds of Europe and North America*. London: John van Voorst.

48 "became convinced that an error had been committed": Pearson, T. 1898. "An Addition and a Correction to the List of North Carolina Birds." *Auk* 15: 275.

48 painted by Robert Ridgway: Deignan, H. 1961. *Type Specimens of Birds in the United States National Museum*. Bulletin of the United States National Museum 221. Washington: Government Printing Office.

48 *History of North American Birds*: Baird, S., et al. 1874. *A History of North American Birds*. Vol. 2. Boston: Little, Brown.

48 pronounced the species "untenable": Ridgway, R. 1881. *Nomenclature of North American Birds*. Bulletin of the United States National Museum 21. Washington: Government Printing Office.

48 "in a plumage hitherto unrecognized": Coues, E. 1883. "Note on 'Passerculus caboti'." *Bulletin of the Nuttall Ornithological Club* 8: 58.

48 "yellow morph" of the species: Rowland, E. 1928. "Abnormal Yellow Coloration of Swamp Sparrows." *Bulletin of the Northeastern Bird-banding Association* 4: 53–56.

50 "an identification game": Peterson, R. 1985. "Memories of Ludlow Griscom." *Birding* 17.2/3: 105–110.

51 as far as Bermuda: Dobson, A. 2004. "Bird Report October to December 2003." *Bermuda Audubon Society Newsletter* 15.1: 6–8.

51 "Although the mid-Atlantic Coast": Greenberg, R., et al. 2008. "Temporal Distribution of the Coastal Plain Swamp Sparrow: The Importance of Field Identification." *Birding* 45.5: 42–49.

51 "very weakly defined" races: Paynter and Storer, *Check-List*.

51 Latham's type: Bond, G., and R. Stewart. 1951. "A New Swamp Sparrow from the Maryland Coastal Plain." *Wilson Bulletin* 63.1: 38–40.

51 The Coastal Plain Swamp Sparrow: Greenberg, et al., "Temporal Distribution."

51 "virtually nonoverlapping": Greenberg, R., and S. Droege. 1990. "Adaptations to Tidal Marshes in Breeding Populations of the Swamp Sparrow." *Condor* 92: 393–404.

LINCOLN SPARROW

52 "Mr. A finished a drawing": Townsend, C. 1924. "A Visit to Tom Lincoln's House with Some Auduboniana." *Auk* 41: 237–242.

52 "These two always go together": Audubon, J. 1899. "The Labrador Journal." In M. Audubon, ed. *Audubon and His Journals*. Vol. 1. New York: Charles Scribner's Sons.

52 "The Caribou flies": Ibid.

52 "surpassing in vigor": Audubon, J. 1834. *Ornithological Biography*. Vol. 2. Edinburgh: Adam Black and Charles Black.

53 "more abundant and less shy": Ibid.

53 others had meanwhile been secured: Audubon, J. 1839. *A Synopsis of the Birds of North America*. Edinburgh: Adam Black and Charles Black.

53 "strange place for it": Audubon, J. 1899. "The Missouri River Journals." In M. Audubon, ed. *Audubon and His Journals*. Vol. 1. New York: Charles Scribner's Sons.

53 In the 1850s: Ridgway, R. 1901. *The Birds of North and Middle America*. Bulletin of the United States National Museum, Bulletin 50, pt. 1. Washington: Government Printing Office.

53 Heinrich von Kittlitz: Kittlitz, F. H. von. 1858. *Denkwürdigkeiten einer Reise nach dem russischen Amerika, nach Mikronesien und durch Kamtschatka*. Vol. 1. Gotha: Justus Perthes.

53 "readily recognizable": Finsch, O. 1872. "Zur Ornithologie Nordwest-Amerikas." *Abhandlungen herausgegeben vom naturwissenschaftlichen Vereine zu Bremen* 3: 17–86.

53 the scientific name properly to be applied: Oberholser, H. 1906. "An Earlier Name for *Melospiza lincolnii striata*." *Proceedings of the Biological Society of Washington* 19: 42.

54 "easily known among the American sparrows": Baird, S., et al. 1860. *The Birds of North America*. Vol. 1. New York: Appleton and Co.

54 "the immature Swamp Sparrow": Peterson, R. 1947. *A Field Guide to the Birds*. 2nd ed. Boston: Houghton Mifflin.

55 " a skulker": Ibid.

56 Vagrants have strayed: Ammon, E. 1995. "Lincoln's Sparrow (*Melospiza lincolnii*)." *Birds of North America Online*. Ithaca: Cornell Lab of Ornithology.

56 stripes "crowd together": Miller, A., and T. McCabe. 1935. "Racial Differentiation in *Passerella (Melospiza) lincolnii*." *Condor* 37: 144–160.

56 mixing of populations: Ibid.

SIERRA MADRE SPARROW

57 "had not planned on collecting": Bailey, A., and H. Conover. 1935. "Notes from the State of Durango, Mexico." *Auk* 52: 421–424.

57 Bangs named the species: Bangs, O. 1931. "A New Genus and Species of American Buntings." *Proceedings of the New England Zoölogical Club* 12: 85–88.

57 "Ridgway wrote saying": Ibid.

58 "to doubt that this sparrow should be segregated": Pitelka, F. 1947. "Taxonomy and Distribution of the Mexican Sparrow *Xenospiza baileyi*." *Condor* 49: 199–203.

58 La Cima population: Dickerman, R., et al. 1967. "On the Sierra Madre Sparrow, *Xenospiza baileyi*, of Mexico." *Auk* 84: 49–60.

58 a greatly enlarged *Ammodramus*: Paynter, R., and R. Storer. 1970. *Check-List of Birds of the World: A Continuation of the Work of James L. Peters*. Vol. 13. Cambridge: Museum of Comparative Zoology.

58 "better placed in *Ammodramus*": Howell, S., and S. Webb. 1995. *A Guide to the Birds of Mexico and Northern Central America*. Oxford: Oxford University Press.

58 Recent molecular study: Klicka et al. "Comprehensive Multilocus Assessment."

58 a vastly expanded *Passerculus*: Klicka, J., et al. 2007. "A Molecular Evaluation of the North American 'Grassland' Sparrow Clade." *Auk* 124: 537–551.

58 The "curious little" Sierra Madre Sparrow: Hellmayr, C. 1938. *Catalogue of Birds of the Americas and the Adjacent Islands in Field Museum of Natural History*. Zoological Series 13, pt. 11. Chicago: Field Museum of Natural History.

59 Savannah Sparrow: Bailey and Conover, "Notes."

60 bill is fairly small: Pitelka, "Taxonomy and Distribution."

60 Newly hatched Sierra Madre Sparrows: Dickerman et al., "Sierra Madre Sparrow."

60 Helmuth Wagner: Pitelka, "Taxonomy and Distribution."

60 three breeding pairs: Oliveras de Ita, A., and O. Rogas-Soto. 2006. "A Survey for the Sierra Madre Sparrow (*Xenospiza baileyi*), With Its Rediscovery in the State of Durango, Mexico." *Bird Conservation International* 16: 25–32.

60 total global population: Oliveras de Ita, A., et al. 2002. "El Gorrión serrano (*Xenospiza baileyi*)." In Gómez de Silva, H., and A. Oliveras de Ita, eds. *Conservación de aves: Experiencias en México*. Mexico City: Museo de Historia Natural.

60 southeastern subspecies *sierrae*: Pitelka, "Taxonomy and Distribution."

60 could not be distinguished: Dickerman et al., "Sierra Madre Sparrow."

BAIRD SPARROW

61 late July of 1843: Marks, J., and T. Nordhagen. 2005. "Type Locality of *Ammodramus bairdii* (Audubon)." *Auk* 122.1: 350.

61 They had much difficulty in raising them: Audubon, J. 1844. *The Birds of America*. 2nd ed. Vol. 7. Philadelphia: E.G. Dorsey.

61 they succeeded in shooting three birds: Marks and Nordhagen, "Type Locality."

61 "the glorious Audubonian sun": Coues, E. 1878. *Birds of the Colorado Valley*. United States Geological Survey of the Territories Miscellaneous Publications 11. Washington: Government Printing Office.

61 "faded specimen in worn plumage": Coues, E. 1874. *Birds of the Northwest*. Miscellaneous Publications of the Geological Survey of the Territories 3. Washington: Government Printing Office.

61 "evidently closely related to *C. Bairdii*": Aiken, C. 1873. "A New Species of Sparrow." *American Naturalist* 7: 236–237.

62 "this nominal species": Scott, D. 1873. "*Centronyx 'ochrocephalus'* Aiken." *American Naturalist* 7: 564–565.

62 the fresh, autumn plumage of *bairdii*: Coues, E. 1873. "Notes on Two Little-known Birds of the United States." *American Naturalist* 7: 695–697.

62 "*Centronyx bairdii* (Aud.) = *C. Ochrocephalus* Aiken.": Ridgway, R. 1873. "The Birds of Colorado." *Bulletin of the Essex Institute* 5: 174–195.

62 he returned with some 75 skins: Coues, *Birds of the Northwest*.

62 "in some places outnumbering all the other birds": Coues, "Two Little-known Birds."

62 first nest and eggs: Allen, J. 1874. "Notes on the Natural History of Portions of Dakota and Montana Territories." *Proceedings of the Boston Society of Natural History* 17: 33–86.

62 "in immense numbers": Henshaw, H. 1874. "On a Hummingbird New to Our Fauna, With Certain Other Facts Ornithological." *American Naturalist* 8: 241–243.

62 at least 18 specimens: Henshaw, H. 1875. "Report Upon the Ornithological Collections Made in Portions of Nevada, Utah, California, Colorado, New Mexico, and Arizona, During the Years 1871, 1872, 1873, and 1874." *Report Upon United States Geological Surveys West of the One Hundredth Meridian* 5: 133–507.

62 Within little more than a decade: Brewster, W. 1885. "Additional Notes on Some Birds Collected In Arizona and the Adjoining Province of Sonora, Mexico, by Mr. F. Stephens in 1884; with a Description of a New Species of *Ortyx*." *Auk* 2: 196–200.

62 Ernest Thompson Seton: Seton, E. 1885. "Notes on Manitoban Birds." *Auk* 2: 267–271.

62 the obvious question: Coues, *Birds of the Northwest*.

62 "distinguish this species very readily": Baird, S., with J. Cassin and G. Lawrence. 1858. *Birds*. Explorations and Surveys for a Railroad Route vol. 9. Washington: Beverly Tucker.

62 "so much like a savanna[h] sparrow": Coues, "Two Little-known Birds."

62 "entailed a considerable mortality": Wetmore, A. 1920. "An Erroneous Kansas Record for Baird's Sparrow." *Auk* 37: 457–458.

62 "Many sight records": American Ornithologists' Union. 1998. *Check-list of North American Birds*. 7th ed. Washington: American Ornithologists' Union.

63 Baird Sparrows in flight away: Zimmer, K. 2000. *Birding in the American West*. Ithaca: Cornell University Press.

63 color and pattern of the nape: Zimmer, K. 1985. *The Western Bird Watcher*. Englewood Cliffs: Prentice-Hall.

64 northern Durango and Zacatecas: Green, M., et al. 2002. "Baird's Sparrow (*Ammodramus bairdii*)." *Birds of North America Online*. Ithaca: Cornell Lab of Ornithology.

64 specimen record from Missouri: Easterla, D. 1966. "The Baird's Sparrow and Burrowing Owl in Missouri." *Condor* 69: 88–89.

HENSLOW SPARROW

65 declaration of bankruptcy in July 1819: Ford, A. 1988. *John James Audubon: A Biography*. New York: Abbeville Press.

65 a sparrow "on the ground, amongst tall grass": Audubon, J. 1831. *Ornithological Biography*. Vol. 1. Edinburgh: Adam Black and Charles Black.

65 "many kind attentions": Jackson, C. n.d. [2014]. *John James Laforest Audubon: An English Perspective*. Winslow: privately published.

65 "procured a great number": Audubon, J. 1839. *Ornithological Biography*. Vol. 5. Edinburgh: Adam Black and Charles Black.

65 numerous in the agricultural fields of New Jersey: Ibid.

65 the range of the "abundant" Henslow Sparrow: Audubon, J. 1839. *A Synopsis of the Birds of North America*. Edinburgh: Adam and Charles Black.

66 Ferdinand V. Hayden collected one: Hayden, F. 1862. "On the Geology and Natural History of the Upper Missouri." *Transactions of the American Philosophical Society* 12: 1–218.

66 Witmer Stone examined the birds: Stone, W. 1904. "Henslow's Sparrow at Bethlehem, Pa.—A Correction." *Auk* 21: 386–387.

67 "Something surely went amiss here": Peterson, R. 1934. *A Field Guide to the Birds*. 1st ed. Boston: Houghton Mifflin.

69 northernmost record: Heagy, A. 2011. *COSEWIC Assessment and Status Report on the Henslow's Sparrow Ammodramus henslowii in Canada*. Ottawa: Committee on the Status of Endangered Wildlife in Canada.

69 phenological data available from the old eastern breeding range: Herkert, J., et al. 2002. "Henslow's Sparrow (*Ammodramus henslowii*)." Birds of North America Online. Ithaca: Cornell Lab of Ornithology.

69 Wintering Henslow Sparrows: Ibid.

69 Apparent vagrants have been reported: Alderfer, J., and J. Dunn, eds. 2014. *National Geographic Complete Birds of North America*. 2nd edition. Washington, D.C.: National Geographic.

69 and the Bahamas: Connor, H., and R. Loftin. 1985. "The Birds of Eleuthera Island, Bahamas." *Florida Field Naturalist* 14: 77–93.

69 The new subspecies, which he named *susurrans*: Brewster, W. 1918. "An Undescribed Race of Henslow's Sparrow." *Proceedings of the New England Zoölogical Club* 6: 77–79.

69 "an ultra-typical example": Ibid.

69 "Indeed, whether the eastern race": Herkert et al., "Henslow's Sparrow (*Ammodramus henslowii*)."

LARGE-BILLED SPARROW

70 an undescribed finch Heermann had collected: Stone, W. 1907. "Adolphus L. Heermann, M.D." *Cassinia* 11: 1–6.

70 "this plain-plumaged bird": Cassin, J. 1852. "Descriptions of New Species of Birds, Specimens of Which are in the Collection of the Academy of Natural Sciences of Philadelphia." *Proceedings of the Academy of Natural Sciences of Philadelphia* 6: 184–188.

70 "strange as it may seem": Grinnell, J. 1905. "Where Does the Large-billed Sparrow Spend the Summer?" *Auk* 22: 16–21.

70 "That the Large-billed Sparrow breeds somewhere": Ibid.

70 "also solves the migration mystery": Oberholser, H. 1919. "A Revision of the Subspecies of *Passerculus rostratus* (Cassin)." *Ohio Journal of Science* 19: 344–354.

70 initial comparisons of the specimens: Cassin, "Descriptions of New Species."

70 share the genus *Ammodramus*: Cassin, J. 1853–1855. *Illustrations of the Birds of California, Texas, Oregon, British and Russian America*. Philadelphia: J.B. Lippincott and Co.

70 "With some hesitation": Baird, S., with J. Cassin and G. Lawrence. 1858. *Birds*. Explorations and Surveys for a Railroad Route. Vol. 9. Washington: Beverly Tucker.

70 Heermann himself adopted *Passerculus*: Heermann, A. 1859. "Report Upon Birds Collected on the Survey." in *Reports of Explorations and Surveys to Ascertain the Most Practicable and Economical Route for a Railroad*. Vol. 10, 29–77. Washington: Beverley Tucker.

70 a single species, the Savannah Sparrow, with 16 subspecies: American Ornithologists' Union. 1944. "Nineteenth Supplement to the American Ornithologists' Union *Check-list of North American Birds*." *Auk* 61: 441–464.

71 rejected proposals to resplit that bird: AOS Committee on Taxonomy and Nomenclature (North and Middle America). 2011. "Votes on Proposals 2011-C." tinyurl.com/NACC2011.

71 three distinct clades: Zink, R., et al. 2005. "Mitochondrial DNA Variation, Species Limits, and Rapid Evolution of Plumage Coloration and Size in the Savannah Sparrow." *Condor* 107: 21–28.

71 plumage pattern and color: Rising, J. 2009. "Geographic Variation in Plumage Pattern and Coloration of Savannah Sparrows." *Wilson Bulletin* 121: 253–264.

71 size and structure: Rising, J. 2001. *Geographic Variation in Size and Shape of Savannah Sparrows (*Passerculus sandwichensis*)*. Studies in Avian Biology 23. Lawrence: Cooper Ornithological Society.

71 "consistency of treatment is impossible": Grinnell, J. 1939. "Proposed Shifts of Names in *Passerculus*—A Protest." *Condor* 41: 112–119.

72 two subspecies: Van Rossem, A. 1947. "A Synopsis of the Savannah Sparrows of Northwestern Mexico." *Condor* 49: 97–107.

72 California's Salton Basin: Small, A. 1994. *California Birds: Their Status and Distribution*. Vista, CA: Ibis.

72 Colorado River in Arizona: Rosenberg, K., et al. 1991. *Birds of the Lower Colorado River Valley*. Tucson: University of Arizona Press.

BELDING SPARROW

73 "I don't enjoy these birds": Fisher, W. 1918. "In Memoriam: Lyman Belding." *Condor* 20: 51–61.

73 expedition led by A.W. Anthony: McGregor, R. 1898. "Description of a New *Ammodramus* from Lower California." *Auk* 15: 265–266.

73 a distinct subspecies of the Large-billed Sparrow: Ridgway, R. 1901. *The Birds of North and Middle America*. Bulletin of the United States National Museum, Bulletin 50, pt. 1. Washington: Government Printing Office.

73 a different, much older puzzle: Brewster, W. 1902. "Birds of the Cape Region of Lower California." *Bulletin of the Museum of Comparative Zoology* 41: 1–243.

73 George N. Lawrence described it as new: Lawrence, G. 1867. "Descriptions of New Species of American Birds." *Annals of the Lyceum of Natural History of New York* 8: 466–482.

74 the two "species" were probably identical: Brewster, "Birds of the Cape Region."

74 eliminating McGregor's Lagoon Sparrow: American Ornithologists' Union. 1904. "Thirteenth Supplement to the American Ornithologists' Union *Check-list of North American Birds*." *Auk* 21: 411–424.

74 synonymized once again in 1949: American Ornithologists' Union. 1949. "Twenty-fourth Supplement to the American Ornithologists' Union *Check-list of North American Birds*." *Auk* 66: 281–285.

74 *Passerculus anthinus*: Baird, S., with J. Cassin and G. Lawrence. 1858. *Birds*. Explorations and Surveys for a Railroad Route. Vol. 9. Washington: Beverly Tucker.

74 "two quite different birds": Ridgway, R. 1885. "On Two Hitherto Unnamed Sparrows from the Coast of California." *Proceedings of the United States National Museum* 7: 516–518.

74 the species entered the AOU *Check-list*: American Ornithologists' Union. 1886. *The Code of Nomenclature and Check-list of North American Birds*. 1st ed. New York: American Ornithologists' Union.

74 a mere subspecies of the Savannah Sparrow: American Ornithologists' Union. 1944. "Nineteenth Supplement to the American Ornithologists' Union *Check-list of North American Birds*." *Auk* 61: 441–464.

74 *beldingi* and *guttatus* are more closely related to each other: Wheelwright, N., and J. Rising. 2008. "Savannah Sparrow (*Passerculus sandwichensis*)." Birds of North America Online. Ithaca: Cornell Lab of Ornithology.

75 "a naturalist of the old school": Fisher, "In Memoriam: Lyman Belding."

75 "Mr. Ridgway was very patient and prompt": Ibid.

75 Belding shot the two sparrows: Ridgway, "Two Hitherto Unnamed Sparrows."

75 the Belding Sparrow's legs and feet: Ibid.

77 In addition to the nominate *guttatus* and *beldingi*: Huey, L. 1930. "Comment on the Marsh Sparrows of Southern and Lower California, With the Description of a New Race." *Transactions of the San Diego Society of Natural History* 6: 203–206.

77 Van Rossem considered that distinction "minor": Van Rossem, A. 1947. "A Synopsis of the Savannah Sparrows of Northwestern Mexico." *Condor* 49: 97–107.

77 Van Rossem himself described the fourth: Ibid.

SAN BENITO SPARROW

78 On the label of the first known specimen: Oberholser, H. 1919. "A Revision of the Subspecies of *Passerculus rostratus* (Cassin)." *Ohio Journal of Science* 19: 344–354.

78 the published report appeared under Streets's name: Streets, T. 1877. "Contributions to the Natural History of the Hawaiian and Fanning Islands and Lower California." *Bulletin of the United States National Museum* 7: 1–172.

78 the skins brought by Streets "were not in good order": Coues, E. 1897. "*Ammodramus (Passerculus) sanctorum*." *Auk* 14: 92–93.

78 "mummified": McGregor, R. 1897. "Nest and Eggs of the San Benito Sparrow." *Osprey* 2: 42.

78 a new species, which he named, rather tentatively: Coues, E. 1884. *Key to North American Birds*. 2nd ed. Boston: Estes and Lauriat.

78 He later explained the unusual name: Coues, E. 1903. *Key to North American Birds*. 5th ed. Vol. 1. Boston: Dana Estes and Company.

78 He found the controverted specimens "essentially identical": Ridgway, R. 1883. "Catalogue of a Collection of Birds Made Near the Southern Extremity of the Peninsula of Lower California by L. Belding." *Proceedings of the United States National Museum* 5: 532–550.

79 Coues complained: Coues, "*Ammodramus (Passerculus) sanctorum*."

79 "promptly accepted" the species: American Ornithologists' Union. 1897. "Eighth Supplement to the American Ornithologists' Union *Check-list of North American Birds*." *Auk* 14: 117–135.

79 the AOU committee reversed itself: Ridgway, R. 1901. *The Birds of North and Middle America*. Bulletin of the United States National Museum, Bulletin 50, pt. 1. Washington: Government Printing Office.

79 the name was accordingly changed: American Ornithologists' Union. 1902. "Eleventh Supplement to the American Ornithologists' Union *Check-list of North American Birds*." *Auk* 19: 315–342.

79 Coues's role in the story: Rising, J. 2014. "San Benito Sparrow (*Passerculus sanctorum*)." In del Hoyo, J., et al., eds. *Handbook of the Birds of the World*. Vol. 16. Barcelona: Lynx Edicions.

79 highest densities: Salinas-Ortiz, Q., et al. 2015. "Éxito reproductivo del gorrión sabanero (*Passerculus sandwichensis sanctorum*) en el archipiélago San Benito, México." *Revista Mexicana de Biodiversidad* 86: 196–201.

79 some limited seasonal movement: Rising, "San Benito Sparrow (*Passerculus sanctorum*)."

SAVANNAH SPARROW

80 "The very slight distinctions of colour": Wilson, A. 1811. *American Ornithology, or the Natural History of the Birds of the United States*. Vol. 3. Philadelphia: Bradford and Inskeep.

80 access to all the relevant handbooks: Burtt, E., and W. Davis. 2013. *Alexander Wilson: The Scot Who Founded American Ornithology*. Cambridge, MA: Harvard University Press.

80 which Latham called the Sandwich Bunting: Latham, J. 1783. *A General Synopsis of Birds*. Vol. 2. London: Benjamin White.

80 Gmelin named it: Gmelin, J. 1788. *Caroli a Linné . . . Systema naturae per regna tria naturae*. 13th ed. Vol. 1. Leipzig: Georg Emanuel Beer.

80 Wilson used the English translation: Burtt and Davis, *Alexander Wilson*.

80   Otto Finsch, for example, adopted the old name *arctica*: Finsch, O. 1872. "Zur Ornithologie Nordwest-Amerikas." *Abhandlungen herausgegeben vom naturwissenschaftlichen Vereine zu Bremen* 3: 17–86.

80   abiding by the principle of priority: Baird, S., with J. Cassin and G. Lawrence. 1858. *Birds*. Explorations and Surveys for a Railroad Route. Vol. 9. Washington: Beverly Tucker.

80   Baird was also the first to point out: Baird, *Birds*.

80   "the *Emberiza sandwichensis* of Gmelin": Allen, J. 1870. "On the Mammals and Winter Birds of East Florida, With an Examination of Certain Assumed Specific Characters in Birds, and a Sketch of the Bird Faunae of Eastern North America." *Bulletin of the Museum of Comparative Zoology* 2: 161–450.

80   specimens he had "shot about Washington": Coues, E. 1866. "From Arizona to the Pacific." *Ibis* (n.s.) 2: 259–275.

80   mere subspecies of *sandwichensis*: American Ornithologists' Union. 1886. *The Code of Nomenclature and Check-list of North American Birds*. 1st ed. New York: American Ornithologists' Union.

80   "a particularly awkward and unlucky matter": Coues, E. 1897. "*Ammodramus (Passerculus) sanctorum*." *Auk* 14: 92–93.

83   Japan in the winter: Brazil, M. 2009. *Birds of East Asia*. London: Christopher Helm.

83   "I must candidly confess": Coues, "From Arizona to the Pacific."

83   "These specimens are separable to some extent": Allen, J. 1870. "On the Mammals and Winter Birds of East Florida, With an Examination of Certain Assumed Specific Characters in Birds, and a Sketch of the Bird Faunae of Eastern North America." *Bulletin of the Museum of Comparative Zoology* 2: 161–450.

83   28 described races: Wheelwright, N., and J. Rising. 2008. "Savannah Sparrow (*Passerculus sandwichensis*)." *Birds of North America Online*. Ithaca: Cornell Lab of Ornithology.

83   The descriptive tradition: Peters, J., and L. Griscom. 1938. "Geographical Variation in the Savannah Sparrow." *Bulletin of the Museum of Comparative Zoology* 80: 445–478.

84   there was contemporary—almost immediate—criticism: Grinnell, J. 1939. "Proposed Shifts of Names in *Passerculus*—A Protest." *Condor* 41: 112–119.

84   the senior author convinced himself: Peters and Griscom, "Geographical Variation."

84   *Passerculus sandwichensis rufofuscus*: Camras, S. 1940. "A New Savannah Sparrow from Mexico." *Zoological Series of Field Museum of Natural History* 24: 159–160.

84   "ornithologist, artist, and brave defender of his country": Bishop, L. 1915. "Description of a New Race of Savannah Sparrow and Suggestions on Some California Birds." *Condor* 17: 185–189.

84   "a mounted bird, worn and faded": Grinnell, "Proposed Shifts of Names."

84   the "transition" between the coastal Savannah Sparrows and the Belding Sparrow: Wheelwright and Rising, "Savannah Sparrow (*Passerculus sandwichensis*)."

85   The race *rufofuscus*: Camras, "New Savannah Sparrow."

85   "argue against the latter's diagnosibility": Wheelwright and Rising, "Savannah Sparrow (*Passerculus sandwichensis*)."

85   When *mediogriseus* is admitted as a distinct subspecies: Aldrich, J. 1940. "Geographic Variation in Eastern North American Savannah Sparrows (*Passerculus sandwichensis*)." *Ohio Journal of Science* 40: 1–8.

85   could represent any one of the four eastern-wintering subspecies: Parkes, K., and R. Panza. 1991. "The Type Locality of *Fringilla savanna* Wilson." *Auk* 108: 185–186.

86   affirmed the correctness: Parkes, K, and R. Panza. 1991. "The Type Locality of *Fringilla savanna* Wilson." *Auk* 108: 185–186.

86   Butler collected half a dozen Savannah Sparrows: Butler, A. 1888. "On a New Subspecies of *Ammodramus sandwichensis* from Mexico." *Auk* 5: 264–266.

86   *rufofuscus* inseparable from *brunnescens*: Hubbard, J. 1974. "Geographic Variation in the Savannah Sparrows of the Inland Southwest, Mexico, and Guatemala." *Nemouria* 12: 1–21.

86   rediscovered and its song and breeding habits described: Eisermann, K., et al. 2017. "Nesting Evidence, Density and Vocalisations in a Resident Population of Savannah Sparrow *Passerculus sandwichensis wetmorei* in Guatemala." *Bulletin of the British Ornithologists' Club* 137: 37–45.

86   Earlier records of the species in Guatemala: Eisermann, K., and C. Avendaño. 2007. *Lista comentada de las aves de Guatemala / Annotated Checklist of the Birds of Guatemala*. Barcelona: Lynx Edicions.

86   specimens taken by W.B. Richardson: Van Rossem, A. 1938. "Descriptions of Twenty-one New Races of Fringillidae and Icteridae from Mexico and Guatemala." *Bulletin of the British Ornithologists' Club* 58: 124–139.

86   adaptations to habitat conditions: Rising, J. 2009. "Geographic Variation in Plumage Pattern and Coloration of Savannah Sparrows." *Wilson Bulletin* 121: 253–264.

86   not "much value in delimiting subspecies": Rising, J. 2001. *Geographic Variation in Size and Shape of Savannah Sparrows (*Passerculus sandwichensis*)*. Studies in Avian Biology 23. Lawrence: Cooper Ornithological Society.

86   thus making the species monotypic: Rising, J., et al. 2014. "Savannah Sparrow (*Passerculus sandwichensis*)." In del Hoyo, J., Elliott, A., Sargatal, J., Christie, D.A., & de Juana, E., eds. *Handbook of the Birds of the World Alive*. Barcelona: Lynx Edicions.

IPSWICH SPARROW

87   "While the Ipswich Sparrow has been maintained as a distinct species": Peters, J., and L. Griscom. 1938. "Geographical Variation in the Savannah Sparrow." *Bulletin of the Museum of Comparative Zoology* 80: 445–478.

87   "the gray bird": Dwight, J. 1895. *The Ipswich Sparrow (Ammodramus princeps) and Its Summer Home*. Memoirs of the Nuttall Ornithological Club 2. Cambridge: Nuttall Club.

87   first specimen of his "Large Barren Ground Sparrow": Maynard, C. 1869. "The Capture of the *Centronyx Bairdii* at Ipswich." *American Naturalist* 3: 554–555.

87   "Previous to the capture of this": Maynard, C. 1870. *Naturalist's Guide In Collecting and Preserving Objects of Natural History, With a Complete Catalogue of the Birds of Eastern Massachusetts*. 1st ed. Boston: Fields, Osgood, and Co.

88   "in all essential points it seems to be the same": Ibid.

88   a full description of his specimen: Ibid.

88   "regular winter visitants from the North": Brewster, W. 1872. "Birds New to Massachusetts Fauna." *American Naturalist* 6: 306–307.

88   they were not identical after all: Maynard, C. 1872. "A New Species of *Passerculus* from Eastern Massachusetts." *American Naturalist* 6: 637–638.

88 "a full investigation will reveal something": Coues, E. 1872. *Key to North American Birds*. 1st ed. New York: Dodd and Mead.

88 formal description of the new sparrow: Maynard, "New Species of *Passerculus*."

88 the two could not "justly be compared": Maynard, *Naturalist's Guide*. 1st ed.

88 he had in fact discovered a new species: Maynard, "New Species of *Passerculus*."

88 ratified Maynard's diagnosis: Baird, S., et al. 1874. *A History of North America Birds*. Vol. 1. Boston: Little, Brown.

89 "but in justice to myself": Maynard, C. 1877. *Naturalist's Guide in Collecting and Preserving Objects of Natural History, With a Complete Catalogue of the Birds of Eastern Massachusetts*. 2nd ed. Salem: Naturalists' Agency.

89 "the once prized Ipswich Sparrow": Dwight, *Ipswich Sparrow* (Ammodramus princeps).

89 "a considerable series of eggs" Ridgway, R. 1884. "The Probable Breeding Place of *Passerculus princeps*." *Auk* 1: 292–293.

89 "an unquestionable Ipswich Sparrow": Merriam, C. 1884. "Breeding of *Passerculus princeps* on Sable Island." *Auk* 1: 390.

89 The first nests: Dwight, *Ipswich Sparrow* (Ammodramus princeps).

89 Isaac Norris De Haven informally addressed a meeting: De Haven, I. 1898. *Abstract of the Proceedings of the Delaware Valley Ornithological Club of Philadelphia: For the Years 1892-1897*. Philadelphia: Delaware Valley Ornithological Club.

89 "The drawing of this bird was in the hands of the engraver": Wilson, A. 1811. *American Ornithology, or the Natural History of the Birds of the United States*. Vol. 3. Philadelphia: Bradford and Inskeep.

89 "small pointed spots of brown": Wilson, A. 1811. *American Ornithology, or the Natural History of the Birds of the United States*. Vol. 4. Philadelphia: Bradford and Inskeep.

90 "ribbon-like crest of a submerged bank": Dwight, *The Ipswich Sparrow* (Ammodramus princeps) *and Its Summer Home*.

90 infrequent cases of mixed pairings: McLaren, I., and A. Horn. 2006. "The Ipswich Sparrow: Past, Present, and Future." *Birding* 38: 52–59.

90 large numbers sometimes linger: Ibid.

90 inland reports from New Hampshire and Texas: Dwight, *Ipswich Sparrow* (Ammodramus princeps).

90 Northerly strays: McLaren and Horn, "Ipswich Sparrow."

90 An April bird in southwest England: Broyd, S. 1985. "Savannah Sparrow: New to the Western Palearctic." *British Birds* 78: 647–646.

VESPER SPARROW

91 "rich museum of American birds": Pennant, T. 1784. *Arctic Zoology*. Vol. 1. London: Henry Hughs.

91 beautiful American warbler: Pennant, T. 1785. *Arctic Zoology*. Vol. 2. London: Henry Hughs.

91 "the Gray Grass-bird": Ibid.

91 Pennant reports: Ibid.

91 Gmelin simply rendered the English: Gmelin, J. 1788. *Caroli a Linné . . . Systema naturae per regna tria naturae*. Vol. 1. Leipzig: Georg Emanuel Beer.

91 adjured since the late nineteenth century: Wright, M. 1895. *Birdcraft: A Field Book of 200 Song, Game, and Water Birds*. New York: Macmillan.

91 "the white outer tail-feathers flashing conspicuously": Peterson, R. 1934. *A Field Guide to the Birds*. 1st ed. Boston: Houghton Mifflin.

91 "the next tipt and edged": Wilson, A. 1811. *American Ornithology, or the Natural History of the Birds of the United States*. Vol. 4. Philadelphia: Bradford and Inskeep.

92 "grass finch" survived: Coues, E. 1903. *Key to North American Birds*. 5th ed. Vol. 1. Boston: Dana Estes and Company.

92 John James Audubon: Audubon, J. 1831. *Ornithological Biography*. Vol. 1. Philadelphia: Judah Dobson.

92 Thomas Nuttall: Nuttall, *Manual*. 1st ed.

92 Spencer Baird: Baird, S., with J. Cassin and G. Lawrence. 1858. *Birds*. Explorations and Surveys for a Railroad Route. Vol. 9. Washington: Beverly Tucker.

92 *History* of 1874: Baird, S., et al. 1874. *A History of North America Birds*. Vol. 1. Boston: Little, Brown.

92 "astonishing numbers" in the Southeast: Audubon, J. 1839. *Ornithological Biography*. Vol. 5. Edinburgh: Adam Black and Charles Black.

92 "They sing with a clear and agreeable note": Nuttall, *Manual*. 1st ed.

92 Wilson Flagg: Flagg, W. 1858. "The Birds of the Garden and Orchard." *Atlantic Monthly* 1: 593–605.

93 Ralph Hoffmann: Hoffmann, R. 1904. *Guide to the Birds of New England and Eastern New York*. Boston: Houghton, Mifflin and Company

94 "inclined to look upon it as a resident": Audubon, J. 1831. *Ornithological Biography*. Vol. 1. Edinburgh: Adam Black and Charles Black.

95 occasional cold-season records north: Campbell, R., et al. 2001. *The Birds of British Columbia*. Vol. 4. Vancouver: UBC Press.

95 rare inland: Jones, S., and J. Cornely. 2002. "Vesper Sparrow (*Pooecetes gramineus*)." *Birds of North America Online*. Ithaca: Cornell Lab of Ornithology.

95 *affinis* is now largely restricted: Ibid.

95 "a good deal of difference in specimens": Baird, *Birds*.

95 he named the subspecies *affinis*: Miller, G. 1880. "Description of an Apparently New *Poocætes* from Oregon." *Auk* 5: 404–405.

LECONTE SPARROW

96 described in 1790 by John Latham: Latham, J. 1790. *Index ornithologicus*. Vol. 1. London: privately published.

96 "It has yet, however, escaped": Nuttall, T. 1832. *A Manual of the Ornithology of the United States and of Canada*. 1st ed. Vol. 1. Cambridge: Hilliard and Brown.

96 Bonaparte suggested: Ibid.

96 "nearly allied if not identic": Nuttall, T. 1840. *A Manual of the Ornithology of the United States and Canada*. 2nd ed. Boston: Hilliard, Gray, and Co.

96 specimens "of this pretty little Sharp-tailed Finch": Audubon, J. 1844. *The Birds of America*. 2nd ed. Vol. 7. Philadelphia: E.G. Dorsey.

96 "young friend Dr. [John Lawrence] Le Conte": Ibid.

96 seven large crates of specimens and notes: Witte, S., and M. Gallagher, eds.; D. Karch, transl. 2012. *The North American Journals of Prince Maximilian von Wied*. Norman: University of Oklahoma Press.

96 "the extremely incomplete notes": Maximilian zu Wied-Neuwied. 1858. "Verzeichniss der Vögel, welche auf einer Reise in Nord-America beobachtet wurden." *Journal für Ornithologie* 6: 1–29, 97–124, 177–204, 257–284, 337–354, 417–444.

97 "On the ground we believed we saw a mouse": Witte and Gallagher, eds., *Journals of Prince Maximilian.*

97 Maximilian duly and properly cited that authority: Maximilian zu Wied-Neuwied, "Verzeichniss der Vögel."

97 Maximilian's skin, the oldest in existence: Allen, J. 1886. "Three Interesting Birds in the American Museum of Natural History: *Ammodramus leconteii, Helinaia swainsonii,* and *Saxicola oenanthe.*" *Auk* 3: 489–490.

97 "it has somehow been mislaid": Baird, S., with J. Cassin and G. Lawrence. 1858. *Birds.* Explorations and Surveys for a Railroad Route. Vol. 9. Washington: Beverly Tucker.

97 "long remained an extreme rarity": Coues, E. 1899. "Zoological and Other Notes." In Audubon, M. 1899, *Audubon and His Journals.* Vol. 1. New York: Charles Scribner's Sons.

97 Lincecum sent a specimen from Texas: Coues, E. 1873. "Notice of a Rare Bird." *American Naturalist* 7: 748–750.

97 skin was in very poor condition: Baird, S., et al. 1874. *A History of North America Birds.* Vol. 1. Boston: Little, Brown.

97 so "very fumble-fisted" a preparator: Lincecum, J., et al., eds. 1997. *Science on the Texas Frontier: Observations of Dr. Gideon Lincecum.* College Station: Texas A&M University Press.

97 "This long-lost species": Coues, E. 1874. *Birds of the Northwest.* Miscellaneous Publications of the Geological Survey of the Territories 3. Washington: Government Printing Office.

97 "found the bird to be not uncommon": Baird et al., *History of North America Birds.*

97 "not without difficulty . . . the only chance was a snap shot": Coues, "Rare Bird."

98 "skulking in rank herbage": Thompson, E. 1890. "The Birds of Manitoba." *Proceedings of the United States National Museum* 13: 457–643.

98 first nest and eggs: Seton, E. 1885. "Manitoban Notes." *Auk* 2: 21–24.

99 an infrequently witnessed longer song: Murray, B. 1969. "A Comparative Study of the Le Conte's and Sharp-tailed Sparrows." *Auk* 86: 199–231.

99 the "aerial trill": Lowther, P. 2005. "Le Conte's Sparrow (*Ammodramus leconteii*)." *Birds of North America Online.* Ithaca: Cornell Lab of Ornithology.

99 described the known range of the LeConte Sparrow: Coues, E. 1872. *Key to North American Birds.* 1st ed. Salem: Naturalists' Agency.

99 The LeConte Sparrow breeds: Lowther, "Le Conte's Sparrow (*Ammodramus leconteii*)."

99 Migrants move south: Ibid.

99 nearly annual autumn rarity in California: Hamilton, R., et al., eds. 2007. *Rare Birds of California.* Camarillo: Western Field Ornithologists.

99 north along the Pacific Coast: Lowther, "Le Conte's Sparrow (*Ammodramus leconteii*)."

SEASIDE SPARROW

100 discovered by Arthur H. Howell in February of 1918: Howell, A. 1919. "Description of a New Seaside Sparrow from Florida." *Auk* 36: 86–87.

100 lumping of this and other putative species: American Ornithologists' Union. 1973. "Thirty-second Supplement to the *Check-list of North American Birds.*" *Auk* 90: 411–419.

100 the Seaside Sparrow in the kitchen: Wilson, A. 1811. *American Ornithology, or the Natural History of the*

Birds of the United States. Vol. 4. Philadelphia: Bradford and Inskeep.

101 "monotonous chirpings": Audubon, J. 1831. *Ornithological Biography.* Vol. 1. Philadelphia: Judah Dobson.

101 readily distinguishable from both: Audubon, J. 1834. *Ornithological Biography.* Vol. 2. Edinburgh: Adam and Charles Black.

101 "very abundant in the Texas": Audubon, J. 1838. *Ornithological Biography.* Vol. 4. Philadelphia: Judah Dobson.

101 discovery of a "very remarkable," "very striking" bird: Ridgway, R. 1873. "On Some New Forms of American Birds." *American Naturalist* 7: 602–619.

101 Dusky Seaside Sparrow a species of its own: Ridgway, R. 1880. "Revisions of Nomenclature of Certain North American Birds." *Proceedings of the United States National Museum* 3: 1–16.

101 a new subspecies of the former, *peninsulae*: Allen, J. 1888. "Descriptions of Two New Subspecies of the Seaside Sparrow (*Ammodramus maritimus*)." *Auk* 5: 284–287.

102 "In the light of this new material": Scott, W. 1889. "A Summary of Observations on the Birds of the Gulf Coast of Florida." *Auk* 6: 318–326.

102 Allen's new concept: Allen, "Two New Subspecies."

102 "evidently entitled to recognition": Ibid.

102 Frank Chapman's revision: Chapman, F. 1899. "The Distribution and Relationships of *Ammodramus maritimus* and Its Allies." *Auk* 16: 1–12.

102 a new understanding of the Seaside Sparrows: Griscom, L., and J. Nichols. 1920. "A Revision of the Seaside Sparrows." *Abstract of the Proceedings of the Linnaean Society of New York.* 32: 18–30.

102 Oberholser crowded the subspecific field: Oberholser, H. 1931. "The Atlantic Coast Races of *Thryospiza maritima* (Wilson)." *Proceedings of the Biological Society of Washington* 44: 123–128.

103 uncertainty as to whether Bailey's *shannoni* and Oberholser's *pelonota*: Hubbard, J., and R. Banks. 1970. "The Types and Taxa of Harold H. Bailey." *Proceedings of the Biological Society of Washington* 83: 321–332.

103 have been extinct from Jacksonville, Florida, south: Ibid.

103 a cogent explanation for its strange zoogeographic patterns: Tompkins, I. 1934. "Hurricanes and Subspecific Variation." *Wilson Bulletin* 46: 238–240.

103 "A certain Finch which I discovered in Florida": Maynard, C. 1875. "A New Species of Finch from Florida." *American Sportsman* 5: 248.

103 lumped with the more widespread Atlantic Coast forms: American Ornithologists' Union, "Thirty-second Supplement."

103 The last Dusky Seaside Sparrow: Greenlaw, J., et al. 2014. *The Robertson and Woolfenden Florida Bird Species: An Annotated List.* Florida Ornithological Society Special Publication 8. Gainesville: Florida Ornithological Society.

105 variability and complexity inaudible to human ears: Hardy, J. 1983. "Geographic Variation in Primary Song of the Seaside Sparrow." In Quay, T., et al., eds. *The Seaside Sparrow, Its Biology and Management.* Raleigh: North Carolina Biological Survey.

106 considerable variation in songs: Ibid.

106 quite unlike the broadly modulated, burry buzzes of coastal birds: Ibid.

106 a longer, more complex song given primarily in flight: McDonald, M. 1983. "Vocalization Repertoire of a Marked Population of Seaside Sparrows." In Quay, T.,

et al., eds. *The Seaside Sparrow, Its Biology and Management.* Raleigh: North Carolina Biological Survey.

106 "a whirling, quavering" call: Ibid.

106 south to Duval and Flagler Counties, Florida: Greenlaw et al., *Florida Bird Species.*

106 Virginia's Chesapeake Bay Islands: National Audubon Society. 2016. "Chesapeake Bay Islands: Accomack County." tinyurl.com/VASeaside.

106 northeasterly post-breeding dispersal: Robbins, C. 1983. "Distribution and Migration of Seaside Sparrows." In Quay, T., et al., eds. *The Seaside Sparrow, Its Biology and Management.* Raleigh: North Carolina Biological Survey.

106 an overland route through the Carolinas: Ibid.

107 The resident subspecies *mirabilis*: Pimm, S., et al. 2002. *Sparrow in the Grass: A Report on the First Ten Years of Research on the Cape Sable Seaside Sparrow* (Ammodramus maritimus mirabilis). Homestead: National Park Service.

107 it now breeds, in small and declining numbers: Greenlaw et al., *Florida Bird Species.*

107 "mixed prairie, often with a substantial percentage of muhly grass": Pimm et al., *Sparrow in the Grass.*

107 Pasco County, Florida: Greenlaw et al., *Florida Bird Species.*

107 the mouth of the Rio Grande: Lockwood, M., and B. Freeman. 2014. *The TOS Handbook of Texas Birds.* College Station: Texas A&M University Press.

107 "wanders to [the] adjacent bank": Howell, S., and S. Webb. 1995. *A Guide to the Birds of Mexico and Northern Central America.* Oxford: Oxford University Press.

107 McLennan County, Texas: Lockwood and Freeman, *TOS Handbook.*

107 interbred near Corpus Christi: Woltmann, S., et al. 2014. "Population Genetics of Seaside Sparrow (*Ammodramus maritimus*) Subspecies Along the Gulf of Mexico." *PLoS ONE* 9: e112739.

NELSON SPARROW

108 "One of the keenest naturalists we have ever had": Goldman, E. 1935. "Edward William Nelson—Naturalist." *Auk* 52: 135–148.

108 more than 3,000 bird skins and eggs: Willard, S. 1877. *A Directory of the Ornithologists of the United States.* Utica: The Oologist.

108 "very markedly" different: Allen, J. 1874. "A Specimen of the Sharp-tailed Finch." *Proceedings of the Boston Society of Natural History* 17: 292–294.

108 breeding in the marshes of south Chicago: Nelson, E. 1876. "Birds of North-eastern Illinois." *Bulletin of the Essex Institute* 8: 90–155.

108 reported in the breeding season from Wisconsin, Texas, Maine, and Kansas: Ridgway, R. 1901. *The Birds of North and Middle America.* Bulletin of the United States National Museum, Bulletin 50, pt. 1. Washington: Government Printing Office.

108 Dakota Territory: Cooke, W. 1888. *Report on Bird Migration in the Mississippi Valley.* Division of Economic Ornithology Bulletin 2. Washington: Government Printing Office.

108 Manitoba: Thompson, E. 1893. "Additions to the List of Manitoba Birds." *Auk* 10: 49–50.

108 "the inland representative of its strictly littoral relatives": Dwight, J. 1896. "The Sharp-tailed Sparrow (*Ammodramus caudacutus*) and Its Geographical Races." *Auk* 13: 271–278.

109 "passes gradually into *nelsoni*": Dwight, J. 1887. "A New Race of the Sharp-tailed Sparrow (*Ammodramus caudacutus*)." *Auk* 4: 232–239.

109 "there appears to be a wide gap": Dwight, "Sharp-tailed Sparrow (*Ammodramus caudacutus*)."

109 "presumed intergrades between *nelsoni* and *subvirgata*": Todd, W. 1938. "Two New Races of North American Birds." *Auk* 55: 116–118.

109 "if it should be proved": Dwight, "Sharp-tailed Sparrow (*Ammodramus caudacutus*)."

109 the two were indeed separate species: Norton, A. 1897. "The Sharp-tailed Sparrows of Maine with Remarks on Their Distribution and Relationship." *Proceedings of the Portland Society of Natural History* 2: 97–102.

109 The American Ornithologists' Union adopted that view: American Ornithologists' Union. 1899. "Ninth Supplement to the A.O.U. *Check-list.*" *Auk* 16: 97–133.

109 reaffirmed it three years later: American Ornithologists' Union. 1902. "Eleventh Supplement to the American Ornithologists' Union *Check-list of North American Birds.*" *Auk* 19: 315–342.

109 *Ammodramus* "belonged" to the Grasshopper Sparrow: Oberholser, H. 1905. "Notes on the Nomenclature of Certain Genera of Birds." *Smithsonian Miscellaneous Collections* 48: 59–68.

109 Maynard's resurrected *Passerherbulus*: American Ornithologists' Union. 1909. "Fifteenth Supplement to the American Ornithologists' Union *Check-list of North American Birds.*" *Auk* 26: 294–303.

109 *Ammospiza*, a name coined by Oberholser: Oberholser, "Nomenclature of Certain Genera."

109 "*Ammospiza nelsoni* being now regarded as only subspecifically different": American Ornithologists' Union. 1931. *Check-list of North American Birds.* 4th ed. Lancaster: American Ornithologists' Union.

110 Greenlaw took up the investigation: Greenlaw, J. 1993. "Behavioral and Morphological Diversification in Sharp-tailed Sparrows (*Ammodramus caudacutus*) of the Atlantic Coast." *Auk* 110: 286–303.

110 distinct species was accepted by the AOU in 1995: American Ornithologists' Union. 1995. "Fortieth Supplement to the A.O.U. *Check-list.*" *Auk* 112: 819–830.

110 formally simplified in 2009 to Nelson's Sparrow: American Ornithologists' Union. 2009. "Fiftieth Supplement to the A.O.U. *Check-list.*" *Auk* 126: 705–714.

111 "Genetic data revealed": Walsh, J., et al. 2015. "Relationship of Phenotypic Variation and Genetic Admixture in the Saltmarsh–Nelson's Sparrow Hybrid Zone." *Auk* 132: 704–716.

111 "a challenge for accurate hybrid identification": Ibid.

111 a diagnostic feature in areas where both may be present: Greenlaw, "Behavioral and Morphological Diversification."

112 apparent flight display: Ibid.

112 lacks the "whisper" song: Ibid.

112 numbers can vary considerably: Campbell, R., et al. 2001. *The Birds of British Columbia.* Vol. 4. Vancouver: UBC Press.

112 from the Eastmain River in Quebec west to Churchill: Greenlaw, J., and J. Rising. 1994. "Saltmarsh Sparrow (*Ammodramus caudacutus*)." *Birds of North America Online.* Ithaca: Cornell Lab of Ornithology.

113 the winter status of this subspecies: Lockwood, M., and B. Freeman. 2014. *The TOS Handbook of Texas Birds.* College Station: Texas A&M University Press.

113 do not appear to be any adequately documented records: Greenlaw, J., and G. Woolfenden. 2007. "Wintering Distributions and Migration of Saltmarsh and Nelson's Sharp-tailed Sparrows." *Wilson Bulletin* 119: 361–377.

113 before continuing south into the usual wintering areas: Ibid.

113 rare fall migrant and rare winter visitor: Hamilton, R., et al., eds. 2007. *Rare Birds of California*. Camarillo: Western Field Ornithologists.

113 an overland route south: Greenlaw and Woolfenden, "Wintering Distributions and Migration."

113 migrants stay close to the coast: Ibid.

114 straight north to the prairie breeding grounds: Lockwood and Freeman, *TOS Handbook*.

114 "Almost yearly the Linnaean Society of New York held seminars": Peterson, R. 1947. *A Field Guide to the Birds*. 2nd ed. Boston: Houghton Mifflin.

114 The James Bay subspecies *altera*: Smith, F. 2011. "Photo Essay: Subspecies of Saltmarsh Sparrow and Nelson's Sparrow." *North American Birds* 65: 368–377.

SALTMARSH SPARROW

115 "seen more than forty": Audubon, J. 1834. *Ornithological Biography*. Vol. 2. Edinburgh: Adam and Charles Black.

115 "The name of Quail-head": De Kay, J. 1844. *Zoology of New-York*. Part 2: Birds. Albany: Carroll and Cook.

115 "This new (as I apprehend it) and beautiful species": Wilson, A. 1811. *American Ornithology, or the Natural History of the Birds of the United States*. Vol. 4. Philadelphia: Bradford and Inskeep.

115 "A bird of this denomination": Ibid.

115 William Turton had published: Turton, W. 1802. *A General System of Nature . . . Translated from Gmelin's Last Edition of the Celebrated* Systema naturae *by Sir Charles Linné*. Vol. 1. London: Lackington, Allen, and Co.

115 entomologist Thomas Say: Burtt, E., and W. Davis. 2013. *Alexander Wilson: The Scot Who Founded American Ornithology*. Cambridge, MA: Harvard University Press.

116 Thomas Pennant, the Welsh collector: Latham, J. 1782. *A General Synopsis of Birds*. Vol. 1, pt. 2. London: Benjamin White.

116 Wilson cited Pennant: Burtt and Davis, *Alexander Wilson*.

116 *Arctic Zoology*: Pennant, T. 1785. *Arctic Zoology*. Vol. 2. London: Henry Hughs.

116 "slope off on each side": Ibid.

116 as Jonathan Dwight would later point out: Dwight, J. 1896. "The Sharp-tailed Sparrow (*Ammodramus caudacutus*) and Its Geographical Races." *Auk* 13: 271–278.

116 "almost willing to believe they are identical": Wilson, *American Ornithology*.

116 "allied in form and habits": Audubon, *Ornithological Biography*.

116 hybridization between Saltmarsh Sparrows and *subvirgata* Nelson Sparrows: Walsh, J., et al. 2015. "Relationship of Phenotypic Variation and Genetic Admixture in the Saltmarsh–Nelson's Sparrow Hybrid Zone." *Auk* 132: 704–716.

116 song is a helpful way: Greenlaw, J. 1993. "Behavioral and Morphological Diversification in Sharp-tailed Sparrows (*Ammodramus caudacutus*) of the Atlantic Coast." *Auk* 110: 286–303.

117 Fall migration: Greenlaw, J., and J. Rising. 1994. "Saltmarsh Sparrow (*Ammodramus caudacutus*)." In A.

Poole, ed. *Birds of North America Online*. Ithaca: Cornell Lab of Ornithology.

117 All Texas records: Lockwood, M., and B. Freeman. 2014. *The TOS Handbook of Texas Birds*. College Station: Texas A&M University Press.

117 in Nova Scotia: McLaren, I. 2012. *All the Birds of Nova Scotia: Status and Critical Identification*. Kentville: Gaspereau Press.

117 northbound flight in spring: Greenlaw and Rising, "Saltmarsh Sparrow (*Ammodramus caudacutus*)."

117 Massachusetts breeding grounds: Veit, R., and W. Petersen. 1993. *Birds of Massachusetts*. Lincoln: Massachusetts Audubon Society.

118 in the hand, more than 90 percent: Smith, F. 2011. "Photo Essay: Subspecies of Saltmarsh Sparrow and Nelson's Sparrow." *North American Birds* 65: 368–377.

118 "very distinct" southern-breeding *diversa*: Bishop, L. 1901. "A New Sharp-tailed Finch from North Carolina." *Auk* 18: 269–270.

118 *diversa* is darker: Smith, "Photo Essay."

118 Bishop noted: Bishop, "New Sharp-tailed Finch."

118 The darkness of the upper parts: Ibid.

118 intergrades or backcrosses: Smith, "Photo Essay."

BELL SPARROW

119 Cassin at the Academy of Natural Sciences: Cassin, J. 1850. "Descriptions of New Species of Birds of the Genera *Parus*, Linn.; *Emberiza*, Linn.; *Carduelis*, Briss.; *Myiothera*, Ill.; and *Leuconerpes*, Sw., Specimens of Which Are in the Collection of the Academy of Natural Sciences of Philadelphia." *Proceedings of the Academy of Natural Sciences of Philadelphia* 5: 103–106.

119 a new subspecies: Ridgway, R. 1873. "The Birds of Colorado." *Bulletin of the Essex Institute* 5: 174–195.

119 collected in California: Belding, L. 1890. *Land Birds of the Pacific District*. Occasional Papers of the California Academy of Sciences 2. San Francisco: California Academy of Sciences.

119 breeding side by side: Grinnell, J. 1898. "Rank of the Sage Sparrow." *Auk* 15: 58–59.

119 refuted by A.K. Fisher: Fisher, A. 1898. "Rank of the Sage Sparrow." *Auk* 15: 190.

119 deserved a name of their own: Grinnell, J. 1905. "The California Sage Sparrow." *Condor* 7: 18–19.

119 "gaps" in color and size: Ibid.

120 recognition in 2013: American Ornithologists' Union. 2013. "Fifty-fourth Supplement to the American Ornithologists' Union *Check-list of North American Birds*." *Auk* 130: 1–14.

120 distinguishable by phenotype and DNA: Cicero, C., and M. Koo. 2012. "The Role of Niche Divergence and Phenotypica Adaptation in Promoting Lineage Diversification in the Sage Sparrow (*Artemisiospiza belli*, Aves: Emberizidae)." *Biological Journal of the Linnean Society* 107: 332–354.

120 plumage characters used to differentiate these species: Pyle, P. 2013. "On Separating Sagebrush Sparrow (*Artemisiospiza nevadensis*) from Bell's Sparrow (*A. belli*), With Particular Reference to *A.b. canescens*." tinyurl.com/PyleSAGS.

122 Juveniles, like adults: Ridgway, R. 1901. *The Birds of North and Middle America*. Bulletin of the United States National Museum, Bulletin 50, pt. 1. Washington: Government Printing Office.

122 "confusing intermediate group": Szeliga, W. 2011. "Sage Sparrow Subspecies." earbirding.com/blog/archives/3040.

123 "isolated pocket": Small, A. 1994. *California Birds: Their Status and Distribution*. Vista, CA: Ibis.

123 Vagrant individuals of *belli*: Iliff, M. 2003. "A Vagrant Bell's Sparrow in Baja California Sur." *Western Birds* 34: 256–259.

123 Mojave Sparrow, *canescens*, is a breeder: Martin, J., and B. Carlson. 1998. "Sage Sparrow (*Artemisiospiza belli*)." In A. Poole, ed. *Birds of North America Online*. Ithaca: Cornell Lab of Ornithology.

123 Nevada's Grapevine Mountains: American Ornithologists' Union. 1957. *Check-list of North American Birds*. 5th ed. Ithaca: American Ornithologists' Union.

123 regular wintering in southwestern Arizona: Rosenberg, K., et al. 1991. *Birds of the Lower Colorado River Valley*. Tucson: University of Arizona Press.

123 "nearly as common": Phillips, A., et al. 1964. *The Birds of Arizona*. Tucson: University of Arizona Press.

123 southern and western Arizona: McCreedy, C., et al. 2014. "Winter Distribution and Plumage Characteristics of Sagebrush and Bell's Sparrows in Arizona." Abstract: Eighth Annual Meeting of the Arizona Field Ornithologists.

123 Nova Scotia: McLaren, I. 2012. *All the Birds of Nova Scotia: Status and Critical Identification*. Kentville: Gaspereau Press.

123 "exactly like *A. belli*": Ridgway, R. 1898. "Descriptions of Supposed New Genera, Species, and Subspecies of American Birds. I: Fringillidae." *Auk* 15: 223–230.

123 retracted the claim: Ridgway, *North and Middle America*.

123 visited San Clemente to collect Bell Sparrows: Van Rossem, A. 1932. "On the Validity of the San Clemente Island Bell's Sparrow." *Auk* 49: 490–491.

123 the race *cinerea*: Townsend, C. 1890. "Birds from the Coasts of Western North America and Adjacent Islands, Collected in 1888-'89, with Descriptions of New Species." *Proceedings of the United States National Museum* 13: 131–142.

123 narrower and more often incomplete: Ridgway, *North and Middle America*.

123 began his description of *cinerea*: Townsend, C. 1890. "Birds from the Coasts of Western North America and Adjacent Islands, Collected in 1888-'89, with Descriptions of New Species." *Proceedings of the United States National Museum* 13: 131–142.

123 a distinct subspecies *xerophilus*: Huey, L. 1930. "A New Race of Bell Sparrows from Lower California, Mexico." *Transactions of the San Diego Society of Natural History* 6: 229–230.

124 south-central Arizona: Monson, G. and A. Phillips. 1981. *Annotated Checklist of the Birds of Arizona*. 2nd ed. Tucson: University of Arizona Press.

124 "very much paler": Grinnell, "California Sage Sparrow."

SAGEBRUSH SPARROW

125 the type specimen: Baird, S., and R. Ridgway. 1873. "On Some New Forms of American Birds." *Bulletin of the Essex Institute* 5: 195–201.

125 collected at Fort Thorn: Cassin, J. 1853–1855. *Illustrations of the Birds of California, Texas, Oregon, British and Russian America*. Philadelphia: J.B. Lippincott and Co.

125 Henry named two new species: Hume, E. 1942. *Ornithologists of the United States Army Medical Corps*. Baltimore: Johns Hopkins Press.

125 Neither Henry nor Spencer Baird: Henry, T. 1859. "Catalogue of the Birds of New Mexico as Compiled from Notes and Observations Made While in that Territory, During a Residence of Six Years." *Proceedings*

of the Academy of Natural Sciences of Philadelphia 11: 104–109.

125 published without further comment: Baird, S., with J. Cassin and G. Lawrence. 1858. *Birds*. Explorations and Surveys for a Railroad Route. Vol. 9. Washington: Beverly Tucker.

125 birds of the inland West: Ridgway, R. 1873. "The Birds of Colorado." *Bulletin of the Essex Institute* 5: 174–195.

125 "a zealous naturalist": Coues, E. 1878. *Birds of the Colorado Valley*. United States Geological Survey of the Territories Miscellaneous Publications 11. Washington: Government Printing Office.

125 "difference in size": Baird, S., et al. 1874. *A History of North America Birds*. Vol. 1. Boston: Little, Brown.

126 Artemisia Sparrow: Ridgway, R. 1877. *Ornithology*. United States Geological Exploration of the Fortieth Parallel. Vol. 4, pt. 3. Washington, D.C.: Government Printing Office.

126 a polytypic Bell Sparrow: Ridgway, R. 1880. "A Catalogue of the Birds of North America." *Proceedings of the United States National Museum* 3: 163–246.

126 specifically distinct: Grinnell, J. 1898. "Rank of the Sage Sparrow." *Auk* 15: 58–59.

126 the race *canescens*: Grinnell, J. 1905. "The California Sage Sparrow." *Condor* 7: 18–19.

126 Sage Sparrow comprised three subspecies: American Ornithologists' Union. 1908. "Fourteenth Supplement to the American Ornithologists' Union *Check-list of North American Birds*." *Auk* 25: 343–399.

126 a single species: Hellmayr, C. 1938. *Catalogue of Birds of the Americas and the Adjacent Islands in Field Museum of Natural History*. Zoological Series 13, pt. 11. Chicago: Field Museum of Natural History.

126 the AOU followed Hellmayr: American Ornithologists' Union. 1944. "Nineteenth Supplement to the American Ornithologists' Union *Check-list of North American Birds*." *Auk* 61: 441–464.

127 "the Sage Sparrow is a relief": Phillips, A., et al. 1964. *The Birds of Arizona*. Tucson: University of Arizona Press.

127 more obscure in spring: Pyle, P. 2013. "On Separating Sagebrush Sparrow (*Artemisiospiza nevadensis*) from Bell's Sparrow (*A. belli*), With Particular Reference to *A.b. canescens*." tinyurl.com/PyleSAGS.

128 grayer streaks: American Ornithologists' Union. 2013. "Fifty-fourth Supplement to the American Ornithologists' Union *Check-list of North American Birds*." *Auk* 130: 1–14.

128 "The song of this bird": Ridgway, *Ornithology*.

128 that "confusing intermediate group": Szeliga, W. 2011. "Sage Sparrow Subspecies." earbirding.com/blog/archives/3040.

129 In Arizona: Corman, T., and C. Wise-Gervais, eds. 2005. *Arizona Breeding Bird Atlas*. Albuquerque: University of New Mexico Press.

129 western Colorado: Andrews, R., and R. Righter. 1992. *Colorado Birds: A Reference to Their Distribution and Habitat*. Denver: Denver Museum of Natural History.

129 Modoc Plateau and from the Mono Basin east: Martin, J., and B. Carlson. 1998. "Sage Sparrow (*Artemisiospiza belli*)." In A. Poole, ed. *Birds of North America Online*. Ithaca: Cornell Lab of Ornithology.

129 southbound migration: Ibid.

129 Wintering birds: Ibid.

129 at least 15 *Artemisiospiza* records for British Columbia: Toochin, R. 2013. "Status and Occurrence of the Sagebrush Sparrow (*Artemisiospiza nevadensis*) in British Columbia." tinyurl.com/ToochinArtemisiospizaBC.

129 photographed in Warren County, Kentucky: Hulsey, A. 2008. "Sage Sparrow in Warren County." *Kentucky Warbler* 84: 77–79.

129 photographed in Nova Scotia: McLaren, I. 2012. *All the Birds of Nova Scotia: Status and Critical Identification.* Kentville: Gaspereau Press.

129 the northernmost population: Oberholser, H. 1946. "Three New North American Birds." *Journal of the Washington Academy of Sciences* 36: 388–389.

CANYON TOWHEE

130 the type specimen: Baird, S., et al. 1874. *A History of North American Birds.* Vol. 2. Boston: Little, Brown.

130 the Cañon Towhee: Baird, S. 1854. "Descriptions of New Birds Collected Between Albuquerque, N.M., and San Francisco, California, During the Winter of 1853-54, by Dr. C.B.R. Kennerly and H.B. Möllhausen, Naturalists Attached to the Survey of the Pacific R.R. Route, Under Lt. A.W. Whipple." *Proceedings of the Academy of Natural Sciences of Philadelphia* 7: 118–120.

130 "restless" and "retiring": Baird et al., *History.*

130 behavior quite unlike: Coues, E. 1866. "List of the Birds of Fort Whipple, Arizona." *Proceedings of the Academy of Natural Sciences of Philadelphia* 18: 39–133.

130 "the localities it selects": Henshaw, H. 1873. "Report Upon the Ornithological Collections made in Portions of Nevada, Utah, California, Colorado, New Mexico, and Arizona During the Years 1871, 1872, 1873, and 1874." In G. Wheeler, ed. *Geographical and Geological Explorations and Surveys West of the One Hundredth Meridian.* Vol. 5. Washington: Government Printing Office.

130 "No one of the gentlemen": Baird, S. 1861. "Appendix to the Report of the Secretary." *Annual Report of the Board of Regents of the Smithsonian Institution.* 1861: 49–67.

130 by Nicholas Vigors in 1839: Richardson, J., et al. 1839. *The Zoology of Captain Beechey's Voyage.* London: Henry G. Bohn.

130 declared it a subspecies: Coues, E. 1872. *Key to North American Birds.* 1st ed. New York: Dodd and Mead.

130 synonymized the brown towhees: Belding, L., and R. Ridgway. 1879. "A Partial List of the Birds of Central California." *Proceedings of the United States National Museum* 1: 388–449.

130 his original concept of two separate species: Ridgway, R. 1901. *The Birds of North and Middle America.* Bulletin of the United States National Museum, Bulletin 50, pt. 1. Washington: Government Printing Office.

131 a committee to investigate: American Ornithologists' Union. 1902. "Eleventh Supplement to the American Ornithologists' Union *Check-list of North American Birds.*" *Auk* 19: 315–342.

131 the AOU removed the subspecies: American Ornithologists' Union. 1908. "Fourteenth Supplement to the American Ornithologists' Union *Check-list of North American Birds.*" *Auk* 25: 343–399.

131 a new subspecies of brown towhee: Oberholser, H. 1919. "Description of a New Subspecies of *Pipilo fuscus.*" *Condor* 21: 210–211.

131 "not each other's nearest relative": Zink, R. 1988. "Evolution of Brown Towhees: Allozymes, Morphometrics and Species Limits." *Condor* 90: 72–82.

131 different vocal repertoires: American Ornithologists' Union. 1989. "Thirty-seventh Supplement to the American Ornithologists' Union *Check-list of North American Birds.*" *Auk* 106: 532–538.

132 a reunion greeting: Marshall, J. 1964. "Voice in Communication and Relationships Among Brown Towhees." *Condor* 66: 345–356.

133 In Arizona: Corman, T., and C. Wise-Gervais, eds. 2005. *Arizona Breeding Bird Atlas.* Albuquerque: University of New Mexico Press.

133 in Texas: Lockwood, M., and B. Freeman. 2014. *The TOS Handbook of Texas Birds.* College Station: Texas A&M University Press.

133 In the south: Howell, S., and S. Webb. 1995. *A Guide to the Birds of Mexico and Northern Central America.* Oxford: Oxford University Press.

133 A 1975 report from Nebraska: Sharpe, R., et al. 2001. *Birds of Nebraska: Their Distribution and Temporal Occurrence.* Lincoln: University of Nebraska Press.

133 ten geographic races: Johnson, R., and L. Haight. 1996. "Canyon Towhee (*Melozone fusca*)." In A. Poole, ed. *Birds of North America Online.* Ithaca: Cornell Lab of Ornithology.

133 northernmost race: Oberholser, H. 1937. "Descriptions of Two New Passerine Birds from the Western United States." *Proceedings of the Biological Society of Washington* 50: 117–119.

133 Panhandle of Texas: Lockwood and Freeman, *TOS Handbook.*

133 easternmost race of the species in the United States: Van Rossem, A. 1934. "A Subspecies of the Brown Towhee from South-central Texas." *Transactions of the San Diego Society of Natural History* 7: 371–372.

133 In Texas, this race is resident: Lockwood and Freeman, *TOS Handbook.*

133 the subspecies *intermedia*: Nelson, E. 1899. "Descriptions of New Birds from Northwestern Mexico." *Proceedings of the Biological Society of Washington* 13: 25–31.

133 The race *jamesi*: Townsend, C. 1923. "Birds Collected in Lower California." *Bulletin of the American Museum of Natural History* 2: 1–26.

134 "desert pocket": Van Rossem, A. 1934. "Critical Notes on Middle American Birds." *Bulletin of the Museum of Comparative Zoology* 77: 385–490.

134 Mexico's central plateau: Johnson and Haight, "Canyon Towhee (*Melozone fusca*)."

134 August Sallé: Sclater, P. 1856. "Catalogue of the Birds Collected by M. Auguste Sallé in Southern Mexico, with Descriptions of New Species." *Proceedings of the Zoological Society of London* 24: 283–311.

134 Ridgway described it: Ridgway, R. 1899. "New Species, etc., of American Birds." *Auk* 16: 254–256.

134 nominate *Melozone fusca fusca*: Johnson and Haight, "Canyon Towhee (*Melozone fusca*)."

134 Hidalgo Canyon Towhee: Moore, R. 1949. "A New Race of *Pipilo fuscus* from Mexico." *Proceedings of the Biological Society of Washington* 62: 101–102.

134 southernmost Canyon Towhees: Moore, R. 1942. "Notes on *Pipilo fuscus* of Mexico and Description of a New Form." *Proceedings of the Biological Society of Washington* 55: 45–48.

CALIFORNIA TOWHEE

135 "quite feebly": Hartlaub, G. 1855. "Zwei unbeschriebene Vögel des Leydner Museums." *Journal für Ornithologie* 3: 361–362.

135 swift and withering: Bonaparte, C. 1856. "Excursions dans les divers musées d'Allemagne, de Hollande et de Belgique, et Tableaux paralléliques de l'ordre des Échassiers." *Comptes rendus hébdomadaires des séances de l'Académie des sciences* 43: 410–421.

135 corrections to his earlier account: Bonaparte, C. 1850 *Conspectus generum avium.* Leiden: Brill.

135 "How did Hartlaub not notice": Bonaparte, "Excursions."

135 named 15 years before: Vigors, N. 1839. "Ornithology." In F. Beechey, ed. *The Zoology of Captain Beechey's Voyage*. London: H.G. Bohn.

135 George Robert Gray at last gave Vigors credit: Gray, G. 1870. *Hand-list of Genera and Species of Birds*. Part 2. London: Trustees of the British Museum.

135 "publication has suffered": Beechey, F., ed. *The Zoology of Captain Beechey's Voyage*. London: H.G. Bohn.

136 lumped Vigors's bird with the Canyon Towhee: Coues, E. 1872. *Key to North American Birds*. 1st ed. New York: Dodd and Mead.

136 the brown towhees as separate species: Ridgway, R. 1877. *Ornithology*. United States Geological Exploration of the Fortieth Parallel. Vol. 4, pt. 3. Washington, D.C.: Government Printing Office.

136 *senicula* of southern California: American Ornithologists' Union. 1895. "Seventh Supplement to the American Ornithologists' Union *Check-list of North American Birds*." *Auk* 12: 163–168.

136 *carolae* of the state's central interior: American Ornithologists' Union. 1901. "Tenth Supplement to the American Ornithologists' Union *Check-list of North American Birds*." *Auk* 18: 295–321.

136 Ridgway's treatment of the brown towhees: Ridgway, R. 1901. *The Birds of North and Middle America*. Bulletin of the United States National Museum, Bulletin 50, pt. 1. Washington: Government Printing Office.

136 the AOU to reexamine: American Ornithologists' Union. 1902. "Eleventh Supplement to the American Ornithologists' Union *Check-list of North American Birds*." *Auk* 19: 315–342.

136 three questions: American Ornithologists' Union. 1903. "Twelfth Supplement to the American Ornithologists' Union *Check-list of North American Birds*." *Auk* 20: 331–368.

136 "alleged form": American Ornithologists' Union. 1904. "Thirteenth Supplement to the American Ornithologists' Union *Check-list of North American Birds*." *Auk* 21: 411–424.

136 The other issues: American Ornithologists' Union. 1908. "Fourteenth Supplement to the American Ornithologists' Union *Check-list of North American Birds*." *Auk* 25: 343–399.

136 Oberholser described a new population: Oberholser, H. 1919. "Description of a New Subspecies of *Pipilo fuscus*." *Condor* 21: 210–211.

136 new race *petulans*: Grinnell, J., and H. Swarth. 1926. "Systematic Review of the Pacific Coast Brown Towhees." *University of California Publications in Zoology* 21: 427–433.

136 ratified by Davis's morphological study: Davis, J. 1951. "Distribution and Variation of the Brown Towhees." *University of California Publications in Zoology* 52: 1–120.

137 "not each other's nearest relative": Zink, R. 1988. "Evolution of Brown Towhees: Allozymes, Morphometrics and Species Limits." *Condor* 90: 72–82.

137 different vocal repertoires: American Ornithologists' Union. 1989. "Thirty-seventh Supplement to the American Ornithologists' Union *Check-list of North American Birds*." *Auk* 106: 532–538.

138 as long as 25 minutes: Benedict, L., et al. 2011. "California Towhee (*Melozone crissalis*)." In A. Poole, ed., *Birds of North America Online*. Ithaca: Cornell Lab of Ornithology.

138 range of this widespread species: Small, A. 1994. *California Birds: Their Status and Distribution*. Vista, CA: Ibis.

138 strays have been recorded: Benedict et al., "California Towhee (*Melozone crissalis*)."

138 Russian Far East: Ibid.

138 "too many subspecies": Patten, M., et al. 2003. *Birds of the Salton Sea: Status, Biogeography, and Ecology*. Berkeley: University of California Press.

138 reduce the number of recognized subspecies: Benedict et al., "California Towhee (*Melozone crissalis*)."

138 revision carried out by Patten: Patten, et al., *Salton Sea*.

139 "extreme of ruddiness": Grinnell and Swarth, "Pacific Coast Brown Towhees."

139 deep chestnut of the crown: Benedict, L., et al. 2011. "California Towhee (*Melozone crissalis*)." In A. Poole, ed., *Birds of North America Online*. Ithaca: Cornell Lab of Ornithology.

139 As defined by Patten: Patten et al., *Salton Sea*.

139 resident of chaparral habitats: Benedict et al., "California Towhee (*Melozone crissalis*)."

139 population of the Argus Mountains: Recce, S. 1987. "Determination of Threatened Status and Critical Habitat Designation for the Inyo Brown Towhee." *Federal Register* 52: 28780–28781.

139 *viejecita*: Anthony, A. 1895. "Description of a New *Pipilo* from Southern and Lower California." *Auk* 12: 109–112.

139 December specimen from the Salton Sea: Patten et al., *Salton Sea*.

139 small, small-billed birds: Benedict et al., "California Towhee (*Melozone crissalis*)."

139 *aripolia* shows a paler belly: Ibid.

139 "gradation in several phenotypic characters": Zink, "Evolution of Brown Towhees."

ABERT TOWHEE

140 "exceedingly interesting collection": Baird, S. 1852. "Birds." Pp. 314–335 in Stansbury, H. *Exploration and Survey of the Valley of the Great Salt Lake of Utah*. Philadelphia: Lippincott, Grambo, and Co.

140 red fox: Larivière, S., and M. Pasitschniak-Arts. 1996. "*Vulpes vulpes*." *Mammalian Species* 537: 1–11.

140 The bird's "characteristic features": Baird, "Birds."

140 slipped into Abert's shipment by a colleague: Mearns, B., and R. Mearns. 1992. *Audubon to Xantus: The Lives of Those Commemorated in North American Bird Names*. London: Academic Press.

140 only a trace: Baird, "Birds."

140 "highly divergent": DaCosta, J., et al. 2009. "A Molecular Systematic Revision of Two Historically Problematic Songbird Clades: *Aimophila* and *Pipilo*." *Journal of Avian Biology* 40: 206–216.

140 a genus of their own: DaCosta et al., "Molecular Systematic Revision."

140 AOU split the genus: American Ornithologists' Union. 2010. "Fifty-first Supplement to the American Ornithologists' Union *Check-list of North American Birds*." *Auk* 127: 726–744.

141 Juvenile Abert Towhees: Tweit, R. C., and D. M. Finch. 1994. "Abert's Towhee (*Melozone aberti*)." In A. Poole, ed., *Birds of North America Online*. Ithaca: Cornell Lab of Ornithology.

141 In Utah: Behle, W. 1976. "Mohave Desert Avifauna in the Virgin River Valley of Utah, Nevada, and Arizona." *Condor* 78: 40–48.

141 into Nevada: Floyd, T., et al. 2007. *Atlas of the Breeding Birds of Nevada*. Reno: University of Nevada Press.

141 In California: Small, A. 1994. *California Birds: Their Status and Distribution*. Vista, CA: Ibis.

141 Arizona: Corman, T., and C. Wise-Gervais, eds. 2005. *Arizona Breeding Bird Atlas*. Albuquerque: University of New Mexico Press.

142 Nebraska: Sharpe, R., et al. 2001. *Birds of Nebraska: Their Distribution and Temporal Occurrence*. Lincoln: University of Nebraska Press.

142 Colorado: Henshaw, W. 1875. "Report Upon the Ornithological Collections Made in Portions of Nevada, Utah, California, Colorado, New Mexico, and Arizona During the Years 1871, 1872, 1873, and 1874." In *Report Upon Geographical and Geological Explorations and Surveys West of the One Hundredth Meridian, In Charge of First Lieut. Geo. M. Wheeler*. Vol. 5. Washington: Government Printing Office.

142 subspecies *dumeticola*: Van Rossem, A. 1946. "Two New Races of Birds from the Lower Colorado River Valley." *Condor* 48: 80–82.

142 Phillips also recognized two subspecies: Phillips, A. 1962. "Notas sistematicas sobre aves mexicanas II." *Anales del Instituto de Biología Universidad de México* 33: 331–372.

WHITE-THROATED TOWHEE

143 Ferdinand Deppe: Stresemann, E. 1954. "Ferdinand Deppe's Travels in Mexico, 1824-1829." *Condor* 56: 86–92.

143 two and a half Prussian thalers: Cabanis, J. 1863. "Lichtenstein's Preis-Verzeichniss mexicanischer Vögel etc. vom Jahre 1830." *Journal für Ornithologie* 11: 54–60.

143 half a week's rent: Engelsing, R. 1978. *Zur Sozialgeschichte deutscher Mittel- und Unterschichten*. 2nd ed. Kritische Studien zur Geschichtswissenschaft 4. Göttingen: Vandenhoeck & Ruprecht.

143 *Tangara rutila* on specimen labels: Van Rossem, A. 1934. "Critical Notes on Middle American Birds." *Bulletin of the Museum of Comparative Zoology* 77: 385–490.

143 Stresemann, sitting in Lichtenstein's chair: Stresemann, "Ferdinand Deppe's Travels."

143 *nomen dubium*: Dickinson, E., and L. Christidis, eds. 2014. *The Howard and Moore Complete Checklist of the Birds of the World*. 4th ed. Vol. 2. Eastbourne: Aves Press.

144 clear description of the White-throated Towhee: Sclater, P. 1856. "Catalogue of the Birds Collected by M. Auguste Sallé in Southern Mexico, with Descriptions of New Species." *Proceedings of the Zoological Society of London* 24: 283–311.

144 in Baird, Brewer, and Ridgway's influential *History*: Baird, S., et al. 1874. *A History of North American Birds*. Vol. 2. Boston: Little, Brown.

144 priority over Sclater's new name: Ridgway, R. 1901. *The Birds of North and Middle America*. Bulletin of the United States National Museum, Bulletin 50, pt. 1. Washington: Government Printing Office.

144 a warning: American Ornithologists' Union. 1998. *Check-list of North American Birds*. 7th ed. Washington: American Ornithologists' Union.

144 "the most positive character": Baird et al., *History*.

144 central breast spot: Ridgway, *North and Middle America*.

144 southeasternmost localities: Binford, L. 1989. *A Distributional Survey of the Birds of the Mexican State of Oaxaca*. Ornithological Monographs 43. Lawrence, KS: American Ornithologists' Union.

144 the subspecies *parvirostris*: Davis, J. 1951. "Distribution and Variation of the Brown Towhees." *University of California Publications in Zoology* 52: 1–120.

144 a matter of seasonal change: Davis, J. 1954. "Seasonal Changes in Bill Length of Certain Passerine Birds." *Condor* 56: 142–149.

144 Guerrero, Puebla, and northernmost Oaxaca: Parkes, K. 1974. "Systematics of the White-throated Towhee (*Pipilo albicollis*)." *Condor* 76: 457–459.

RUSTY-CROWNED GROUND SPARROW

145 Pierre Louis Jouy: Jouy, P. 1894. "Notes on Birds of Central Mexico, With Descriptions of Forms Believed to Be New." *Proceedings of the United States National Museum* 16: 771–791.

145 Bonaparte described the species: Bonaparte, C. 1850. *Conspectus generum avium*. Leiden: Brill.

145 the French warship *Danaïde*: Sclater, P., and O. Salvin. 1868. "Descriptions of New or Little-known American Birds of the Families Fringillidae, Oxyrhamphidae, Bucconidae, and Strigidae." *Proceedings of the Zoological Society of London* 1868: 321–329.

145 deemed one of a kind: Salvin, O., and F. Godman. 1886. *Biologia centrali-americana: Aves*. Vol. 1. London: R.H. Porter.

145 Ridgway could only speculate: Ridgway, R. 1901. *The Birds of North and Middle America*. Bulletin of the United States National Museum, Bulletin 50, pt. 1. Washington: Government Printing Office.

145 specimen in the Berlin museum: Cabanis, J. 1851. *Museum Heineanum: Verzeichniss der ornithologischen Sammlung des Oberamtmann Ferdinand Heine*. Vol. 1. Halberstadt: R. Frantz.

145 represented by a single skin: Lichtenstein, H. 1854. *Nomenclator avium Musei Zoologici Berolinensis*. Berlin: Königliche Akademie der Wissenschaften.

145 collected by Ferdinand Deppe: Salvin and Godman, *Biologia centrali-americana*.

145 some doubt about the bird's specific status: Cabanis, J. 1860. "Übersicht der im Berliner Museum befindlichen Vögel von Costa Rica." *Journal für Ornithologie* 10: 401–416.

146 "clearly different": Cabanis, J. 1866. "Über neue oder weniger bekannte exotische Vögel." *Journal für Ornithologie* 14: 231–235.

146 Lawrence published a description: Lawrence, G. 1867. "Description of New Species of American Birds." *Annals of the Lyceum of Natural History of New York* 8: 466–482.

146 after all identical to Cabanis's: Lawrence, G. 1874. "The Birds of Western and Northwestern Mexico, Based Upon Collections Made by Col. A.J. Grayson, Capt. J. Xantus and Ferd. Bischoff, Now in the Museum of the Smithsonian Institution, at Washington, D.C." *Memoirs Read Before the Boston Society of Natural History* 2: 265–319.

146 lack of complete series: Ridgway, *North and Middle America*.

146 gathered the evidence in 1898: Nelson, E. 1898. "Notes on Certain Species of Mexican Birds." *Auk* 15: 155–161.

146 engraving of Bonaparte's type: Sclater, P., and O. Salvin. 1869. *Exotic Ornithology: Containing Figures and Descriptions of New or Rare Species of American Birds*. London: Bernard Quaritch.

146 "series of true *P. rubricatum*": Ridgway, *North and Middle America*.

146 by John Cassin: Cassin, J. 1865. "On Some Conirostral Birds from Costa Rica in the Collection of the Smithso-

nian Institution." *Proceedings of the Academy of Natural Sciences of Philadelphia* 17: 169–172.

146 by Jean Cabanis: Cabanis, "Exotische Vögel."

146 "trenchant features" separating them: Hellmayr, C. 1938. *Catalogue of Birds of the Americas and the Adjacent Islands in Field Museum of Natural History.* Zoological Series 13, pt. 11. Chicago: Field Museum of Natural History.

146 distinct at the species level: Miller, A., et al. 1957. *Distributional Checklist of the Birds of Mexico.* Part 2. Pacific Coast Avifauna 33. Berkeley: Cooper Ornithological Society.

146 "until the type of *kieneri* can be examined": Van Rossem, A. 1933. "A Northern Race of *Melozone rubricatum* (Cabanis)." *Transactions of the San Diego Society of Natural History* 7: 283–284.

146 "rumpled and stained" bird: Van Rossem, A. 1934. "Critical Notes on Middle American Birds." *Bulletin of the Museum of Comparative Zoology* 77: 385–490.

146 coined originally by Johann Georg Wagler: Wagler, J. 1831. "Einige Mittheilungen über Thiere Mexicos. *Isis* 24: 510–535.

146 "the genus combines characters": Ibid.

146 the genus *Embernagra*: Gray, G. 1870. *Hand-list of Genera and Species of Birds.* Part 2. London: Trustees of the British Museum.

146 returned the ground sparrows to *Pyrgisoma*: Sclater and Salvin, "New or Little-known American Birds."

146 Reichenbach had erected the genus *Melozone*: Reichenbach, L. 1850. *Avium systema naturale: Das natürliche System der Vögel.* Dresden: Expedition der vollständigsten Naturgeschichte.

147 Juveniles are even less frequently observed: Howell, S., and S. Webb. 1995. *A Guide to the Birds of Mexico and Northern Central America.* Oxford: Oxford University Press.

147 sputtering labial trill: Ibid.

147 endemic to thorn forest and dry brush: Rising, J. (2011). Rusty-crowned Ground-sparrow (*Pyrgisoma kieneri*). In del Hoyo, J., Elliott, A., Sargatal, J., Christie, D.A. & de Juana, E. (eds.) (2015). *Handbook of the Birds of the World Alive.* Lynx Edicions, Barcelona.

148 three subspecies: Rising, " Rusty-crowned Ground-sparrow (*Pyrgisoma kieneri*)."

RUSTY SPARROW

149 "indefatigable explorer Auguste Sallé": Bonaparte, C. 1856. "Note sur les Tableaux des Gallinacés." *Comptes rendus hebdomadaires des séances de l'Académie des sciences* 42: 953–957.

149 Philip Lutley Sclater, examining the specimen: Sclater, P. 1856. "Catalogue of the Birds Collected by M. Auguste Sallé in Southern Mexico, with Descriptions of New Species." *Proceedings of the Zoological Society of London* 24: 283–311.

149 Swainson had described the adult: Swainson, W. 1827. "A Synopsis of the Birds Discovered in Mexico by W. Bullock, F.L.S. and H.S., and Mr. William Bullock, jun." *The Philosophical Magazine* 2nd ser., 1: 364–369 and 433–442.

149 a specimen of his own: Swainson, W. 1838. *Animals in Menageries.* London: Longman, Orme, Brown, Green & Longmans, and John Taylor.

150 Reichenbach "corrected" Swainson's original: Reichenbach, L. 1850. *Avium systema naturale: Das natürliche System der Vögel.* Dresden: Expedition der vollständigsten Naturgeschichte.

150 emended spelling prevailed: Ridgway, R. 1901. *The Birds of North and Middle America.* Bulletin of the United States National Museum, Bulletin 50, pt. 1. Washington: Government Printing Office.

150 notable exception was Elliott Coues: Coues, E. 1903. *Key to North American Birds.* 5th ed. Vol. 1. Boston: Dana Estes and Company.

150 overlap in range: Howell, S., and S. Webb. 1995. *A Guide to the Birds of Mexico and Northern Central America.* Oxford: Oxford University Press.

150 Oaxaca Sparrow is smaller: Storer, R. 1955. "A Preliminary Survey of the Sparrows of the Genus *Aimophila*." *Condor* 57: 192–201.

151 "strong yellow wash": Ibid.

151 "reunion duet": Ibid.

152 West Mexican birds: Van Rossem, A. 1942. "A New Race of the Rusty Sparrow from North Central Sonora, Mexico." *Transactions of the San Diego Society of Natural History* 9: 435–436.

152 divided into a dozen subspecies: Rising, J. 2011. Rusty Sparrow (*Aimophila rufescens*). In J. del Hoyo, et al., eds. *Handbook of the Birds of the World Alive.* Barcelona: Lynx Edicions.

152 Brewster in 1888: Brewster, W. 1888. "Descriptions of Supposed New Birds from Lower California, Sonora, and Chihuahua, Mexico, and the Bahamas." *Auk* 5: 82–95.

152 John C. Cahoon: Webster, F. 1891. "John C. Cahoon Meets Death at Newfoundland." *Ornithologist and Oölogist* 16: 73–75.

152 Brewster suspected: Brewster, "Supposed New Birds."

152 again named and described as new: Salvin, O., and F. Godman. 1889. "Notes on Mexican Birds." *Ibis* ser. 6, 1: 232–243.

152 this subspecies is resident: Rising, "Rusty Sparrow (*Aimophila rufescens*)."

152 slightly smaller than *rufescens*: Ridgway, *North and Middle America.*

152 95 miles south of the United States border: Russell, S., and G. Monson. 1998. *The Birds of Sonora.* Tucson: University of Arizona Press.

152 paler subspecies *antonensis*: Van Rossem, "A New Race."

152 *pyrgitoides* is an eastern subspecies: Rising, "Rusty Sparrow (*Aimophila rufescens*)."

RUFOUS-CROWNED SPARROW

153 *Birds of the Pacific States*: Hoffmann, R. 1927. *Birds of the Pacific States.* Boston: Houghton Mifflin.

153 Harriet William Myers: Stone, W. 1923. "Mrs. Myers' 'Western Birds.'" *Auk* 40: 350.

153 the simply titled *Western Birds*: Myers, H. 1922. *Western Birds.* New York: MacMillan.

153 coined the familiar verbalization: Myers, H. 1909. "Nesting Habits of the Rufous-crowned Sparrow." *Condor* 11: 131–133.

153 The first specimens: Cassin, J. 1853–1855. *Illustrations of the Birds of California, Texas, Oregon, British and Russian America.* Philadelphia: J.B. Lippincott and Co.

154 Cooper, following Spencer Baird: Baird, S., with J. Cassin and G. Lawrence. 1858. *Birds.* Explorations and Surveys for a Railroad Route. Vol. 9. Washington: Beverly Tucker.

154 "their favorite resort": Baird, S. ed. 1870. *Ornithology: From the Manuscript and Notes of J.G. Cooper.* Vol. 1. Cambridge: University Press.

154 Even in 1874: Baird, S., et al. 1874. *A History of North American Birds.* Vol. 2. Boston: Little, Brown.

154 Charles A. Allen: Brewster, W. 1879. "Notes on the Habits and Distribution of the Rufous-crowned Sparrow (*Peucaea ruficeps*)." *Bulletin of the Nuttall Ornithological Club* 4: 47–48.

154 "is thought to convey mild alertness": Collins, P. 1999. "Rufous-crowned Sparrow (*Aimophila ruficeps*)." In A. Poole, ed. *Birds of North America Online*. Ithaca: Cornell Lab of Ornithology.

155 An infrequent variant: Ibid.

155 On the Pacific Slope: Ibid.

155 southernmost Nevada and Utah: Floyd, T., et al. 2007. *Atlas of the Breeding Birds of Nevada*. Reno: University of Nevada Press.

156 On the Mexican mainland: Collins, "Rufous-crowned Sparrow (*Aimophila ruficeps*)."

156 elevational movement after breeding: Rising, J. 2015. "Rufous-crowned Sparrow (*Aimophila ruficeps*)." In del Hoyo, J., Elliott, A., Sargatal, J., Christie, D.A., & de Juana, E. eds. *Handbook of the Birds of the World Alive*. Barcelona: Lynx Edicions.

156 coast of Texas: Lockwood, M., and B. Freeman. 2014. *The TOS Handbook of Texas Birds*. College Station: Texas A&M University Press.

156 Hubbard's 1975 review: Hubbard, J. 1975. "Geographic Variation in Non-California Populations of the Rufous-crowned Sparrow." *Nemouria: Occasional Papers of the Delaware Museum of Natural History* 15: 1–28.

RUFOUS-CAPPED BRUSH FINCH

157 accidental death: Glaw, F. 2001. "Johann Georg Wagler (1800-1832)." In W. Rieck et al., eds. *Die Geschichte der Herpetologie und Terrarienkunde im deutschsprachigen Raume*. Mertensiella 12: 633–637.

157 "a number of new species": Wagler, J. 1831. "Einige Mittheilungen über Thiere Mexicos." *Isis* 24: 510–535.

157 The scientific name Wagler coined: Cabanis, J. 1851. *Museum Heineanum: Verzeichniss der ornithologischen Sammlung des Oberamtmann Ferdinand Heine*. Vol. 1. Halberstadt: R. Frantz.

157 described it as combining characteristics: Wagler, "Thiere Mexicos."

157 the list of genera recalled: Sclater, P. 1856. "Catalogue of the Birds Collected by M. Auguste Sallé in Southern Mexico, with Descriptions of New Species." *Proceedings of the Zoological Society of London* 24: 283–311.

157 greatly expanded *Embernagra*: Gray, G. 1870. *Hand-list of Genera and Species of Birds*. Part 2. London: Trustees of the British Museum.

157 catholic generic concept: Giebel, C. 1875. *Thesaurus ornithologiae: Repertorium der gesammten ornithologischen Literatur und Nomenclator sämmtlicher Gattungen und Arten der Vögel*. Vol. 2. Leipzig: Brockhaus.

158 brown tips to the wing coverts: Rising, J. 2016. "Rufous-capped Brush-finch (*Atlapetes pileatus*)." In J. del Hoyo, et al., eds. *Handbook of the Birds of the World*. Barcelona: Lynx Edicions.

158 Nominate *pileatus . . . dilutus*: Ibid.

158 "these differences are easily distinguished": Paynter, R. 1978. "Biology and Evolution of the Avian Genus *Atlapetes* (Emberizinae)." *Bulletin of the Museum of Comparative Zoology* 148: 323–369.

WHITE-NAPED BRUSH FINCH

159 de Candé: Lauro, F., et al., eds. 2003. *Etat général des fonds privés de la Marine*. Vol. 1. Vincennes: Service historique de la Défense.

159 described the exotic new species: de Lafresnaye, F., and A. d'Orbigny. 1838. "Notice sur quelques oiseaux de Carthagène et de la partie du Mexique la plus voisine,

rapportés par M. Ferdinand de Candé, officier de la marine royale." *Revue zoologique* 1: 164–166.

159 introduced it to science again: Lesson, R. 1839. "Oiseaux rares ou nouveaux de la collection du Docteur Abbeillé, à Bordeaux." *Revue zoologique* 2: 40–43.

159 priority of the name: Gray, G. 1849. *The Genera of Birds*. Vol. 2. London: Longman, Brown, Green, and Longmans.

159 the *Didon*'s stops in Mexico: Paynter, R. 1964. "The Type Locality of *Atlapetes albinucha*." *Auk* 81: 223–224.

159 described a few years later: de Lafresnaye, F. 1843. "Quelques nouveaux espèces d'Oiseaux." *Revue zoologique* 6: 97–104.

160 Lesson erected the genus: Lesson, R. 1831. *Traité d'ornithologie*. Vol. 1. Brussels: F.G. Levrault.

160 the new genus *Atlapetes*: Cabanis, J. 1851. *Museum Heineanum: Verzeichniss der ornithologischen Sammlung des Oberamtmann Ferdinand Heine*. Vol. 1. Halberstadt: R. Frantz.

160 cast the group's lot with the tanagers: Sclater, P. 1856. "Synopsis avium tanagrinarum: A Descriptive Catalogue of the Known Species of Tanagers." *Proceedings of the Zoological Society of London* 24: 64–94.

160 moved to the genus *Buarremon*: Bonaparte, C. 1850. *Conspectus generum avium*. Leiden: Brill.

160 still accounted a tanager: Salvin, O., and F. Godman. 1886. *Biologia centrali-americana: Aves*. Vol. 1. London: R.H. Porter.

160 definitively reassigned that genus: Ridgway, R. 1901. *The Birds of North and Middle America*. Bulletin of the United States National Museum, Bulletin 50, pt. 1. Washington: Government Printing Office.

160 Atlantic slope of Mexico: Howell, S., and S. Webb. 1995. *A Guide to the Birds of Mexico and Northern Central America*. Oxford: Oxford University Press.

160 southerly race: Miller, A., et al. 1957. *Distributional Checklist of the Birds of Mexico*. Part 2. Pacific Coast Avifauna 33.

160 The northern, nominate race: Rising, J. 2013. "White-naped Brush-finch (*Atlapetes albinucha*)." In J. del Hoyo et al., eds. *Handbook of the Birds of the World*. Vol. 16. Barcelona: Lynx Edicions.

160 "lighter yellow and more extensive": Dwight, J., and L. Griscom. 1921. *A Revision of* Atlapetes *with Descriptions of Three New Races*. American Museum Novitates 16. New York: American Museum of Natural History.

EASTERN TOWHEE

161 depicted by a European painter: Bescoby, J., et al. 2007. "New Visions of a New World: The Conservation and Analysis of the John White Watercolours." *British Museum Technical Research Bulletin* 1: 1–22.

161 by Edward Topsell: Christy, B. 1933. "Topsell's 'Fowles of Heauen.'" *Auk* 50: 275–283.

161 for Hans Sloane: Croft-Murray, E., and P. Hulton. 1960. *Catalogue of British Drawings in the British Museum, XVI and XVII Centuries*. London: Trustees of the British Museum.

161 Mark Catesby published his account: Catesby, M. 1731. *The Natural History of Carolina, Florida and the Bahama Islands*. Vol. 1. London: privately published.

161 Linnaeus in 1758 simply translated: Linnaeus, C. 1758. *Systema naturae per regna tria naturae*. 10th ed. Vol. 1. Stockholm: Laurentius Salvius.

161 John Abbott and Steven Elliot: Snavely, I. 2015. "Biographer of the Feathered Tribes: Alexander Wilson and *American Ornithology*." *Pennsylvania Heritage Magazine* 41: 16–25.

161 "examined a great number": Wilson, A. 1810. *American Ornithology, or the Natural History of the Birds of the United States.* Vol. 2. Philadelphia: Bradford and Inskeep.

162 "color of the iris": Burling, G. 1870. "American Birds." *Appleton's Journal* 4: 256–258.

162 "a local race": Allen, J. 1870. "On the Mammals and Winter Birds of East Florida, With an Examination of Certain Assumed Specific Characters in Birds, and a Sketch of the Bird Faunae of Eastern North America." *Bulletin of the Museum of Comparative Zoology* 2: 161–450.

162 "the curious little *Pipilo*": Coues, E. 1871. "Progress of American Ornithology." *American Naturalist* 5: 364–373.

162 revised his opinion: Coues, E. 1872. *Key to North American Birds.* 1st ed. Salem: Naturalists' Agency.

162 misquote his own nomenclatural act: Coues, E. 1874. *Birds of the Northwest.* Miscellaneous Publications of the Geological Survey of the Territories 3. Washington: Government Printing Office.

162 "A few words of explanation": Maynard, C. 1881. *The Birds of Eastern North America.* Newtonville: C.J. Maynard and Co.

163 "absurd classifications": Lewis, D. 2012. *The Feathery Tribe: Robert Ridgway and the Modern Study of Birds.* New Haven: Yale University Press.

163 "the case of *Pipilo alleni*": Henshaw, H. 1880. "Maynard's *Birds of Eastern North America.*" *Bulletin of the Nuttall Ornithological Club* 5: 170–173.

163 white spotting and streaking: Coues, E. 1878. "*Pipilo erythophthalmus* with Spotted Scapulars." *Bulletin of the Nuttall Ornithological Club* 3: 41–42.

163 "a great or less number of minute white spots": Baird, S., et al. 1874. *A History of North American Birds.* 3 vols. Boston: Little, Brown.

163 considered conspecific: Sibley, C. 1950. "Species Formation in the Red-eyed Towhees of Mexico." *University of California Publications in Zoology.* 50: 109–194.

163 "particularly pleasurable": Dickinson, J. 1951. "Review of 'Species Formation in the Red-eyed Towhees of Mexico.'" *Wilson Bulletin* 63: 349–350.

163 ratified Sibley's view: American Ornithologists' Union. 1954. "Twenty-ninth Supplement to the American Ornithologists' Union *Check-list of North American Birds.*" *Auk* 71: 310–312.

163 More than 500 specimens: Sibley, C., and D. West. 1959. "Hybridization in the Rufous-sided Towhees of the Great Plains." *Auk* 76: 326–338.

163 "strongly suggest assortative mating": American Ornithologists' Union. 1995. "Fortieth Supplement to the A.O.U. Check-list." *Auk* 112: 819–830.

163 Study of mitochondrial DNA: Ball, R., and J. Avise. 1992. "Mitochondrial DNA Phylogeographic Differentation Among Avian Populations and the Evolutionary Significance of Subspecies." *Auk* 109: 626–636.

163 coining the new English name Eastern towhee: American Ornithologists' Union, "Fortieth Supplement."

164 white spots or bars: Coues, "Spotted Scapulars."

164 "due to 'ancestral' genes": Sibley and West, "Hybridization."

164 "scoring" the plumage: Ibid.

164 apparent hybrids and backcrosses: Ibid.

165 a contact note: Greenlaw, J. 2015. "Eastern Towhee (*Pipilo erythophthalmus*)." In A. Poole, ed. *Birds of North America Online.* Ithaca: Cornell Lab of Ornithology.

165 "somewhat hoarser": Ibid.

165 song of *alleni*: Ibid.

165 "complex quiet song": Ibid.

165 Breeding birds occur in warm brushy habitats: Ibid.

165 Piney Woods of Texas: Lockwood, M., and B. Freeman. 2014. *The TOS Handbook of Texas Birds.* College Station: Texas A&M University Press.

166 arriving in the breeding areas: Greenlaw, "Eastern Towhee (*Pipilo erythrophthalmus*)."

166 regularly overshoots in spring: McLaren, I. 2012. *All the Birds of Nova Scotia: Status and Critical Identification.* Kentville: Gaspereau Press.

166 on Lundy Island: Waller, C. 1970. "Rufous-sided Towhee on Lundy." *British Birds* 63: 147–149.

166 In Colorado: Andrews, R., and R. Righter. 1992. *Colorado Birds: A Reference to Their Distribution and Habitat.* Denver: Denver Museum of Natural History.

166 report from California: Hamilton, R., et al., eds. 2007. *Rare Birds of California.* Camarillo: Western Field Ornithologists.

166 Four subspecies: Dickinson, J. 1952. "Geographic Variation in the Red-eyed Towhees of the Eastern United States." *Bulletin of the Museum of Comparative Zoology* 107: 273–352.

166 Joseph Harvey Rilcy: Wetmore, A. 1943. "In Memoriam: Joseph Harvey Riley." *Auk* 60: 1–16.

SPOTTED TOWHEE

167 "The black series of *Pipilo*": Coues, E. 1884. *Key to North American Birds.* 2nd ed. Boston: Estes and Lauriat.

167 Brewster warned: Brewster, W. 1882. "On a Collection of Birds Lately Made by Mr. F. Stephens in Arizona." *Bulletin of the Nuttall Ornithological Club* 7: 193–212.

167 "a convenient receptacle": Allen, J., and W. Brewster. 1883. "List of Birds Observed in the Vicinity of Colorado Springs, Colorado, During March, April, and May, 1882." *Bulletin of the Nuttall Ornithological Club* 8: 189–198.

167 "constant tangible differences": Coues, E. 1866. "List of the Birds of Fort Whipple, Arizona." *Proceedings of the Academy of Natural Sciences of Philadelphia* 18: 39–133.

167 "strongly marked . . . varieties": Baird, S., with J. Cassin and G. Lawrence. 1858. *Birds.* Explorations and Surveys for a Railroad Route. Vol. 9. Washington: Beverly Tucker.

167 raised a white flag: Cabanis, J. 1862. "Zur Synonymie einiger *Pipilo*-Arten." *Journal für Ornithologie* 10: 473–474.

167 "one of the worst muddles": Allen and Brewster, "Colorado Springs."

167 Swainson's descriptions: Richardson, J., and W. Swainson. 1831. *Fauna boreali-americana: Or the Zoology of the Northern Parts of British America.* Part 2. London: John Murray.

167 "Aesthetically pleasing": Farber, P. 1982. *Discovering Birds: The Emergence of Ornithology as a Scientific Discipline, 1760-1850.* Baltimore: Johns Hopkins University Press.

167 "was supposed to consist": Richardson and Swainson, *Fauna boreali-americana.*

168 1973 Supplement: American Ornithologists' Union. 1973. "Thirty-second Supplement to the American Ornithologists' Union *Check-list of North American Birds.*" *Auk* 90: 411–419.

168 two separate species: Swainson, W. 1827. "A Synopsis of the Birds Discovered in Mexico by W. Bullock, F.L.S. and H.S., and Mr. William Bullock, jun." *The Philosophical Magazine* 2nd ser., 1: 364–369 and 433–442.

168 Misinformed about the localities: Sibley, C. 1950. "Species Formation in the Red-eyed Towhees of Mexico." *University of California Publications in Zoology*. 50: 109–194.

168 affirmed through the nineteenth century: Ridgway, R. 1901. *The Birds of North and Middle America*. Bulletin of the United States National Museum, Bulletin 50, pt. 1. Washington: Government Printing Office.

168 critical review: Sibley, "Species Formation."

168 "confounded" two distinct towhee taxa: Bell, J. 1852. "On the *Pipilo Oregonus*, as Distinguished from the *Pipilo Arcticus* of Swainson." *Annals of the Lyceum of Natural History of New York* 5: 6–8.

168 "shades into": Coues, E. 1872. Key to North American Birds. 1st ed. Salem: Naturalists' Agency.

168 "This form, if not a distinct species": Baird, *Birds*.

168 Coues expressed his doubts: Coues, "Fort Whipple, Arizona."

168 The *Key* treated them as conspecific: Coues, *Key*. 1st ed.

169 specimens collected by Andrew Jackson Grayson: Lawrence, G. 1874. "Descriptions of New Species of Birds from Mexico, Central America, and South America, With a Note on *Rallus longirostris*." *Annals of the Lyceum of Natural History of New York* 10: 1–21.

169 Lawrence reconsidered: Lawrence, G. 1871. "List of the Birds of Western Mexico." *Memoirs Read Before the Boston Society of Natural History* 2: 265–319.

169 the dignity of a species: Ridgway, R. 1887. *A Manual of North American Birds*. Philadelphia: J.B. Lippincott.

169 a position it continued to hold: Ridgway, *North and Middle America*.

169 Hellmayr lumped *carmani*: Hellmayr, C. 1938. *Catalogue of Birds of the Americas and the Adjacent Islands in Field Museum of Natural History*. Zoological Series 13, pt. 11. Chicago: Field Museum of Natural History.

169 "This dwarf form": Miller, A., et al. 1957. *Distributional Checklist of the Birds of Mexico*. Part 2. Pacific Coast Avifauna 33.

169 list the Socorro Towhee as a full species: Clements, J. 2007. *The Clements Checklist of Birds of the World*. 6th ed. Ithaca: Cornell University Press.

169 the last Spotted Towhee: Grinnell, J. 1897. "Description of a New Towhee From California." *Auk* 14: 294–296.

169 large-billed subspecies: American Ornithologists' Union. 1899. "Ninth Supplement to the A.O.U. Checklist." *Auk* 16: 97–133.

169 Ridgway described this "not abundant" bird: Ridgway, R. 1876. "Ornithology of Gaudeloupe Island, Based on Notes and Collections Made by Dr. Edward Palmer." *Bulletin of the United States Geological and Geographical Survey of the Territories* 2: 183–195.

169 a "close affinity": Ridgway, R. 1877. "The Birds of Guadalupe Island, Discussed with Reference to the Present Genesis of Species." *Bulletin of the Nuttall Ornithological Club* 2: 58–66.

169 Richard Bowdler Sharpe demurred: Sharpe, R. 1888. *Catalogue of the Passeriformes or Perching Birds, in the Collection of the British Museum: Fringilliformes, Part III*. Catalogue of the Birds in the British Museum 12. London: Trustees of the British Museum.

170 distinctness of the Guadalupe Towhee until 1944: American Ornithologists' Union. 1944. "Nineteenth Supplement to the American Ornithologists' Union Check-list of North American Birds." *Auk* 61: 441–464.

170 Hellmayr's observation: Hellmayr, *Birds of the Americas*.

170 made explicit in 1981: Monson, G., and A. Phillips. 1981. *Annotated Checklist of the Birds of Arizona*. 2nd ed. Tucson: University of Arizona Press.

170 "scoring" the plumage: Sibley and West, "Hybridization."

170 neither Spotted nor Collared: Sibley, "Species Formation."

171 the most distinctive vocalization: Audubon, J. 1839. *Ornithological Biography*. Vol. 5. Edinburgh: Adam Black and Charles Black.

171 states of both calm and alarm: Bartos Smith, S., and J. Greenlaw. 2015. "Spotted Towhee (*Pipilo maculatus*)." In A. Poole, ed. *Birds of North America Online*. Ithaca: Cornell Lab of Ornithology.

172 "often occurs in social contexts": Bartos Smith and Greenlaw, "Spotted Towhee (*Pipilo maculatus*)."

172 songs of the Spotted Towhee: Ibid.

172 Inland in western Canada: Ibid.

172 Riske Creek, British Columbia: Campbell, R., et al. 2001. *The Birds of British Columbia*. Vol. 4. Vancouver: UBC Press.

172 Baja California: Miller, A., et al. 1957. *Distributional Checklist of the Birds of Mexico*. Part 2. Pacific Coast Avifauna 33.

172 the migrant populations: Bartos Smith and Greenlaw, "Spotted Towhee (*Pipilo maculatus*)."

172 southeasternmost records: Greenlaw, J., et al. 2014. *The Robertson and Woolfenden Florida Bird Species: An Annotated List*. Florida Ornithological Society Special Publication 8. Gainesville: Florida Ornithological Society.

172 Yorkshire coast of the North Sea: O'Sullivan, J., et al. 1977. "Report on Rare Birds in Great Britain in 1976." *British Birds* 70: 405–453.

173 "white spots in the scapulars": Mather, J. 1976. *Yorkshire Naturalists' Union Ornithological Report for 1975*. Leeds: University Printing Service.

173 escape from captivity: Lewington, I., et al. 1991. *A Field Guide to the Rare Birds of Britain and Europe*. London: HarperCollins.

173 published photograph: Speight, G. 1976. "The Spurn Towhee." *Yorkshire Birding* 1: 23–24.

173 most of these differences "occur in a mosaic pattern": Bartos Smith and Greenlaw, "Spotted Towhee (*Pipilo maculatus*)."

173 *curtatus* was named for: Grinnell, J. 1911. "Description of a New Spotted Towhee from the Great Basin." *University of California Publications in Zoology* 7: 309–311.

173 in Arizona: Monson and Phillips, *Annotated Checklist*.

174 winter specimens from eastern Arizona and Utah: Phillips, A., et al. 1964. *The Birds of Arizona*. Tucson: University of Arizona Press.

174 Nova Scotia: McLaren, I. 2012. *All the Birds of Nova Scotia: Status and Critical Identification*. Kentville: Gaspereau Press.

174 the subspecies *megalonyx*: Bartos Smith and Greenlaw, "Spotted Towhee (*Pipilo maculatus*)."

174 the subspecies *umbraticola*: Grinnell, J., and H. Swarth. 1926. "An Additional Subspecies of Spotted Towhee from Lower California." *Condor* 28: 130–133.

174 southernmost Baja California Sur: Ibid.

174 Mexican subspecies: Bartos Smith and Greenlaw, "Spotted Towhee (*Pipilo maculatus*)."

174 *griseipygius* has a gray rump: Van Rossem, A. 1934. "Critical Notes on Middle American Birds." *Bulletin of the Museum of Comparative Zoology* 77: 385–490.

175 the race *gaigei*: Van Tyne, J., and G. Sutton. 1937. *The Birds of Brewster County, Texas*. University of Michigan Museum of Zoology Miscellaneous Publications 37. Ann Arbor: University of Michigan Press.

175 northernmost subspecies in eastern Mexico: Bartos Smith and Greenlaw, "Spotted Towhee (*Pipilo maculatus*)."

175 less extensively spotted white above: Sibley, "Species Formation."

175 highlands of central Mexico are inhabited by three subspecies: Ibid.

175 "the western slope of the Volcán de Toluca": Sibley, C. 1950. "Species formation in the red-eyed towhees of Mexico." University of California Publications in Zoology 50: 109–194.

175 Middle America south of Mexico's Transvolcanic Belt: Ibid.

GUADALUPE TOWHEE

176 botanist and naturalist Edward Palmer: Blake, S. 1961. "Edward Palmer's Visit to Guadalupe Island, Mexico, in 1875." *Madroño* 16: 1–4.

176 mistrusted Coues: Cutright, P., and M. Brodhead. 1981. *Elliott Coues: Naturalist and Frontier Historian*. Urbana: University of Illinois Press.

176 "very interesting" Guadalupe birds: Ridgway, R. 1876. "Ornithology of Guadeloupe Island, Based on Notes and Collections Made by Dr. Edward Palmer." *Bulletin of the United States Geological and Geographical Survey of the Territories* 2: 183–195.

176 "notes accompanying the specimens": Ibid.

177 no ship arrived: Blake, "Edward Palmer's Visit."

177 a single Guadalupe Towhee: Thoburn, W. 1899. "Report of an Expedition in Search of the Fur Seal of Guadalupe Island, Lower California, June, 1897." In D. Jordan, ed. *The Fur Seals and Fur Sea Islands of the North Pacific Ocean* 3. Washington: Government Printing Office: 275–283.

177 Guadalupe suffered dramatically: Howell, T., and T. Cade. 1954. "The Birds of Guadalupe Island in 1953." *Condor* 56: 283–294.

177 The most striking distinctions: Ridgway, R. 1901. *The Birds of North and Middle America*. Bulletin of the United States National Museum, Bulletin 50, pt. 1. Washington: Government Printing Office.

177 "when startled": Ridgway, "Ornithology of Guadeloupe Island."

SOCORRO TOWHEE

178 put in at Socorro Island: Grayson, A. 1871. "Exploring Expedition to the Island of Socorro, from Mazatlan, Mexico." *Proceedings of the Boston Society of Natural History* 14: 287–302.

178 credit was not his alone: Lawrence, G. 1874. "The Birds of Western and Northwestern Mexico, Based Upon Collections Made by Col. A.J. Grayson, Capt. J. Xantus and Ferd. Bischoff, Now in the Museum of the Smithsonian Institution, at Washington, D.C." *Memoirs Read Before the Boston Society of Natural History* 2: 265–319.

178 "For this providential service": Grayson, A. 1871. "Exploring Expedition to the Island of Socorro, from Mazatlan, Mexico." *Proceedings of the Boston Society of Natural History* 14: 287–302.

178 collected by snaring: Ibid.

178 representatives of a new sparrow species: Taylor, L. 1951. "Prior Descriptions of Two Mexican Birds by Andrew Jackson Grayson." *Condor* 53: 194–197.

179 Grayson's own death: Taylor, L. 1949. "Andrew Jackson Grayson." *Condor* 51: 49–51.

179 "abundant" or "very common": Anthony, A. 1898. "Avifauna of the Revillagigedo Islands." *Auk* 15: 311–318.

179 fairly common: Howell, S., and S. Webb. 1995. *A Guide to the Birds of Mexico and Northern Central America*. Oxford: Oxford University Press.

179 non-native vertebrates: Brattstrom, B. 1990. "Biogeography of the Islas Revillagigedo, Mexico." *Journal of Biogeography* 17: 177–183.

179 a pair of domestic hogs: Grayson, "Exploring Expedition."

179 eradication efforts: Aguirre-Muñoz, A., et al. 2011. "Island Restoration in Mexico: Ecological Outcomes After Systematic Eradications of Invasive Mammals." Pp. 250–258 in Veitch, C., et al., eds. *Island Invasives: Eradication and Management*. Gland, Switzerland: IUCN.

179 similarity in "general appearance": Grayson, "Exploring Expedition."

COLLARED TOWHEE

180 exotically handsome bird: Du Bus de Gisignies, B. 1847. "Note sur quelques espèces nouvelles d'oiseaux d'Amérique." *Bulletins de l'Académie royale des sciences, des lettres et des beaux-arts de Belgique* 14: 101–108.

180 colored engraving: Du Bus de Gisignies, B. 1851. *Esquisses ornithologiques. Description et figures d'oiseaux nouveaux et peu connus*. Part 4. Brussels: Van Dale.

180 confused the towhee with another species: Salvin, O. 1866. "A Further Contribution to the Ornithology of Guatemala." *Ibis* (n.s.) 2: 188–206.

180 error was at first overlooked: Salvin, O. 1874. "A Visit to the Principal Museums of the United States, With Notes on Some of the Birds Contained Therein." *Ibis* ser. 3, 4: 305–329.

181 male sparrow from the vicinity of Xalapa: Lawrence, G. 1865. "Descriptions of New Species of Birds of the Families Tanagridae, Dendrocolaptidae, Formicaridae, Tyrannidae, and Trochilidae." *Annals of the Lyceum of Natural History of New York* 8: 126–135.

181 Lawrence had been misled: Salvin, "Visit to the Principal Museums."

181 "original diagnosis and plate": Van Rossem, A. 1940. "Du Bus' Type of the Collared Towhee, *Pipilo torquatus*." *Wilson Bulletin* 52: 173–174.

181 "an obvious hybrid": Ibid.

181 elevations above 3,000 feet: Howell, S., and S. Webb. 1995. *A Guide to the Birds of Mexico and Northern Central America*. Oxford: Oxford University Press.

182 hybrids and intergrades with Spotted Towhees: Sibley, C. 1950. "Species Formation in the Red-eyed Towhees of Mexico." *University of California Publications in Zoology*. 50: 109–194.

182 without "the slightest hint of any crossing": Ibid.

182 on Mount Orizaba: Ibid.

182 Ridgway described two other specimens: Ridgway, R. 1886. "Catalogue of Animals Collected by the Geographical and Exploring Commission of the Republic of Mexico: Birds." *Proceedings of the United States National Museum* 9: 130–182.

182 later synonymized both: Ridgway, R. 1901. *The Birds of North and Middle America*. Bulletin of the United States National Museum, Bulletin 50, pt. 1. Washington: Government Printing Office.

182 "so near the precise intermediate point": Sibley, "Species Formation."

182 *Pipilo orizabae*: Cox, U. 1894. "Description of a New Species of *Pipilo* from Mount Orizaba, Mexico." *Auk* 11: 161–162.

182 "exceptionally abundant": Sibley, "Species Formation."

182 the two species display a "gradient": Ibid.

182 Other, similar gradients: Sibley, C. 1954. "Hybridization in the Red-eyed Towhees of Mexico." *Evolution* 8: 252–290.

182 shorter and sharper: Sibley, "Species Formation."

183 "high, clear, usually ascending whistle": Howell and Webb, *Mexico and Northern Central America.*

183 a wide variety of songs: Sibley, "Species Formation."

183 five subspecies: Ibid.

183 The race *alticola*: Salvin, O., and F. Godman. 1886. *Biologia centrali-americana: Aves.* Vol. 1. London: R.H. Porter.

183 dark subspecies of the Collared Towhee: Ridgway, *North and Middle America.*

183 initially considered a separate species: Salvin and Godman, *Biologia centrali-americana.*

183 or a hybrid population: Ridgway, *North and Middle America.*

183 relationship to the other Collared Towhees: Blake, E., and H. Hanson. 1942. "Notes on a Collection of Birds from Michoacán, Mexico." *Field Museum of Natural History Zoological Series* 22: 513–551.

183 "effects of infiltration": Sibley, "Species Formation."

GREEN-TAILED TOWHEE

184 "the handsomest bird": Cassin, J. 1853–1855. *Illustrations of the Birds of California, Texas, Oregon, British and Russian America.* Philadelphia: J.B. Lippincott and Co.

184 "new and singularly marked species": Gambel, W. 1843. "Descriptions of Some New and Rare Birds of the Rocky Mountains and California." *Proceedings of the Academy of Natural Sciences of Philadelphia* 1: 258–262.

184 William Blanding: Anon. 1857. "Obituary Notice of Dr. William Blanding." *New England Journal of Medicine* 57: 295–296.

184 first published illustration: Gambel, W. 1847. "Remarks on the Birds Observed in Upper California, With Descriptions of New Species." *Journal of the Academy of Natural Sciences of Philadelphia* (n.s.) 1: 25–56.

184 George White's drawing: Cassin, *Illustrations.*

184 The next Green-tailed Towhee: Woodhouse, S. 1853. "Report on the Natural History." Pp. 31–105 in Sitgreaves, L. 1853. *Report of an Expedition Down the Zuni and Colorado Rivers.* Washington, DC: R. Armstrong.

184 west to California: Heermann, A. 1859. "Report Upon Birds Collected on the Survey." *Reports of Explorations and Surveys to Ascertain the Most Practicable and Economical Route for a Railroad.* Vol. 10, 29–77. Washington: Beverley Tucker.

184 Gambel raised for the first time the possibility: Gambel, "Upper California."

184 described by Audubon: Audubon, J. 1839. *Ornithological Biography.* Vol. 5. Edinburgh: Adam Black and Charles Black.

184 Woodhouse included Audubon's name: Woodhouse, S. 1853. "Birds." Pp. 58–105 in L. Sitgreaves, ed. *Report of an Expedition Down the Zuni and Colorado Rivers.* Washington, DC: Robert Armstrong.

185 Heermann asserted their identity: Heermann, "Report Upon Birds Collected."

185 Cassin, though, would have none of it: Cassin, *Illustrations.*

185 shot his bird: Audubon, *Ornithological Biography.*

185 skin finally reemerged: Baird, "Birds."

185 the missing link: Baird, S., with J. Cassin and G. Lawrence. 1858. *Birds.* Explorations and Surveys for a Railroad Route. Vol. 9. Washington: Beverly Tucker.

185 even in flight: Dobbs, R., P. Martin, and T. Martin. 2012. "Green-tailed Towhee (*Pipilo chlorurus*)." In A. Poole, ed. *Birds of North America Online.* Ithaca: Cornell Lab of Ornithology.

185 "loud and distinct *mew-wée*": Merrill, J. 1888. "Notes on the Birds of Fort Klamath, Oregon." *Auk* 4: 357–366.

185 a sharp *tick*: Dobbs, Martin, and Martin, "Green-tailed Towhee (*Pipilo chlorurus*)."

186 50 distinguishable song types: Ibid.

186 Breeding Green-tailed Towhees: Ibid.

186 fall migration: Ibid.

186 wintering distribution: Ibid.

186 Texas winterers: Lockwood, M., and B. Freeman. 2014. *The TOS Handbook of Texas Birds.* College Station: Texas A&M University Press.

186 in Mexico: Dobbs, Martin, and Martin, "Green-tailed Towhee (*Pipilo chlorurus*)."

187 "irruption" years: Lockwood and Freeman, *TOS Handbook.*

187 Florida: Greenlaw, J., et al. 2014. *The Robertson and Woolfenden Florida Bird Species: An Annotated List.* Florida Ornithological Society Special Publication 8. Gainesville: Florida Ornithological Society.

187 southernmost records for the species: American Ornithologists' Union. 1998. *Check-list of North American Birds.* 7th ed. Washington: American Ornithologists' Union.

187 migration in spring: Dobbs, Martin, and Martin, "Green-tailed Towhee (*Pipilo chlorurus*)."

187 Nova Scotia: McLaren, I. 2012. *All the Birds of Nova Scotia: Status and Critical Identification.* Kentville: Gaspereau Press.

187 Oberholser determined that those birds differed: Oberholser, H. 1932. "Descriptions of New Birds from Oregon, Chiefly from the Warner Valley Region." *Scientific Publications of the Cleveland Museum of Natural History* 4: 1–12.

187 likely due to seasonal wear: Miller, A. 1941. "A Review of Centers of Differentiation for Birds in the Western Great Basin Region." *Condor* 43: 257–267.

AMERICAN TREE SPARROW

188 "a misnomer": Naugler, Christopher T. 2014. "American Tree Sparrow (*Spizella arborea*)." In A. Poole, ed. *Birds of North America Online.* Ithaca: Cornell Lab of Ornithology.

188 he named the Mountain Sparrow: Edwards, G. 1760. *Gleanings of Natural History.* Pt. 2. London: Royal College of Physicians.

188 "one of the very few mistakes": Richardson, J., and W. Swainson. 1831. *Fauna boreali-americana: Or the Zoology of the Northern Parts of British America.* Part 2. London: John Murray.

188 French ornithologist Mathurin Brisson: Brisson, M. 1760. *Ornithologie, ou Méthode contenant la division des oiseaux.* Vol. 3. Paris: Jean-Baptiste Bauche.

188 Forster reluctantly identified it: Forster, J. 1772. "An Account of the Birds Sent from Hudson's Bay, With Observations Relative to Their Natural History, and Latin Descriptions of Some of the Most Uncommon." *Philosophical Transactions* 62: 382–433.

188 "that judicious and excellent naturalist": Wilson, A. 1810. *American Ornithology, or the Natural History of the Birds of the United States.* Vol. 2. Philadelphia: Bradford and Inskeep.

188　a species distinct from Edwards's: Pennant, T. 1785. *Arctic Zoology*. Vol. 2. London: Henry Hughs.

188　renamed the European bird: Pennant, T. 1770. *British Zoology*. Part 4. London: Benjamin White.

188　rechristened the New World species: Pennant, *Arctic Zoology*.

188　Alexander Wilson Americanized Pennant's Tree Finch: Wilson, *American Ornithology*.

189　diminished, degenerate reflex: Buffon, Georges Louis Leclerc, Comte de. 1775. *Histoire naturelle, générale et particulière*. Vol. 18. Paris: Imprimerie Royale.

189　Brisson's *canadensis*: Brisson, *Ornithologie*.

189　"some of our own naturalists": Wilson, *American Ornithology*.

191　individual males sing only a single type: Naugler, "American Tree Sparrow (*Spizella arborea*)."

191　across Alaska and the northern Canadian territories: Ibid.

191　south of the Stikine: Campbell, R., et al. 2001. *The Birds of British Columbia*. Vol. 4. Vancouver: UBC Press.

191　"small, isolated population" in north-central Alberta: Naugler, "American Tree Sparrow (*Spizella arborea*)."

191　northernmost breeding areas: Campbel et al., *British Columbia*.

191　St. Lawrence Island: Lehman, P. 2005. "Fall Bird Migration at Gambell, St. Lawrence Island, Alaska." *Western Birds* 36: 2–55.

191　Wintering birds are most abundant: Naugler, "American Tree Sparrow (*Spizella arborea*)."

191　In California: Small, A. 1994. *California Birds: Their Status and Distribution*. Vista, CA: Ibis.

192　Texas: Lockwood, M., and B. Freeman. 2014. *The TOS Handbook of Texas Birds*. College Station: Texas A&M University Press.

192　Florida: Greenlaw, J., et al. 2014. *The Robertson and Woolfenden Florida Bird Species: An Annotated List*. Florida Ornithological Society Special Publication 8. Gainesville: Florida Ornithological Society.

192　southern tier of Canadian provinces: Naugler, "American Tree Sparrow (*Spizella arborea*)."

192　Russia: Brazil, M. 2009. *Birds of East Asia*. London: Christopher Helm.

192　as in the western subspecies: Brewster, W. 1882. "Notes on Some Birds Collected by Capt. Charles Bendire at Fort Walla Walla, Washington Territory." *Bulletin of the Nuttall Ornithological Club* 7: 225–232.

192　smallest female *arborea*: Ridgway, R. 1901. *The Birds of North and Middle America*. Bulletin of the United States National Museum, Bulletin 50, pt. 1. Washington: Government Printing Office.

192　The western subspecies was first collected: Brewster, "Birds Collected by Capt. Charles Bendire."

192　breeding range: Naugler, "American Tree Sparrow (*Spizella arborea*)."

192　eastern Great Plains: American Ornithologists' Union. 1957. *Check-list of North American Birds*. 5th ed. Ithaca: American Ornithologists' Union.

192　east to the Mississippi: Naugler, "American Tree Sparrow (*Spizella arborea*)."

192　in Nebraska: Sharpe, R., et al. 2001. *Birds of Nebraska: Their Distribution and Temporal Occurrence*. Lincoln: University of Nebraska Press.

RED FOX SPARROW

193　Blasius Merrem: Bohle, H. 2015. *Von der Naturgeschichte zur Zoologie: Blasius Merrem und die Entwicklung der Zoologie an der Universität Marburg im 19. Jahrhundert*. Münster: Waxmann.

193　"new species of finch": Merrem, B. 1786. *Avium rariorum et minus cognitarum icones et descriptiones*. Leipzig: Müller.

193　"was a winter specimen": Swarth, H. 1920. "Revision of the Avian Genus *Passerella*, with Special Reference to the Distribution and Migration of the Races in California." *University of California Publications in Zoology* 21: 75–224.

193　Oberholser restricted the type locality: Oberholser, H. 1946. "Three New North American Birds." *Journal of the Washington Academy of Sciences* 36: 388–389.

193　"It is the size of our Corn Bunting": Brisson, M. 1760. *Ornithologie, ou Méthode contenant la division des oiseaux*. Vol. 3. Paris: Jean-Baptiste Bauche.

194　*cul-rousset*, the red-tail: Buffon, Georges Louis Leclerc, Comte de. 1778. *Histoire naturelle, générale et particulière, avec la description du Cabinet du Roi*. Vol. 19. Paris: Imprimerie royale.

195　males and females are known to sing: Weckstein, J., et al. 2002. "Fox Sparrow (*Passerella iliaca*)." In A. Poole, ed. *Birds of North America Online*. Ithaca: Cornell Lab of Ornithology.

195　one of the most widespread breeding birds: Ibid.

196　more than 1,200 miles: Ibid.

196　wintering grounds in Texas: Lockwood, M., and B. Freeman. 2014. *The TOS Handbook of Texas Birds*. College Station: Texas A&M University Press.

196　two Greenland records: Boertmann, D. 1994. "An Annotated Checklist to the Birds of Greenland." *Meddelelser om Grønland: Bioscience* 38: 3–28.

196　Iceland: Weckstein et al., "Fox Sparrow (*Passerella iliaca*)."

196　Estonia: "Fox Sparrow in Estonia." http://www.surf-birds.com/forum/forum/rare-bird-information/european-rare-birds/10575-fox-sparrow-in-estonia.

196　Wintering birds: Weckstein et al., "Fox Sparrow (*Passerella iliaca*)."

196　Sierra Nevada and in northern Baja California: Miller, A., et al. 1957. *Distributional Checklist of the Birds of Mexico*. Part 2. Pacific Coast Avifauna 33.

196　Chihuahua and northern Sonora: Russell, S., and G. Monson. 1998. *The Birds of Sonora*. Tucson: University of Arizona Press.

196　eastern two-thirds of Texas: Lockwood and Freeman, *TOS Handbook*.

196　Florida records: Greenlaw, J., et al. 2014. *The Robertson and Woolfenden Florida Bird Species: An Annotated List*. Florida Ornithological Society Special Publication 8. Gainesville: Florida Ornithological Society.

196　northbound movement: Weckstein et al., "Fox Sparrow (*Passerella iliaca*)."

197　this species' song that "predominated": Peterson, R., and J. Fisher. 1955. *Wild America*. Boston: Houghton Mifflin.

197　Ireland and Germany: Lewington, I., et al. 1991. *A Field Guide to the Rare Birds of Britain and Europe*. London: HarperCollins.

197　"average larger than eastern examples": Ridgway, R. 1901. *The Birds of North and Middle America*. Bulletin of the United States National Museum, Bulletin 50, pt. 1. Washington: Government Printing Office.

197　consistently overall darker: Oberholser, "Three New North American Birds."

197 named by J. H. Riley: Riley, J. 1911. "Descriptions of Three New Birds from Canada." *Proceedings of the Biological Society of Washington* 24: 233–236.

197 Swarth allocated: Swarth, "Avian Genus *Passerella*."

197 a subspecies of both: Beadle, D., and J. Rising. 2003. *Sparrows of the United States and Canada: The Photographic Guide.* Princeton: Princeton University Press.

197 *altivagans* is slightly smaller: Swarth, "Avian Genus *Passerella*."

197 are probably intergrades: Weckstein et al., "Fox Sparrow (*Passerella iliaca*)."

197 British Columbia: Campbell, R., et al. 2001. *The Birds of British Columbia.* Vol. 4. Vancouver: UBC Press.

197 specimen record in Arizona: Phillips, A., et al. 1964. *The Birds of Arizona.* Tucson: University of Arizona Press.

SOOTY FOX SPARROW

198 Baron von Kittlitz: Gebhardt, L. 1977. "Kittlitz, Heinrich Freiherr von." In Historische Kommission bei der Bayerischen Akademi der Wissenschaften, ed. *Neue Deutsche Biographie* 11: 694–695. Berlin: Duncker and Humblot.

198 "the most musical of any": Kittlitz, F. H. von. 1858. *Denkwürdigkeiten einer Reise nach dem russischen Amerika, nach Mikronesien und durch Kamtschatka.* Vol. 1. Gotha: Justus Perthes.

198 a distinct species: Ibid. Kittlitz, *Denkwürdigkeiten einer Reise.*

198 name was preoccupied: Ridgway, R. 1901. *The Birds of North and Middle America.* Bulletin of the United States National Museum, Bulletin 50, pt. 1. Washington: Government Printing Office.

198 named almost 50 years before: Latham, J. 1783. *A General Synopsis of Birds.* Vol. 2. London: Benjamin White.

198 Gmelin took the bird over: Gmelin, J. 1788. *Caroli a Linné . . . Systema naturae per regna tria naturae.* 13th ed. Vol. 1. Leipzig: Georg Emanuel Beer.

198 "brownish above, with rusty wings": Vigors, N. 1839. "Ornithology." In F. Beechey, ed. *The Zoology of Captain Beechey's Voyage.* London: H.G. Bohn.

198 In February 1836, John Townsend killed: Audubon, J. 1839. *Ornithological Biography.* Vol. 5. Edinburgh: Adam Black and Charles Black.

198 a provisional resolution: Baird, S., with J. Cassin and G. Lawrence. 1858. *Birds.* Explorations and Surveys for a Railroad Route. Vol. 9. Washington: Beverly Tucker.

199 affirmed in 1872 by Otto Finsch: Finsch, O. 1872. "Zur Ornithologie Nordwest-Amerikas." *Abhandlungen herausgegeben vom naturwissenschaftlichen Vereine zu Bremen* 3: 17–86.

199 "forms are but modifications": Henshaw, H. 1878. "On the Species of the Genus *Passerella*." *Bulletin of the Nuttall Ornithological Club* 3: 3–7.

199 The northerly races: Mlodinow, S., et al. 2012. "The Sooty Fox Sparrows of Washington's Puget Trough." *Birding* 44.2: 46–52.

200 Hybridization: McCarthy, E. 2006. *Handbook of Avian Hybrids of the World.* Oxford: Oxford University Press.

201 "most combinations of hybridization": Zink, R. 1994. "The Geography of Mitochondrial DNA Variation, Population Structure, Hybridization, and Species Limits in the Fox Sparrow (*Passerella iliaca*)." *Evolution* 48: 96–111.

201 call is very similar: Garrett, K., et al. 2000. "Call Note and Winter Distribution in the Fox Sparrow Complex." *Birding* 32: 412–417.

201 loud, melodious song: Ibid.

201 breeds from the eastern Aleutians south: Weckstein, J., et al. 2002. "Fox Sparrow (*Passerella iliaca*)." In A. Poole, ed. *Birds of North America Online.* Ithaca: Cornell Lab of Ornithology.

202 southwestern British Columbia: Campbell, R., et al. 2001. *The Birds of British Columbia.* Vol. 4. Vancouver: UBC Press.

202 "leap-frog migration": Bell, C. 1997. "Leap-frog Migration in the Fox Sparrow: Minimizing the Cost of Spring Migration." *Condor* 99: 470–477.

202 western Alaska subspecies: Weckstein et al., "Fox Sparrow (*Passerella iliaca*)."

202 southeast Arizona: Garrett et al., "Call Note and Winter Distribution."

202 Chukchi Peninsula: Arkhopiv, V., and Ł. Łuwacki. 2016. "Nearctic Passerines in Russia." *Dutch Birding* 38: 201–214.

202 winters regularly: Weckstein et al., "Fox Sparrow (*Passerella iliaca*)."

202 In California: Small, A. 1994. *California Birds: Their Status and Distribution.* Vista, CA: Ibis.

202 San Diego County: Unitt, P. 2004. *San Diego County Bird Atlas.* Proceedings of the San Diego Society of Natural History 39. San Diego: San Diego Natural History Museum.

202 Salton Sea: Patten, M., et al. 2003. *Birds of the Salton Sea: Status, Biogeography, and Ecology.* Berkeley: University of California Press.

202 North Dakota: Garrett et al., "Call Note and Winter Distribution."

202 New Hampshire: Keith, A., and R. Fox. 2013. *The Birds of New Hampshire.* Memoirs of the Nuttall Ornithological Club 19. Cambridge: Nuttall Ornithological Club.

202 Japan: Brazil, M. 2009. *Birds of East Asia.* London: Christopher Helm.

202 movement in spring: Weckstein et al., "Fox Sparrow (*Passerella iliaca*)."

202 Russian Far East: Arkhopiv and Łuwacki, "Nearctic Passerines in Russia."

202 Central Park: Chang, S. 2011. "A 'Sooty' Fox Sparrow in Central Park, New York City." *Kingbird* 61: 203–205.

202 two "categories": Swarth, H. 1920. "Revision of the Avian Genus *Passerella*, with Special Reference to the Distribution and Migration of the Races in California." *University of California Publications in Zoology* 21: 75–224.

202 remains unsettled: Weckstein et al., "Fox Sparrow (*Passerella iliaca*)."

202 Breeding from Unalaska: Ibid.

202 probably intergrades: Swarth, "Avian Genus *Passerella*."

202 "rather more reddish": Grinnell, J. 1910. "Birds of the 1908 Alexander Alaska Expedition." *University of California Publications in Zoology* 5: 361–428.

202 plumage tone is intermediate: Swarth, "Avian Genus *Passerella*."

202 Kodiak Island: Ibid.

203 enlarged genus *Zonotrichia*: Paynter, R., and R. Storer. 1970. *Check-List of Birds of the World: A Continuation of the Work of James L. Peters.* Vol. 13. Cambridge: Museum of Comparative Zoology.

203 *annectens* ("linking"): Swarth, "Avian Genus *Passerella*."

203 Beechey's *meruloides*: Grinnell, J. 1902. "The Monterey Fox Sparrow." *Condor* 4: 44–45.

203 "one of the least known": Swarth, "Avian Genus *Passerella*."

203 Darkest of all: Ridgway, R. 1899. "New Species, Etc., of American Birds. III: Fringillidae." *Auk* 16: 35–37.

203 considered "non-typical": Swarth, "Avian Genus *Passerella*."

203 a distinct subspecies, *chilcatensis*: Webster, J. 1983. "A New Subspecies of Fox Sparrow from Alaska." *Proceedings of the Biological Society of Washington* 96: 664–668.

203 not recognized by all: Weckstein et al., "Fox Sparrow (*Passerella iliaca*)."

SLATE-COLORED FOX SPARROW

204 "readily distinguished": Baird, S., with J. Cassin and G. Lawrence. 1858. *Birds*. Explorations and Surveys for a Railroad Route. Vol. 9. Washington: Beverly Tucker.

204 Bryan's party: Bryan, F. 1945. "Report of Lieut. F.T. Bryan Concerning His Operations in Locating a Practicable Road Between Fort Riley to Bridger's Pass 1856." Repr. *Annals of Wyoming* 17: 24–55.

204 "the specimen in question": Cooke, W. 1897. *The Birds of Colorado*. Fort Collins: Colorado State Agricultural College.

204 "south fork of the Platte": American Ornithologists' Union. 1910. *Check-list of North American Birds*. 3rd ed. New York: American Ornithologists' Union.

205 Bryan described the habitat: Bryan, "Report of Lieut. F.T. Bryan."

205 "it may be that this bird had strayed": Swarth, H. 1920. "Revision of the Avian Genus *Passerella*, with Special Reference to the Distribution and Migration of the Races in California." *University of California Publications in Zoology* 21: 75–224.

205 lower-pitched and less metallic: Garrett, K., et al. 2000. "Call Note and Winter Distribution in the Fox Sparrow Complex." *Birding* 32: 412–417.

205 several different song types: Ibid.

205 the interior of British Columbia: Campbell, R., et al. 2001. *The Birds of British Columbia*. Vol. 4. Vancouver: UBC Press.

206 from southwest Alberta south: Weckstein, J., et al. 2002. "Fox Sparrow (*Passerella iliaca*)." In A. Poole, ed. *Birds of North America Online*. Ithaca: Cornell Lab of Ornithology.

206 dark gray *swarthi*: Weckstein et al., "Fox Sparrow (*Passerella iliaca*)."

206 southern British Columbia by September: Campbell et al., *British Columbia*.

206 arriving on the Colorado River: Weckstein et al., "Fox Sparrow (*Passerella iliaca*)."

206 northeastern Arizona: Monson, G., and A. Phillips. 1981. *Annotated Checklist of the Birds of Arizona*. 2nd ed. Tucson: University of Arizona Press.

206 eastern plains of Colorado: Andrews, R., and R. Righter. 1992. *Colorado Birds: A Reference to Their Distribution and Habitat*. Denver: Denver Museum of Natural History.

206 El Paso County, Texas: Lockwood, M., and B. Freeman. 2014. *The TOS Handbook of Texas Birds*. College Station: Texas A&M University Press.

206 California's foothills and lowlands: Small, A. 1994. *California Birds: Their Status and Distribution*. Vista, CA: Ibid.

206 deserts of the southeast: Patten, M., et al. 2003. *Birds of the Salton Sea: Status, Biogeography, and Ecology*. Berkeley: University of California Press.

206 lower Colorado River: Rosenberg, K., et al. 1991. *Birds of the Lower Colorado River Valley*. Tucson: University of Arizona Press.

206 northern Baja California: Miller, A., et al. 1957. *Distributional Checklist of the Birds of Mexico*. Part 2. Pacific Coast Avifauna 33.

207 migration north: Weckstein et al., "Fox Sparrow (*Passerella iliaca*)."

207 "failed to confirm": Zink, R. 1986. *Patterns and Evolutionary Significance of Geographic Variation in the* schistacea *Group of the Fox Sparrow* (Passerella iliaca). Ornithological Monographs 40. Washington, DC: American Ornithologists' Union.

207 smallest and thickest-billed is *canescens*: Swarth, H. 1918. "Three New Subspecies of *Passerella iliaca*." *Proceedings of the Biological Society of Washington* 31: 161–164.

207 Breeding in the White and Inyo Mountains: Floyd, T., et al. 2007. *Atlas of the Breeding Birds of Nevada*. Reno: University of Nevada Press.

207 winter in southern California: Weckstein et al., "Fox Sparrow (*Passerella iliaca*)."

207 "still grayer," so much so: Behle, W., and R. Selander. 1951. "The Systematic Relationships of the Fox Sparrows (*Passerella iliaca*) of the Wasatch Mountains, Utah, and the Great Basin." *Journal of the Washington Academy of Sciences* 41: 364–367.

207 Apparent intermediates: Ibid.

207 wintering range remains unknown: Weckstein et al., "Fox Sparrow (*Passerella iliaca*)."

207 winters in California: Weckstein et al., "Fox Sparrow (*Passerella iliaca*)."

207 subspecies *olivacea*: Aldrich, J. 1943. "A New Fox Sparrow from the Northwestern United States." *Proceedings of the Biological Society of Washington* 56: 163–166.

207 comes into contact and intergrades: Ibid.

THICK-BILLED FOX SPARROW

208 John Xantus: Zwinger, A. ed. *John Xántus: The Fort Tejon Letters 1857-1859*. Tucson: University of Arizona Press.

208 "all unhesitatingly concur": Ibid.

208 "I undertook a trip": Ibid.

208 his new Slate-colored Fox Sparrow: Baird, S., with J. Cassin and G. Lawrence. 1858. *Birds*. Explorations and Surveys for a Railroad Route. Vol. 9. Washington: Beverly Tucker.

208 "solved satisfactorily": Zwinger, *John Xántus*.

209 "the bird of Fort Tejon": Baird, *Birds*.

209 "where individuals of each [species] co-occur": Zink, R. 1994. "The Geography of Mitochondrial DNA Variation, Population Structure, Hybridization, and Species Limits in the Fox Sparrow (*Passerella iliaca*)." *Evolution* 48: 96–111.

209 color of the lower mandible: Rising, J. 2016. "Thick-billed Fox-sparrow (*Passerella megarhyncha*)." In J. del Hoyo et al., eds. *Handbook of the Birds of the World Alive*. Lynx Edicions, Barcelona.

209 distinction lies in the contact call: Pieplow, M. 2011. "Fox Sparrow: Alarm and Contact Calls." *Earbirding.com: Recording, Identifying, and Interpreting Bird Sounds*. earbirding.com.

209 "spectrographically distinct": Ibid.

209 "second to none": Baird, S., et al. 1874. *A History of North American Birds*. Vol. 2. Boston: Little, Brown.

210 southern Washington: Mlodinow, S., and M. Bartels. 2016. "Tenth Report of the Washington Bird Records Committee (2010-2013)." *Western Birds* 47: 86–119.

210 northern and central Cascades of Oregon: Gillson, G. 2003. "Breeding Fox Sparrows in Oregon." thebirdguide.com/fox/fox.htm.

210 In California: Weckstein, J., et al. 2002. "Fox Sparrow (*Passerella iliaca*)." In A. Poole, ed. *Birds of North America Online*. Ithaca: Cornell Lab of Ornithology.

210 Nevada: Floyd, T., et al. 2007. *Atlas of the Breeding Birds of Nevada.* Reno: University of Nevada Press.

210 Baja California: Miller, A., et al. 1957. *Distributional Checklist of the Birds of Mexico.* Part 2. Pacific Coast Avifauna 33.

211 commonest of the fox sparrows in San Diego County: Unitt, P. 2004. *San Diego County Bird Atlas.* Proceedings of the San Diego Society of Natural History 39. San Diego: San Diego Natural History Museum.

211 Arizona specimens from October and December: Monson, G., and A. Phillips. 1981. *Annotated Checklist of the Birds of Arizona.* 2nd ed. Tucson: University of Arizona Press.

211 nominate subspecies: Weckstein et al., "Fox Sparrow (*Passerella iliaca*)."

211 the very heavy-billed *stephensi*: Ibid.

211 "a geographic designation": Mailliard, J. 1918. "The Yolla Bolly Fox Sparrow." *Condor* 20: 138–139.

211 Mono Fox Sparrow: Grinnell, J., and T. Storer. 1917. "A New Race of Fox Sparrow, from the Vicinity of Mono Lake, California." *Condor* 19: 165–166.

211 The subspecies *mariposae*: Swarth, H. 1918. "Three New Subspecies of *Passerella iliaca*." *Proceedings of the Biological Society of Washington* 31: 161–164.

211 The uninformatively named subspecies *fulva*: Ibid.

211 sometimes allocated to that species: Rising, "Thick-billed Fox-sparrow (*Passerella megarhyncha*)."

GUADALUPE JUNCO

212 a collecting party: Gaylord, H. 1897. "Notes from Guadalupe Island." *Nidologist* 4: 41–43.

212 "remarkable confidence": Gaylord, H. 1897. "Remarkable Confidence of the Guadalupe Junco." *Osprey* 1: 98.

212 "They are the most abundant birds": Ridgway, R. 1876. "Ornithology of Guadeloupe Island, Based on Notes and Collections Made by Dr. Edward Palmer." *Bulletin of the United States Geological and Geographical Survey of the Territories* 2: 183–195.

212 "nearly starved to death": Fisher, W. 1905. "In Memoriam: Walter E. Bryant." *Condor* 7: 129–131.

213 Bryant found the juncos less abundant: Bryant, W. 1887. "Additions to the Ornithology of Guadalupe Island." *Bulletin of the California Academy of Sciences* 2: 269–319.

213 "probably not uncommon": American Ornithologists' Union. 2014. "Fifty-fifth Supplement to the American Ornithologists' Union *Check-list of North American Birds.*" *Auk* 131: CSi–CSxv.

213 tail feathers are less extensively white: Miller, A. 1941. "Speciation in the Avian Genus *Junco.*" *University of California Publications in Zoology* 44: 173–434.

213 simple loose trill: Bryant, "Additions to the Ornithology."

213 considerably more complex: Mirsky, E. 1976. "Song Divergence in Hummingbird and Junco Populations on Guadalupe Island." *Condor* 78: 230–235.

213 500 acres: American Ornithologists' Union, "Fifty-fifth Supplement."

BAIRD JUNCO

214 excavating middens and graves: Ten Kate, H. 2004. *Travels and Researches in Native North America, 1882–1883.* Transl. and ed. P. Hovens et al. Albuquerque: University of New Mexico Press.

214 skeletons they discovered: Tyson, R. 1977. "Human Skeletal Material from the Cape Region of Baja California, Mexico: The American Collections." *Journal de la Société des américanistes* 64: 167–174.

214 Belding set off alone: Deignan, H. 1961. *Type Specimens of Birds in the United States National Museum.* Bulletin of the United States National Muscum 221. Washington: Government Printing Office.

214 "the trail leading to Laguna": Belding, L. 1883. "Second Catalogue of a Collection of Birds Made Near the Southern Extremity of Lower California." *Proceedings of the United States National Museum* 6: 344–352.

214 "pretty and very distinct": Ridgway, R. 1883. "Descriptions of Some New Birds from Lower California, Collected by Mr. L. Belding." *Proceedings of the United States National Museum* 6: 154–156.

214 Purely casual: Erickson, R., et al. 2013. "Annotated Checklist of the Birds of Baja California and Baja California Sur, Second Edition." *North American Birds* 66: 582–613.

214 "a pale, dwarfed representative": Miller, A. 1941. "Speciation in the Avian Genus *Junco.*" *University of California Publications in Zoology* 44: 173–434.

215 buffy patch on the nape: Ibid.

215 tail feathers average more extensively white: Ibid.

216 high number of unique syllables: Pieplow, N., and C. Francis. 2011. "Song Differences Among Subspecies of Yellow-eyed Juncos (*Junco phaeonotus*)." *Wilson Journal of Ornithology* 123: 464–471.

216 compared to that of a House Wren: Howell, S., and S. Webb. 1995. *A Guide to the Birds of Mexico and Northern Central America.* Oxford: Oxford University Press.

216 Rufous-crowned Sparrow: Pieplow, N. 2011. "The Next Junco." *Earbirding.com: Recording, Identifying, and Interpreting Bird Sounds.* earbirding.com.

216 a strongly sedentary resident: Miller, "Avian Genus *Junco.*"

YELLOW-EYED JUNCO

217 left England for a newly independent Mexico: Costeloe, M. 2006. "William Bullock and the Mexican Connection." *Mexican Studies / Estudios Mexicanos* 22: 275–309.

217 an exhaustive catalog: Swainson, W. 1827. "A Synopsis of the Birds Discovered in Mexico by W. Bullock, F.L.S. and H.S., and Mr. William Bullock, jun." *The Philosophical Magazine* 2nd ser., 1: 364–369 and 433–442.

217 Swainson named it: Swainson, "Birds Discovered in Mexico."

217 "the smaller bird collection": Costeloe, "William Bullock."

218 approach each other most closely in Arizona: Miller, A. 1941. "Speciation in the Avian Genus *Junco.*" *University of California Publications in Zoology* 44: 173–434.

218 "Its plumage was most like that of Yellow-eyed": Hoyer, R. 2005. "Birding the Pinal Mountains." *Arizona Birds Online* 1: 8–9. azfo.org.

218 Robert Ridgway's illustration: Henshaw, H. 1873. "Report Upon the Ornithological Collections made in Portions of Nevada, Utah, California, Colorado, New Mexico, and Arizona During the Years 1871, 1872, 1873, and 1874." In G. Wheeler, ed. *Geographical and Geological Explorations and Surveys West of the One Hundredth Meridian.* Vol. 5. Washington: Government Printing Office.

219 humid pine, oak, and spruce forests and agricultural habitats: Miller, A., et al. 1957. *Distributional Checklist of the Birds of Mexico.* Part 2. Pacific Coast Avifauna 33.

219 Mexican range of this subspecies: Miller, "Avian Genus *Junco.*"

219 Chiapas Junco: Ibid.

219 mountains of southern Mexico: Eisermann, K., and C. Avendaño. 2007. *Lista comentada de las aves de Guatemala / Annotated Checklist of the Birds of Guatemala*. Barcelona: Lynx Edicions.

219 mixed and coniferous forests: Miller et al., *Distributional Checklist*.

219 Two isolated populations: Miller, "Avian Genus *Junco*."

219 "constitutes an even gradient": Ibid.

219 southeastern Arizona: Corman, T., and C. Wise-Gervais, eds. 2005. *Arizona Breeding Bird Atlas*. Albuquerque: University of New Mexico Press.

219 New Mexico and Arizona mountain ranges: Monson, G., and A. Phillips. 1981. *Annotated Checklist of the Birds of Arizona*. 2nd ed. Tucson: University of Arizona Press.

219 west Texas: Lockwood, M., and B. Freeman. 2014. *The TOS Handbook of Texas Birds*. College Station: Texas A&M University Press.

220 Two recognizable subspecies: Miller, "Avian Genus *Junco*."

RED-BACKED JUNCO

221 Thomas Charlton Henry: Hume, E. 1942. *Ornithologists of the United States Army Medical Corps*. Baltimore: Johns Hopkins Press.

221 published by others: Brodhead, M. 1989. "Contributions of Medical Officers of the Regular Army to Natural History in the Pre-Civil War Era." Pp. 3–14 in F. Hartigan, ed. *History and Humanities: Essays in Honor of Wilbur S. Shepperson*. Reno: University of Nevada Press.

221 near Fort Stanton in southwestern New Mexico: Miller, A. 1941. "Speciation in the Avian Genus *Junco*." *University of California Publications in Zoology* 44: 173–434.

221 Henry named *Junco dorsalis*: Henry, T. 1858. "Descriptions of New Birds from Fort Thorn, New Mexico." *Proceedings of the Academy of Natural Sciences of Philadelphia* 10: 117–118.

221 Baird affirmed: Baird, S., with J. Cassin and G. Lawrence. 1858. *Birds*. Explorations and Surveys for a Railroad Route. Vol. 9. Washington: Beverly Tucker.

221 specimen with yellow eyes: Baird, S., et al. 1860. *The Birds of North America: Atlas*. Philadelphia: J.B. Lippincott and Co.

221 "only separable as varieties": Henshaw, H. 1873. "Report Upon the Ornithological Collections made in Portions of Nevada, Utah, California, Colorado, New Mexico, and Arizona During the Years 1871, 1872, 1873, and 1874." In G. Wheeler, ed. *Geographical and Geological Explorations and Surveys West of the One Hundredth Meridian*. Vol. 5. Washington: Government Printing Office.

222 "hybrid between *caniceps* and *cinereus*": Baird, S., et al. 1874. *A History of North America Birds*. Vol. 1. Boston: Little, Brown.

222 an even more extreme solution: Coues, E. 1884. *Key to North American Birds*. 2nd ed. Boston: Estes and Lauriat.

222 "they replace each other geographically": Miller, "Avian Genus *Junco*."

222 adopted by the AOU in 1945: American Ornithologists' Union. 1945. "Twentieth Supplement to the American Ornithologists' Union *Check-list of North American Birds*." *Auk* 62: 436–449.

222 lumping of all other brown-eyed juncos: American Ornithologists' Union. 1973. "Thirty-second Supplement to the American Ornithologists' Union *Check-list of North American Birds*." *Auk* 90: 411–419.

222 constituting a subspecies group: American Ornithologists' Union. 1982. "Thirty-fourth Supplement to the American Ornithologists' Union *Check-list of North American Birds*." *Auk* 99 (supplement): 1CC–16CC.

222 shares with the closely related Gray-headed: Miller, "Avian Genus *Junco*."

223 "interbreed freely": Ibid.

224 In Arizona, breeders are found: Nolan, Jr., V., et al. 2002. "Dark-eyed Junco (*Junco hyemalis*)." In A. Poole, ed. *Birds of North America Online*. Ithaca: Cornell Lab of Ornithology.

224 recently discovered populations: Corman, T., and C. Wise-Gervais, eds. 2005. *Arizona Breeding Bird Atlas*. Albuquerque: University of New Mexico Press.

224 intergrades with the Gray-headed Junco: Nolan et al., "Dark-eyed Junco (*Junco hyemalis*)."

224 in the Pinal Mountains: Hoyer, R. 2005. "Birding the Pinal Mountains." *Arizona Birds Online* 1: 8–9. azfo.org.

224 In New Mexico: Nolan et al., "Dark-eyed Junco (*Junco hyemalis*)."

224 Datil and San Mateo Mountains: Miller, "Avian Genus *Junco*."

224 the Guadalupe Mountains of Texas: Lockwood, M., and B. Freeman. 2014. *The TOS Handbook of Texas Birds*. College Station: Texas A&M University Press.

224 northwestern Mexico: Miller, "Avian Genus *Junco*."

224 central Colorado: Brinkley, E., ed. 2012. "Pictorial Highlights." *North American Birds* 66: 373.

224 plumage or morphological variations: Miller, "Avian Genus *Junco*."

GRAY-HEADED JUNCO

225 The first published image: Henshaw, H. 1873. "Report Upon the Ornithological Collections made in Portions of Nevada, Utah, California, Colorado, New Mexico, and Arizona During the Years 1871, 1872, 1873, and 1874." In G. Wheeler, ed. *Geographical and Geological Explorations and Surveys West of the One Hundredth Meridian*. Vol. 5. Washington: Government Printing Office.

225 Red-backed Junco was portrayed for the first time: Baird, S., et al. 1860. *The Birds of North America: Atlas*. Philadelphia: J.B. Lippincott and Co.

225 Gray-headed Junco first appeared in print: Ibid.

225 who had secured one of the birds: Woodhouse, S. 1853. "Description of a New Snow Finch of the Genus *Struthus*, Boie." *Proceedings of the Academy of Natural Sciences of Philadelphia* 6: 202–203.

225 "dark brown, almost black": Ibid. Woodhouse, "New Snow Finch."

226 unwittingly based: Merriam, C. 1890. "Annotated List of Birds of the San Francisco Mountain Plateau and the Desert of the Little Colorado River, Arizona." *North American Fauna* 3: 87–101.

226 "male birds have now vanished": Deignan, H. 1961. *Type Specimens of Birds in the United States National Museum*. Bulletin of the United States National Museum 221. Washington: Government Printing Office.

226 "probably due to blood stain": Miller, "Avian Genus *Junco*."

226 "yellowish, black at the tip": Baird, S., with J. Cassin and G. Lawrence. 1858. *Birds*. Explorations and Surveys for a Railroad Route. Vol. 9. Washington: Beverly Tucker.

226 Intergrades with the Red-backed Junco: Miller, "Avian Genus *Junco*."

227 interbreeding of Gray-headed Juncos with Oregon Juncos: Van Rossem, A. 1931. "Descriptions of New Birds from the Mountains of Southern Nevada." *Transactions of the San Diego Society of Natural History* 6: 325–332.

227 "the uniformly red back": Ibid.

227 "some mixture" of red and brown: Miller, "Avian Genus *Junco*."

227 Similar intermediates are also found: Ibid.

227 an even more complex situation: Ibid.

227 "the only case": Hamilton, R., and P. Gaede. 2015. "Pink-sided x Gray-headed Juncos." *Western Birds* 36: 150–152.

227 "having the sides reddish": Baird, *Birds*.

227 named it *Junco annectens*: Baird, *Birds*.

227 again listed it as a hybrid: Baird, S., et al. 1874. *A History of North America Birds*. Vol. 1. Boston: Little, Brown.

227 Mearns reported his discovery: Mearns, E. 1890. "Descriptions of a New Species and Three New Subspecies of Birds from Arizona." *Auk* 7: 243–251.

228 "without a name": Ridgway, R. 1897. "Note on *Junco annectens* Baird and *J. ridgwayi* Mearns." *Auk* 14: 94.

228 product of interbreeding: Nolan, Jr., V., et al. 2002. "Dark-eyed Junco (*Junco hyemalis*)." In A. Poole, ed. *Birds of North America Online*. Ithaca: Cornell Lab of Ornithology.

228 Mearns's description: Mearns, "New Species."

228 variability of plumage: Miller, "Avian Genus *Junco*."

228 Collecting in Utah's Wasatch Mountains: Ibid.

228 "The acoustic properties of song": Cardoso, G., and D. Reichard. 2016. "Dark-eyed Junco Song: Linking Ontogeny and Function with a Potential Role in Reproductive Isolation." In E. Ketterson and J. Atwell, eds. *Snowbird: Integrative Biology and Evolutionary Diversity in the Junco*. Chicago: University of Chicago Press.

229 "short-range" song: Nolan et al., "Dark-eyed Junco (*Junco hyemalis*)."

229 In California, this subspecies breeds: Small, A. 1994. *California Birds: Their Status and Distribution*. Vista, CA: Ibis.

229 Volcan Mountain: Unitt, P. 2004. *San Diego County Bird Atlas*. Proceedings of the San Diego Society of Natural History 39. San Diego: San Diego Natural History Museum.

229 mountains of Nevada: Floyd, T., et al. 2007. *Atlas of the Breeding Birds of Nevada*. Reno: University of Nevada Press.

229 Kaibab Plateau and in the Zuni Mountains: American Ornithologists' Union. 1957. *Check-list of North American Birds*. 5th ed. Ithaca: American Ornithologists' Union.

229 Interbreeding with Pink-sided Juncos: Nolan et al., "Dark-eyed Junco (*Junco hyemalis*)."

229 San Diego County: Unitt, *San Diego County Bird Atlas*.

229 Salton Sea area: Patten, M., et al. 2003. *Birds of the Salton Sea: Status, Biogeography, and Ecology*. Berkeley: University of California Press.

229 lower Colorado River: Rosenberg, K., et al. 1991. *Birds of the Lower Colorado River Valley*. Tucson: University of Arizona Press.

229 regions of Texas: Lockwood, M., and B. Freeman. 2014. *The TOS Handbook of Texas Birds*. College Station: Texas A&M University Press.

229 reliable records east to Illinois: Stotz, D., and D. Johnson. 2003. "The Ninth Report of the Illinois Ornithological Records Committee." *Meadowlark* 12: 55–61.

229 darker-headed than normal: Miller, "Avian Genus *Junco*."

PINK-SIDED JUNCO

230 Constantin Charles Drexler: Dos Passos, C., ed. 1951. "The Entomological Reminiscences of William Henry Edwards." *Journal of the New York Entomological Society* 59: 129–186.

230 taxidermist at the Smithsonian: Burnett, W. 1931. *Life-History Studies of the Wyoming Ground Squirrel (Citellus elegans elegans) in Colorado*. Colorado Agricultural Experiment Station Bulletin 373. Fort Collins: Colorado Agricultural College.

230 Baird described him: Binnema, T. 2014. *Enlightened Zeal: The Hudson's Bay Company and Scientific Networks, 1670-1870*. Toronto: University of Toronto Press.

230 assistant to James Graham Cooper: Taylor, W. 1919. "Notes on Mammals Collected Principally in Washington and California Between the Years 1853 and 1874 by Dr. James Graham Cooper." *Proceedings of the California Academy of Sciences*. 4th ser. 9: 69–121.

230 Fort Bridger: Burnett, *Wyoming Ground Squirrel (Citellus elegans elegans)*.

230 labeled as an Oregon Junco: Miller, "Avian Genus *Junco*."

230 Ridgway pointed out: Ridgway, R. 1897. "Note on *Junco annectens* Baird and *J. ridgwayi* Mearns." *Auk* 14: 94.

231 more white in the tail: Miller, "Avian Genus *Junco*."

232 inaccurate illustrations: Dunn, J. 2002. "The Identification of Pink-sided Juncos, With Cautionary Notes about Plumage Variation and Hybridization." *Birding* 34: 433–443.

232 Ridgway Junco: Miller, A. 1941. "Avian Genus *Junco*."

232 four "conclusive" hybrids: Ibid.

232 most subtly marked of Miller's hybrids: Ibid.

232 gradient between the Oregon Junco: Ibid.

233 Idaho birds: Ibid.

233 "The acoustic properties of song": Cardoso, G., and D. Reichard. 2016. "Dark-eyed Junco Song: Linking Ontogeny and Function with a Potential Role in Reproductive Isolation." In E. Ketterson and J. Atwell, eds. *Snowbird: Integrative Biology and Evolutionary Diversity in the Junco*. Chicago: University of Chicago Press.

233 "short-range" song: Nolan, Jr., V., et al. 2002. "Dark-eyed Junco (*Junco hyemalis*)." In A. Poole, ed. *Birds of North America Online*. Ithaca: Cornell Lab of Ornithology.

233 from Banff National Park: Semenchuk, G., ed. 1992. *The Atlas of Breeding Birds of Alberta*. Edmonton: Federation of Alberta Naturalists.

233 south to Montana: Marks, J., et al. 2016. *Birds of Montana*. Arrington, Virginia: Buteo Books.

233 Idaho and the Teton and Wind River Mountains: American Ornithologists' Union. 1957. *Check-list of North American Birds*. 5th ed. Ithaca: American Ornithologists' Union.

233 "the most common birds in Yellowstone": Phelps, J. 1968. "*Junco oreganus mearnsi* Ridgway: Pink-sided Oregon Junco." In A. Bent. 1968. *Life Histories of North American Cardinals, Grosbeak, Buntings, Towhees, Finches, Sparrows, and Allies*. Part 2. United States National Museum Bulletin 237. Washington: Smithsonian University Press.

233 Autumn flocks: Ibid.

233 a decidedly western junco: Ibid.

234 plains of eastern Colorado: Andrews, R., and R. Righter. 1992. *Colorado Birds: A Reference to Their Distribution and Habitat*. Denver: Denver Museum of Natural History.

234 west Texas: Lockwood, M., and B. Freeman. 2014. *The TOS Handbook of Texas Birds*. College Station: Texas A&M University Press.

234 Nebraska: Sharpe, R., et al. 2001. *Birds of Nebraska: Their Distribution and Temporal Occurrence.* Lincoln: University of Nebraska Press.

234 Virginia, and . . . Michigan: Dunn, "Identification of Pink-sided Juncos."

234 New York's first: Burgiel, J., et al. 2000. "Hudson-Delaware." *North American Birds* 54: 158–162.

234 four reports in Nova Scotia: McLaren, I. 2012. *All the Birds of Nova Scotia: Status and Critical Identification.* Kentville: Gaspereau Press.

234 Fire Island: Buckley, P., and S. Mitra. 2003. "Williamson's Sapsucker, Cordilleran Flycatcher, and Other Long-distance Vagrants at a Long Island, New York, Stopover Site." *North American Birds* 57: 292–304.

234 to the mountains of San Diego County: Unitt, P. 2004. *San Diego County Bird Atlas.* Proceedings of the San Diego Society of Natural History 39. San Diego: San Diego Natural History Museum.

234 April and May: Cooke, W. 1915. "The Migration of North American Sparrows." *Bird-Lore* 17: 18–19.

234 "appear suddenly in March": Cardoso and Reichard, "Dark-eyed Junco Song."

234 no clearly demonstrable geographic variation: Miller, "Avian Genus *Junco*."

OREGON JUNCO

235 Coues described and named *connectens*: Coues, E. 1884. *Key to North American Birds.* 2nd ed. Boston: Estes and Lauriat.

235 "before any specimens were received": Coues, E. 1884. "On the Application of Trinomial Nomenclature to Zoology." *Zoologist* 3rd ser., 8: 241–244.

235 shot in Colorado by William Brewster: Ridgway, R. 1901. *The Birds of North and Middle America.* Bulletin of the United States National Museum, Bulletin 50, pt. 1. Washington: Government Printing Office.

235 flock of White-crowned Sparrows: Brewster, W. 1882. "Ornithology 207046, *Junco hyemalis*." Unpubl. ms. ledger, Museum of Comparative Zoology. ids.lib .harvard.edu/ids/view/44820287.

235 "accidentally overlooked": Coues, E. 1897. "Rectifications of Synonymy in the Genus *Junco*." *Auk* 14: 94–95.

236 Eighth Supplement: American Ornithologists' Union. 1897. "Eighth Supplement to the American Ornithologists' Union *Check-list of North American Birds*." *Auk* 14: 117–135.

236 Ridgway "carefully examined and compared": Ridgway, *North and Middle America.*

236 twice declined to judge: American Ornithologists' Union. 1902. "Eleventh Supplement to the American Ornithologists' Union *Check-list of North American Birds*." *Auk* 19: 315–342.

American Ornithologists' Union. 1903. "Twelfth Supplement to the American Ornithologists' Union *Check-list of North American Birds*." *Auk* 20: 331–368.

236 resolution it finally offered: American Ornithologists' Union. 1910. *Check-list of North American Birds.* 3rd ed. New York: American Ornithologists' Union.

236 "dispose of the '*Junco hyemalis connectens*'": Dwight, J. 1918. "The Geographical Distribution of Color and of Other Variable Characters in the Genus *Junco*: A New Aspect of Specific and Subspecific Values." *Bulletin of the American Museum of Natural History* 38: 269–314.

237 "a contrary opinion": Swarth, H. 1922. "Birds and Mammals of the Stikine River Region of Northern British Columbia and Southeastern Alaska." *University of California Publications in Zoology* 24: 125–314.

237 justified its decision twice: American Ornithologists' Union. 1931. *Check-list of North American Birds.* 4th ed. Lancaster: American Ornithologists' Union.

237 "with much interest": Miller, A. 1941. "Speciation in the Avian Genus *Junco*." *University of California Publications in Zoology* 44: 173–434.

238 "short-range" song: Nolan, Jr., V., et al. 2002. "Dark-eyed Junco (*Junco hyemalis*)." In A. Poole, ed. *Birds of North America Online.* Ithaca: Cornell Lab of Ornithology.

238 The various Oregon Juncos: American Ornithologists' Union. 1957. *Check-list of North American Birds.* 5th ed. Ithaca: American Ornithologists' Union.

238 British Columbia has some of: Campbell, R., et al. 2001. *The Birds of British Columbia.* Vol. 4. Vancouver: UBC Press.

239 western third of Montana: Marks, J., et al. 2016. *Birds of Montana.* Arrington, Virginia: Buteo Books.

239 in northern Nevada: Floyd, T., et al. 2007. *Atlas of the Breeding Birds of Nevada.* Reno: University of Nevada Press.

239 In California: American Ornithologists' Union, *Check-list.* 5th ed.

239 Salton Sea: Patten, M., et al. 2003. *Birds of the Salton Sea: Status, Biogeography, and Ecology.* Berkeley: University of California Press.

239 lower Colorado River: Rosenberg, K., et al. 1991. *Birds of the Lower Colorado River Valley.* Tucson: University of Arizona Press.

239 Sonora in mid-October: Russell, S., and G. Monson. 1998. *The Birds of Sonora.* Tucson: University of Arizona Press.

239 Wrangel Island: Arkhopiv, V., and Ł. Łuwacki. 2016. "Nearctic Passerines in Russia." *Dutch Birding* 38: 201–214.

239 desert habitats of the east and southeast: Small, A. 1994. *California Birds: Their Status and Distribution.* Vista, CA: Ibis.

239 winters in western Texas: Lockwood, M., and B. Freeman. 2014. *The TOS Handbook of Texas Birds.* College Station: Texas A&M University Press.

239 In Nebraska: Sharpe, R., et al. 2001. *Birds of Nebraska: Their Distribution and Temporal Occurrence.* Lincoln: University of Nebraska Press.

239 easternmost records are from Nova Scotia: McLaren, I. 2012. *All the Birds of Nova Scotia: Status and Critical Identification.* Kentville: Gaspereau Press.

239 spring migration: Phillips, A., et al. 1964. *The Birds of Arizona.* Tucson: University of Arizona Press.

239 Howard and Moore: Dickinson, E., and L. Christidis, eds. 2014. *The Howard and Moore Complete Checklist of the Birds of the World.* 4th ed. Vol. 2. Eastbourne: Aves Press.

239 type specimen of the nominate race: Campbell et al., *British Columbia.*

239 plumage varies significantly: Miller, "Avian Genus *Junco*."

239 Robert W. Shufeldt: Lambrecht, K. 1935. "In Memoriam: Robert Wilson Shufeldt, 1850-1934." *Auk* 52: 359–361.

239 vagrants in eastern North America: Nolan et al., "Dark-eyed Junco (*Junco hyemalis*)."

239 *thurberi*: Small, *California Birds.*

240 both slopes of the Sierra Nevada: Beedy, C., and E. Pandolfino. 2013. *Birds of the Sierra Nevada: Their Natural History, Status, and Distribution.* Berkeley: University of California Press.

240 coastal lowlands of San Diego County: Atwell, J., et al. 2016. "Shifts in Hormonal, Morphological, and Behavioral Traits in a Novel Environment: Comparing Recently Diverged *Junco* Populations." In E. Ketterson and J. Atwell, eds. *Snowbird: Integrative Biology and Evolutionary Diversity in the Junco*. Chicago: University of Chicago Press.

240 "lighter, more pinkish back": Miller, "Avian Genus *Junco*."

240 *pinosus*: Loomis, L. 1893. "Description of a New Junco from California." *Auk* 10: 47–48.

240 Leverett M. Loomis: Bishop, L. 1929. "In Memoriam: Leverett Mills Loomis." *Auk* 46: 1–13.

240 "distinguished at a glance": Loomis, "New Junco."

240 overlap in the head color: Miller, "Avian Genus *Junco*."

240 smaller and longer-billed: Ridgway, *North and Middle America*.

240 Entirely nonmigratory: Nolan et al., "Dark-eyed Junco (*Junco hyemalis*)."

240 the race *pontilis*: Oberholser, H. 1919. "Description of an Interesting New Junco from Lower California." *Condor* 21: 119–120.

240 Townsend Junco: Nolan et al., "Dark-eyed Junco (*Junco hyemalis*)."

240 "bridging": Oberholser, "Interesting New Junco."

240 Townsend Junco was described: Anthony, A. 1889. "New Birds from Lower California and Mexico." *Proceedings of The California Academy of Sciences* 2nd ser., 2: 73–82.

240 lumped by the AOU: American Ornithologists' Union. 1908. "Fourteenth Supplement to the American Ornithologists' Union *Check-list of North American Birds*." *Auk* 25: 343–399.

240 tail averages slightly more white: Miller, "Avian Genus *Junco*."

240 led Dwight to propose: Dwight, "Characters in the Genus *Junco*."

CASSIAR JUNCO

241 first account of this bird: Dwight, J. 1918. "The Geographical Distribution of Color and of Other Variable Characters in the Genus *Junco*: A New Aspect of Specific and Subspecific Values." *Bulletin of the American Museum of Natural History* 38: 269–314.

241 breeding along the Stikine River: Swarth, H. 1922. "Birds and Mammals of the Stikine River Region of Northern British Columbia and Southeastern Alaska." *University of California Publications in Zoology* 24: 125–314.

241 "a 'good subspecies'": Ibid.

241 Miller affirmed Swarth's assessment: Miller, A. 1941. "Speciation in the Avian Genus *Junco*." *University of California Publications in Zoology* 44: 173–434.

241 lectotype: LeCroy, M. 2012. *Type Specimens of Birds in the American Museum of Natural History. Part 10. Passeriformes: Emberizidae: Emberizinae, Catamblyrhynchinae, Cardinalinae, Thraupinae, and Tersininae.* Bulletin of the American Museum of Natural History 368. New York: American Museum of Natural History.

242 the type of Coues's *connectens*: Miller, "Avian Genus *Junco*."

242 the Sibley painting: Sibley, D. 2014. *The Sibley Guide to Birds*. 2nd ed. New York: Alfred A. Knopf.

242 "of hybrid origin": Miller, "Avian Genus *Junco*."

243 Any pairing of Slate-colored and Oregon: Miller, "Avian Genus *Junco*."

243 such areas include: Miller, "Avian Genus *Junco*."

243 southeast Alaska: Gibson, D., and B. Kessel. 1997. "Inventory of the Species and Subspecies of Alaska Birds." *Western Birds* 28: 45–95.

243 Nova Scotia: McLaren, I. 2012. *All the Birds of Nova Scotia: Status and Critical Identification*. Kentville: Gaspereau Press.

243 "to all purposes disappear": Miller, "Avian Genus *Junco*."

SLATE-COLORED JUNCO

244 "companion of every child": Audubon, J. 1831. *Ornithological Biography*. Vol. 1. Edinburgh: Adam Black and Charles Black.

244 whose names for the species: Dall, W., and H. Bannister. 1869. "List of the Birds of Alaska, With Biographical Notes." *Transactions of the Chicago Academy of Sciences* 1: 267–310.

244 White first documented the species: Sloan, K. ed. 2007. *A New World: England's First View of America*. Chapel Hill: University of North Carolina Press.

244 "a kind of small bird": Kalm, P. 1761. *En resa til Norra America*. Vol. 3. Stockholm: Konglige Swenska Wetenskaps-Academien.

244 the name "snowbird": Ibid.

244 Catesby's day: Catesby, M. 1731. *The Natural History of Carolina, Florida and the Bahama Islands*. Vol. 1. London: privately published.

244 "by far the most numerous": Wilson, A. 1810. *American Ornithology, or the Natural History of the Birds of the United States*. Vol. 2. Philadelphia: Bradford and Inskeep.

244 "pretty general[ly]" believed: Ibid.

245 "Jacobin bunting": Buffon, Georges Louis Leclerc, Comte de. 1775. *Histoire naturelle, générale et particulière*. Vol. 18. Paris: Imprimerie Royale.

245 unequivocally refuted: Vieillot, L. 1817. *Nouveau dictionnaire d'histoire naturelle*. Vol. 25. Paris: Deterville.

245 Thomas Nuttall: Nuttall, *Manual*. 1st ed.

245 Bonaparte's endorsement: Bonaparte, C. 1826. *Observations on the Nomenclature of Wilson's Ornithology*. Philadelphia: Anthony Finley.

245 Wilson's repeating: Wilson, *American Ornithology*.

245 invalid name *nivalis*: Catesby, *Natural History of Carolina*.

245 the scorn of Maximilian: Maximilian zu Wied-Neuwied. 1858. "Verzeichniss der Vögel, welche auf einer Reise in Nord-America beobachtet wurden." *Journal für Ornithologie* 6: 1–29, 97–124, 177–204, 257–284, 337–354, 417–444.

245 "Black Bunting": Latham, J. 1783. *A General Synopsis of Birds*. Vol. 2. London: Benjamin White.

245 "Winter Finch": Gould, J. 1837. *The Birds of Europe*. Vol. 3. London: privately published.

245 "the head, neck, and upperparts": Wilson, *American Ornithology*.

245 "amateur bird students": Griscom, L. 1947. "Common Sense in Common Names." *Wilson Bulletin* 59: 131–138.

246 "short-range" song: Nolan, Jr., V., et al. 2002. "Dark-eyed Junco (*Junco hyemalis*)." In A. Poole, ed. *Birds of North America Online*. Ithaca: Cornell Lab of Ornithology.

246 lower Stikine River: Campbell, R., et al. 2001. *The Birds of British Columbia*. Vol. 4. Vancouver: UBC Press.

247 Carolina Juncos: American Ornithologists' Union. 1957. *Check-list of North American Birds*. 5th ed. Ithaca: American Ornithologists' Union.

247 altitudinal migrants: Ibid.

247 autumn movement: Nolan et al., "Dark-eyed Junco (*Junco hyemalis*)."

247 uncommon in San Diego County: Unitt, P. 2004. *San Diego County Bird Atlas*. Proceedings of the San Diego Society of Natural History 39. San Diego: San Diego Natural History Museum.

247 southern Arizona: Monson, G., and A. Phillips. 1981. *Annotated Checklist of the Birds of Arizona*. 2nd ed. Tucson: University of Arizona Press.

247 specimen from Sonora: Phillips, A., and D. Amadon. 1952. "Some Birds of Northwestern Sonora, Mexico." *Condor* 54: 163–168.

247 Texas: Lockwood, M., and B. Freeman. 2014. *The TOS Handbook of Texas Birds*. College Station: Texas A&M University Press.

247 Florida: Greenlaw, J., et al. 2014. *The Robertson and Woolfenden Florida Bird Species: An Annotated List*. Florida Ornithological Society Special Publication 8. Gainesville: Florida Ornithological Society.

248 spring migration: Nolan et al., "Dark-eyed Junco (*Junco hyemalis*)."

248 Iceland: Temminck, C. 1835. *Manuel d'ornithologie*. 2nd ed. Vol. 3. Paris: Edmond d'Ocagne.

248 Greenland: Boertmann, D. 1994. "An Annotated Checklist to the Birds of Greenland." *Meddelelser om Grønland: Bioscience* 38: 3–28.

248 50 European records: Lewington, I., et al. 1991. *A Field Guide to the Rare Birds of Britain and Europe*. London: HarperCollins.

248 Ireland and Italy: Polder, J., and K. Voous. 1969. "De Grijze Junco (*Junco hyemalis*), een nieuw vogel voor de Nederlandse avifauna.z." *Limosa* 42: 198–200.

248 seagoing Slate-colored Juncos: Raffaele, H., et al. 2003. *Birds of the West Indies*. Princeton: Princeton University Press.

248 In the Northwest: Nolan et al., "Dark-eyed Junco (*Junco hyemalis*)."

248 first specimen from Asia: Blyth, E. 1849. *Catalogue of the Birds in the Museum [of the] Asiatic Society*. Calcutta: Baptist Mission Press.

248 "had never been encountered before": Palmén, J. 1887. *Bidrag till kännedomen om Sibiriska Ishafskustens fogelfauna*. Stockholm: n.p.

248 eastern Russia: Arkhopiv, V., and Ł. Łuwacki. 2016. "Nearctic Passerines in Russia." *Dutch Birding* 38: 201–214.

248 larger, paler, and more uniformly blue-gray: Brewster, W. 1886. "An Ornithological Reconnaissance in Western North Carolina." *Auk* 3: 94–112.

248 a full species: Brewster, W. 1886. "*Junco hyemalis* Nesting in a Bush." *Auk* 3: 277–278.

248 affirmed without comment: American Ornithologists' Union. 1889. *Supplement to the* Code of Nomenclature and Check-list of North American Birds. New York: American Ornithologists' Union.

248 "expressly with a view": Dwight, J. 1891. "*Junco carolinensis* Shown to Be a Subspecies." *Auk* 8: 290–292.

248 "an excellent text for a sermon": Ibid.

248 followed suit the next year: American Ornithologists' Union. 1892. "Fourth Supplement to the American Ornithologists' Union *Check-list of North American Birds*." *Auk* 9: 105–108.

248 Carolina Junco is slightly larger: Ridgway, R. 1901. *The Birds of North and Middle America*. Bulletin of the United States National Museum, Bulletin 50, pt. 1. Washington: Government Printing Office.

WHITE-WINGED JUNCO

249 first collected: Holden, C., and C. Aiken. 1872. "Notes on the Birds of Wyoming and Colorado Territories." *Proceedings of the Boston Society of Natural History* 15: 193–210.

249 Black Hills of Wyoming: Allen, J. 1873. "Recent Contributions to American Geographical Ornithology." *American Naturalist* 7: 361–364.

249 "Colorado's pioneer ornithologist": Warren, E. "Charles Edward Howard Aiken." *Condor* 38: 234–238.

249 to see the junco himself: Allen, "American Geographical Ornithology."

249 combined their results: Holden and Aiken, "Wyoming and Colorado Territories."

249 "always genial and courteous": Ridgway, R. 1901. *The Birds of North and Middle America*. Bulletin of the United States National Museum, Bulletin 50, pt. 1. Washington: Government Printing Office.

249 Ridgway removed: Ridgway, R. 1873. "On Some New Forms of American Birds." *American Naturalist* 7: 602–619.

249 not a distinct species: Ibid.

249 a full species, *Junco aikeni*: Baird, S., et al. 1874. *A History of North America Birds*. Vol. 1. Boston: Little, Brown.

249 T. Martin Trippe: Coues, E. 1874. *Birds of the Northwest*. Miscellaneous Publications of the Geological Survey of the Territories 3. Washington: Government Printing Office.

250 newly minted Dark-eyed Junco: American Ornithologists' Union. 1973. "Thirty-second Supplement to the *Check-list of North American Birds*." *Auk* 90: 411–419.

250 "even at gunshot range": Coues, E. 1897. "Rectifications of Synonymy in the Genus *Junco*." *Auk* 14: 94–95.

250 "yet to examine an equivocal specimen": Miller, "Avian Genus *Junco*."

250 "much the largest of the juncos": Dwight, J. 1918. "The Geographical Distribution of Color and of Other Variable Characters in the Genus *Junco*: A New Aspect of Specific and Subspecific Values." *Bulletin of the American Museum of Natural History* 38: 269–314.

250 a "giant species": Miller, "Avian Genus *Junco*."

250 longer than Slate-colored Juncos: Ridgway, *North and Middle America*.

250 outweigh Oregon Juncos: Nolan et al., "Dark-eyed Junco (*Junco hyemalis*)."

250 applies most clearly to females: Ridgway, *North and Middle America*.

250 or older birds: Miller, "Avian Genus *Junco*."

250 Miller speculated: Ibid.

251 white wing bars: Ibid.

251 never show white: Ibid.

251 the tail pattern: Ibid.

252 a new subspecies of the Slate-colored Junco: Coues, E. 1895. "Letter from Sylvan Lake, S. Dak." *Nidiologist* 3: 14–15.

252 should "not be mistaken for *hyemalis*": Coues, "Rectifications of Synonymy."

252 "conclusive" hybrids: Miller, "Avian Genus *Junco*."

252 "chiefly" White-winged: Ibid.

252 closer visually to Pink-sided: Appleton, D. 2014. "Dark-eyed Junco Intergrades: Pink-sided Junco x White-winged Junco." *Bird Hybrids: A Collaborative Project to Improve Understanding of Bird Hybrids*. http://birdhybrids.blogspot.com/2014/11/dark-eyed-junco-intergrades-pink-sided.html.

252 Dominance by race: Yaukey, P. 1994. "Variation in Racial Dominance Within the Winter Range of the Dark-eyed Junco (*Junco hyemalis* L.)." *Journal of Biogeography* 21: 359–368.

252 Pine Ridge in Nebraska: Mollhoff, W. 2016. *The Second Nebraska Breeding Bird Atlas*. Bulletin of the University of Nebraska State Museum 29. Lincoln: University of Nebraska State Museum.

252 that land feature in South Dakota: Peterson, R. 1995. *The South Dakota Breeding Bird Atlas*. Brookings: South Dakota Ornithologists' Union.

253 June specimen from Clear Creek, Colorado: Andrews, R., and R. Righter. 1992. *Colorado Birds: A Reference to Their Distribution and Habitat*. Denver: Denver Museum of Natural History.

253 In southeastern Montana: Marks, J., et al. 2016. *Birds of Montana*. Arrington, Virginia: Buteo Books.

253 flocks of up to 30: Whitney, N. 1968. "*Junco aikeni* Ridgway: White-winged Junco." In Bent, A., ed. *Life Histories of North American Cardinals, Grosbeaks, Buntings, Towhees, Finches, Sparrows, and Allies*. Part 2. *Bulletin of the United States National Museum* 237. Washington: Smithsonian Institution Press.

253 on the breeding range in winter: Marks et al., *Birds of Montana*.

253 New Mexico and extreme western Kansas: American Ornithologists' Union. 1957. *Check-list of North American Birds*. 5th ed. Ithaca: American Ornithologists' Union.

253 casual to very rare in west Texas: Lockwood, M., and B. Freeman. 2014. *The TOS Handbook of Texas Birds*. College Station: Texas A&M University Press.

253 Yuma County: Patten, M., et al. 1998. "First Records of the White-winged Junco for California." *Western Birds* 29: 41–48.

253 Grand Canyon: Gatlin, B. 2013. "White-winged Dark-eyed Junco (*Junco hyemalis aikeni*), Hearst Tanks, Grand Canyon South Rim, Coconino County." *Arizona Field Ornithologists: Photo Documentation*. azfo.net.

253 Flagstaff: Monson, G., and A. Phillips. 1981. *Annotated Checklist of the Birds of Arizona*. 2nd ed. Tucson: University of Arizona Press.

253 Pima County: Rosenberg, G., et al. 2007. "Arizona Bird Committee Report: 2000-2004 Records." *Western Birds* 38: 74–101.

253 In California: Patten et al., "White-winged Junco for California."

253 "extraordinary" winter of 2000–2001: Engilis, A., and M. Biddlecomb. 2003. "First Record of White-winged Junco (*Junco hyemalis aikeni*) for the Central Valley, California." *Central Valley Bird Club Bulletin* 6: 1–3.

253 photographed in Deschutes County, Oregon: Crabtree, T. 1987. "Oregon's First 'White-winged' Junco." *Oregon Birds* 13: 296–300.

253 northbound movement: Whitney, "*Junco aikeni* Ridgway: White-winged Junco."

253 Most birds have departed: Andrews and Righter, *Colorado Birds*.

253 eggs in South Dakota: Johnsgard, P. 1979. *Birds of the Great Plains*. Lincoln: University of Nebraska Press.

253 Lincoln County, Nebraska: Sharpe, R., et al. 2001. *Birds of Nebraska: Their Distribution and Temporal Occurrence*. Lincoln: University of Nebraska Press.

253 Plymouth County, Massachusetts: Rines, M. 2009. "Thirteenth Annual Report of the Massachusetts Avian Records Committee (MARC)." *Bird Observer of Eastern Massachusetts* 37: 85–97.

GOLDEN-CROWNED SPARROW

254 sparrow from the south Pacific: Bonaparte, C. 1850. *Conspectus generum avium*. Leiden: Brill.

254 the Galápagos sparrow: Gray, G. 1870. *Hand-list of Genera and Species of Birds*. Part 2. London: Trustees of the BritishMuseum.

254 "only a specimen of the Californian *Z. coronata*": Salvin, O. 1875. "On the Avifauna of the Galapagos Archipelago." *Transactions of the Zoological Society of London* 9: 447–510.

254 at anchor in Prince William Sound: Stresemann, E. 1949. "Birds Collected in the North Pacific Area During Captain James Cook's Last Voyage (1778 and 1779)." *Ibis* 91: 244–255.

254 "a small land bird, of the finch kind": Cook, J. 1784. *A Voyage to the Pacific Ocean*. Vol. 2. London: W. and A. Strahan.

254 painted by William W. Ellis: Sharpe, R. 1906. *The History of the Collections Contained in the Natural History Departments of the British Museum*. Vol. 2, pt. 3. London: Trustees of the British Museum.

254 described by John Latham: Latham, J. 1783. *A General Synopsis of Birds*. Vol. 2. London: Benjamin White.

254 female from Nootka Sound: Pennant, T. 1785. *Arctic Zoology*. Vol. 2. London: Henry Hughs.

255 long-lived uncertainty: Latham, *General Synopsis*.

255 a geographic error: Stresemann, "North Pacific Area."

255 his most important sources: Stresemann, "North Pacific Area."

255 combining the range descriptions: Gmelin, J. 1788. *Caroli a Linné . . . Systema naturae per regna tria naturae*. Vol. 1. Leipzig: Georg Emanuel Beer.

255 name the bird anew: Pallas, P. 1811. *Zoographia rosso-asiatica*. Vol. 2. St. Petersburg: Imperial Academy of Sciences.

255 Bonaparte's *Geographical and Comparative List*: Bonaparte, C. 1838. *A Geographic and Comparative List of the Birds of Europe and North America*. London: John van Voorst.

255 Audubon's *Birds of America*: Audubon, J. 1839. *Ornithological Biography*. Vol. 5. Edinburgh: Adam Black and Charles Black.

255 *Synopsis* of 1839: Audubon, J. 1839. *A Synopsis of the Birds of North America*. Edinburgh: Adam Black and Charles Black.

255 Thomas Nuttall: Nuttall, T. 1840. *A Manual of the Ornithology of the United States and Canada*. 2nd ed. Boston: Hilliard, Gray, and Co.

255 in occasional use: Newberry, J. 1857. *Report Upon the Zoology of the Route*. Reports of Explorations and Surveys to Ascertain the Most Practicable and Economical Route for a Railroad. Vol. 6. Washington: Beverley Tucker.

255 the ornithological establishment's discomfort: Baird, S., with J. Cassin and G. Lawrence. 1858. *Birds*. Explorations and Surveys for a Railroad Route. Vol. 9. Washington: Beverly Tucker.

255 "essentially a totally different bird": Ridgway, R. 1901. *The Birds of North and Middle America*. Bulletin of the United States National Museum, Bulletin 50, pt. 1. Washington: Government Printing Office.

255 "obviously [belonged] to the same species": Stresemann, "North Pacific Area."

255 changed the scientific species name: American Ornithologists' Union. 1952. "Twenty-seventh Supplement to the American Ornithologists' Union *Check-list of North American Birds*." *Auk* 69: 308–312.

256 "like large female House Sparrows": Peterson, R. 1934. *A Field Guide to the Birds*. 1st ed. Boston: Houghton Mifflin.

257 "an ashy streak above the eye": Baird, *Birds*.

258 hybrid wintering in Michigan: Payne, R. 1979. "Two Apparent Hybrid *Zonotrichia* Sparrows." *Auk* 96: 595–599.

258 "a broader supercilium": Appleton, D. 2007. "Golden-crowned Sparrow x White-crowned Sparrow." Flickr Group Hybrid Birds. Tinyurl.com/gcwcsparrow.

258 spring specimens from California: Morton, M., and L. Mewaldt. 1960. "Further Evidence of Hybridization Between *Zonotrichia atricapilla* and *Zonotrichia leucophrys*." *Condor* 62: 485–486.

258 "suppressed or obscured": Miller, A. 1940. "A Hybrid Between *Zonotrichia coronata* and *Zonotrichia leucophrys*." *Condor* 42: 45–48.

258 Feeding birds call: Norment, C., et al. 1998. "Golden-crowned Sparrow (*Zonotrichia atricapilla*)." In A. Poole, ed. *Birds of North America*. Ithaca: Cornell Lab of Ornithology.

258 three notes descending by whole steps: Hoffmann, R. 1927. *Birds of the Pacific States*. Boston: Houghton Mifflin.

258 "To the miners carrying their packs": Kelly, J. 1964. "Golden-crowned Sparrow." In A. Bent, ed. *Life Histories of North American Cardinals, Grosbeaks, Buntings, Towhees, Finches, Sparrows, and Allies*. Pt. 3. Bulletin of the United States National Museum 237. Washington: Smithsonian Institution Press.

258 "No gold here": Petersen, M., et al. 1991. *Birds of the Kilbuck and Ahklun Mountain Region, Alaska*. North American Fauna 76. Washington: U.S. Fish and Wildlife Service.

258 "running on the ground": Audubon, *Ornithological Biography*.

259 "on the Rocky Mountains": Ibid.

259 "occasionally breeds in California": Heermann, "Birds Collected on the Survey."

259 another nest, again with a clutch of four: Brewer, T. 1878. "Nest and Eggs of *Zonotrichia coronata*." *Bulletin of the Nuttall Ornithological Club* 3: 42–43.

259 "probably erroneous": Ridgway, *North and Middle America*.

259 Alaska: Gibson, D., and B. Kessel. 1997. "Inventory of the Species and Subspecies of Alaska Birds." *Western Birds* 28: 45–95.

259 Yukon breeders: Norment et al., "Golden-crowned Sparrow (*Zonotrichia atricapilla*)."

259 Northwest Territories . . . Alberta: Ibid.

259 British Columbia: Campbell, R., et al. 2001. *The Birds of British Columbia*. Vol. 4. Vancouver: UBC Press.

259 northern Cascades of Washington: Paulson, D. 2013. "Birds of Washington." tinyurl.com/PaulsonBWashington.

259 1987–1996 breeding bird atlas surveys: Patuxent Wildlife Research Center. 2016. "BBA (Breeding Bird Atlas) Explorer." www.pwrc.usgs.gov/bba/.

259 leave the breeding range: Norment et al., "Golden-crowned Sparrow (*Zonotrichia atricapilla*)."

260 Pallas anticipated: Pallas, *Zoographia rosso-asiatica*.

260 Russian Far East: Arkhopiv, V., and Ł. Łuwacki. 2016. "Nearctic Passerines in Russia." *Dutch Birding* 38: 201–214.

260 In Canada: Campbell et al., *British Columbia*.

260 In Washington and Oregon: Norment et al., "Golden-crowned Sparrow (*Zonotrichia atricapilla*)."

260 western half of California: Small, A. 1994. *California Birds: Their Status and Distribution*. Vista, CA: Ibis.

260 East of the Sierra Nevada: Beedy, C., and E. Pandolfino. 2013. *Birds of the Sierra Nevada: Their Natural History, Status, and Distribution*. Berkeley: University of California Press.

260 San Diego County: Unitt, P. 2004. *San Diego County Bird Atlas*. Proceedings of the San Diego Society of Natural History 39. San Diego: San Diego Natural History Museum.

260 eastern and southeastern deserts: Small, *California Birds*.

260 Baja California: Erickson, R., et al. 2013. "Annotated Checklist of the Birds of Baja California and Baja California Sur, Second Edition." *North American Birds* 66: 582–613.

260 off the Baja coast: Miller, A., et al. 1957. *Distributional Checklist of the Birds of Mexico*. Part 2. Pacific Coast Avifauna 33.

260 Sonora: Russell, S., and G. Monson. 1998. *The Birds of Sonora*. Tucson: University of Arizona Press.

260 Nova Scotia's ten records: McLaren, I. 2012. *All the Birds of Nova Scotia: Status and Critical Identification*. Kentville: Gaspereau Press.

260 Orange County, Florida: Greenlaw, J., et al. 2014. *The Robertson and Woolfenden Florida Bird Species: An Annotated List*. Florida Ornithological Society Special Publication 8. Gainesville: Florida Ornithological Society.

260 northbound migration: Norment et al., "Golden-crowned Sparrow (*Zonotrichia atricapilla*)."

260 Russian Far East: Arkhopiv and Łuwacki, "Nearctic Passerines in Russia."

260 Three of Nova Scotia's records: McLaren, *Nova Scotia*.

260 "detected surprisingly often in spring": Veit, R., and W. Petersen. 1993. *Birds of Massachusetts*. Lincoln: Massachusetts Audubon Society.

260 New Hampshire: Keith, A., and R. Fox. 2013. *The Birds of New Hampshire*. Memoirs of the Nuttall Ornithological Club 19. Cambridge: Nuttall Ornithological Club.

WHITE-CROWNED SPARROW

261 first scientific description: Brisson, M. 1760. *Ornithologie, ou Méthode contenant la division des oiseaux*. Vol. 3. Paris: Jean-Baptiste Bauche.

261 Jean-François Gaultier: Boivin, B. 1974. "Gaultier (Gautier, Gauthier ou Gaulthier), Jean-François." *Dictionnaire biographique du Canada*. Vol. 3. Toronto: University of Toronto Press.

261 transferred to the Académie: Roger, J. 1997. *Buffon: A Life in Natural History*. Ithaca: Cornell University Press.

261 "We have named it the 'soulciet'": Buffon, Georges Louis Leclerc, Comte de. 1778. *Histoire naturelle, générale et particulière, avec la description du Cabinet du Roi*. Vol. 19. Paris: Imprimerie royale.

261 "elegant little" bird was an adult: Forster, J. 1772. "An Account of the Birds Sent from Hudson's Bay, With Observations Relative to Their Natural History, and Latin Descriptions of Some of the Most Uncommon." *Philosophical Transactions* 62: 382–433.

262 only three individuals of "this beautifully marked species": Wilson, A. 1810. *American Ornithology, or the Natural History of the Birds of the United States*. Vol. 2. Philadelphia: Bradford and Inskeep.

262 "the wild regions of Labrador": Audubon, J. 1834. *Ornithological Biography*. Vol. 2. Edinburgh: Adam and Charles Black.

262 Thomas Nuttall: Nuttall, T. 1832. *A Manual of the Ornithology of the United States and of Canada.* 1st ed. Vol. 1. Cambridge: Hilliard and Brown.

262 "convinced that these birds lose the white":Audubon, *Ornithological Biography.*

262 Nuttall encountered his first: Nuttall, T. 1840. *A Manual of the Ornithology of the United States and Canada.* 2nd ed. Boston: Hilliard, Gray, and Co.

262 "seen in almost every hedge": Gambel, W. 1843. "Descriptions of Some New and Rare Birds of the Rocky Mountains and California." *Proceedings of the Academy of Natural Sciences of Philadelphia* 1: 258–262.

262 "my friend Nuttall": Gambel, W. 1847. "Remarks on the Birds Observed in Upper California, With Descriptions of New Species." *Journal of the Academy of Natural Sciences of Philadelphia* (n.s.) 1: 25–56.

263 Heermann could declare "conclusively": Heermann, A. 1852. "Notes on the Birds of California, Observed During a Residence of Three Years in That Country." *Journal of the Academy of Natural Sciences of Philadelphia* ser. 2, 2: 259–272.

263 Baird's great *Birds*: Baird, S., with J. Cassin and G. Lawrence. 1858. *Birds.* Explorations and Surveys for a Railroad Route. Vol. 9. Washington: Beverly Tucker.

263 nine different call behaviors: Hill, B., and M. Lein. 1985. "The Non-song Vocal Repertoire of the White-crowned Sparrow." *Condor* 87: 327–335.

263 "high arousal": Ibid.

263 flatter in the populations breeding on the Pacific Coast: Rising, J. 2016. "White-crowned Sparrow (*Zonotrichia leucophrys*)." In J. del Hoyo et al., eds. *Handbook of the Birds of the World Alive.* Barcelona: Lynx Edicions.

264 common flight note: Hill and Lein, "Repertoire of the White-crowned Sparrow."

264 Other calls identified by Hill and Lein: Ibid.

264 Vocal differences among subspecies: Derryberry, E. 2007. *Song Evolution in White-crowned Sparrows (Zonotrichia leucophrys): Patterns and Mechanisms.* Dissertation. Duke University. UMI 328381.

264 Sierra Nevada: Orejuela, J., and M. Morton. 1975. "Song Dialects in Several Populations of Mountain White-crowned Sparrows (*Zonotrichia leucophrys oriantha*) in the Sierra Nevada." *Condor* 77: 145–153.

264 number of dialects: Ibid.

264 vast northern breeding range: Chilton, G., et al. 1995. "White-crowned Sparrow (*Zonotrichia leucophrys*)." In A. Poole, ed. *Birds of North America Online.* Ithaca: Cornell Lab of Ornithology.

264 British Columbia: Campbell, R., et al. 2001. *The Birds of British Columbia.* Vol. 4. Vancouver: UBC Press.

264 western half of Montana: Marks, J., et al. 2016. *Birds of Montana.* Arrington, Virginia: Buteo Books.

265 Arizona: Corman, T., and C. Wise-Gervais, eds. 2005. *Arizona Breeding Bird Atlas.* Albuquerque: University of New Mexico Press.

265 Nevada: Floyd, T., et al. 2007. *Atlas of the Breeding Birds of Nevada.* Reno: University of Nevada Press.

265 California: Beedy, C., and E. Pandolfino. 2013. *Birds of the Sierra Nevada: Their Natural History, Status, and Distribution.* Berkeley: University of California Press.

265 San Diego County: Unitt, P. 2004. *San Diego County Bird Atlas.* Proceedings of the San Diego Society of Natural History 39. San Diego: San Diego Natural History Museum.

265 Del Norte and Humboldt Counties: Small, A. 1994. *California Birds: Their Status and Distribution.* Vista, CA: Ibis.

265 Willamette Valley: Chilton et al., "White-crowned Sparrow (*Zonotrichia leucophrys*)."

265 Mendocino County south: Small, *California Birds.*

265 contact zone between the two breeding populations: Chilton et al., "White-crowned Sparrow (*Zonotrichia leucophrys*)."

265 strongly and conspicuously migratory: Ibid.

265 Belize: Howell, S., and S. Webb. 1995. *A Guide to the Birds of Mexico and Northern Central America.* Oxford: Oxford University Press.

265 small numbers now winter in Montana: Marks et al., *Montana.*

266 Mexico: Howell and Webb, *Mexico and Northern Central America.*

266 Florida: Greenlaw, J., et al. 2014. *The Robertson and Woolfenden Florida Bird Species: An Annotated List.* Florida Ornithological Society Special Publication 8. Gainesville: Florida Ornithological Society.

266 Caribbean: Chilton et al., "White-crowned Sparrow (*Zonotrichia leucophrys*)."

266 the Netherlands: Lewington, I., et al. 1991. *A Field Guide to the Rare Birds of Britain and Europe.* London: HarperCollins.

266 records of the species from Europe: Arkhopiv, V., and Ł. Łuwacki. 2016. "Nearctic Passerines in Russia." *Dutch Birding* 38: 201–214.

266 Russian Far East: Arkhopiv and Łuwacki, "Nearctic Passerines in Russia."

266 South Korea and Japan: Brazil, M. 2009. *Birds of East Asia.* London: Christopher Helm.

266 In spring: Chilton et al., "White-crowned Sparrow (*Zonotrichia leucophrys*)."

266 "Intermediate Sparrow": Baird, S., and R. Ridgway. 1873. "On Some New Forms of American Birds." *Bulletin of the Essex Institute* 5: 197–201.

266 "merely geographical races of one species": Ridgway, R. 1890. "Intergradation Between *Zonotrichia leucophrys* and *Z. intermedia,* and Between the Latter and *Z. gambeli.*" *Auk* 7: 96.

266 ratified by the AOU: American Ornithologists' Union. 1890. "Second Supplement to the American Ornithologists' Union Check-list of North American Birds." *Auk* 7: 60–66.

266 the bird Nuttall had described: Nuttall, *Manual.* 2nd ed.

266 a new name, *nuttalli*: Ridgway, R. 1899. "New Species, Etc., of American Birds. III: Fringillidae." *Auk* 16: 35–37.

266 removed, by the AOU: American Ornithologists' Union. 1901. "Tenth Supplement to the American Ornithologists' Union Check-list of North American Birds." *Auk* 18: 295–321.

266 Ridgway's *Birds of North and Middle America*: Ridgway, R. 1901. *The Birds of North and Middle America.* Bulletin of the United States National Museum, Bulletin 50, pt. 1. Washington: Government Printing Office.

266 Grinnell discovered: Grinnell, J. 1928. "Notes on the Systematics of West American Birds." *Condor* 30: 185–189.

266 "There is that approximate degree of uniformity": Ibid.

267 AOU adopted Grinnell's view: American Ornithologists' Union. 1931. *Check-list of North American Birds.* 4th ed. Lancaster: American Ornithologists' Union.

267 Oberholser split the nominate subspecies: Oberholser, H. 1932. "Descriptions of New Birds from Oregon, Chiefly from the Warner Valley Region." *Scientific Publications of the Cleveland Museum of Natural History* 4: 1–12.

267 a valid race: American Ornithologists' Union. 1944. "Nineteenth Supplement to the American Ornithologists' Union *Check-list of North American Birds.*" *Auk* 61: 441–464.

267 "unfortunate nomenclatural complications": Todd, W. 1948. "Systematics of the White-crowned Sparrow." *Proceedings of the Biological Society of Washington* 61: 19–20.

267 rejected Todd's proposal: Wetmore, A. 1953. "The Application of the Name *leucophrys* Forster." *Auk* 70: 372–373.

267 Not all authorities still recognize *oriantha*: Chilton et al., "White-crowned Sparrow (*Zonotrichia leucophrys*)."

267 Banks's lump of the two: Banks, R. 1964. "Geographic Variation in the White-crowned Sparrow *Zonotrichia leucophrys.*" *University of California Publications in Zoology* 70: 1–123.

268 intergrade commonly: Chilton et al., "White-crowned Sparrow (*Zonotrichia leucophrys*)."

268 "biologically . . . dissimilar": Rising, "White-crowned Sparrow (*Zonotrichia leucophrys*)."

268 significant differences: Mewaldt, R., et al. 1968. "Comparative Biology of Pacific Coastal White-crowned Sparrows." *Condor* 70: 14–30.

268 upperparts color of these Pacific Coast birds: Banks, "Geographic Variation."

268 separable in the hand: Ibid.

HARRIS SPARROW

269 unfamiliar finches: Audubon, J. 1899. "The Missouri River Journals." In M. Audubon, ed. *Audubon and His Journals.* Vol. 1. New York: Charles Scribner's Sons.

269 Nuttall named it: Nuttall, T. 1840. *A Manual of the Ornithology of the United States and Canada.* 2nd ed. Boston: Hilliard, Gray, and Co.

269 "reserv[ing] this discovery for his own book": Harris, H. 1919. "Historical Notes on Harris's Sparrow (*Zonotrichia querula*)." *Auk* 36: 180–190.

270 "why Audubon and his coworkers were in ignorance": Ibid.

270 proved to be males: Audubon, "Missouri River Journals."

270 Harris wrote to his brother-in-law: Morris, G. 1895. "Notes and Extracts from a Letter of Edward Harris." *Auk* 12: 225–231.

270 the identity of the Townsend Bunting: Holt, J. "Notes on Audubon's 'Mystery' Birds." *Cassinia* 70: 22–24.

270 female "of the large new Finches": Audubon, "Missouri River Journals."

270 announcing his "new" species: Audubon, J. 1844. *The Birds of America.* 2nd ed. Vol. 7. Philadelphia: E.G. Dorsey.

270 occasional notices: Cooke, W. 1884. "The Distribution and Migration of *Zonotrichia querula.*" *Auk* 1: 332–337.

270 a virtually affectionate account: Coues, E. 1874. *Birds of the Northwest.* Miscellaneous Publications of the Geological Survey of the Territories 3. Washington: Government Printing Office.

270 "bird of imposing appearance": Coues, E. 1872. *Key to North American Birds.* 1st ed. Boston: Estes and Lauriat.

270 "readily destroyed": Coues, E. 1878. "Field-Notes on Birds Observed in Dakota and Montana Along the Forty-ninth Parallel During the Seasons of 1873 and 1874." *Bulletin of the Geological and Geographical Survey of the Territories* 4: 545–661.

270 a "fine series of specimens": Ibid.

270 "I presume the bird has some special": Coues, *Birds of the Northwest.*

270 bred in the thickets along the Missouri: Maximilian zu Wied-Neuwied. 1841. *Reise in das Innere Nord-America in den Jahren 1832 bis 1834.* Vol. 2. Coblenz: J. Hoelscher.

270 "the summer home of Harris's Sparrow": Bendire, C. 1889. "Description of the Supposed Nest and Eggs of *Zonotrichia querula*, Harris's Sparrow." *Auk* 6: 150–152.

270 "young just from the nest": Preble, E. 1902. *A Biological Investigation of the Hudson Bay Region.* North American Fauna 22. Washington: Government Printing Office.

271 "on the ground under a dwarf birch": Seton, E. 1908. "Bird Records from Great Slave Lake Region." *Auk* 25: 68–74.

271 he was not present: Preble, E. 1908. *A Biological Investigation of the Athabask-Mackenzie Region.* North American Fauna 27. Washington: Government Printing Office.

271 "lots of things": McAtee, W. 1962. "Memorial: Edward Alexander Preble." *Auk* 79: 730–742.

271 "the completion of the Hudson Bay Canadian Government Railway": Semple, J., and G. Sutton. 1932. "Nesting of Harris's Sparrow (*Zonotrichia querula*) at Churchill, Manitoba." *Auk* 49: 166–183.

272 blackish crown feathers: Rohwer, S., et al. 1981. "Variation in Size, Appearance, and Dominance Within and Among the Sex and Age Classes of Harris' Sparrows." *Journal of Field Ornithology* 52: 291–303.

272 "studliness scale": Rohwer, S. 1977. "Status Signaling in Harris Sparrows: Some Experiments in Deception." *Behaviour* 61: 107–129.

272 available online: tinyurl.com/StudlyZonos.

272 "badge of subordination": Rohwer, "Status Signaling."

272 more extensively black: Ibid.

272 somewhat more lightly marked: Ibid.

273 even in alternate plumage: Ibid.

273 An apparent hybrid: Payne, R. 1979. "Two Apparent Hybrid *Zonotrichia* Sparrows." *Auk* 96: 595–599.

273 "a queer chuckling note": Coues, *Key.* 1st ed.

273 "a warble": Thompson, E. 1890. "The Birds of Manitoba." *Proceedings of the United States National Museum* 13: 457–643.

273 at the forest-tundra edge: Norment, C., et al. 2008. "Harris's Sparrow (*Zonotrichia querula*)." In A. Poole, ed. *Birds of North America.* Ithaca: Cornell Lab of Ornithology.

273 in the Yukon: Sinclair, P., et al., eds. 2003. *Birds of the Yukon Territory.* Vancouver: University of British Columbia Press.

273 leave their breeding range: Norment et al., "Harris's Sparrow (*Zonotrichia querula*)."

273 "swarm across the prairie": Baumgartner, A. 1968. "Harris' Sparrow *Zonotrichia querula* (Nuttall)." In A. Bent, ed. *Life Histories of North American Cardinals, Grosbeaks, Buntings, Towhees, Finches, Sparrows, and Allies.* Pt. 3. Bulletin of the United States National Museum 237. Washington: Smithsonian Institution Press.

273 "linger in abundance": Swenk, M., and O. Stevens. 1929. "Harris's Sparrow and the Study of It By Trapping." *Wilson Bulletin* 41: 129–177.

273 "area of greatest abundance": Cooke, "Distribution and Migration."

273 Iowa: Kent, T., and J. Dinsmore. 1996. *Birds in Iowa.* Iowa City: privately published.

274 Hendry County, Florida: Greenlaw, J., et al. 2014. *The Robertson and Woolfenden Florida Bird Species: An Annotated List*. Florida Ornithological Society Special Publication 8. Gainesville: Florida Ornithological Society.

274 spring migration: Swenk and Stevens, "Harris's Sparrow."

WHITE-THROATED SPARROW

275 Edwards engraved the drawing: Edwards, G. 1760. *Gleanings of Natural History*. Pt. 2. London: Royal College of Physicians.

275 "the large brown white throat": Bartram, W. 1791. *Travels Through North and South Carolina, Georgia, East and West Florida, the Cherokee Country, the Extensive Territories of the Muscogulges, or Creek Confederacy, and the Country of the Chactaws*. Philadelphia: James and Johnson.

275 Brisson, however: Brisson, M. 1760. *Supplementum ornithologiae*. Paris: Jean-Baptiste Bauche.

275 Gmelin knew and cited: Gmelin, J. 1788. *Caroli a Linné . . . Systema naturae per regna tria naturae*. 13th ed. Vol. 1. Leipzig: Georg Emanuel Beer.

275 Wilson had chosen to use Gmelin's binomial: Wilson, A. 1811. *American Ornithology, or the Natural History of the Birds of the United States*. Vol. 3. Philadelphia: Bradford and Inskeep.

275 subsequent edition of Wilson's work: Wilson, A. 1832. *American Ornithology, or The Natural History of the Birds of the United States, With a Continuation by Charles Lucian Bonaparte, Prince of Musignano; the Illustrative Notes, and Life of Wilson, by Sir William Jardine, bart.* Vol. 1. London: Whittaker, Treacher, and Arnot.

275 priority of *pensylvanica*: Bonaparte, C. 1826. *Observations on the Nomenclature of Wilson's Ornithology*. Philadelphia: Anthony Finley.

275 definitively adopting *albicollis*: Bonaparte, C. 1850 *Conspectus generum avium*. Leiden: Brill.

275 Audubon asserted the validity of *pennsylvanica*: Ridgway, R. 1901. *The Birds of North and Middle America*. Bulletin of the United States National Museum, Bulletin 50, pt. 1. Washington: Government Printing Office.

275 Maximilian used it as late as 1858: Maximilian zu Wied-Neuwied. 1858. "Verzeichniss der Vögel, welche auf einer Reise in Nord-America beobachtet wurden." *Journal für Ornithologie* 6: 1–29, 97–124, 177–204, 257–284, 337–354, 417–444.

275 As Brisson's nomenclature is not binomial: Baird, S., with J. Cassin and G. Lawrence. 1858. *Birds*. Explorations and Surveys for a Railroad Route. Vol. 9. Washington: Beverly Tucker.

275 Brisson's generic names: Hemming, F. 1955. "*Direction 16: Validation Under the Plenary Power of Brisson (M.J.), 1760, Ornithologia sive Synopsis methodica sistens avium divisionem in ordines* ("*Direction*" in Replacement of *Opinion* 37)." *Opinions and Declarations Rendered by the International Commission of Zoological Nomenclature* 1.C, pt. C.6: 81–88.

276 Thomas Pennant reported: Pennant, T. 1785. *Arctic Zoology*. Vol. 2. London: Henry Hughs.

276 "all the parts that are white": Wilson, *American Ornithology*.

276 Audubon 20 years on: Audubon, J. 1831. *Ornithological Biography*. Vol. 1. Edinburgh: Adam Black and Charles Black.

277 "the males do not attain their mature colors": Allen, J. 1866. "Catalogue of the Birds Found at Springfield, Mass., With Notes on Their Migrations, Habits, etc.; Together With a List of Those Birds Found in the State Not Yet Observed at Springfield." *Proceedings of the Essex Institute* 4: 48–98.

277 Coues offered an alternative explanation: Coues, E. 1874. *Birds of the Northwest*. Miscellaneous Publications of the Geological Survey of the Territories 3. Washington: Government Printing Office.

277 gene inversion on the second chromosome: Huynh, L., et al. 2010. "Contrasting Population Genetic Patterns Within the White-throated Sparrow Genome (*Zonotrichia albicollis*)." *BCM Genetics* 11: 96.

277 "supergene": Tuttle, E., et al. 2016. "Divergence and Functional Degradation of a Sex Chromosome-like Supergene." *Current Biology* 26: 344–350.

277 "Regardless of sex": Lowther, J., and J. Falls. 1968. "White-throated Sparrow *Zonotrichia albicollis* (Gmelin)." In A. Bent, ed. *Life Histories of North American Cardinals, Grosbeaks, Buntings, Towhees, Finches, Sparrows, and Allies*. Pt. 3. Bulletin of the United States National Museum 237. Washington: Smithsonian Institution Press.

277 prealternate molt: Falls, J., and J. Kopachena. 2010. "White-throated Sparrow (*Zonotrichia alibicollis*)." In P. Rodewald, ed. *Birds of North America*. Ithaca: Cornell Lab of Ornithology.

277 even in basic plumage: Piper, W., and R. Wiley. 1989. "Distinguishing Morphs of the White-throated Sparrow in Basic Plumage." *Journal of Field Ornithology* 60: 73–83.

278 head pattern provides signals: Horton, B., et al. 2012. "Morph Matters: Aggression Bias in a Polymorphic Sparrow." *PLoS ONE* 7: e48705.

278 different behavioral strengths: Tuttle, E. 2003. "Alternative Reproductive Strategies in the White-throated Sparrow: Behavioral and Genetic Evidence." *Behavioral Ecology* 14: 425–432.

278 males and females sing: Falls and Kopachena, "White-throated Sparrow (*Zonotrichia alibicollis*)."

278 slurred up or down: Borror, D., and W. Gunn. 1965. "Variation in White-throated Sparrow Songs." *Auk* 82: 26–47.

278 15 different patterns: Ibid.

278 changes in the frequence and distribution of song patterns: Ibid.

278 early deglaciation: Cannings, R., and S. Cannings. 2004. *British Columbia: A Natural History*. 2nd ed. Vancouver: Greystone Books.

278 Yukon: Sinclair, P., et al., eds. 2003. *Birds of the Yukon Territory*. Vancouver: University of British Columbia Press.

279 Nunavut: Richards, J., and T. White. 2008. *Birds of Nunavut: A Checklist*. 2nd ed. Yellowknife: privately published.

279 British Columbia: Campbell, R., et al. 2001. *The Birds of British Columbia*. Vol. 4. Vancouver: UBC Press.

279 Alberta: Federation of Alberta Naturalists. 2007. *The Atlas of Breeding Birds of Alberta: A Second Look*. Edmonton: Federation of Alberta Naturalists.

279 southern limits of the breeding range: American Ornithologists' Union. 1998. *Check-list of North American Birds*. 7th ed. Washington: American Ornithologists' Union.

279 New Jersey: Walsh, J., et al. 1999. *Birds of New Jersey*. Bernardsville: New Jersey Audubon Society.

279 Ohio: Rodewald, P., et al., eds. 2016. *The Second Atlas of Breeding Birds in Ohio*. University Park: Pennsylvania State University Press.

279 leave their breeding grounds: Falls and Kopachena, "White-throated Sparrow (*Zonotrichia albicollis*)."

279 Rio Grande Valley: Lockwood, M., and B. Freeman. 2014. *The TOS Handbook of Texas Birds*. College Station: Texas A&M University Press.

279 Florida: Greenlaw, J., et al. 2014. *The Robertson and Woolfenden Florida Bird Species: An Annotated List*. Florida Ornithological Society Special Publication 8. Gainesville: Florida Ornithological Society.

279 Aruba: Prins, T., et al. 2009. "Checklist of the Birds of Aruba, Curaçao and Bonaire, South Caribbean." *Ardea* 97: 137–268.

279 San Diego County: Unitt, P. 2004. *San Diego County Bird Atlas*. Proceedings of the San Diego Society of Natural History 39. San Diego: San Diego Natural History Museum.

279 Britain and Ireland: Lewington, I., et al. 1991. *A Field Guide to the Rare Birds of Britain and Europe*. London: HarperCollins.

279 Scotland: Angus, C. 1869. "Notice of the Occurrence of the White-throated Sparrow (*Zonotrichia albicollis*) in Aberdeenshire." *Proceedings of the Natural History Society of Glasgow* 1: 209–211.

279 Outer Hebrides: Records Committee of the British Ornithologists' Union. 1960. "Additions to the British and Irish List: White-throated Sparrow, Black-and-white Warbler and Olive-backed Thrush." *British Birds* 53: 97–99.

279 more than 50 accepted records: Lewington et al., *Rare Birds*.

279 spring migration: Falls and Kopachena, "White-throated Sparrow (*Zonotrichia albicollis*)."

RUFOUS-COLLARED SPARROW

280 "pretty little bird": Audubon, J. 1839. *Ornithological Biography*. Vol. 5. Edinburgh: Adam Black and Charles Black.

280 "frantic" to publish the novelties: Mearns, B., and R. Mearns. 1992. *Audubon to Xantus: The Lives of Those Commemorated in North American Bird Names*. London: Academic Press.

280 the "infamous Morton's Finch": Mearns, R., and B. Mearns. 2007. *John Kirk Townsend: Collector of Audubon's Western Birds and Mammals*. Dumfries: privately published.

280 "nothing more than a Chilian specimen": Sclater, P. 1857. "Notes on the Birds in the Museum of the Academy of Natural Sciences of Philadelphia, and Other Collections in the United States of America." *Proceedings of the Zoological Society of London* 25: 1–8.

281 Cape of Good Hope Bunting: Buffon, Georges Louis Leclerc, Comte de. 1778. *Histoire naturelle, générale et particulière, avec la description du Cabinet du Roi*. Vol. 19. Paris: Imprimerie royale.

281 Cape Finch: Boddaert, P. 1772. *Kortbegrip van het zamenstel der natuur, van den Heer C. Linnaeus*. Utrecht: J. can Schoonhoven.

281 putative range in southernmost Africa: Müller, P. 1776. *Des Ritters Carl von Linné . . . vollständigen Natursystems Supplements- und Register-Band*. Nuremberg: Gabriel Nicolaus Raspe.

281 "good-morning bird": Buffon, *Histoire naturelle*.

281 Boddaert eventually agreed: Boddaert, P. 1783. *Table des planches enluminéez*. Utrecht: n.p.

281 "geographically erroneous name": Allen, J. 1891. "On a Collection of Birds from Chapada, Matto Grosso, Brazil, Made by Mr. Herbert H. Smith." *Bulletin of the American Museum of Natural History* 3: 337–380.

281 "The song typically comprises": Handford, P. 2005. "Latin Accents: Song Dialects of a South American Sparrow." *Birding* 37: 510–519.

282 frequently heard calls: Fagan, J., and O. Komar. 2016. *Peterson Field Guide to Birds of Northern Central America*. Boston: Houghton Mifflin Harcourt.

282 "This protean species": Stephens, J. 1815. *Aves*. Vol. 9 in G. Shaw, ed. *General Zoology, or Systematic Natural History*. London: G. Wilkie et al.

282 range and physical diversity: Lougheed, S., et al. 2013. "Continental Phylogeography of an Ecologically and Morphologically Diverse Neotropical Songbird, *Zonotrichia capensis*." *BMC Evolutionary Biology* 13: 58–61.

282 published online: worldbirdnames.org.

282 tropical breeders: Rising, J., and A. Jaramillo. 2016. "Rufous-collared Sparrow (*Zonotrichia capensis*)." In J. del Hoyo et al., eds. *Handbook of the Birds of the World*. Barcelona: Lynx Edicions.

282 black patch at the side of the upper breast: Ibid.

282 hindneck collar: Ibid.

282 a large Rufous-collared Sparrow: Griscom, L. 1930. "Studies From the Dwight Collection of Guatemala Birds III." *American Museum Novitates* 438: 1–18.

282 in Chiapas: Miller, A., et al. 1957. *Distributional Checklist of the Birds of Mexico*. Part 2. Pacific Coast Avifauna 33.

282 To the south: Gallardo, J. 2014. *Guide to the Birds of Honduras*. Tegucigalpa: privately published.

282 El Salvador: Griscom, "Dwight Collection."

282 Clear Creek County, Colorado: Faulkner, D. 2012. "The 63rd Report of the Colorado Bird Records Committee." *Colorado Birds* 46: 188–197.

282 sound recordings: Pieplow, N. 2011. "Rufous-collared Origins." *Earbirding.com: Recording, Identifying, and Interpreting Bird Sounds*. earbirding.com.

CHESTNUT-CAPPED BRUSH FINCH

283 "beautiful ornithological holdings": de Lafresnaye, F. 1835. "Sur le genre *Grimpic* (Picolaptes, Lesson)." *Magasin de zoologie* 5: n.p.

283 bristles for brush manufacturers: Garnaud, N. 2008. *L'émergence du monde ouvrier en milieu rural dans l'ancienne province du Poitou au XIXe siècle*. Dissertation. Université de Poitiers.

283 "the very special enthusiasm": de Lafresnaye, F. 1839. "Quelques oiseaux nouveaux de la collection de M. Charles Brelay, à Bordeaux." *Revue zoologique* 2: 97–100.

283 "giving new birds the names of women": Ibid.

283 another bird from the Brelays' collection: Ibid.

283 widely separated localities: Ridgway, R. 1901. *The Birds of North and Middle America*. Bulletin of the United States National Museum, Bulletin 50, pt. 1. Washington: Government Printing Office.

284 southern and western Mexico: Miller, A., et al. 1957. *Distributional Checklist of the Birds of Mexico*. Part 2. Pacific Coast Avifauna 33.

284 separated into nine or ten subspecies: Parkes, K. 1954. "A Revision of the Neotropical Finch *Atlapetes brunnei-nucha*." *Condor* 56: 129–138.

284 the race *apertus*: Wetmore, A. 1942. "Descriptions of Three Additional Birds from Southern Vera Cruz." *Proceedings of the Biological Society of Washington* 55: 105–108.

284 "no characters uniquely its own": Parkes, "Neotropical Finch *Atlapetes brunnei-nucha*."

284 This race, *suttoni*: Ibid.

284 forehead pattern: Ibid.

284 generally considered indistinguishable: Rising, J. 2016. "Chestnut-capped Brush-finch (*Arremon brunneinucha*)." In del Hoyo, J., et al., eds. *Handbook of the Birds of the World Alive*. Barcelona: Lynx Edicions.

GREEN-STRIPED BRUSH FINCH

285 "To the modest number of species in the genus": Bonaparte, C. 1855. "Les principales espèces nouvelles qu'il vient d'observer dans son récent voyage en Ecosse et Angleterre." *Comptes rendus hebdomadaires des séances de l'Académie des Sciences* 41: 651–660.

285 in the royal museum of Berlin: Sclater, P. 1856. "Synopsis avium tanagrinarum: A Descriptive Catalogue of the Known Species of Tanagers." *Proceedings of the Zoological Society of London* 24: 64–94.

286 most frequent call: Howell, S., and S. Webb. 1995. *A Guide to the Birds of Mexico and Northern Central America*. Oxford: Oxford University Press.

286 only in western and central Mexico: Ibid.

286 Guanajuato: Miller, A., et al. 1957. *Distributional Checklist of the Birds of Mexico*. Part 2. Pacific Coast Avifauna 33.

286 Sinaloa: Moore, R. 1938. "New Races in the Genera of *Vireo* and *Buarremon* from Sinaloa." *Proceedings of the Biological Society of Washington* 51: 69–71.

286 40,000 Mexican specimens: Davis, J. 1974. "Chester Converse Lamb." *Auk* 91: 479–480.

286 described the brush finch in 1938: Moore, "Genera of *Vireo* and *Buarremon*."

286 "wide overlap": Hardy, J., and T. Webber. 1975. *A Critical List of Type Specimens of Birds in the Moore Laboratory of Zoology at Occidental College*. Contributions in Science 273. Los Angeles: Natural History Museum of Los Angeles County.

286 Volcán de Colima: Van Rossem, A. 1938. "Descriptions of Twenty-one New Races of Fringillidae and Icteridae from Mexico and Guatemala." *Bulletin of the British Ornithologists' Club* 58: 124–139.

BLACK-THROATED SPARROW

287 "at least $20,000": Ford, A. 1988. *John James Audubon: A Biography*. New York: Abbeville Press.

287 "one of the most remarkable finches": Cassin, J. 1850. "Descriptions of New Species of Birds of the Genera *Parus*, Linn.; *Emberiza*, Linn.; *Carduelis*, Briss.; *Myiothera*, Ill.; and *Leuconerpes*, Sw., Specimens of Which Are in the Collection of the Academy of Natural Sciences of Philadelphia." *Proceedings of the Academy of Natural Sciences of Philadelphia* 5: 103–106.

287 "this curious little Finch": Cassin, J. 1853–1855. *Illustrations of the Birds of California, Texas, Oregon, British and Russian America*. Philadelphia: J.B. Lippincott and Co.

288 retain juvenile characters: Johnson, M., et al. 2002. "Black-throated Sparrow (*Amphispiza bilineata*)." In P. Rodewald, ed. *Birds of North America*. Ithaca: Cornell Lab of Ornithology.

288 song of the Black-throated Sparrow: Ibid.

288 "sing in flight": Ibid.

288 Texas: Lockwood, M., and B. Freeman. 2014. *The TOS Handbook of Texas Birds*. College Station: Texas A&M University Press.

288 Colorado: Andrews, R., and R. Righter. 1992. *Colorado Birds: A Reference to Their Distribution and Habitat*. Denver: Denver Museum of Natural History.

288 Wyoming: Johnson et al., "Black-throated Sparrow (*Amphispiza bilineata*)."

289 breed throughout the deserts: Ibid.

289 California: Small, A. 1994. *California Birds: Their Status and Distribution*. Vista, CA: Ibis.

289 Oregon: Ibid.

289 northernmost breeders are in Washington: Campbell, R., et al. 2001. *The Birds of British Columbia*. Vol. 4. Vancouver: UBC Press.

289 rare anywhere in Washington: Paulson, D. 2013. "Birds of Washington." tinyurl.com/PaulsonBWashington.

289 British Columbia: Toochin, R. 2013. "Status and Occurrence of the Black-throated Sparrow (*Amphispiza bilineata*) in British Columbia." ibis.geog.ubc.ca/biodiversity.

289 known to be migratory: Johnson et al., "Black-throated Sparrow (*Amphispiza bilineata*)."

289 status is uncertain in Texas: Lockwood and Freeman, *TOS Handbook*.

289 Spring migrants: Ibid.

289 nearly annual in British Columbia: Toochin, "Status and Occurrence."

289 Florida: Greenlaw, J., et al. 2014. *The Robertson and Woolfenden Florida Bird Species: An Annotated List*. Florida Ornithological Society Special Publication 8. Gainesville: Florida Ornithological Society.

289 most widespread and to many birders the most familiar: Johnson et al., "Black-throated Sparrow (*Amphispiza bilineata*)."

289 larger than the gray-backed nominate race: Ridgway, R. 1898. "Descriptions of Supposed New Genera, Species, and Subspecies of American Birds. I: Fringillidae." *Auk* 15: 223–230.

289 nominate race: Johnson et al., "Black-throated Sparrow (*Amphispiza bilineata*)."

289 averages nearly twice as large: Ridgway, R. 1901. *The Birds of North and Middle America*. Bulletin of the United States National Museum, Bulletin 50, pt. 1. Washington: Government Printing Office.

289 from north-central Texas east: Johnson et al., "Black-throated Sparrow (*Amphispiza bilineata*)."

289 from southeast Colorado: Ibid.

289 back is tinged brown: Burleigh, T., and G. Lowery. 1939. "Description of Two New Birds from Western Texas." *Occasional Papers of the Museum of Zoology, Louisiana State University* 6: 67–68.

289 the race *grisea*: Nelson, E. 1898. "Description of New Birds from Mexico, with a Revision of the Genus *Dactylortyx*." *Proceedings of the Biological Society of Washington* 12: 57–68.

290 from central Chihuahua: Miller, A., et al. 1957. *Distributional Checklist of the Birds of Mexico*. Part 2. Pacific Coast Avifauna 33.

290 Mexican resident, *bangsi*: Miller et al., *Birds of Mexico*.

290 paler above: Grinnell, J. 1927. "Six New Subspecies of Birds from Lower Baja California." *Auk* 44: 67–72.

290 *pacifica* is a resident: Miller et al., *Birds of Mexico*.

290 darker upperparts and smaller size: Nelson, E. 1900. "Descriptions of Thirty New North America Birds in the Biological Survey Collection." *Auk* 17: 253–270.

290 San Esteban: Van Rossem, A. 1930. "Four New Birds from Northwestern Mexico." *Transactions of the San Diego Society of Natural History* 6: 213–226.

290 Tortuga Island: Ibid.

290 Cerralvo (Jacques Cousteau) Island: Banks, R. 1963. "New Birds from Cerralvo Island, Baja California, Mexico." *Occasional Papers of the California Academy of Sciences* 37: 1–5.

FIVE-STRIPED SPARROW

291 "This well-marked species": Sclater, P., and O. Salvin. 1868. "Descriptions of New or Little-known American Birds of the Families Fringillidae, Oxyrhamphidae, Bucconidae, and Strigidae." *Proceedings of the Zoological Society of London* 1868: 321–329.

291 "Judging from the preparation of the skin": Salvin, O., and F. Godman. 1886. *Biologia centrali-americana: Aves*. Vol. 1. London: R.H. Porter.

291 Damiano Floresi d'Arcais: Palmer, T. 1928. "Notes on Persons Whose Names Appear in the Nomenclature of California Birds." *Condor* 30: 261–307.

291 sent to Panama: Wagner, M. 1870. "Über die hydrographischen Verhältnisse und das Vorkommen der Süßwasserfische in den Staaten Panama und Ecuador." *Abhandlungen der Königlichen Akademie der Wissenschaften* 10: 1–61.

291 without the use of mercury: Burkart, 1859. "Über den Bergwerkstbetrieb in den Revieren von Pachuca und Real del Monte in Mexico." *Zeitschrift für das Berg-, Hütten- und Salinenwesen in dem Preußischen Staate* 7: 101–168.

291 Imperial Woodpecker: Prŷs-Jones, R. 2011. "Type Specimens of the Imperial Woodpecker *Campephilus imperialis* (Gould, 1832)." *Bulletin of the British Ornithologists' Club* 131: 256–260.

291 "the type was received by Gould": Van Rossem, A. 1934. "Critical Notes on Middle American Birds." *Bulletin of the Museum of Comparative Zoology* 77: 385–490.

292 "Jalisco (Bolaños)": Miller, A., et al. 1957. *Distributional Checklist of the Birds of Mexico*. Part 2. Pacific Coast Avifauna 33.

292 browner than the male: Van Rossem, "Middle American Birds."

292 Wolf discovered similarities: Wolf, L. 1977. *Species Relationships in the Avian Genus* Aimophila. Ornithological Monographs 23. Lawrence: American Ornithologists' Union.

292 "an introductory note": Groschupf, K., and G. Mills. 1982. "Singing Behavior of the Five-striped Sparrow." *Condor* 84: 226–236.

292 variety of note groups: Ibid.

293 usual contact call: Ibid.

293 "slurred chatter": Wolf, *Species Relationships*.

293 series of rushed notes: Groschupf and Mills, "Singing Behavior."

293 collected in June 1957: Binford, L. 1958. "First Record of the Five-striped Sparrow in the United States." *Auk* 75: 103.

293 a systematic search: Mills, G. 1977. "New Locations for the Five-striped Sparrow in the United States." *Western Birds* 8: 121–130.

293 occupy a site for a few years: Groschupf, K. 1994. "Current Status of the Five-striped Sparrow in Arizona." *Western Birds* 25: 192–197.

293 the northerly subspecies: Van Rossem, "Middle American Birds."

293 Nominate *quinquestriata*: Howell, S., and S. Webb. 1995. *A Guide to the Birds of Mexico and Northern Central America*. Oxford: Oxford University Press.

293 southward movements in November: Phillips, A., and R. Phillips Farfan. 1993. "Distribution, Migration, Ecology, and Relationships of the Five-striped Sparrow, *Aimophila quinquestriata*." *Western Birds* 24: 65–72.

293 migrants and winterers: Phillips and Phillips Farfan, "Distribution, Migration, Ecology, and Relationships."

LARK SPARROW

294 "this beautiful species": Audubon, J. 1839. *Ornithological Biography*. Vol. 5. Edinburgh: Adam Black and Charles Black.

294 "they run upon the ground": James, E., ed. 1832. *Account of an Expedition From Pittsburgh to the Rocky Mountains*. Vol. 1. Philadelphia: H.C. Carey and I. Lea.

294 "badly selected": Coues, E. 1903. *Key to North American Birds*. 5th ed. Vol. 1. Boston: Dana Estes and Company.

294 helped Charles Bonaparte: Stroud, P. 1992. *Thomas Say: New World Naturalist*. Philadelphia: University of Pennsylvania Press.

294 the Lark Finch: Bonaparte, C. 1825. *American Ornithology, or, The Natural History of Birds Inhabiting the United States, Not Given by Wilson*. Vol. 1. Philadelphia: Carey, Lea and Carey.

294 "stripe-headed bunting": Maximilian zu Wied-Neuwied. 1858. "Verzeichniss der Vögel, welche auf einer Reise in Nord-America beobachtet wurden." *Journal für Ornithologie* 6: 1–29, 97–124, 177–204, 257–284, 337–354, 417–444.

295 in coastal Massachusetts: Putnam, F. 1856. "Catalogue of the Birds of Essex County, Massachusetts." *Proceedings of the Essex Institute* 1: 201–231.

295 "this species seems to be gradually extending": Ridgway, R. 1877. *Ornithology*. United States Geological Exploration of the Fortieth Parallel. Vol. 4, pt. 3. Washington, D.C.: Government Printing Office.

295 "a large number of the birds are annually destroyed": Coale, H. 1887. "Geographical Variations between *Chondestes grammacus* (Say) and *Chondestes grammacus strigatus* (Swains.)." *Bulletin of the Ridgway Ornithological Club* 2: 24–25.

295 without comment or qualification: Peterson, R. 1934. *A Field Guide to the Birds*. 1st ed. Boston: Houghton Mifflin.

295 "weedy fields": Peterson, R. 1947. *A Field Guide to the Birds*. 2nd ed. Boston: Houghton Mifflin.

295 last recorded breeding in Pennsylvania: Wilson, A., et al., eds. 2012. *Second Atlas of Breeding Birds in Pennsylvania*. University Park: Pennsylvania State University Press.

295 more than 40 Ohio counties: Rodewald, P., et al., eds. 2016. *The Second Atlas of Breeding Birds in Ohio*. University Park: Pennsylvania State University Press.

295 Indiana: Keller, C. 1979. *Indiana Birds and Their Haunts*. Bloomington: Indiana University Press.

295 "in only 2 percent": Brock, K. 2006. *Brock's Birds of Indiana*. Indianapolis: Amos Butler Audubon Society.

295 appropriate habitats in Illinois: Chapman, M. 2004. "Lark Sparrow Nests in DuPage County After 80-year Absence." *Meadowlark* 13: 16–17.

295 "If you are going to introduce a beginner": Beedy, C., and E. Pandolfino. 2013. *Birds of the Sierra Nevada: Their Natural History, Status, and Distribution*. Berkeley: University of California Press.

296 "quail head": Bailey, F. 1898. *Birds of Village and Field: A Bird Book for Beginners*. Boston: Houghton, Mifflin and Company.

296 "snake-bird": Baird, S., et al. 1874. *A History of North America Birds*. Vol. 1. Boston: Little, Brown.

296 second molt: Howell, S. 2010. *Peterson Reference Guide to Molt in North American Birds*. Boston: Houghton Mifflin Harcourt.

296 the English name: Bonaparte, *American Ornithology*.

296 greatly desired cage bird: Baird et al., *History*.

296 "the delightful song of this bird": Ibid.

296 differences in the sequence: Martin, J., and J. Parrish. 2000. "Lark Sparrow (*Chondestes grammacus*). In P. Rodewald, ed. *Birds of North America*. Ithaca: Cornell Lab of Ornithology.

297 bring to mind a wood warbler: Ibid.

297 North Carolina and northern South Carolina: McNair, D. 1990. "Lark Sparrows Breed at Rhine-Luzon Drop Zone, Camp MacKall, Scotland County, N.C." *The Chat* 54: 16–20.

297 low-density breeder in Kentucky: Palmer-Ball, B., ed. 1996. *The Kentucky Breeding Bird Atlas*. Lexington: University Press of Kentucky.

297 through central Tennessee: McNair, D. 2000. "Summary of Historical Breeding and Breeding-season Records of the Lark Sparrow in Tennessee." *The Migrant* 71: 73–78.

297 Alabama and Mississippi: Martin and Parrish, "Lark Sparrow (*Chondestes grammacus*).

297 Minnesota: Erickson, L. 2016. *American Birding Association Field Guide to Birds of Minnesota*. New York: Scott and Nix.

297 British Columbia: Campbell, R., et al. 2001. *The Birds of British Columbia*. Vol. 4. Vancouver: UBC Press.

297 in most of Texas: Lockwood, M., and B. Freeman. 2014. *The TOS Handbook of Texas Birds*. College Station: Texas A&M University Press.

297 Mexico: Howell, S., and S. Webb. 1995. *A Guide to the Birds of Mexico and Northern Central America*. Oxford: Oxford University Press.

297 Nevada: Floyd, T., et al. 2007. *Atlas of the Breeding Birds of Nevada*. Reno: University of Nevada Press.

297 lower Colorado River: Rosenberg, K., et al. 1991. *Birds of the Lower Colorado River Valley*. Tucson: University of Arizona Press.

297 Coachella and Imperial Valleys: Patten, M., et al. 2003. *Birds of the Salton Sea: Status, Biogeography, and Ecology*. Berkeley: University of California Press.

297 California: Small, A. 1994. *California Birds: Their Status and Distribution*. Vista, CA: Ibis.

297 east slope of the Sierra Nevada: Beedy and Pandolfino, *Sierra Nevada*.

297 San Diego County: Unitt, P. 2004. *San Diego County Bird Atlas*. Proceedings of the San Diego Society of Natural History 39. San Diego: San Diego Natural History Museum.

297 northern Baja California: Howell and Webb, *Mexico and Northern Central America*.

297 southbound migration: Martin and Parrish, "Lark Sparrow (*Chondestes grammacus*).

297 "nomadic tendencies": Ibid.

297 Florida peninsula: Greenlaw, J., et al. 2014. *The Robertson and Woolfenden Florida Bird Species: An Annotated List*. Florida Ornithological Society Special Publication 8. Gainesville: Florida Ornithological Society.

297 Great Basin: Martin and Parrish, "Lark Sparrow (*Chondestes grammacus*).

297 northbound movement: Ibid.

297 southern England: Lewington, I., et al. 1991. *A Field Guide to the Rare Birds of Britain and Europe*. London: HarperCollins.

297 by ship: Martin and Parrish, "Lark Sparrow (*Chondestes grammacus*).

297 consistent differences: Ridgway, R. 1880. "A Catalogue of the Birds of North America." *Proceedings of the United States National Museum* 3: 163–246.

297 on the Atlantic coast: American Ornithologists' Union. 1957. *Check-list of North American Birds*. 5th ed. Ithaca: American Ornithologists' Union.

LARK BUNTING

298 "The singing of meadow-larks": Bourke, J. G. 2003. *The Diaries of John Gregory Bourke*. Vol. 1. C. M. Robinson, ed. Denton, TX: University of North Texas Press.

298 The first European scientist: Maximilian zu Wied-Neuwied. 1839. *Reise in das innere Nord-America in den Jahren 1832 bis 1834*. Vol .1. Coblenz: J. Hölscher.

298 stridently point out: Maximilian zu Wied-Neuwied. 1858. "Verzeichniss der Vögel, welche auf einer Reise in Nord-America beobachtet wurden." *Journal für Ornithologie* 6: 1–29, 97–124, 177–204, 257–284, 337–354, 417–444.

298 *Fringilla bicolor*: Townsend, J. 1836. "Description of Twelve New Species of Birds Chiefly from the Vicinity of the Columbia River." *Proceedings of the Academy of Natural Sciences of Philadelphia* 7: 187–193.

298 "in its habits and behaviors": Maximilian zu Wied-Neuwied, *Reise*.

298 a single genus, *Dolichonyx*: Nuttall, T. 1840. *A Manual of the Ornithology of the United States and Canada*. 2nd ed. Boston: Hilliard, Gray, and Co.

298 "presents many of the habits": Townsend, "Twelve New Species."

298 "somewhat allied": Townsend, J. 1839. *Narrative of a Journey Across the Rocky Mountains to the Columbia River, and a Visit to the Sandwich Islands, Chili, etc.* Philadelphia: Henry Perkins.

298 to include only this species: Bonaparte, C. 1838. *A Geographic and Comparative List of the Birds of Europe and North America*. London: John van Voorst.

298 a genus of his own: Audubon, J. 1839. *A Synopsis of the Birds of North America*. Edinburgh: Adam and Charles Black.

298 new epithet for the bunting: Stejneger, L. 1885. "Analecta ornithologica: Fourth Series." *Auk* 2: 43–52.

301 celebrated for its flight song: Townsend, *Narrative of a Journey*.

301 "one of the sweetest songsters": Nuttall, *Manual*. 2nd ed.

301 another gifted singer: Coues, E. 1874. *Birds of the Northwest*. Miscellaneous Publications of the Geological Survey of the Territories 3. Washington: Government Printing Office.

301 "naturally applied to it the cognomen": Allen, J. 1872. "Notes of an Ornithological Reconnaissance of portions of Kansas, Colorado, Wyoming, and Utah." *Bulletin of the Museum of Comparative Zoology* 3: 113–183.

301 The phrase types: Stillwell, J., and N. Stillwell. 1955. "Notes on the Songs of Lark Buntings." *Wilson Bulletin* 67: 138–139.

301 summer range: Rising, J. 2016. "Lark Bunting (*Calamospiza melanocorys*)." In J. del Hoyo, et al., eds. *Handbook of the Birds of the World Alive*. Barcelona: Lynx Edicions.

301 where Nuttall shot the type specimens: Deignan, H. 1961. *Type Specimens of Birds in the United States National Museum*. Bulletin of the United States National Museum 221. Washington: Government Printing Office.

302 Texas: Lockwood, M., and B. Freeman. 2014. *The TOS Handbook of Texas Birds*. College Station: Texas A&M University Press.

302 Montana: Marks, J., et al. 2016. *Birds of Montana*. Arrington, Virginia: Buteo Books.

302 Colorado: Andrews, R., and R. Righter. 1992. *Colorado Birds: A Reference to Their Distribution and Habitat*. Denver: Denver Museum of Natural History.

302 eastern edge of the normal breeding range: Shane, T. 2000. "Lark Bunting (*Calamospiza melanocorys*)." In P. Rodewald, ed. *Birds of North America*. Ithaca: Cornell Lab of Ornithology.

302 irregularly as far west: Ibid.

302 northeastern Arizona: Corman, T., and C. Wise-Gervais, eds. 2005. *Arizona Breeding Bird Atlas*. Albuquerque: University of New Mexico Press.

302 San Bernardino County: Small, A. 1994. *California Birds: Their Status and Distribution*. Vista, CA: Ibis.

302 first arriving adults: Lockwood and Freeman, *TOS Handbook*.

302 peak southbound migration: Shane, "Lark Bunting (*Calamospiza melanocorys*)."

302 Males are the first: Ibid.

302 deserts of the American Southwest: Ibid.

302 interior southern California: Small, *California Birds*.

302 California coast: Ibid.

302 San Diego County: Unitt, P. 2004. *San Diego County Bird Atlas*. Proceedings of the San Diego Society of Natural History 39. San Diego: San Diego Natural History Museum.

302 Rio Grande Valley: Lockwood and Freeman, *TOS Handbook*.

302 "heaviest migration": Shane, "Lark Bunting (*Calamospiza melanocorys*)."

CHIPPING SPARROW

303 "the little house sparrow": Bartram, W. 1791. *Travels Through North and South Carolina, Georgia, East and West Florida, the Cherokee Country, the Extensive Territories of the Muscogulges, or Creek Confederacy, and the Country of the Chactaws*. Philadelphia: James and Johnson.

303 this "sociable" species: Wilson, A. 1810. *American Ornithology, or the Natural History of the Birds of the United States*. Vol. 2. Philadelphia: Bradford and Inskeep.

303 little more than a compilation: Audubon, J. 1834. *Ornithological Biography*. Vol. 2. Edinburgh: Adam and Charles Black.

303 Louis Pierre Vieillot: Oehser, P. 1948. "Louis Jean Pierre Vieillot (1748-1831 [sic])." *Auk* 65: 568–576.

303 survive in manuscript: n.a. 1989. *The Library of H. Bradley Martin: Highly Important Illustrated and Scientific Ornithology*. Sale 5953. New York: Sotheby's, n.a.

303 "This is one of the first": Vieillot, L. 1817. "La passerine des vergers ou le titit." *Nouveau dictionnaire d'histoire naturelle*. Vol. 25. Paris: Deterville.

304 broad gray collar: Lethaby, N. 2014. "Hind-neck Pattern: An Additional Aid in Identifying *Spizella* Sparrows in Fall and Winter." *Birding* 46: 36–39.

306 usual call note: Middleton, A. 1998. "Chipping Sparrow (*Spizella passerina*)." In P. Rodewald, ed. *Birds of North America*. Ithaca: Cornell Lab of Ornithology.

306 simple trilling song: Liu, Wan-Chun. 2001. "Development, Variation, and Use of Songs by Chipping Sparrows." Dissertation, University of Massachusetts–Amherst.

306 east of the Missouri River and north of Texas: Audubon, J. 1839. *Ornithological Biography*. Vol. 5. Edinburgh: Adam Black and Charles Black.

306 Canada: Middleton, "Chipping Sparrow (*Spizella passerina*)."

306 Nova Scotia: McLaren, I. 2012. *All the Birds of Nova Scotia: Status and Critical Identification*. Kentville: Gaspereau Press.

306 Alaska: Kessel, B., and D. Gibson. 1978. *Status and Distribution of Alaska Birds*. Studies in Avian Biology 1. Los Angeles: Cooper Ornithological Society.

306 Washington: Opperman, H. 2003. *A Birder's Guide to Washington*. Colorado Springs: American Birding Association.

306 California: Small, A. 1994. *California Birds: Their Status and Distribution*. Vista, CA: Ibis.

306 Piney Woods region: Lockwood, M., and B. Freeman. 2014. *The TOS Handbook of Texas Birds*. College Station: Texas A&M University Press.

306 Florida: Greenlaw, J., et al. 2014. *The Robertson and Woolfenden Florida Bird Species: An Annotated List*. Florida Ornithological Society Special Publication 8. Gainesville: Florida Ornithological Society.

306 west Texas: Lockwood and Freeman, *TOS Handbook*.

306 Mexico: Miller, A., et al. 1957. *Distributional Checklist of the Birds of Mexico*. Part 2. Pacific Coast Avifauna 33.

306 Belize and Honduras: Fagan, J., and O. Komar. 2016. *Peterson Field Guide to Birds of Northern Central America*. Boston: Houghton Mifflin Harcourt.

306 Guatemala: Eisermann and Avendaño, *Birds of Guatemala*.

306 "molt migration": Carlisle, J., et al. 2005. "Molt Strategies and Age Differences in Migration Timing Among Autumn Landbird Migrants in Southwestern Idaho." *Auk* 122: 1070–1085.

306 begins later in the East: Middleton, "Chipping Sparrow (*Spizella passerina*)."

306 Bering Sea islands: Lehman, P. 2005. "Fall Bird Migration at Gambell, St. Lawrence Island, Alaska." *Western Birds* 36: 2–55.

306 Wrangel Island: Arkhopiv, V., and Ł. Łuwacki. 2016. "Nearctic Passerines in Russia." *Dutch Birding* 38: 201–214.

306 winter north into the breeding range: Middleton, "Chipping Sparrow (*Spizella passerina*)."

307 south into Mexico: Miller, *Birds of Mexico*.

307 movement north: Ibid.

307 mitochondrial DNA: Zink, R., and D. Dittmann. 1993. "Population Structure and Gene Flow in the Chipping Sparrow and a Hypothesis for Evolution in the Genus *Spizella*." *Wilson Bulletin* 105: 399–413.

307 type locality of "Canada": Bechstein, J. 1798. *Johann Lathams allgemeine Uebersicht der Vögel*. Vol. 3, pt. 2. Nuremberg: A.C. Schneider und Wegel.

307 vicinity of Quebec City: Oberholser, H. 1955. "Description of a New Chipping Sparrow from Canada." *Journal of the Washington Academy of Sciences* 45: 59–60.

307 from Minnesota and Newfoundland: Middleton, "Chipping Sparrow (*Spizella passerina*)."

307 arizonae is larger and paler: Ridgway, R. 1901. *The Birds of North and Middle America*. Bulletin of the United States National Museum, Bulletin 50, pt. 1. Washington: Government Printing Office.

307 "curious" new subspecies: Coues, E. 1872. *Key to North American Birds*. 1st ed. Boston: Estes and Lauriat.

307 ear coverts of the western bird: Ridgway, *North and Middle America*.

307 meet on the eastern Great Plains: Coues, *Key*. 1st ed.

307 Birds of the Pacific Coast: Grinnell, J. 1927. "Designation of a Pacific Coast Subspecies of Chipping Sparrow." *Condor* 29: 81–82.

307 on the western Great Plains: Oberholser, "New Chipping Sparrow."

307 To the south: Moore, R. 1937. "New Races of *Myadestes, Spizella,* and *Turdus* from Northwestern Mexico." *Proceedings of the Biological Society of Washington* 50: 201–205.

307 race is resident: Middleton, "Chipping Sparrow (*Spizella passerina*)."

307 northern Transvolcanic Belt: Phillips, A. 1966. "Further Systematic Notes on Mexican Birds." *Bulletin of the British Ornithologists' Club* 86: 148–159.

307 across the Mexican highlands: Middleton, "Chipping Sparrow (*Spizella passerina*)."

307 more rufous feathering: Nelson, E. 1899. "Descriptions of New Birds from Mexico." *Auk* 16: 25–31.

307 Oaxaca and Guerrero: Phillips, "Further Systematic Notes."

307 southernmost of all: Salvin, O. 1863. "Descriptions of Thirteen New Species of Birds Discovered in Central America by Frederick Godman and Osbert Salvin." *Proceedings of the Zoological Society of London* 1863: 186–192.

307 with the Field: Baird, S., et al. 1874. *A History of North American Birds.* Vol. 2. Boston: Little, Brown.

307 in fact a Chipping Sparrow: Ridgway, R. 1884. "Notes on Three Guatemalan Birds." *Ibis* ser. 5, 2: 43–44.

307 "the color of the bill": Ridgway, *North and Middle America.*

307 northern El Salvador: Middleton, "Chipping Sparrow (*Spizella passerina*)."

CLAY-COLORED SPARROW

308 "puzzled by a singular song": Thompson, E. 1890. "The Birds of Manitoba." *Proceedings of the United States National Museum* 13: 457–643.

308 "rambling and foraging": Ibid.

308 "as familiar and confident": Richardson, J., and W. Swainson. 1831. *Fauna boreali-americana: Or the Zoology of the Northern Parts of British America.* Part 2. London: John Murray.

308 Audubon was able to depict: Audubon, J. 1839. *Ornithological Biography.* Vol. 5. Edinburgh: Adam Black and Charles Black.

309 "a mere synopsis": Audubon, "Missouri River Journals."

309 Audubon expressed some uncertainty: Ibid.

309 "handsome little species": Audubon, J. 1844. *The Birds of America.* 2nd ed. Vol. 7. Philadelphia: E.G. Dorsey.

309 the then undescribed Brewer Sparrow: Cassin, J. 1856. "Notes on North American Birds in the Collection of the Academy of Natural Sciences, Philadelphia, and National Museum, Washington." *Proceedings of the Academy of Natural Sciences of Philadelphia* 8: 39–42.

309 broad gray collar: Lethaby, N. 2014. "Hind-neck Pattern: An Additional Aid in Identifying *Spizella* Sparrows in Fall and Winter." *Birding* 46: 36–39.

310 other characters: Pyle, P., and S. Howell. 1996. "*Spizella* Sparrows: Intraspecific Variation and Identification." *Birding*: 374–387.

310 "dull" Clay-colored Sparrows: Ibid.

311 clearly "zoned" nape: Lethaby, "Hind-neck Pattern."

311 first pointed out by Peter Pyle: Pyle and Howell, "*Spizella* Sparrows."

311 British Columbia: Campbell, R., et al. 2001. *The Birds of British Columbia.* Vol. 4. Vancouver: UBC Press.

311 breeding range continues: Grant, T., and R. Knapton. 2012. "Clay-colored Sparrow (*Spizella pallida*)." In P. Rodewald, ed. *Birds of the North America.* Ithaca: Cornell Lab of Ornithology.

311 northern Ontario: Warburton, F. 1952. "Nesting of Clay-colored Sparrow, *Spizella pallida,* in Northern Ontario." *Auk* 69: 314–316.

312 southern Quebec: David, N. 1996. *Liste commentée des oiseaux du Québec.* Montréal: Association québécoise des groupes d'ornithologues.

312 New Brunswick or Nova Scotia: Stewart, R., ed. 2015. *Second Atlas of Breeding Birds of the Maritime Provinces.* Port Rowan: Bird Studies Canada.

312 eastern Washington: Opperman, H. 2003. *A Birder's Guide to Washington.* Colorado Springs: American Birding Association.

312 Idaho: Sturts, S. 2016. "All Records for 'Review Species.'" IdahoBirds.net.

312 New York: Smith, C. 2008. "Clay-colored Sparrow (*Spizella pallida*)." In K. McGowan and K. Corwin, eds. *The Second Atlas of Breeding Birds in New York State.* Ithaca: Cornell University Press.

312 New England: Keith, A., and R. Fox. 2013. *The Birds of New Hampshire.* Memoirs of the Nuttall Ornithological Club 19. Cambridge: Nuttall Ornithological Club.

312 Pennsylvania: Wilson, A., et al., eds. 2012. *Second Atlas of Breeding Birds in Pennsylvania.* University Park: Pennsylvania State University Press.

312 eastern Ohio: Rodewald, P., et al., eds. 2016. *The Second Atlas of Breeding Birds in Ohio.* University Park: Pennsylvania State University Press.

312 Indiana and Illinois: Grant and Knapton, "Clay-colored Sparrow (*Spizella pallida*)."

312 northern Iowa: Kent, T., and J. Dinsmore. 1996. *Birds in Iowa.* Iowa City: privately published.

312 Wyoming: Sharpe, R., et al. 2001. *Birds of Nebraska: Their Distribution and Temporal Occurrence.* Lincoln: University of Nebraska Press.

312 Colorado: Andrews, R., and R. Righter. 1992. *Colorado Birds: A Reference to Their Distribution and Habitat.* Denver: Denver Museum of Natural History.

312 autumn migration: Grant and Knapton, "Clay-colored Sparrow (*Spizella pallida*)."

312 Dry Tortugas: Greenlaw, J., et al. 2014. *The Robertson and Woolfenden Florida Bird Species: An Annotated List.* Florida Ornithological Society Special Publication 8. Gainesville: Florida Ornithological Society.

312 California: Hamilton, R., et al., eds. 2007. *Rare Birds of California.* Camarillo: Western Field Ornithologists.

312 Texas: Lockwood, M., and B. Freeman. 2014. *The TOS Handbook of Texas Birds.* College Station: Texas A&M University Press.

312 widespread in Mexico: Grant and Knapton, "Clay-colored Sparrow (*Spizella pallida*)."

312 Guatemala: Eisermann, K., and C. Avendaño. 2007. *Lista comentada de las aves de Guatemala / Annotated Checklist of the Birds of Guatemala.* Barcelona: Lynx Edicions.

312 Belize: Fagan, J., and O. Komar. 2016. *Peterson Field Guide to Birds of Northern Central America.* Boston: Houghton Mifflin Harcourt.

312 Cuba and the Bahamas: Raffaele, H., et al. 1998. *A Guide to the Birds of the West Indies.* Princeton: Princeton University Press.

312 "moving in waves": Grant and Knapton, "Clay-colored Sparrow (*Spizella pallida*)."

312 flocks numbering in the thousands: Ibid.

BLACK-CHINNED SPARROW

313 "In 1851, while on an expedition": Newman, J. 1968. "*Spizella atrogularis evura* Coues: Arizona Black-chinned Sparrow." In A. Bent, ed. *Life Histories of North American Cardinals, Grosbeaks, Buntings, Towhees, Finches, Sparrows, and Allies*. Pt. 2. United States National Museum Bulletin 237. Washington: Smithsonian Institution.

313 half a world away: Schalow, H. 1906. "Jean Cabanis." *Journal für Ornithologie* 54: 329–358.

313 "adult in abraded plumage": Van Rossem, A. 1935. "Notes on the Forms of *Spizella atrogularis*." *Condor* 37: 282–284.

313 "in very fair condition": Ibid.

313 "Thus, in the year 1838": Aschenborn, H. 1843. *De iure protimiseos et retractus secundum ius commune Borussicum*. Wrocław: Grassius, Barthius, et Socius.

313 "In my mind I hear you asking": Aschenborn, H. 1841. "Atlantisches." *Sundine: Unterhaltungsblatt für Neu-Vorpommern und Rügen* 15: 100–102.

314 label had been attached: Van Rossem, "Forms of *Spizella atrogularis*."

314 Couch's allocation of the species to *Struthus*: Couch, D. 1854. "Descriptions of New Birds of Northern Mexico." *Proceedings of the Academy of Natural Sciences of Philadelphia* 7: 66–67.

314 considerable geographic variation: Tenney, C. 1997. "Black-chinned Sparrow (*Spizella atrogularis*)." In P. Rodewald, ed. *Birds of North America*. Ithaca: Cornell Lab of Ornithology.

314 Another song: Ibid.

315 Oregon: Tweit, B., et al. 1999. "Oregon-Washington Region." *North American Birds* 53: 425–428.

315 locally in California's: Small, A. 1994. *California Birds: Their Status and Distribution*. Vista, CA: Ibis.

315 San Diego County: Unitt, P. 2004. *San Diego County Bird Atlas*. Proceedings of the San Diego Society of Natural History 39. San Diego: San Diego Natural History Museum.

315 the Sierra Nevada: Beedy, C., and E. Pandolfino. 2013. *Birds of the Sierra Nevada: Their Natural History, Status, and Distribution*. Berkeley: University of California Press.

315 northern Baja California: Miller, A., et al. 1957. *Distributional Checklist of the Birds of Mexico*. Part 2. Pacific Coast Avifauna 33.

315 Isla Cerralvo: Ruizcampos, G. 2013. "Annotated Checklist of the Birds of Baja California and Baja California Sur, Second Edition." *North American Birds* 66: 582–613.

315 "a sporadic, southerly distribution" in Nevada: Floyd, T., et al. 2007. *Atlas of the Breeding Birds of Nevada*. Reno: University of Nevada Press.

315 Colorado: Spencer, A. 2013. Black-chinned Sparrow in "Pictorial Highlights." *North American Birds* 66: 748.

315 New Mexico: Parmeter, J., et al. 2002. *New Mexico Bird Finding Guide*. Albuquerque: New Mexico Ornithological Society.

316 Arizona: Corman, T., and C. Wise-Gervais, eds. 2005. *Arizona Breeding Bird Atlas*. Albuquerque: University of New Mexico Press.

316 Texas's central Trans-Pecos region: Lockwood, M., and B. Freeman. 2014. *The TOS Handbook of Texas Birds*. College Station: Texas A&M University Press.

316 Sonora: Russell, S., and G. Monson. 1998. *The Birds of Sonora*. Tucson: University of Arizona Press.

316 Mexican Plateau: Miller et al., *Birds of Mexico*.

316 Oaxaca: Binford, L. 1989. *A Distributional Survey of the Birds of the Mexican State of Oaxaca*. Ornithological Monographs 43. Lawrence, KS: American Ornithologists' Union.

316 resident within its range: Tenney, "Black-chinned Sparrow (*Spizella atrogularis*)."

316 Texas records: Lockwood and Freeman, *TOS Handbook*.

316 California coast and islands: Tenney, "Black-chinned Sparrow (*Spizella atrogularis*)."

316 winter: Ibid.

316 Arizona's winterers: Monson, G., and A. Phillips. 1981. *Annotated Checklist of the Birds of Arizona*. 2nd ed. Tucson: University of Arizona Press.

316 lower Colorado River Valley: Rosenberg, K., et al. 1991. *Birds of the Lower Colorado River Valley*. Tucson: University of Arizona Press.

316 northbound migration: Tenney, "Black-chinned Sparrow (*Spizella atrogularis*)."

316 nominate subspecies: Van Rossem, "Forms of *Spizella atrogularis*."

317 females show a black chin: Tenney, "Black-chinned Sparrow (*Spizella atrogularis*)."

317 *caurina* ("northwestern"): Miller, A. 1929. "A New Race of Black-chinned Sparrow from the San Francisco Bay District." *Condor* 31: 205–207.

317 Pacific race, *cana*: Ibid.

317 washed with brownish: Tenney, "Black-chinned Sparrow (*Spizella atrogularis*)."

317 fourth race, *evura*: Ibid.

317 overlap in western Arizona: Monson and Phillips, *Annotated Checklist*. 2nd ed.

WORTHEN SPARROW

318 "few of our birds have a briefer history": Chapman, F. 1914. "Notes on the Plumage of North American Sparrows." *Bird-Lore* 16: 352.

318 sent professional collectors: Thayer, J. 1925. "The Nesting of the Worthen Sparrow in Tamaulipas, Mexico." *Condor* 27: 34.

318 "the most successful collector in Texas": Casto, S. 1994. "The Ornithological Collections of Frank B. Armstrong." *Bulletin of the Texas Ornithological Society* 27: 8–18.

318 western race *arenacea* of the Field Sparrow: Chadbourne, A. 1886. "On a New Race of the Field Sparrow from Texas." *Auk* 3: 248–249.

318 eight specimens known at the time: Wege, C., et al. 1993. "The Distribution and Status of Worthen's Sparrow *Spizella entheni*: A Review." *Bird Conservation International* 3: 211–220.

318 "on the flat near the town": Marsh, C. 1885. "A New Species of Field Sparrow." *Ornithologist and Oölogist* 10: 5.

318 "Territorial Taxidermist": Marsh, C. 1885. Classified ad in *The Agassiz Association Journal* 1: 55.

318 "any sign of the Worthen Sparrow": Thayer, "Nesting of the Worthen Sparrow."

318 essential unanimity today: Hubbard, J., and C. Dove. 2013. "Our Reevaluation of a Worthen's Sparrow and Three Other Anomalous Specimens from Charles H. Marsh's 1884-1886 Collections of Birds from Southwestern New Mexico." tinyurl.com/MarshWorthen.

318 "his tastes led him": Allen, J. 1909. "Charles K. Worthen." *Auk* 26: 332.

318 exchanged 118 bird specimens: Ridgway, R. 1885. "Report Upon the Department of Birds in the U.S. Museum, 1884." *Annual Report of the Board of Regents of the Smithsonian Institution*. Pt. 2. Washington: Government Printing Office.

318   the type of the new sparrow: Ridgway, R. 1884.
      "Description of a New Species of Field-Sparrow from
      New Mexico." *Proceedings of the United States National
      Museum* 7: 259.

318   "always intelligent and trustworthy": Allen, "Charles K.
      Worthen."

319   removed the original tags: Hubbard and Dove. 2013.
      "Reevaluation of a Worthen's Sparrow.

319   labels bearing his own name: Hoffmeister, D. 2002.
      Mammals of Illinois. Urbana: University of Illinois
      Press.

319   original collector of the sparrow: Hubbard and Dove,
      "Reevaluation of a Worthen's Sparrow."

319   "kindly presented": Ridgway, "New Species of
      Field-Sparrow."

319   his own notice of the new species: Marsh, "New Species
      of Field Sparrow."

319   accusation of "plagiarism": Hubbard and Dove, "Reeval-
      uation of a Worthen's Sparrow."

319   "very glad to give Mr. Marsh due credit": Ridgway, R.
      1885. "Editor's Notes." *Ornithologist and Oölogist* 10:
      24.

319   "no bad feelings in the matter": Hubbard and Dove,
      "Reevaluation of a Worthen's Sparrow."

319   the two species are "quite distinct": Coues, E. 1903. *Key
      to North American Birds*. 5th ed. Vol. 1. Boston: Dana
      Estes and Company.

319   so similar as to be indistinguishable: Burleigh, T., and G.
      Lowery. 1942. *Notes on the Birds of Southeastern Coa-
      huila*. Occasional Papers of the Museum of Zoology.
      Baton Rouge: Louisiana State University Press.

320   only weak wing bars: Webster, J., and R. Orr. 1954.
      "Summering Birds of Zacatecas, Mexico, with a
      Description of a New Race of Worthen Sparrow."
      Condor 56: 155–160.

320   tarsus and toes: Behrstock, R., et al. 1997. "First Nest-
      ing Records of Worthen's Sparrow *Spizella wortheni* for
      Nuevo Léon, Mexico, with a Habitat Characterisation of
      the Nest Site and Notes on Ecology, Voice, Additional
      Recent Sightings, and Leg Coloration." *Cotinga* 8:
      27–33.

320   "consists of a thin, introductory *seep*": Ibid.

320   slur the introductory note: Webster and Orr, "Summer-
      ing Birds of Zacatecas."

320   similarity to juncos: Howell, S., and S. Webb. 1995.
      *A Guide to the Birds of Mexico and Northern Central
      America*. Oxford: Oxford University Press.

320   steady, even rhythm: Webster and Orr, "Summering
      Birds of Zacatecas."

320   "a high, thin, fairly dry *tssip*": Howell and Webb, *Mexico
      and Northern Central America*.

320   once present in Chihuahua: Wege et al., "Distribution
      and Status."

320   "no adequate evidence" of long-distance migration:
      Ibid.

320   the race *browni*: Webster and Orr, "Summering Birds of
      Zacatecas."

TIMBERLINE SPARROW

321   "almost impossible": Kaufman, K. 2011. *Kaufman Field
      Guide to Advanced Birding*. Boston: Houghton Mifflin
      Harcourt.

321   "In 1924, Allan Brooks joined Harry Swarth": Candy,
      R., and R. Campbell. 2012. "Allan Brooks: Naturalist
      and Wildlife Illustrator (1869-1946)." *Wildlife Afield* 9:
      88–106.

321   described the Timberline Sparrow: Swarth, H., and A.
      Brooks. 1925. "The Timberline Sparrow: A New Species
      from Northwestern Canada." *Condor* 27: 67–69.

321   first migrant specimen: Grinnell, J. 1932. "An [sic]
      United States Record of the Timberline Sparrow." *Con-
      dor* 35: 231–232.

321   "a previously unheard sparrow": Rea, A. 1967. "Some
      Bird Records from San Diego, California." *Condor* 69:
      316–318.

322   slightly larger than the Brewer Sparrow: Swarth and
      Brooks, "Timberline Sparrow."

322   head markings: Ibid.

322   eye-ring: Pyle, P., and S. Howell. 1996. "*Spizella* Spar-
      rows: Intraspecific Variation and Identification."
      Birding: 374–387.

322   "no specimen of equivocal character": Swarth and
      Brooks, "Timberline Sparrow."

322   On average: Pyle and Howell, "*Spizella* Sparrows."

322   "the possibility of overlap": Lethaby, N. 2014. "Hind-
      neck Pattern: An Additional Aid in Identifying *Spizella*
      Sparrows in Fall and Winter." *Birding* 46: 36–39.

322   Juvenile Timberline Sparrows: Swarth and Brooks,
      "Timberline Sparrow."

322   two song "types": Rich, T. 2002. "The Short Song of
      Brewer's Sparrow: Individual and Geographic Variation
      in Southern Idaho." *Western North American Naturalist*
      62: 288–299.

322   "more musical and tinkling": Spencer, A. 2014. "How to
      Identify a Timberline Sparrow." *Earbirding*. earbirding.
      com/blog/archives/4764.

322   "a series of descending sweet notes": Doyle, T. 1997.
      "The Timberline Sparrow, *Spizella (breweri) taverneri*,
      in Alaska, with Notes on Breeding Habitat and Vocal-
      izations." *Western Birds* 28: 1–12.

323   Alaska: Ibid.

323   first Timberline Sparrows: Swarth and Brooks, "Tim-
      berline Sparrow."

323   British Columbia: Campbell, R., et al. 2001. *The Birds of
      British Columbia*. Vol. 4. Vancouver: UBC Press.

323   Alberta: Semenchuk, G., ed. 1992. *The Atlas of Breed-
      ing Birds of Alberta*. Edmonton: Federation of Alberta
      Naturalists.

323   Montana: Marks, J., et al. 2016. *Birds of Montana*.
      Arrington, Virginia: Buteo Books.

323   Wyoming: Hansley, P., and G. Beauvais. 2004. *Species
      Assessment for Brewer's Sparrow (Spizella breweri) in
      Wyoming*. Cheyenne: Bureau of Land Management.

323   Colorado: Semo, L., and B. Percival. 2004. "News From
      the Field." *Colorado Birds* 38: 35–52.

323   northern California: Pyle and Howell, "*Spizella*
      Sparrows."

323   leave British Columbia: Doyle, "Timberline Sparrow,
      Spizella (breweri) taverneri."

323   New Mexico specimen: Grinnell, "Record of the Tim-
      berline Sparrow."

323   portions of west Texas: Lockwood, M., and B. Freeman.
      2014. *The TOS Handbook of Texas Birds*. College Sta-
      tion: Texas A&M University Press.

323   San Diego County: Unitt, P. 2004. *San Diego County
      Bird Atlas*. Proceedings of the San Diego Society of
      Natural History 39. San Diego: San Diego Natural His-
      tory Museum.

323   The only Arizona specimens: Monson, G., and A. Phil-
      lips. 1981. *Annotated Checklist of the Birds of Arizona*.
      2nd ed. Tucson: University of Arizona Press.

323 Washington in April and early May: Opperman, H. 2003. *A Birder's Guide to Washington*. Colorado Springs: American Birding Association.

323 arrive on the Alaskan breeding grounds: Campbell et al., *British Columbia*.

323 Brewer Sparrows are fledging young: Klicka, J., et al. 1999. "Evidence Supporting the Recent Origin and Species Status of the Timberline Sparrow." *Condor* 101: 577–588.

BREWER SPARROW

324 "Sparrow Wars": Coates, P. 2006. *American Perceptions of Immigrant and Native Species: Strangers on the Land*. Berkeley: University of California Press.

324 "Everybody knows": Coues, E. 1897. "The Documents in the Bendire Business." *Osprey* 2: 22–23.

324 acrimonious climax: Coates, *Immigrant and Native Species*.

324 praised his work: Coues, E. 1884. *Key to North American Birds*. 2nd ed. Boston: Estes and Lauriat.

324 "long the leading oölogist": Coues, E. 1882. *The Coues Check List of North American Birds*. 2nd ed. Boston: Estes and Lauriat.

324 "the highest abilities and social qualities": Cassin, J. 1856. "Notes on North American Birds in the Collection of the Academy of Natural Sciences, Philadelphia, and National Museum, Washington." *Proceedings of the Academy of Natural Sciences of Philadelphia* 8: 39–42.

324 "Dr. T.M. Brewer": Coues, E. 1903. *Key to North American Birds*. 5th ed. Boston: D. Estes.

324 Coues fell into the same error: Marks, J., et al. 2016. *Birds of Montana*. Arrington, Virginia: Buteo Books.

325 gray on the neck sides and nape: Lethaby, N. 2014. "Hind-neck Pattern: An Additional Aid in Identifying *Spizella* Sparrows in Fall and Winter." *Birding* 46: 36–39.

326 "soft twittering": Rotenberry, T., et al. 1999. "Brewer's Sparrow (*Spizella breweri*)." In P. Rodewald, ed. *Birds of North America*. Ithaca: Cornell Lab of Ornithology.

326 "The song of Brewer's Sparrow": Baird, S., et al. 1874. *A History of North American Birds*. Vol. 2. Boston: Little, Brown.

326 two song "types": Rich, T. 2002. "The Short Song of Brewer's Sparrow: Individual and Geographic Variation in Southern Idaho." *Western North American Naturalist* 62: 288–299.

326 "a series of sweet descending notes": Doyle, T. 1997. "The Timberline Sparrow, *Spizella (breweri) taverneri*, in Alaska, with Notes on Breeding Habitat and Vocalizations." *Western Birds* 28: 1–12.

326 short song: Rotenberry et al., "Brewer's Sparrow (*Spizella breweri*)."

326 British Columbia's breeding Brewer Sparrows: Campbell, R., et al. 2001. *The Birds of British Columbia*. Vol. 4. Vancouver: UBC Press.

327 White Lake and Richter Pass: Cannings, R., et al. 1987 *Birds of the Okanagan Valley, British Columbia*. Victoria: Royal British Columbia Museum.

327 Alberta: Semenchuk, G., ed. 1992. *The Atlas of Breeding Birds of Alberta*. Edmonton: Federation of Alberta Naturalists.

327 Washington: Opperman, H. 2003. *A Birder's Guide to Washington*. Colorado Springs: American Birding Association.

327 Oregon: Oregon Department of Fish and Wildlife. 2016. "Oregon Wildlife Species." Tinyurl.com/oregonbrsp.

327 Arizona: Corman, T., and C. Wise-Gervais, eds. 2005. *Arizona Breeding Bird Atlas*. Albuquerque: University of New Mexico Press.

327 California: Small, A. 1994. *California Birds: Their Status and Distribution*. Vista, CA: Ibis.

327 West of the Sierra crest: Beedy, C., and E. Pandolfino. 2013. *Birds of the Sierra Nevada: Their Natural History, Status, and Distribution*. Berkeley: University of California Press.

327 San Diego County: Unitt, P. 2004. *San Diego County Bird Atlas*. Proceedings of the San Diego Society of Natural History 39. San Diego: San Diego Natural History Museum.

327 Colorado: Wickersham, L., ed. 2016. *The Second Colorado Breeding Bird Atlas*. Denver: Colorado Division of Wildlife.

328 North Dakota: North Dakota Game and Fish Department. 2016. "North Dakota Grassland Birds." gf.nd.gov/wildlife/id/grasslandbirds.

328 South Dakota: Peterson, R. 1995. *The South Dakota Breeding Bird Atlas*. Brookings: South Dakota Ornithologists' Union.

328 Nebraska: Mollhoff, W. 2016. *The Second Nebraska Breeding Bird Atlas*. Bulletin of the University of Nebraska State Museum 29. Lincoln: University of Nebraska State Museum.

328 in central Nebraska is not credible: Sharpe, R., et al. 2001. *Birds of Nebraska: Their Distribution and Temporal Occurrence*. Lincoln: University of Nebraska Press.

328 Kansas: Holmes, J., and M. Johnson. 2005. *Brewer's Sparrow* (Spizella breweri*): A Technical Conservation Assessment*. Denver: USDA Forest Service, Rocky Mountain Region.

328 Texas Panhandle: Tweit, R. 2008. "Brewer's Sparrow." In K. Benson and K. Arnold, eds. *The Texas Breeding Bird Atlas*. College Station: Texas A&M University Press. txtbba.tamu.edu.

328 Autumn migration: Rotenberry et al., "Brewer's Sparrow (*Spizella breweri*)."

328 California coast: Unitt, *San Diego County Bird Atlas*.

328 Nova Scotia: McLaren, I. 2012. *All the Birds of Nova Scotia: Status and Critical Identification*. Kentville: Gaspereau Press.

328 lower Rio Grande Valley: Holmes and Johnson, *Brewer's Sparrow* (Spizella breweri*).

328 Mexico: Miller, A., et al. 1957. *Distributional Checklist of the Birds of Mexico*. Part 2. Pacific Coast Avifauna 33.

328 Massachusetts: Veit, R., and W. Petersen. 1993. *Birds of Massachusetts*. Lincoln: Massachusetts Audubon Society.

328 eastern Virginia: Virginia Avian Records Committee. 2016. "Database for Ornithological Verification and Submission." vsodoves.org.

328 Great Basin breeding grounds: Rotenberry et al., "Brewer's Sparrow (*Spizella breweri*)."

328 Alberta: Ibid.

328 more heavily marked: Pyle, P., and S. Howell. 1996. "*Spizella* Sparrows: Intraspecific Variation and Identification." *Birding*: 374–387.

328 "contentious": Rotenberry et al., "Brewer's Sparrow (*Spizella breweri*)."

328 an undescribed race: Rea, A. 1983. *Once a River: Bird Life and Habitat Changes on the Middle Gila*. Tucson: University of Arizona Press.

FIELD SPARROW

329 "surely a fine singer": Cheney, S. 1892. *Wood Notes Wild: Notations of Bird Music*. Boston: Lee and Shepard.

329 Mark Catesby: Catesby, M. 1731. *The Natural History of Carolina, Florida and the Bahama Islands*. Vol. 1. London: privately published.

329 Vieillot: Bonnaterre, P., and L. Vieillot. 1823. *Tableau encyclopédique et méthodique des trois règnes de la nature.* Pt. 3, Ornithologie. Paris: Agasse.

329 "has no song": Wilson, A. 1810. *American Ornithology, or the Natural History of the Birds of the United States.* Vol. 2. Philadelphia: Bradford and Inskeep.

329 Bonaparte: Bonaparte, C. 1826. *Observations on the Nomenclature of Wilson's Ornithology.* Philadelphia: Anthony Finley.

329 "remarkable, although not fine": Audubon, J. 1834. *Ornithological Biography.* Vol. 2. Edinburgh: Adam and Charles Black.

329 "Our little bird": Nuttall, T. 1832. *A Manual of the Ornithology of the United States and of Canada.* 1st ed. Vol. 1. Cambridge: Hilliard and Brown.

330 "a very varied and fine singer": Baird, S., et al. 1874. *A History of North American Birds.* Vol. 2. Boston: Little, Brown.

330 color of tarsus and toes: Behrstock, R., et al. 1997. "First Nesting Records of Worthen's Sparrow *Spizella wortheni* for Nuevo Léon, Mexico, with a Habitat Characterisation of the Nest Site and Notes on Ecology, Voice, Additional Recent Sightings, and Leg Coloration." *Cotinga* 8: 27–33.

331 Field Sparrows sing two songs: Carey, M., et al. 2008. "Field Sparrow (*Spizella pusilla*)." In P. Rodewald, ed. *Birds of North America.* Ithaca: Cornell Lab of Ornithology.

331 "signal[ing] heightened aggressive tendencies": Ibid.

331 Alberta: Semenchuk, G., ed. 2007. *The Atlas of Breeding Birds of Alberta: A Second Look.* Calgary: Federation of Alberta Naturalists.

331 Montana: Marks, J., et al. 2016. *Birds of Montana.* Arrington, Virginia: Buteo Books.

331 Colorado: Andrews, R., and R. Righter. 1992. *Colorado Birds: A Reference to Their Distribution and Habitat.* Denver: Denver Museum of Natural History.

331 Texas: Lockwood, M., and B. Freeman. 2014. *The TOS Handbook of Texas Birds.* College Station: Texas A&M University Press.

331 northern edge: Carey et al., "Field Sparrow (*Spizella pusilla*)."

331 Ontario: Cadmon, M., et al., eds. *Atlas of the Breeding Birds of Ontario 2001-2005.* Toronto: Bird Studies Canada.

331 Quebec: Carey et al., "Field Sparrow (*Spizella pusilla*)."

331 New Brunswick: McLaren, I. 2012. *All the Birds of Nova Scotia: Status and Critical Identification.* Kentville: Gaspereau Press.

331 Florida: Greenlaw, J., et al. 2014. *The Robertson and Woolfenden Florida Bird Species: An Annotated List.* Florida Ornithological Society Special Publication 8. Gainesville: Florida Ornithological Society.

331 lingering through the winter: Hamilton, R., et al., eds. 2007. *Rare Birds of California.* Camarillo: Western Field Ornithologists.

332 Wintering Field Sparrows: Carey et al., "Field Sparrow (*Spizella pusilla*)."

332 The southern limit: Miller, A., et al. 1957. *Distributional Checklist of the Birds of Mexico.* Part 2. Pacific Coast Avifauna 33.

332 Spring arrivals: Carey et al., "Field Sparrow (*Spizella pusilla*)."

332 Chadbourne formally described the new subspecies: Chadbourne, A. 1886. "On a New Race of the Field Sparrow from Texas." *Auk* 3: 248–249.

332 clinal in size and in plumage tone: Carey et al., "Field Sparrow (*Spizella pusilla*)."

GRASSHOPPER SPARROW

333 "lukewarm" nature of "the slender countenance": Grosart, A. 1876. *Memoir and Remains of Alexander Wilson, the American Ornithologist.* Vol. 1. Paisley: Alexander Gardner.

333 "This small species": Wilson, A. 1811. *American Ornithology, or the Natural History of the Birds of the United States.* Vol. 3. Philadelphia: Bradford and Inskeep.

333 Hans Sloane: Sloane, H. 1725. *A Voyage to the Islands Madera, Barbadoes, Nieves, St. Christophers, and Jamaica.* Vol. 2. London: privately published.

333 Ray's *Synopsis methodica*: Ray, J. 1713. *Synopsis methodica avium et piscium.* Vol. 1. London: William Innys.

333 "They sit on the Ground in the Plains": Sloane, *Voyage to the Islands.*

333 Louis Pierre Vieillot: Vieillot, L. 1817. "La Passerine des prés, *Passerina pratensis.*" *Nouveau dictionnaire d'histoire naturelle.* Vol. 25. Paris: Déterville.

334 "He is . . . censurable": Bonaparte, C. 1826. *Observations on the Nomenclature of Wilson's Ornithology.* Philadelphia: Anthony Finley.

334 Bonaparte was himself uncertain: Ibid.

334 "a rather small lark": Sloane, *Voyage to the Islands.*

334 "it is not a lark": Klein, J. 1750. *Historiae avium prodromus.* Lübeck: Jonas Schmidt.

334 Mathurin Brisson retained the bird: Brisson, M. 1760. *Ornithologie, ou Méthode contenant la division des oiseaux.* Vol. 3. Paris: Jean-Baptiste Bauche.

334 name coined by Johann Friedrich Gmelin: Gmelin, J. 1788. *Caroli a Linné . . . Systema naturae per regna tria naturae.* Vol. 1. Leipzig: Georg Emanuel Beer.

334 Thomas Nuttall: Nuttall, T. 1832. *A Manual of the Ornithology of the United States and of Canada.* 1st ed. Vol. 1. Cambridge: Hilliard and Brown.

334 adopted by the AOU: Ridgway, R. 1885. "Some Emended Names of North American Birds." *Proceedings of the United States National Museum* 8: 354–356.

335 only the high call notes: Vieillot, "Passerine des prés."

335 "It has a short, weak interrupted chirrup": Wilson, *American Ornithology.*

335 "They perch in sheltered trees": Nuttall, *Manual.* 1st ed.

335 "an unmusical ditty": Audubon, J. 1834. *Ornithological Biography.* Vol. 2. Edinburgh: Adam and Charles Black.

335 Spencer Trotter: Trotter, S. 1910. "An Inquiry into the History of the Current English Names of North American Land Birds." *Annual Report of the Board of Regents of the Smithsonian Institution for the Year 1909:* 505–519.

335 "a humble effort": Coues, E. 1874. *Birds of the Northwest.* Miscellaneous Publications of the Geological Survey of the Territories 3. Washington: Government Printing Office.

335 "a close resemblance to the note of a grasshopper": Baird, S., et al. 1874. *A History of North America Birds.* Vol. 1. Boston: Little, Brown.

336 the English name "Grasshopper Sparrow": Coues, E. 1884. *Key to North American Birds.* 2nd ed. Boston: Estes and Lauriat.

336 "sustained song": Vickery, P. 1996. "Grasshopper Sparrow (*Ammodramus savannarum*)." In P. Rodewald, ed. *Birds of North America.* Ithaca: Cornell Lab of Ornithology.

336 a short trill: Ibid.

336 Widespread but very local: Ibid.

336 British Columbia: Campbell, R., et al. 2001. *The Birds of British Columbia.* Vol. 4. Vancouver: UBC Press.

336 eastern Washington and northeastern Oregon: Opperman, H. 2003. *A Birder's Guide to Washington.* Colorado Springs: American Birding Association.

336 west slope of the Sierra Nevada: Beedy, C., and E. Pandolfino. 2013. *Birds of the Sierra Nevada: Their Natural History, Status, and Distribution.* Berkeley: University of California Press.

336 California: Shuford, D., and T. Gardall, eds. 2008. *California Bird Species of Special Concern.* Studies of Western Birds 1. Camarillo: Western Field Ornithologists.

336 Baja California: Vickery, "Grasshopper Sparrow (*Ammodramus savannarum*)."

336 Nevada: Floyd, T., et al. 2007. *Atlas of the Breeding Birds of Nevada.* Reno: University of Nevada Press.

336 Utah, western Wyoming, and northwestern Colorado: Vickery, "Grasshopper Sparrow (*Ammodramus savannarum*)."

336 Montana: Marks, J., et al. 2016. *Birds of Montana.* Arrington, Virginia: Buteo Books.

336 Yavapai County, Arizona: Corman, T., and C. Wise-Gervais, eds. 2005. *Arizona Breeding Bird Atlas.* Albuquerque: University of New Mexico Press.

336 Texas: Lockwood, M., and B. Freeman. 2014. *The TOS Handbook of Texas Birds.* College Station: Texas A&M University Press.

336 and New Hampshire: Keith, A., and R. Fox. 2013. *The Birds of New Hampshire.* Memoirs of the Nuttall Ornithological Club 19. Cambridge: Nuttall Ornithological Club.

336 southwestern subspecies *ammolegus*: Corman and Wise-Gervais, *Arizona Breeding Bird Atlas.*

337 Florida Grasshopper Sparrow: Greenlaw, J., et al. 2014. *The Robertson and Woolfenden Florida Bird Species: An Annotated List.* Florida Ornithological Society Special Publication 8. Gainesville: Florida Ornithological Society.

337 Autumn departures: Vickery, "Grasshopper Sparrow (*Ammodramus savannarum*)."

337 Nova Scotia: McLaren, I. 2012. *All the Birds of Nova Scotia: Status and Critical Identification.* Kentville: Gaspereau Press.

337 Bermuda: Vickery, "Grasshopper Sparrow (*Ammodramus savannarum*)."

337 winterers: Vickery, "Grasshopper Sparrow (*Ammodramus savannarum*)."

337 Costa Rica: Fagan, J., and O. Komar. 2016. *Peterson Field Guide to Birds of Northern Central America.* Boston: Houghton Mifflin Harcourt.

337 Panama: Olson, S. 1980. "The Subspecies of Grasshopper Sparrow (*Ammodramus savannarum*) in Panamá (Aves: Emberizinae)." *Proceedings of the Biological Society of Washington* 93: 757–759.

337 Mexican winterers: Miller, A., et al. 1957. *Distributional Checklist of the Birds of Mexico.* Part 2. Pacific Coast Avifauna 33.

337 May in British Columbia: Campbell et al., *British Columbia.*

337 western Nebraska: Sharpe, R., et al. 2001. *Birds of Nebraska: Their Distribution and Temporal Occurrence.* Lincoln: University of Nebraska Press.

337 New Hampshire: Keith, A., and R. Fox. 2013. *The Birds of New Hampshire.* Memoirs of the Nuttall Ornithological Club 19. Cambridge: Nuttall Ornithological Club.

337 *Handbooks of Birds of the World*: Rising, J. 2011. "Grasshopper Sparrow (*Ammodramus savannarum*)." In J. del Hoyo, et al., eds. *Handbook of Birds of the World.* Vol. 16. Barcelona: Lynx Edicions.

337 "slightly darker": Ridgway, R. 1901. *The Birds of North and Middle America.* Bulletin of the United States National Museum, Bulletin 50, pt. 1. Washington: Government Printing Office.

337 "paler" on the back: Vickery, "Grasshopper Sparrow (*Ammodramus savannarum*)."

337 *borinquensis*: Peters, J. 1917. "The Porto Rican Grasshopper Sparrow." *Proceedings of the Biological Society of Washington* 30: 95–96.

337 *intricatus* is larger-billed and darker: Hartert, E. 1907. "A New Subspecies of *Ammodramus.*" *Bulletin of the British Ornithologists' Club* 19: 73–74.

337 *caribaeus*: Hartert, E. 1902. "Die mit Sicherheit festgestellten Vögel der Inseln Aruba, Curaçao und Bonaire." *Novitates zoologicae* 9: 295–309.

337 *caucae* is dark: Chapman, F. 1912. "Diagnoses of Apparently New Colombian Birds." *Bulletin of the American Museum of Natural History* 31: 139–166.

338 *beatriceae* . . . is the palest: Olson, "Grasshopper Sparrow (*Ammodramus savannarum*) in Panamá."

338 *bimaculatus*, first described: Ridgway, *North and Middle America.*

338 Zacatecas: Miller et al., *Birds of Mexico.*

338 long-winged and long-tailed: Ridgway, *North and Middle America.*

338 "difficult to understand": Van Rossem, A. 1934. "Notes on Some Types of North American Birds." *Transactions of the San Diego Society of Natural History* 7: 349–361.

338 *perpallidus* is longer-winged and thinner-billed: Oberholser, H. 1974. *The Bird Life of Texas.* Vol. 2. Austin: University of Texas Press.

338 Florida Grasshopper Sparrow: Mearns, E. 1902. "Descriptions of Three New Birds from the Southern United States." *Proceedings of the United States National Museum* 24: 915–919.

338 extremely limited south-central Florida range of *floridanus*: Greenlaw et al., *Florida Bird Species.*

338 very locally distributed *ammolegus*: Oberholser, H. 1942. "Description of a New Arizona Race of the Grasshopper Sparrow." *Proceedings of the Biological Society of Washington* 55: 15–16.

OLIVE SPARROW

339 described in 1851: Lawrence, G. 1851. "Descriptions of New Species of Birds of the Genera *Conirostrum,* D'Orb. et Lafr., *Embernagra,* Less., and *Xanthornus,* Briss., Together with a List of Other Species Not Heretofore Noticed As Being Found within the Limits of the United States." *Annals of the Lyceum of Natural History of New York* 5: 112–117.

339 collected at Fort Ringgold: American Ornithologists' Union. 1998. *Check-list of North American Birds.* 7th ed. Ithaca: American Ornithologists' Union.

339 Nuevo León: Baird, S., with J. Cassin and G. Lawrence. 1858. *Birds.* Explorations and Surveys for a Railroad Route vol. 9. Washington: Beverly Tucker.

339 "birds abundant": Chapman, F. 1891. "On the Birds Observed near Corpus Christi, Texas, During Parts of March and April 1891." *Bulletin of the American Museum of Natural History* 3: 315–328.

339 Samuel Nicholson Rhoads: Rhoads, N. 1892. "The Birds of Southeastern Texas and Southern Arizona Observed During May, June, and July 1891." *Proceedings of the Academy of Natural Sciences of Philadelphia* 44: 98.

340 northern Guatemala: Eisermann, K., and C. Avendaño. 2007. *Lista comentada de las aves de Guatemala / Annotated Checklist of the Birds of Guatemala.* Barcelona: Lynx Edicions.

340 Texas: Lockwood, M., and B. Freeman. 2014. *The TOS Handbook of Texas Birds*. College Station: Texas A&M University Press.

341 short-distance movements: Brush, T. 2013. "Olive Sparrow (*Arremonops rufivirgatus*)." In P. Rodewald, ed. *Birds of North America*. Ithaca: Cornell Lab of Ornithology.

341 nominate race: Ibid.

341 *crassirostris*: Ridgway, R. 1901. *The Birds of North and Middle America*. Bulletin of the United States National Museum, Bulletin 50, pt. 1. Washington: Government Printing Office.

341 full specific rank: Salvin, O., and F. Godman. 1886. *Biologia centrali-americana: Aves*. Vol. 1. London: R. H. Porter.

341 intergrades: Brush, "Olive Sparrow (*Arremonops rufivirgatus*)."

341 *sinaloae*: Ridgway, *North and Middle America*.

341 *sumichrasti*: Ibid.

341 *chiapensis*: Nelson, E. 1904. "Descriptions of Four New Birds from Mexico." *Proceedings of the Biological Society of Washington* 17: 151–152.

341 *superciliosus*: Brush, "Olive Sparrow (*Arremonops rufivirgatus*)."

341 Yucatán Peninsula: Ibid.

BRIDLED SPARROW

342 "man's first recorded glimpse": Semple, J., and G. Sutton. 1932. "Nesting of Harris's Sparrow (*Zonotrichia querula*) at Churchill, Manitoba." *Auk* 49: 166–183.

342 Hartlaub described the species: Hartlaub, G. 1852. "Descriptions de quelques nouvelles espèces d'Oiseaux." *Revue et magasin de zoologie pure et appliquée* n.s. 4: 3–9.

342 with a large chick: Binford, L. 1989. *A Distributional Survey of the Birds of the Mexican State of Oaxaca*. Ornithological Monographs 43. Lawrence, KS: American Ornithologists' Union.

342 accident in the mountains: Orr, R. 1971. "John Stuart Rowley." *Auk* 88: 711.

342 eggs were discovered: Palacios-Silva, R., et al. 2005. "Descripción del nido y huevos del gorrión embridado (*Aimophila mystacalis*)." *Ornitologia neotropical* 16: 101–104.

343 juvenile plumage in both species: Wolf, L. 1977. *Species Relationships in the Avian Genus* Aimophila. Ornithological Monographs 23. Lawrence: American Ornithologists' Union.

343 song: Ibid.

343 resident of dry slopes: Ibid.

343 México State and eastern Morelos: Howell, S., and S. Webb. 1995. *A Guide to the Birds of Mexico and Northern Central America*. Oxford: Oxford University Press.

343 occurs in treed habitats: Wolf, *Avian Genus* Aimophila.

BLACK-CHESTED SPARROW

344 Geographical and Exploring Commission: Ferrari-Pérez, F. 1886. "Catalogue of Animals Collected by the Geographical and Exploring Commission of the Republic of Mexico." *Proceedings of the United States National Museum* 9: 125–199.

344 "The considerable expense": Ibid.

345 Ridgway formally described: Ridgway, R. 1883. "Preliminary Descriptions of Some New Species of Birds from Southern Mexico, in the Collection of the Mexican Geographical and Exploring Commission." *Auk* 3: 331–333.

345 skins returned to Mexico: Ibid.

345 "no difficulty in recognizing them": Salvin, O., and F. Godman. 1886. *Biologia centrali-americana: Aves*. Vol. 1. London: R.H. Porter.

345 described 35 years earlier: Cabanis, J. 1851. *Museum Heineanum: Verzeichniss der ornithologischen Sammlung des Oberamtmann Ferdinand Heine*. Vol. 1. Halberstadt: R. Frantz.

345 "unmistakable and striking": Howell, S., and S. Webb. 1995. *A Guide to the Birds of Mexico and Northern Central America*. Oxford: Oxford University Press.

345 Juveniles: Wolf, L. 1977. *Species Relationships in the Avian Genus* Aimophila. Ornithological Monographs 23. Lawrence: American Ornithologists' Union.

345 The formative plumage that results: Howell and Webb, *Mexico and Northern Central America*.

345 The calls of this species: Ibid.

345 Wolf distinguished two song types: Wolf, *Avian Genus* Aimophila.

345 Mexican endemic: Miller, A., et al. 1957. *Distributional Checklist of the Birds of Mexico*. Part 2. Pacific Coast Avifauna 33.

345 Oaxaca: Binford, L. 1989. *A Distributional Survey of the Birds of the Mexican State of Oaxaca*. Ornithological Monographs 43. Lawrence, KS: American Ornithologists' Union.

345 *asticta*: Griscom, L. 1934. "The Ornithology of Guerrero, Mexico." *Bulletin of the Museum of Comparative Zoology* 75: 367–422.

345 "but doubtfully valid": Rising, J. 2012. "Black-chested Sparrow (*Peucaea humeralis*)." In J. del Hoyo, et al., eds. *Handbook of Birds of the World*. Barcelona: Lynx Edicions.

STRIPE-HEADED SPARROW

346 Sumichrast had wearied: Boucard, A. 1884. "Notice sur François Sumichrast." *Bulletin de la Société zoologique de France* 9: 305–312.

346 "had [his] fill": Joseph, J. 2012. *Saussure*. Oxford: Oxford University Press.

346 "an extended exploration": Lawrence, G. 1876. "Catalogue of Birds Collected by Prof. Francis Sumichrast in Southwestern Mexico, and Now in the National Museum at Washington, D.C." *Bulletin of the United States National Museum* 1.4: 5–56.

346 "The specimens sent": Ibid.

346 "I regret to be unable to tell you certainly": Ibid.

347 originally described: Bonaparte, C. 1853. "Notes sur les collections rapportées en 1853, par M. A. Delattre, de son voyage en California et dans le Nicaragua." *Comptes rendus hébdomadaires des séances de l'Académie des sciences* 37: 913–925.

347 Tehuantepec Ground Sparrow: Salvin, O., and F. Godman. 1886. *Biologia centrali-americana: Aves*. Vol. 1. London: R. H. Porter.

347 Sumichrast did not live: Boucard, "François Sumichrast."

347 a subspecies of the Stripe-headed Sparrow: Ridgway, R. 1901. *The Birds of North and Middle America*. Bulletin of the United States National Museum, Bulletin 50, pt. 1. Washington: Government Printing Office.

347 two different songs: Wolf, L. 1977. *Species Relationships in the Avian Genus* Aimophila. Ornithological Monographs 23. Lawrence: American Ornithologists' Union.

347 northern *acuminata*: Rising, J. 2014. "Stripe-headed Sparrow (*Peucaea ruficauda*)." In J. del Hoyo et al., eds. *Handbook of Birds of the World*. Vol. 16. Barcelona: Lynx Edicions.

348 *lawrencii*: Binford, L. 1989. *A Distributional Survey of the Birds of the Mexican State of Oaxaca*. Ornithological Monographs 43. Lawrence, KS: American Ornithologists' Union.

348 two southern races: Rising, "Stripe-headed Sparrow (*Peucaea ruficauda*)."

348 extreme northwestern Guatemala: Eisermann, K., and C. Avendaño. 2007. *Lista comentada de las aves de Guatemala / Annotated Checklist of the Birds of Guatemala*. Barcelona: Lynx Edicions.

348 size increases: Rising, J. "Stripe-headed Sparrow (*Peucaea ruficauda*)."

348 *acuminata*, often known as the Colima Sparrow: Ibid.

348 described the Guatemala race *connectens*: Griscom, L. 1932. *The Distribution of Bird-life in Guatemala*. Bulletin of the American Museum of Natural History 64. New York: American Museum of Natural History.

348 grayer above: Griscom, L. 1930. "Studies From the Dwight Collection of Guatemala Birds III." *American Museum Novitates* 438: 1–18.

BACHMAN SPARROW

349 a catalog of duplicate specimens: Lichtenstein, H. 1823. *Verzeichniss der Doubletten des Zoologischen Museums der Königlichen Universität zu Berlin*. Berlin: T. Trautwein.

349 acquired it from a London dealer: Stresemann, E. 1953. "On a Collection of Birds from Georgia and Carolina Made About 1810 by John Abbot." *Auk* 70: 113–117.

349 "Summer Sparrow": Brunk Auctions. 2009. Lot 0188: Rare John Abbot Watercolor, Summer Sparrow or Grass Sparrow. Auction of May 30, 2009: Griffin Collection. brunkauctions.com/lot-detail-past/?id=47177.

349 three surviving specimens: Stresemann, "Birds from Georgia and Carolina."

349 labeled in Lichtenstein's hand: Van Rossem, A. 1939. "Manuscript Notes on Avian Types." Unpub. ms., Smithsonian Institution Libraries. tinyurl.com/AJVRMS.

349 described by another ornithologist, John Latham: Latham, J. 1823. *A General History of Birds*. Vol. 6. Winchester: privately published.

350 first meeting in Savannah: Burtt and Davis, *Alexander Wilson*.

350 Wilson thanked Abbot: Stone, W. 1906. "Some Unpublished Letters of Alexander Wilson and John Abbot." *Auk* 23: 361–368.

350 "obtained many specimens of it": Audubon, J. 1834. *Ornithological Biography*. Vol. 2. Edinburgh: Adam Black and Charles Black.

350 *Peucaea*: Audubon, J. 1839. *A Synopsis of the Birds of North America*. Edinburgh: Adam Black and Charles Black.

350 Pinewood-Finch: Audubon, *Synopsis*.

350 two races of a single species: Brewster, W. 1885. "*Peucaea aestivalis* and Its Subspecies *illinoensis*." *Auk* 2: 105–106.

351 at times been considered conspecific: Ridgway, R. 1901. *The Birds of North and Middle America*. Bulletin of the United States National Museum, Bulletin 50, pt. 1. Washington: Government Printing Office.

351 might not be reliably distinguishable: Dunning, J. 2006. "Bachman's Sparrow (*Peucaea aestivalis*)." In P. Rodewald, ed. *Birds of North America*. Ithaca: Cornell Lab of Ornithology.

351 Seal Island, Nova Scotia: McLaren, I. 2012. *All the Birds of Nova Scotia: Status and Critical Identification*. Kentville: Gaspereau Press.

351 lumped by ornithologists in the past: Ridgway, *North and Middle America*.

352 a repertoire of three songs: Dunning, "Bachman's Sparrow (*Peucaea aestivalis*)."

352 different primary songs: Wolf, L. 1977. *Species Relationships in the Avian Genus* Aimophila. Ornithological Monographs 23. Lawrence: American Ornithologists' Union.

352 Bachman described it: Audubon, *Ornithological Biography*.

352 songs reverse the sequence: Wolf, *Avian Genus* Aimophila.

352 "barely audible to observers": Dunning, "Bachman's Sparrow (*Peucaea aestivalis*)."

352 "excited" song: Ibid.

352 fast and without a pause: Wolf, *Avian Genus* Aimophila.

352 adult's calls: Dunning, "Bachman's Sparrow (*Peucaea aestivalis*)."

352 "a series of chip calls": Wolf, *Avian Genus* Aimophila.

353 Oklahoma: Carter, W., and M. Duggan. 1998. *Bachman's Sparrow Survey and Habitat Characterization on Southeastern Oklahoma Wildlife Management Areas*. Oklahoma City: Oklahoma Department of Wildlife Conservation.

353 Indiana: Brock, K. 2006. *Brock's Birds of Indiana*. Indianapolis: Amos Butler Audubon Society.

353 Illinois: Bohlen, D. 2017. "Bachman's Sparrow *Aimophila aestivalis*." Champaign: Illinois Natural History Survey. inhs.illinois.edu.

353 Ohio: Rodewald, P., et al., eds. 2016. *The Second Atlas of Breeding Birds in Ohio*. University Park: Pennsylvania State University Press.

353 Pennsylvania: Wilson, A., et al., eds. 2012. *Second Atlas of Breeding Birds in Pennsylvania*. University Park: Pennsylvania State University Press.

353 Arkansas: Dunning, "Bachman's Sparrow (*Peucaea aestivalis*)."

353 Texas: Lockwood, M., and B. Freeman. 2014. *The TOS Handbook of Texas Birds*. College Station: Texas A&M University Press.

353 Louisiana: Fontenot, B., and R. DeMay. S.d. *Louisiana Sparrows*. Thibodaux: Barataria-Terrebonne National Estuary Program.

353 Red-cockaded Woodpeckers: Gibbons, R., et al. 2013. *A Birder's Guide to Louisiana*. Colorado Springs: American Birding Association.

353 Mississippi: Turcotte, W., and D. Watts. 1999. *Birds of Mississippi*. Jackson: University Press of Mississippi.

354 Alabama: Tucker, J. 2014. "Bachman's Sparrow." tinyurl.com/AlabamaBASP.

354 Florida: Greenlaw, J., et al. 2014. *The Robertson and Woolfenden Florida Bird Species: An Annotated List*. Florida Ornithological Society Special Publication 8. Gainesville: Florida Ornithological Society.

354 Georgia: Gobris, N. 2010. "Bachman's Sparrow (*Aimophila aestivalis*)." In T. Schneider, et al., eds. *The Breeding Bird Atlas of Georgia*. Athens: University of Georgia Press.

354 South Carolina: Dunning, "Bachman's Sparrow (*Peucaea aestivalis*)."

354 North Carolina: LeGrand, H. 2016. "Bachman's Sparrow—*Peucaea aestivalis*." In H. LeGrand et al., eds. *Birds of North Carolina: Their Distribution and Abundance*. ncbirds.carolinabirdclub.org.

354 "usually fallow for four years": Weston, F. 1968. "*Aimophila aestivalis bachmani* (Audubon): Bachman's Sparrow." In A. Bent., ed. *Life Histories of North American Cardinals, Grosbeaks, Buntings, Towhees, Finches, Sparrows, and Allies*. Pt. 2. Bulletin of the United States National Museum 237. Washington: Smithsonian Institution.

354 migration is documented: Dunning, "Bachman's Sparrow (*Peucaea aestivalis*)."

354 New Jersey: Boyle, W. 2011. *The Birds of New Jersey*. Princeton: Princeton University Press.

354 New York's three records: Meade, G. 1941. "Bachman's Sparrow in New York." *Auk* 58: 103–104.

354 Long Island: Levine, E., ed. *Bull's Birds of New York State*. Ithaca: Comstock Publishing Associates.

354 Canada: Godfrey, W. 1979. *The Birds of Canada*. Ottawa: National Museum of Natural Sciences.

354 lumped with *bachmani*: Brewster, "Subspecies *illinoensis*."

354 western and northwestern edges of the species' range: Sutton, G. 1938. "Some Findings of the Semple Oklahoma Expedition." *Auk* 55: 501–508.

354 this western race: Wetmore, A. 1939. "Notes on the Birds of Tennessee." *Proceedings of the United States National Museum* 86: 175–243.

354 Audubon's *bachmani*: Ibid.

354 Intergradation: Ibid.

354 intergrade in Georgia: Dunning, "Bachman's Sparrow (*Peucaea aestivalis*)."

CASSIN SPARROW

355 "a peculiar character of marking": Coues, E. 1884. *Key to North American Birds*. 2nd ed. Boston: Estes and Lauriat.

355 "but upon examination": Woodhouse, S. 1852. "Descriptions of New Species of Birds of the Genera *Vireo*, Vieill., and *Zonotrichia*, Swains." *Proceedings of the Academy of Natural Sciences of Philadelphia* 6: 60–61.

355 three additional skins: Baird, S., with J. Cassin and G. Lawrence. 1858. *Birds*. Explorations and Surveys for a Railroad Route. Vol. 9. Washington: Beverly Tucker.

355 "probably considerably older": Baird, *Birds*.

355 "considered as in quite immature plumage": Baird, S., et al. 1874. *A History of North American Birds*. Vol. 2. Boston: Little, Brown.

355 correct the error: Ridgway, R. 1873. "On Some New Forms of American Birds." *American Naturalist* 7: 602–619.

355 not an "adult" Cassin Sparrow but a Botteri Sparrow: Baird et al., *History*.

355 "upon examination of ample material": Coues, E. 1874. *Birds of the Northwest*. Miscellaneous Publications of the Geological Survey of the Territories 3. Washington: Government Printing Office.

355 Ridgway in the pages of the *American Naturalist*: Ridgway, "On Some New Forms."

355 Coues in the first edition of his *Check List*: Coues, E. 1873. *A Check List of North American Birds*. Salem: Naturalists' Agency.

355 only Ridgway was cited: Baird et al., *History*.

355 posthumous edition of his *Key*: Coues, E. 1903. *Key to North American Birds*. 5th ed. Vol. 1. Boston: Dana Estes and Company.

355 "individuals may appear to be molting": Dunning, J., et al. 1999. "Cassin's Sparrow "*Peucaea cassinii*)." In P. Rodewald, ed. *Birds of North America*. Ithaca: Cornell Lab of Ornithology.

355 "looks like its picture": Nash, O. 1957. *You Can't Get There From Here*. Boston: Little, Brown.

357 pattern of the central rectrices: Kaufman, K. 1990. *A Field Guide to Advanced Birding*. Boston: Houghton Mifflin.

358 "have a mixture of older, abraded [formative] feathers": Dunning et al., "Cassin's Sparrow "*Peucaea cassinii*)."

358 definitive prebasic molt: Ibid.

358 "Rising with a tremulous motion": Heermann, A. 1859. "Report Upon Birds Collected on the Survey." *Reports of Explorations and Surveys to Ascertain the Most Practicable and Economical Route for a Railroad*. Vol. 10, 29–77. Washington: Beverley Tucker.

358 "its sweetly modulated song": Baird et al., *History*.

358 "very plaintive, but quite pretty": Henshaw, H. 1873. "Report Upon the Ornithological Collections made in Portions of Nevada, Utah, California, Colorado, New Mexico, and Arizona During the Years 1871, 1872, 1873, and 1874." In G. Wheeler, ed. *Geographical and Geological Explorations and Surveys West of the One Hundredth Meridian*. Vol. 5. Washington: Government Printing Office.

358 a secondary song: Dunning et al., "Cassin's Sparrow "*Peucaea cassinii*)."

358 "whisper song": Ibid.

359 "This opportunistic bird": Luke, D. 1997. *Status of the Cassin's Sparrow*. Denver: U.S. Fish & Wildlife Service.

359 range can be extended west: Rosenberg, K., et al. 1991. *Birds of the Lower Colorado River Valley*. Tucson: University of Arizona Press.

359 Nebraska: Mollhoff, W. 2016. *The Second Nebraska Breeding Bird Atlas*. Bulletin of the University of Nebraska State Museum 29. Lincoln: University of Nebraska State Museum.

359 in Texas, exceptional "irruptions": Lockwood, M., and B. Freeman. 2014. *The TOS Handbook of Texas Birds*. College Station: Texas A&M University Press.

359 Arizona: Corman, T., and C. Wise-Gervais, eds. 2005. *Arizona Breeding Bird Atlas*. Albuquerque: University of New Mexico Press.

359 breeding range extends south: Howell, S., and S. Webb. 1995. *A Guide to the Birds of Mexico and Northern Central America*. Oxford: Oxford University Press.

359 plains of Texas: Lockwood and Freeman, *TOS Handbook*.

359 migratory at least in the northern portion: Dunning et al., "Cassin's Sparrow (*Peucaea cassinii*)."

359 Wintering birds: Lockwood and Freeman, *TOS Handbook*.

359 spring arrivals: Ibid.

359 southwest Nebraska: Sharpe, R., et al. 2001. *Birds of Nebraska: Their Distribution and Temporal Occurrence*. Lincoln: University of Nebraska Press.

359 eastern Colorado: Andrews, R., and R. Righter. 1992. *Colorado Birds: A Reference to Their Distribution and Habitat*. Denver: Denver Museum of Natural History.

359 occurred some 50 times in California: Hamilton, R., et al., eds. 2007. *Rare Birds of California*. Camarillo: Western Field Ornithologists.

359 Curry County, Oregon: Nehls, H. 2015. "The Records of the Oregon Bird Records Committee Through April 2015." orbirds.org.

359 Baja California: Dunning et al., "Cassin's Sparrow (*Peucaea cassinii*)."

359 East of the species' usual range: Ibid.

359 Maryland: Maryland Biodiversity Project. n.d. "Cassin's Sparrow *Peucaea cassinii* (Woodhouse, 1852)." marylandbiodiversity.com.

359 New Jersey: Boyle, W. 2011. *The Birds of New Jersey*. Princeton: Princeton University Press.

359 Long Island: Burke, T. 2001. "First Record of Cassin's Sparrow (*Aimophila cassinii*) for New York State. *Kingbird* 51: 450–451.

359 Nova Scotia: McLaren, I. 2012. *All the Birds of Nova Scotia: Status and Critical Identification*. Kentville: Gaspereau Press.

BOTTERI SPARROW

360 a fresh eye: Ridgway, R. 1898. "Descriptions of Supposed New Genera, Species, and Subspecies of American Birds. I: Fringillidae." *Auk* 15: 223–230.

360 badly worn Botteri Sparrow: Baird, S., et al. 1874. *A History of North American Birds*. Vol. 2. Boston: Little, Brown.

360 "undisputed king of birds": Lewis, D. 2012. *The Feathery Tribe: Robert Ridgway and the Modern Study of Birds*. New Haven: Yale University Press.

360 in fact represented a new species: Ridgway, "Supposed New Genera."

360 a race of the Botteri Sparrow: Ridgway, R. 1901. *The Birds of North and Middle America*. Bulletin of the United States National Museum, Bulletin 50, pt. 1. Washington: Government Printing Office.

360 a revival in the 1960s: Howell, T. 1965. "New Subspecies of Birds from the Lowland Pine Savanna of Northeastern Nicaragua." *Auk* 82: 438–464.

360 one of the Central American races: Hellmayr, C. 1938. *Catalogue of Birds of the Americas and the Adjacent Islands in Field Museum of Natural History*. Zoological Series 13, pt. 11. Chicago: Field Museum of Natural History.

360 an intermediate population: Webster, J. 1959. "A Revision of the Botteri Sparrow." *Condor* 61: 136–146.

360 Sartorius fled his native land: Langman, I. 1949. "Dos figuras casi olvidadas en la historia de la botanica mexicana." *Revista de la Sociedad Méxicana de Historia Natural* 10: 329–336.

361 a large plantation: Scharrer, B. 1982. "Estudie de caso: El grupo familiar de empresarios Stein-Sartorius." Pp. 231–286 in B. von Mentz, et al., eds. *Los Pioneros del imperialismo alemán en México*. Hidalgo: Centro de Investigaciones y Estudios Superiores en Antropología Social.

361 "constantly taking notes": Ibid.

361 collaborator with the Smithsonian: Langman, "Dos figuras casi olvidadas."

361 Otto, the first of the Sartorius children: Baird et al., *History*.

361 The younger son, Florentin: Ibid.

361 a large sparrow: Deignan, H. 1961. *Type Specimens of Birds in the United States National Museum*. Bulletin of the United States National Museum 221. Washington: Government Printing Office.

361 their modest appearance: Bock, C., and J. Bock. 2000. *The View From Bald Hill: Thirty Years in an Arizona Grassland*. Berkeley: University of California Press.

362 the juvenile's preformative molt: Howell, S. 2010. *Peterson Reference Guide to Molt in North American Birds*. Boston: Houghton Mifflin Harcourt.

362 compared to the end of a Black-chinned Sparrow song: Monson, G. 1968. "*Aimophila botterii* (Sclater): Botteri's Sparrow." In A. Bent, ed. *Life Histories of North American Cardinals, Grosbeaks, Buntings, Towhees, Finches, Sparrows, and Allies*. Pt. 2. United States National Museum Bulletin 237. Washington: Smithsonian Institution Press.

362 "short, variable introductory notes": Webb, E., and C. Bock. 2012. "Botteri's Sparrow (*Peucaea botterii*)." In P. Rodewald, ed. *Birds of North America*. Ithaca: Cornell Lab of Ornithology.

363 "piercing, high-pitched, ventriloquial": Ibid.

363 coastal prairies of Texas: Lockwood, M., and B. Freeman. 2014. *The TOS Handbook of Texas Birds*. College Station: Texas A&M University Press.

363 the race *texana*: Webb and Bock, "Botteri's Sparrow (*Peucaea botterii*)."

363 Arizona: Corman, T., and C. Wise-Gervais, eds. 2005. *Arizona Breeding Bird Atlas*. Albuquerque: University of New Mexico Press.

363 northeastern Sonora: Russell, S., and G. Monson. 1998. *The Birds of Sonora*. Tucson: University of Arizona Press.

363 southern subspecies: Webb and Bock, "Botteri's Sparrow (*Peucaea botterii*)."

363 eastern Honduras: Fagan, J., and O. Komar. 2016. *Peterson Field Guide to Birds of Northern Central America*. Boston: Houghton Mifflin Harcourt.

363 Costa Rica: Stiles, G., and A. Skutch. 1989. *A Guide to the Birds of Costa Rica*. Ithaca: Cornell University Press.

363 arrive in south Texas: Lockwood and Freeman, *TOS Handbook*.

363 Arizona and northern Sonora: Russell and Monson, *Birds of Sonora*.

363 Guerrero and Morelos: Webb and Bock, "Botteri's Sparrow (*Peucaea botterii*)."

363 currently recognized subspecies: Rising, J. 2016. "Botteri's Sparrow (*Peucaea botterii*)." In J. del Hoyo, et al., eds. *Handbook of Birds of the World*. Vol. 16. Barcelona: Lynx Edicions.

364 recognize *mexicana*: Webb and Bock, "Botteri's Sparrow (*Peucaea botterii*)."

364 *petenica*: Salvin, O. 1863. "Descriptions of Thirteen New Species of Birds Discovered in Central America by Frederick Godman and Osbert Salvin." *Proceedings of the Zoological Society of London* 1863: 186–192.

364 coverts have brown edges: Mill, W., and L. Griscom. 1925. *Further Notes on Central American Birds, with Descriptions of New Forms*. American Museum Novitates 184. New York: American Museum of Natural History.

364 *tabascensis*: Dickerman, R., and A. Phillips. 1967. "Botteri's Sparrows of the Atlantic Coastal Lowlands of México." *Condor* 69, no. 6: 596–600.

364 synonyms of *petenica*: Webb and Bock, "Botteri's Sparrow (*Peucaea botterii*)."

364 *vantynei*: Webster, "Revision of the Botteri Sparrow."

364 *spadiconigrescens*: Howell, "Northeastern Nicaragua."

364 broad rusty edges: Webb and Bock, "Botteri's Sparrow (*Peucaea botterii*)."

364 *vulcanica*: Mill and Griscom, *Central American Birds*.

364 nominate *botterii*: Ibid.

364 streaking on the back of *mexicana*: Rising, "Botteri's Sparrow (*Peucaea botterii*)."

364 Sonora to Nayarit: Webb and Bock, "Botteri's Sparrow (*Peucaea botterii*)."

364 *goldmani*: Phillips, A. 1943. "Critical Notes on Two Southwestern Sparrows." *Auk* 60: 242–248.

364 more gray on average in *texana*: Mill and Griscom, *Central American Birds*.

RUFOUS-WINGED SPARROW

365 "a homely little bird": Coues, E. 1873. "Some United States Birds, New to Science, and Other Things Ornithological." *American Naturalist* 7: 321–331.

365 "I took my first specimens": Bendire, C. 1882. "The Rufous-winged Sparrow." *Ornithologist and Oölogist* 7: 121–122.

365 "brighter and purer": Coues, "Some United States Birds."

365 "turned loose" among the Rufous-winged Sparrows: Henshaw, H. 1875. "Report Upon the Ornithological Collections Made in Portions of Nevada, Utah, California, Colorado, New Mexico, and Arizona, During the Years 1871, 1872, 1873, and 1874." *Report Upon United States Geological Surveys West of the One Hundredth Meridian* 5: 133–507.

366 "to birds there is no limit": Nelson, E. 1932. "Henry Wetherbee Henshaw—Naturalist: 1850-1930." *Auk* 49: 399–427.

366 "sparingly" in the traditional localities: Brewster, W. 1882. "On a Collection of Birds Lately Made by Mr. F. Stephens in Arizona." *Bulletin of the Nuttall Ornithological Club* 7: 193–212.

366 last found in Arizona: Phillips, A., et al. 1964. *The Birds of Arizona*. Tucson: University of Arizona Press.

366 rediscovered in the Tucson area: Ibid.

366 numbers in the early 1970s: Walters, P., et al. 1984. "Twelve Years of Banding at Tanque Verde Ranch, Tucson, Arizona." *North American Bird Bander* 9: 2–10.

366 a noticeable crash: Lowther, P., et al. 1999. "Rufous-winged Sparrow (*Peucaea carpalis*)." In A. Poole, ed. *Birds of North America*. Ithaca: Cornell Lab of Ornithology.

366 Arizona's Guadalupe Canyon: Wolf, C. 2009. "Rufous-winged Sparrow (*Aimophila carpalis*), Guadalupe Canyon Gate, Cochise County." azfo.org.

368 three dozen different song types: Groschupf, K. 1983. "A Comparative Study of the Vocalizations and Singing Behavior of Four *Aimophila* Sparrows." Dissertation. Virginia Polytechnical Institute and State University.

368 warbling vocalization: Lowther et al., "Rufous-winged Sparrow (*Peucaea carpalis*)."

369 three generally recognized subspecies: Ibid.

369 *bangsi*: Moore, R. 1932. "A New Race of *Aimophila carpalis* from Mexico." *Proceedings of the Biological Society of Washington* 45: 231–234.

369 intergrading with the southernmost subspecies *cohaerens*: Miller, A., et al. 1957. *Distributional Checklist of the Birds of Mexico*. Part 2. Pacific Coast Avifauna 33.

369 darker and browner: Moore, R. 1946. "The Rufous-winged Sparrow, Its Legends and Taxonomic Status." *Condor* 48: 117–123.

INDEXES

INDEX OF PEOPLE NAMES

See also the Notes, not indexed here.

Abbot, John (1751–1840/1841), 161, 349–350
Abeillé, Grégoire (1798–1848),159
Abert, James William (1820–1897), 140
Aiken, Charles E. (1850–1936), 61–62, 249
Allen, Charles A. (1841–1930), 154
Allen, Joel Asaph (1838–1921), 62, 80, 83, 101–102, 108, 162, 163, 277, 301
Anthony, A. W. (1865–1939), 73, 79, 212
Ardis, W. F. (fl. 1931), 57
Armstrong, Frank B. (1863–1915), 318, 332
Aschenborn, Heinrich Alwin (1816–1865), 313–314
Aubry, Jean–Thomas (1714–1785), 193
Audubon, John James (1785–1851), 4, 5, 20, 22, 25, 27, 33, 34–35, 52–53, 61, 65, 87, 92, 96, 97, 101, 115, 168, 171, 184–185, 258–259, 262, 269, 276–277, 280, 298, 303, 308–309, 329, 335, 350
Audubon, John Woodhouse (1812–1862), 52, 287
Avilès, Mario del Toro (fl. 1929/1932), 134

Bachman, John (1790–1874), 65, 101, 350, 352
Bailey, Alfred M. (1894–1978), 57
Bailey, Harold H. (1878–1962), 102
Baird, Spencer Fullerton (1823–1887), 4, 6, 16–17, 25, 43, 47, 48, 61, 70, 74, 80, 87–89, 92, 95, 125–126, 130, 139, 140, 154, 163, 168, 169, 176, 185, 199, 204, 208–209, 214, 221–222, 226, 227, 230, 255, 257–258, 263, 275, 324, 355

Bakewell, John Howard (1825–1849), 287
Bangs, Outram (1863–1932), 57
Banks, Joseph (1743–1820), 34, 80, 198
Banks, Richard C., 267
Bartram, William (1739–1823), 32, 46, 188, 275, 303
Beadle, David, 197
Beechey, Frederick William (1796–1856), 135, 198
Belding, Lyman (1829–1917), 73, 75, 214
Bell, John G. (1812–1899), 61, 96, 119, 168, 269
Bendire, Charles (1836–1897), 270, 365
Benedict, Lauryn, 138
Benson, Seth B. (1905–2005), 321
Best, Robert (1790–1831), 65
Billings, Joseph (1758–1806), 255
Blackburne, Anna (1726–1793), 32, 91, 188
Blackburne, Ashton (ca. 1730—ca. 1780), 32, 91, 116
Blanchan, Neltje (1865–1918), 4, 8
Blanding, William (1773–1857), 184
Blossom, Elizabeth Beardsley B. (1881–1970), 186
Boddaert, Pieter (1730–1795), 281
Bonaparte, Charles Lucien (1803–1857), 5, 16, 20, 25, 31, 33, 35, 46–47, 84, 135, 145, 146, 149, 160, 245, 254, 275, 285, 294, 296, 298, 329, 334, 347
Borror, Donald J. (1907–1988), 278
Boucard, Adolphe (1839–1905), 347
Brandt, Johann Friedrich von (1802–1879), 53
Brelay, Aglaé (1800–1879), 283
Brelay, Charles (1791–1857), 283
Brewer, Thomas Mayo (1814–1880), 6, 48, 88, 92, 163, 221–222, 227, 259, 324, 330, 355
Brewster, William (1851–1919), 53, 57, 69, 73, 74, 88,

130–131, 136, 152, 154, 167, 235, 248, 350, 354
Brisson, Mathurin (1723–1806), 188, 193–194, 261, 275, 334
Brodkorb, Pierce (1908–1992), 148
Brooks, Allan (1869–1946), 84, 241, 321, 322
Brown, Wilmot W. (ca. 1878—ca. 1953), 318, 320
Bryan, Francis T. (1823–1917), 204, 205
Bryant, Walter E. (1861–1905), 212–213
Buffon, Georges–Louis Leclerc, comte de (1707–1788), 189, 261, 281, 303
Bullock, William, Jr. (1796?–1827?), 149, 217
Bullock, William, Sr. (ca. 1773–1849), 149, 217
Burling, Gilbert (fl. 1870), 162
Butler, Amos W. (1860–1937), 86

Cabanis, Jean (1816–1906), 20–21, 145–146, 160, 167, 313–314, 345
Cabot, Samuel, Jr. (1815–1885), 48
Cahoon, John C. (1863–1891), 152
Carman, Benjamin F. (?–1886), 169
Carson, Rebecca J., 27
Cassin, John (1813–1869), 70, 119, 125, 146, 154, 184–185, 225, 287
Catesby, Mark (1683–1749), 4, 5, 32, 161, 193, 244, 329
Chadbourne, Arthur P. (1862–1936), 332
Chapman, Frank M. (1864–1945), 30, 102, 318, 339
Cheney, Simeon Pease (1818–1890), 329
Cicero, Carla, 120
Coale, Henry K. (1858–1926), 235–236, 295, 297
Colen, John H. (fl. 1838/1854), 184
Conover, Henry Boardman (1892–1950), 57

Cook, James (1728–1779), 80, 198, 254

Cooke, Wells (1858–1916), 204, 273

Cooper, James Graham (1830–1902), 154, 230

Crook, George (1830–1890), 298

Couch, Darius N. (1822–1897), 314

Coues, Elliott (1842–1899), 6, 16, 34, 43, 47, 48, 61, 62, 78–79, 83, 88, 97, 99, 130, 136, 150, 162–163, 167, 168, 176, 222, 249–250, 252, 270, 277, 294, 307, 319, 324, 335, 355, 365

Cristobal (last name unrecorded) (1853—?), 178–179

DaCosta, Jeffery M., 27, 140

Danby, Durward B. (1864–1948), 252

Davis, John (1916–1986), 60, 136–137

De Haven, Isaac Norris (1847–1924), 89

DeLattre, Adolphe (1805–1854), 149

Deppe, Ferdinand (1794–1861), 143, 144, 145

Deppe, Wilhelm (1800–1844), 143

Detwiller, John W. (1851–1898), 66

Dickerman, Robert (1926–2015), 58, 60

Donsker, David, 168

Drexler, Constantin Charles (fl. 1858/1860), 230

Du Bus de Gisignies, Bernard (1808–1874), 180

Dugès, Alfredo (1826–1910), 134

Dwight, Jonathan (1858–1929), 89, 90, 108–109, 116, 130–131, 136, 236, 240, 241, 248

Edwards, George (1694–1773), 32, 34, 188–189, 275

Elliott, Henry W. (1846–1930), 7

Elliott, Steven (fl. ca. 1808), 161

Ellis, William W. (?–1785), 254

Emory, William H. (1811–1887), 140

Ferrari-Pérez, Fernando (1857–1933), 344–345

Finsch, Otto (1839–1917), 53, 80, 199

Fisher, A. K. (1856–1948), 57, 119

Fisher, James (1912–1970), 197

Flack, William, 133

Flagg, Wilson (1805–1884), 92

Floresi d'Arcais, Damiano (fl. 1840/1845), 291

Floyd, Ted, 306

Forster, Johann Reinhold (1729–1798), 188, 261

Fremont, John C. (1813–1890), 140

Gaige, Frederick M. (1890–1976), 175

Gambel, William (1823–1849), 184–185, 262

Gaultier, Jean-François (1708–1756), 261

Gaylord, Horace A. (1877–1916), 212

Giebel, Christoph Gottfried (1820–1881), 16, 157

Gmelin, Johann Friedrich (1748–1804), 25, 33, 34, 80, 91, 115, 198, 255, 275, 334

Godman, Frederick DuCane (1834–1919), 30, 152, 291, 345, 347

Goldman, Luther J. (fl. 1908/1915), 70, 131

Gould, John (1804–1881), 245, 291

Gray, George Robert (1808–1872), 31, 135, 157

Grayson, Andrew Jackson (1819–1869), 169, 178–179

Grayson, Edward (1851–1867), 178–179

Greenlaw, Jon S. , 110, 112

Griffin, David, 219, 220, 224

Grinnell, Joseph (1877–1939), 70, 71, 86, 119, 120, 123, 126, 169, 266–267

Griscom, Ludlow (1890–1959), 50, 83–84, 86, 87, 102, 245, 345, 348

Gunn, William (1913–1984), 278

Hall, Harrison (1787–1866), 6

Hammond, William A. (1828–1900), 208

Hardy, John William (1930–2012), 286

Harris, Edward (1799–1863), 22, 61, 65, 269–270, 287

Harris, Harry (1878–1954), 269

Hartlaub, Gustav (1814–1900), 135, 342

Hayden, Ferdinand V. (1829–1887), 66

Heermann, Adolphus L. (1827–1865), 70, 153–154, 184, 259, 263, 358

Heine, Ferdinand (1809–1894), 313

Hellmayr, Charles E. (1878–1944), 126, 146, 169, 170

Henry, Thomas Charlton (1825–1877), 125, 221

Henshaw, Henry W. (1850–1930), 62, 163, 199, 221, 365–366

Henslow, John Stevens (1796–1861), 16, 65

Hill, Brad G., 264

Hitchcock, William E. (ca. 1824—ca. 1880), 72, 287

Hoffmann, Ralph (1870–1932), 93, 153

Holden, Charles H., Jr. (fl. 1869), 249

Howell, Arthur H. (1872–1940), 100

Howell, Steve N. G., 58, 183, 311, 320

Hubbard, John P., 86, 156

Huey, Lawrence M. (1892–1963), 77, 123

Humphrey, George (c. 1745–1830), 46

Iliff, Marshall, 9

Illiger, Karl Wilhelm (1775–1813), 349

James, Arthur Curtiss (1867–1941), 133–134

Jardine, William (1800–1874), 25, 275

Jaurès, Charles (1808–1870), 145

Jouy, Alice Elizabeth Craig (1853–1880), 145

Jouy, Pierre Louis (1856–1894), 145

Kalm, Pehr (1716–1779), 244

Kaufman, Kenn , 9

Keerl, F. W. (fl. 1830), 157

Kennerly, Caleb (1829–1861), 130, 355

Keiner, Louis Charles (1799–1881), 145

Kittlitz, Heinrich von (1799–1874), 53, 198

Klein, Jacob Theodor (1685–1759), 334

Klicka, John, 7, 13, 14, 16, 18

Koo, Michelle S., 120

Krider, John (1819–ca. 1886), 66

Kumlien, Ludovic (1853–1902), 259

La Fresnaye, Frédéric de (1783–1861), 152, 159

Lamb, Chester Converse (1882–1965), 286

Latham, John (1740–1837), 6, 32, 34, 46, 80, 96, 97, 115–116, 198, 245, 254–255, 349–350

Lawrence, George N. (1806–1895), 73, 78, 146, 148, 168–169, 181, 339, 346–347

LeConte, John Eatton (1784–1860), 96

LeConte, John Lawrence (1825–1883), 96

Lein, M. Ross, 264

Lesson, René-Primevère (1794–1849), 146, 159–160,

Lever, Ashton (1729–1788), 188, 254,

Lichtenstein, Hinrich (1780–1857), 46, 143, 285, 313, 349

Lincecum, Gideon E. (1793–1874), 97

Lincoln, Thomas (1812–1883), 52–53

Linnaeus, Carl (1707–1778), 161, 244, 275, 298

Litke, Fyodor (1797–1882), 198

Lloyd, Bert C. , 271

Loomis, Leverett M. (1857–1928), 240

MacGillivray, William (1796–1852), 101

Mailliard, Joseph (1857–1945), 211

Malherbe, Alfred (1804–1865), 248

Marsh, Charles H. (fl. 1880/1899), 318–319

Marshall, Joe T. (1918–2015), 144

Martín del Campo, Rafael (1910–1987), 134

Martinet, François Nicolas (1731–1804), 194

Maximilian see Wied–Neuwied, Maximilian zu

Maussion de Candé, Antoine Marie (1801–1867), 159,

Maynard, Charles Johnson (1845–1929) , 16, 87–89, 101, 103, 162–163

McGregor, Richard C. (1871–1936), 73, 78

McLaren, Ian , 234

McLeod, Richard Randall (fl. 1883/1899), 152

Mearns, Edgar (1856–1916), 227–228, 338

Merrem, Blasius (1761–1824), 193

Merriam, C. Hart (1855–1942), 89, 226, 366

Miller, Alden H. (1906–1965), 169, 187, 213, 214, 219, 224, 226, 227, 228, 232–233, 234, 237, 240, 241, 251, 292, 317

Miller, Gerrit S., Jr. (1869–1956), 95

Monson, Gale (1912–2012), 127

Montbeillard, Philippe Guéneau de (1720–1785), 245

Montes de Oca, Rafael (fl. 1875), 181

Moore, Robert T. (1882–1958), 134, 286

Morton, Samuel George (1799–1851), 280

Müller, Philipp Ludwig Statius (1725–1776), 281

Myers, Harriet Williams (1867–1949), 153

Native Guide, Unnamed, 314

Nelson, Edward W. (1855–1934), 57, 108, 131, 146, 289

Nichols, J. T. (1883–1958), 102

Nickerson, Azor H. (1837–1910), 298

Norton, Arthur (1888–1943), 109

Nunes, Christian, 253

Nuttall, Thomas (1786–1859), 6, 33–34, 47, 92, 96, 168, 171, 245, 255, 258–259, 262, 269, 298, 301, 308–309, 329, 334

Oberholser, Harry C. (1870–1963), 16, 34, 53, 57, 70, 102, 109, 129, 131, 136, 139, 187, 193, 240, 267

d'Orbigny, Alcide (1802–1857), 159

Ord, George (1781–1866), 6

Otto, Adolph Wilhelm (1786–1845), 313

Pallas, Peter Simon (1741–1811), 80, 255, 260

Palmén, J. A. (1845–1919), 248

Palmer, Edward (1829–1911), 176–177, 212

Palmer, T. S. (1868–1955), 204

Panza, R. K. , 86

Pape, Wilhelm Georg (1806–1875), 53

Parkes, Kenneth C. (1922–2007), 86, 284

Patten, Michael A., 40–41, 43, 138–139

Paynter, Raymond A. (1925–2003), 35, 48, 58

Peale, Charles Willson (1741–1827), 80

Pearson, T. Gilbert (1873–1943), 48

Pemberton, John Roy (1884–1968), 123

Pennant, Thomas (1726–1798), 32–33, 34, 80, 91, 116, 188, 254, 275, 276

Peters, James L. (1889–1952), 35, 83–84, 86, 87

Peterson, Roger Tory (1908–1996), 7, 50, 55, 67, 93, 114, 197, 256, 295

Pettingill, Olin Sewall, Jr. (1907–2001), 271

Phillips, Allan R. (1914–1996), 40, 58, 142, 144, 148

Pitelka, Frank (1916–2003), 58, 60

Preble, Alfred (1880—?), 270

Preble, Edward (1871–1957), 270–271

Pruett, Christine L., 40–41, 43

Pyle, Peter, passim, with gratitude

Raleigh, Walter (1554–1618), 161

Rea, Amadeo, 321–322, 328

Réaumur, Antoine Ferchault de (1683–1757), 261

Rhoads, Samuel Nicholson (1862–1952), 339–340

Reichenbach, Ludwig (1793–1879), 146, 150

Richardson, John (1787–1865), 308

Richardson, William B. (1868 1927), 57, 86, 345

Richmond, Charles (1868–1932), 193

Ridgway, Robert (1850–1929), 6, 7, 22–23, 25, 26, 31, 34, 40, 43, 48, 57, 61–62, 73, 74, 78–79, 88, 89, 101, 119, 123, 125, 126, 128, 130, 134, 136, 144, 160, 163, 169, 176, 182, 183, 209, 214, 218, 221–222, 227–228, 236, 249, 255, 259, 266–267, 289, 295, 296, 297, 307, 318–319, 324, 326, 335–336, 344–345, 347, 350, 354, 355, 360

Riley, Joseph Harvey (1873–1941), 166, 197

Rising, James (1942–2018), 197

Rohwer, Sievert , 272

Roosevelt, Theodore (1858–1919), 108

Rowley, J. Stuart (1907–1968), 342

Sada, Andres M., 320

Sallé, Auguste (1820–1896), 134, 149

Salvin, Osbert (1835–1898), 22, 146, 152, 180–181, 254, 291, 345

Sartorius, Carl Christian (1796–1872), 360–361

Sartorius, Florentin (1837—?), 360–361

Saussure, Henri de (1829–1905), 346

Say, Thomas (1787–1834), 115, 294, 296, 297

Schiede, Wilhelm (1798–1836), 143

Schoepf, Johann David (1752–1800), 189

Sclater, Philip Lutley (1829–1913), 22, 30, 144, 146, 149, 157, 280, 285, 291, 347

Scott, D. W. (fl. 1873/1886), 34, 62

Scott, W.E.D. (1852–1910), 101–102

Semple, John Bonner (1869–1947), 271

Sennett, George B. (1840–1900), 102

Seton, Ernest Thompson (1860–1946), 62, 270–271, 308, 309

Shannon, W.E. (fl. 1931), 103

Sharpe, Richard Bowdler (1847–1909), 169–170

Shattuck, George C. (1813–1893), 309

Sheeter, Cathy, 243

Shufeldt, Robert W. (1850–1934), 239

Sibley, Charles G. (1917–1998), 163, 168, 171, 175, 182

Sibley, David, 242

Sieber, Friedrich Wilhelm (fl. 1801/1812), 349

Sloane, Hans (1660–1753), 161, 333–334, 337

Smit, Joseph (1836–1929), 148

Sonnini de Manoncourt, Charles (1751–1812), 281

Spicer, Greg S., 27

Sprague, Isaac (1811–1895), 269, 309

Stansbury, Howard (1806–1863), 140

Stejneger, Leonhard (1851–1943), 298, 344

Stephens, James Francis (1792–1852), 46

Stone, Witmer (1866–1939), 30, 66, 130–131, 136

Storer, Robert W. (1914–2008), 27

Streets, Thomas H. (fl. 1873–1877), 78

Stresemann, Erwin (1889–1972), 143, 255

Sumichrast, François (1828–1882), 346

Sutton, George M. (1898–1982), 109, 271

Swainson, William (1789–1855), 18, 20, 25, 30, 48, 134, 149, 150, 167–168, 170, 175, 217, 297, 308, 338

Swanson, Phil, 232

Swarth, Harry (1878–1935), 197, 202, 203, 205, 207, 211, 237, 241, 321, 322

Taverner, Percy A. (1875–1947), 321

Ten Kate, Herman (1858–1931), 214

Thayer, John Eliot (1862–1933), 318

Thompson, Ernest E. see Seton, Ernest Thompson

Todd, W.E. Clyde (1874–1969), 109, 267

Tompkins, Ivan R. (1893–1966), 103

Topsell, Edward (ca. 1572–1625), 161

Townsend, Charles (1859–1944), 123, 126

Townsend, John K. (1809–1851), 184–185, 198, 239, 255, 259, 262, 269, 280–281, 298, 301, 308

Trippe, T. Martin (fl. 1865/1874), 249–250

Trotter, Spencer (1860–1931), 335

Trudeau, James (1817–1887), 65

Turton, William (1762–1835), 80, 115

Van Rossem, Adriaan J. (1892–1949), 31, 77, 86, 123, 146, 181, 227, 286, 291, 313, 338

Vieillot, Louis Pierre (1748–1830), 5, 20, 25, 245, 303–304, 329, 333–334, 335

Vigors, Nicholas (1785–1840), 130, 135, 136, 198

Wagler, Johann Georg (1800–1832), 146, 157

Wagner, Helmuth (1897–1977), 60

Wangenheim, Friedrich von (1749–1800), 189

Warner, Dwaine W. (1917–2005), 58

Webb, Sophie, 183

Webber, Thomas , 286

Weeks, Edwin Lord (1849–1903) , 88

Wege, David. C. , 320

West, David A. (1934–2015), 164

Wetmore, Alexander (1886–1978), 62

Wetmore, Beatrice (1910–1997), 338

White, George (?–1898), 184

White, John (ca. 1540–ca. 1593), 161, 244

Wied–Neuwied, Maximilian zu (1782–1867), 96–97, 245, 270, 275, 294, 298

Wilson, Alexander (1766–1813), 5, 25, 32–33, 46–47, 80, 85–86, 89, 91–92, 93, 100–101, 115–116, 118, 161–162, 185–186, 189, 198, 244–245, 261–262, 275, 276, 303, 304, 329, 333, 335

Wolf, Larry L. , 27, 292, 345

Wood, William S., Jr. (fl. 1856), 204

Woodhouse, Samuel W. (1821–1904), 184, 225–226, 355

Worthen, Charles K. (1850–1909), 318–319

Wright, Mabel Osgood (1859–1934), 47

Xantus, John (1825–1894), 73, 146, 208–209

Zimmer, Kevin, 63

Zink, Robert M., 131, 136, 139

INDEX OF BIRD NAMES

Page numbers in **bold** indicate the beginning of the full species account. Page numbers in *italics* indicate the Notes for each species. Scientific species names are provided only for taxa given full accounts in this guide.

Abert Towhee, 19, 130, 132, 137, **140**, *389*
Aimophila, 19
Aimophila rufescens, 19, **149**
Aimophila ruficeps, 19, **153**
Ambiguous Sparrow, 6, 12
Amphispiza, 20
Amphispiza bilineata, 20, **287**
Amphispiza quinquestriata, 20, **291**
American Robin, 2
American Tree Sparrow, 20–21, **188**, 367, 396
Ammodramus, 25
Ammodramus savannarum, 25, **333**
Ammospiza, 17
Ammospiza caudacuta, 17, **115**
Ammospiza leconteii, 17, **96**
Ammospiza maritima, 17, **100**
Ammospiza nelsoni, 17, **108**
Arremon, 20
Arremon brunneinucha, 20, **283**
Arremon virenticeps, 20, **285**
Arremonops, 26
Arremonops rufivirgatus, 26, **339**
Artemisiospiza, 19
Artemisiospiza belli, 19, **119**
Artemisiospiza nevadensis, 19, **125**
Atlapetes, 20
Atlapetes albinucha, 20, 159
Atlapetes pileatus, 20, 157

Bachman Sparrow, 26–27, 47, 154, **349**, 361, *423*
Baird Junco, 21–22, **214**, *400*
Baird Sparrow, 4, 16–17, 18, 25–26, **61**, 66–67, 82–83, 335, *379*
Belding Sparrow, 17, **73**, 79, 82, *381*
Belted Kingfisher, 151
Bell Sparrow, 15, 19, 23, **119**, 127–128, 287–288, *386*
Bewick Wren, 219
Bishop, Northern Red, 335
Black Phoebe, 137
Blackbird, Red-winged, 12, 76, 105
Blackbird, Tricolored, 12
Black-capped Chickadee, 358
Black-chested Sparrow, 26–27, **344**, *422*

Black-chinned Sparrow, 25, **313**, 362, *417*
Black-headed Grosbeak, 10
Black-throated Blue Warbler, 264
Black-throated Sparrow, 22–24, 120, 127, **287**, 292, *412*
Blue Grosbeak, 137, 140
Blue-winged Warbler, 311
Bobolink, 13, 298–299
Botteri Sparrow, 26–27, 147, 351, 355, 356–357, **360**, *425*
Brewer Sparrow (*see also* Timberline Sparrow), 6, 25, 304, 310–311, 321, 322–323, **324**, 356, *419*
Bridled Sparrow, 26–17, **342**, *422*
Bronzed Cowbird, 12, 141
Brown Creeper, 191
Brown-headed Cowbird, 6, 12–13, 141, 164
Brown Thrasher, 195
Brown Towhee. *See* Canyon Towhee *and* California Towhee
Brush Finch, Chestnut-capped, 22, 181–182, 185, **283**, 285, *411*
Brush Finch, Green-striped, 22, **285**, *412*
Brush Finch, Rufous-capped, 20, **157**, 160, *392*
Brush Finch, White-naped, 20, **159**, *392*
Brush Finch, Yellow-throated, 159, 160
Bunting, Indigo, 10, 155
Bunting, Lark, 8–9, 24–25, **298**, *414*
Bunting, Lazuli, 10
Bunting, Little, 90
Bunting, McKay, 13–14
Bunting, Painted, 10
Bunting, Snow, 13–14
Bunting, Townsend , 269
Bunting, Varied, 10
Bush Tanagers, 3

Cabanis Ground Sparrow, 146
Cabot Sparrow. *See* Swamp Sparrow
Calamospiza, 24
Calamospiza melanocorys, 24, **298**
California Sparrow. *See* Bell Sparrow
California Towhee, 19, 132, **135**, 141, *388*
Canyon Towhee, 19, **130**, 137–138, 140, 141, 143, 144, 147, 155, 368, *388*
Caracara, Guadalupe, 212
Cardinal, Northern , 263, 273, 368
Carolina Junco. *See* Slate-colored Junco

Cassiar Junco, 1, 21–22, 237–238, **241**, 245–246, 247, *404*
Cassin Finch, 11–12
Cassin Sparrow, 26–27, 351–352, **359**, 361–362, *424*
Centronyx, 16
Centronyx bairdii, 16, **61**
Centronyx henslowii, 16, **65**
Chat, Yellow-breasted, 301
Chestnut-capped Brush Finch, 22, 181–182, 185, **283**, 285, *411*
Chestnut-collared Longspur, 14
Chiapas Junco, 219
Chickadee, Black-capped, 358
Chipping Sparrow, 20, 25, 31, 154, 186, 189–190, 233, 255, 263, **303**, 309–310, 352, 367, 368, *415*
Chondestes, 24
Chondestes grammacus, 24, **294**
Cinnamon-tailed Sparrow, 27, 367
Clay-colored Sparrow, 7, 8, 20, 25, 48, 98, 264, 304, 305–306, **308**, 321, 322–323, 324, 326, *416*
Colima Sparrow, 348
Collared Towhee, 20, 170–171, **180**, 185, 283–284, *395*
Common Yellowthroat, 98
Cowbird, Bronzed, 12, 141
Cowbird, Brown-headed, 6, 12–13, 141, 164
Cowbird, Shiny, 12
Creeper, Brown, 191

Dark-eyed Junco. *See* Red-backed Junco, Gray-headed Junco, Pink-sided Junco, Oregon Junco, Cassiar Junco, Slate-colored Junco, *and* White-winged Junco
Dickcissel, 11, 66, 151
Dove, Socorro, 179
Dunnock, 2, 261

Eastern Towhee, 20, **161**, 170, 172, 352, *392*
Eurasian Tree Sparrow, 2, 10, 188–189

Field Sparrow, 6, 25, 189, 304, 305, 307, 319–320, **329**, 351, *419*
Finch, Cassin, 11–12
Finch, Chestnut-capped Brush, 22, 181–182, 185, **283**, 285, *411*
Finch, Green-striped Brush, 22, **285**, *412*
Finch, House, 11–12
Finch, Plumbeous Sierra, 149
Finch, Purple, 11–12
Finch, Rufous-capped Brush, 20, **157**, 160, *392*

Finch, Rufous-sided Warbling , 119,

Finch, Townsend. *See* Townsend Bunting

Finch, White-naped Brush, 20, **159**, *392*

Finch, Yellow-throated Brush, 159, 160

Five-striped Sparrow, 22–24, 27, **291**, *413*

Fox Sparrow, Red, 21, 39, **193**, *397*

Fox Sparrow, Slate-colored, 21, **204**, *399*

Fox Sparrow, Sooty, 21, 39, **198**, *398*

Fox Sparrow, Thick-billed, 21, **208**, *399*

Fox Sparrow, Trinity. *See* Thick-billed Fox Sparrow

Fox Sparrow, Yolla Bolly. *See* Thick-billed Fox Sparrow

Gambel Sparrow. *See* White-crowned Sparrow

Golden-crowned Sparrow, 22, **254**, 263, 276, *406*

Grasshopper Sparrow, 14, 18, 25–26, 67, 98, **333**, 356, 361, *420*

Grassland Sparrow, 26

Gray-headed Junco, 21–22, 222, **225**, 231, 252, *401*

Grayish Saltator, 285

Green-backed Sparrow, 340

Green-striped Brush Finch, 22, **285**, *412*

Green-tailed Towhee, 20, 150, 157–158, **184**, *396*

Grosbeak, Black-headed, 10

Grosbeak, Blue, 137, 140

Grosbeak, Rose-breasted, 10

Ground Sparrow, Cabanis, 146

Ground Sparrow, Prevost, 146

Ground Sparrow, Rusty-crowned, 19, 140, **145**

Ground Sparrow, Tehuantepec. *See* Stripe-headed Sparrow

Ground Sparrow, White-eared, 180

Guadalupe Caracara, 212

Guadalupe Junco, 21–22, **212**, *400*

Guadalupe Towhee, 169, **176**, *395*

Guatemala Junco, 219

Harris Sparrow, 22, 256, **269**, 287, *409*

Hedge Sparrow, 2, 261

Henslow Sparrow, 16–17, 25–26, 64, **65**, 82–83, 335, *380*

House Finch, 11–12

House Sparrow, 2, 9–10, 256, 271, 287

House Wren, 155, 216, 218

Huatusco Sparrow. *See* Botteri Sparrow

Indigo Bunting, 10, 155

Ipswich Sparrow, **87**, *382*

Jolla Bolly Fox Sparrow. *See* Thick-billed Fox Sparrow

Junco, 21

Junco, Baird, 21–22, **214**, *400*

Junco bairdi, 21, **214**

Junco, Carolina. *See* Slate-colored Junco

Junco, Cassiar, 21–22, 237–238, **241**, 245–246, 247, *404*

Junco, Chiapas, 219

Junco, Dark-eyed. *See* Red-backed Junco, Gray-headed Junco, Pink-sided Junco, Oregon Junco, Cassiar Junco, Slate-colored Junco, *and* White-winged Junco

Junco, Gray-headed, 21–22, 222, **225**, 231, 252, *401*

Junco, Guadalupe, 21–22, **212**, *400*

Junco, Guatemala, 219

Junco hyemalis, 21, **221**, **225**, **230**, **235**, **241**, **244**, **249**

Junco insularis, 21, **212**

Junco, Oregon, 21–22, 213, 227, 229, 231–232, 233, **235**, 242–243, 245, *403*

Junco phaeonotus, 21, **217**

Junco, Pink-sided, 21–22 213, 222, 225, **230**, 238, 252, *402*

Junco, Red-backed, 21–22, 218, **221**, 226–227, *401*

Junco, Ridgway, 222, 227–228, 229, 232

Junco, Slate-colored, 21–22, 161, 232, 237, 242–243, **244**, 250–251, 314, *404*

Junco, Townsend. *See* Oregon Junco

Junco, White-winged, 21–22, 232, **249**, *405*

Junco, Yellow-eyed (*see also* Guadalupe Junco, Baird Junco, and Red-backed Junco), 21–22, 216, **217**, 222, 225–226, *400*

Kingfisher, Belted, 151

Lapland Longspur, 14, 271

Large-billed Sparrow, 3, 17, **70**, 75–76, 79, 82, *380*

Lark Bunting, 8–9, 24–25, **298**, *414*

Lark Sparrow, 7, 8, 24, 66, 292, **294**, 300, *413*

Lazuli Bunting, 10

LeConte Sparrow, 17–19, 25–26, 62, 67, **96**, 110–111, 335, *383*

Lincoln Sparrow, 15, 18, 39, 48, 49–50, **52**, 58–60, 195, 281, *378*

Little Bunting, 90

Longspur, Chestnut-collared , 14

Longspur, Lapland, 14, 271

Longspur, McCown, 14

Longspur, Smith, 14

McCown Longspur, 14

McKay Bunting, 13–14

Meadow Pipit, 90

Melospiza, 15

Melospiza georgiana, 15, **46**

Melospiza lincolnii, 15, **52**

Melospiza melodia, 15, **32**

Mojave Sparrow. *See* Bell Sparrow

Mountain Sparrow. *See* Eurasian Tree Sparrow

Nelson Sparrow, 17–19, 25–26, 67–68, 98, 104–105, **108**, 116–117, 334–335, *385*

Northern Cardinal, 263, 273, 368

Northern Red Bishop, 335

Oaxaca Sparrow, 27, 150

Olive Sparrow, 26, 185, **339**, *421*

Orange-crowned Warbler, 233, 238

Oregon Junco, 21–22, 213, 227, 229, 231–232, 233, **235**, 242–243, 245, *403*

Oriturus, 15

Oriturus superciliosus, 15, **30**

Painted Bunting, 10

Palm Warbler, 98

Passerculus, 17

Passerculus guttatus, 17, **73**

Passerculus rostratus, 17, **70**

Passerculus sanctorum, 17, **78**

Passerculus sandwichensis, 17, **80**

Passerella, 21

Passerella iliaca, 21, **193**

Passerella megarhyncha, 21, **208**

Passerella schistacea, 21, **204**

Passerella unalaschensis, 21, **198**

Petronia, Rock, 189, 261

Peucaea, 26

Peucaea aestivalis, 26, **349**

Peucaea botterii, 26, **360**

Peucaea carpalis, 26, **365**

Peucaea cassinii, 26, **355**

Peucaea humeralis, 26, **344**

Peucaea mystacalis, 26, **342**

Peucaea ruficauda, 26, **346**

Phoebe, Black, 137

Pine Siskin, 11

Pine Warbler , 246, 306, 368

Pine-woods Sparrow. *See*
Bachman Sparrow
Pink-sided Junco, 21–22, 213, 222,
225, **230**, 238, 252, 402
Pin-tailed Whydah, 299
Pipilo, 20
Pipilo chlorurus, 20, **184**
Pipilo erythrophthalmus, 20, **161**
Pipilo maculatus, 20, **167**
Pipilo ocai, 20, **180**
Pipit, Meadow, 90
Pipit, Sprague, 298
Plumbeous Sierra Finch, 149
Pooecetes, 17
Pooecetes gramineus, 17, **91**
Prevost Ground Sparrow, 146
Purple Finch, 11–12

Red-backed Junco, 21–22, 218,
221, 226–227, *401*
Red Bishop, Northern, 335
Red Fox Sparrow, 21, **193**, 199–
200, 205, *397*
Red-winged Blackbird, 12, 76, 105
Ridgway Junco, 222, 227–228,
229, 232
Robin, American, 2
Rock Petronia, 189, 261
Rock Sparrow. *See* Rock Petronia
Rock Wren, 212
Rose-breasted Grosbeak, 10
Rufous-capped Brush Finch, 20,
157, 160, *392*
Rufous-collared Sparrow, 22, **280**,
411
Rufous-crowned Sparrow, 19, 27,
132, 150, **153**, 185, 216, 263,
366–367, *391*
Rufous-sided Towhee. *See* Eastern
Towhee *and* Spotted Towhee
Rufous-sided Warbling Finch , 119
Rufous-winged Sparrow, 26–27,
154, 190, **365**, *426*
Rusty Sparrow, 19, 27, 30, **149**,
154, *391*
Rusty-crowned Ground Sparrow,
19, 140, **145**, *390*

Sagebrush Sparrow, 15, 19, 23,
119–124, **125**, 287–288, *387*
Saltator, Grayish, 285
Saltbush Sparrow. *See* Bell
Sparrow
Saltmarsh Sparrow, 17–19, 25–26,
48, 51, 67–68, 98, 105, 111–
112, **115**, *386*
San Benito Sparrow, 17, **78**, *381*
Savannah Sparrow, 17, 37–38,
48, 60, 62–64, 75, **80**, 89, 90,
92–93, 147, 356, 361, *381*
Seaside Sparrow, 17–19, 25–26, 48,
51, 68, **100**, *384*

Sharp-tailed Sparrow. *See* Nelson
Sparrow *and* Saltmarsh
Sparrow
Shiny Cowbird, 12
Sierra Finch, Plumbeous, 149
Sierra Madre Sparrow, 15–16, 39,
57, *379*
Siskin, Pine, 11
Slate-colored Fox Sparrow, 21,
195, 199–200, **204**, 209, *399*
Slate-colored Junco, 21–22, 161,
232, 237, 242–243, **244**,
250–251, 314, *404*
Smith Longspur, 14
Snow Bunting, 13–14
Socorro Dove, 179
Socorro Towhee, 168, **178**, *395*
Song Sparrow, 4, 15, 18, **32**, 48–49,
53–54, 58–60, 75, 81–82,
92–93, 94, 194, 198, 276, *377*
Sooty Fox Sparrow, 21, 39, **198**,
398
Sparrow, Ambiguous, 6
Sparrow, American Tree, 20–21,
188, 367, *396*
Sparrow, Bachman, 26–27, 47,
154, **349**, *423*
Sparrow, Baird, 4, 16–17, 18,
25–26, **61**, 66–67, 82–83,
335, *379*
Sparrow, Belding, 17, **73**, 79, 82,
381
Sparrow, Bell, 15, 19, 23, **119**,
127–128, 287–288, *386*
Sparrow, Black-chested, 26–27,
344, *422*
Sparrow, Black-chinned, 25, **303**,
362, *417*
Sparrow, Black-throated, 22–24,
120, 127, **287**, 292, *412*
Sparrow, Botteri, 26–27, 147, 351,
356–357, **360**, *425*
Sparrow, Brewer (*see also*
Timberline Sparrow), 6, 25,
304, 310–311, 321, 322–323,
324, 356, *419*
Sparrow, Bridled, 26–27, **342**, *422*
Sparrow, Cabanis Ground, 146
Sparrow, Cabot. *See* Swamp
Sparrow
Sparrow, California. *See* Bell
Sparrow
Sparrow, Cassin, 26–27, 351–352,
355, 361–362, *424*
Sparrow, Chipping, 20, 25, 31,
154, 186, 189–190, 233, 255,
263, **303**, 309–310, 352, 367,
368, *415*
Sparrow, Cinnamon-tailed, 27,
367
Sparrow, Clay-colored, 7, 8, 20,
25, 48, 98, 264, 304, 305–306,

308, 321, 322–323, 324, 326,
416
Sparrow, Colima, 348
Sparrow, Eurasian Tree, 2, 10,
188–189
Sparrow, Field, 6, 25, 189, 304,
305, 307, 319–320, **329**, 351,
419
Sparrow, Five-striped, 22–24, 27,
291, *413*
Sparrow, Gambel. *See* White-
crowned Sparrow
Sparrow, Golden-crowned, 22,
254, 263, 276, *406*
Sparrow, Grasshopper, 14, 18,
25–26, 27, 67, 98, **333**, 356,
361, *420*
Sparrow, Grassland, 26
Sparrow, Green-backed , 340
Sparrow, Harris, 22, 256, **269**, 287,
409
Sparrow, Hedge, 2, 261
Sparrow, Henslow, 16–17, 25–26,
64, **65**, 82–83, 335, *380*
Sparrow, House, 2, 9–10, 256, 271,
287
Sparrow, Huatusco. *See* Botteri
Sparrow
Sparrow, Ipswich, **87**, *382*
Sparrow, Large-billed, 3, 17, **70**,
75–76, 79, 82, *380*
Sparrow, Lark, 7, 8, 24, 66, 292,
294, 300, *413*
Sparrow, LeConte, 17–19, 25–26,
62, 67, **96**, 110–111, 335,
383
Sparrow, Lincoln, 15, 18, 39, 48,
49–50, **52**, 58–60, 195, 281,
378
Sparrow, Mojave. *See* Bell Sparrow
Sparrow, Mountain. *See* Eurasian
Tree Sparrow
Sparrow, Nelson, 17–19, 25–26,
67–68, 98, 104–105, **108**,
116–117, 334–335, *385*
Sparrow, Oaxaca, 27, 150
Sparrow, Olive, 26, 185, **339**, *421*
Sparrow, Pine-woods. *See*
Bachman Sparrow
Sparrow, Prevost Ground, 146
Sparrow, Red Fox, 21, 39, **193**,
199–200, 205, *397*
Sparrow, Rock. *See* Rock Petronia
Sparrow, Rufous-collared, 22, **280**,
411
Sparrow, Rufous-crowned, 19,
27, 132, 150, **153**, 185, 263,
366–367, *391*
Sparrow, Rufous-winged, 26–27,
154, 190, 216, **365**, *426*
Sparrow, Rusty, 19, 27, 30, **149**,
154, *391*

Sparrow, Rusty-crowned Ground, 19, 140, **145**, *390*

Sparrow, Sage. *See* Bell Sparrow *and* Sagebrush Sparrow

Sparrow, Sagebrush, 15, 19, 23, 119–124, **125**, 287–288, *387*

Sparrow, Saltbush. *See* Bell Sparrow

Sparrow, Saltmarsh, 17–19, 25–26, 48, 51, 67–68, 98, 105, 111–112, **115**, *386*

Sparrow, San Benito, 17, **78**, *381*

Sparrow, Savannah, 17, 37–38, 48, 60, 62–64, 75, **80**, 89, 90 92–93, 147, 356, 361, *381*

Sparrow, Seaside, 17–19, 25–26, 48, 51, 68, **100**, *384*

Sparrow, Sharp-tailed. *See* Nelson Sparrow *and* Saltmarsh Sparrow

Sparrow, Sierra Madre, 15–16, 39, **57**, *379*

Sparrow, Slate-colored Fox, 21, 195, 199–200, **204**, 209, *399*

Sparrow, Song, 4, 15, 18, **32**, 48–49, 53–54, 58–60, 75, 81–82, 92–93, 94, 194, 198, 276, *377*

Sparrow, Sooty Fox, 21, 39, **198**, *398*

Sparrow, Stripe-capped, 27

Sparrow, Striped, 15, **30**, 149, *376*

Sparrow, Stripe-headed, 26–27, **346**, *422*

Sparrow, Sumichrast. *See* Cinnamon-tailed Sparrow

Sparrow, Swamp, 15, 18, 39, **46**, 53–54, 60, 154, 185, 190, *378*

Sparrow, Tehuantepec Ground. *See* Stripe-headed Sparrow

Sparrow, Thick-billed Fox, 21, 200, 205, 207, **208**, *399*

Sparrow, Timberline (*see also* Brewer Sparrow), 25, 304, 310–311, **321**, 325, 326, 328, *418*

Sparrow, Tree. *See* Sparrow, Eurasian Tree, *and* Sparrow, American Tree

Sparrow, Tumbes, 27

Sparrow, Vesper, 17, 38–39, 81–82, 90, **91**, *383*

Sparrow, White-crowned, 5, 6, 22, 185, 189, 256, 258, **261**, 273, 276, *407*

Sparrow, White-eared Ground, 180

Sparrow, White-throated, 22, 185, 256, 263, 273, **275**, 314, *410*

Sparrow, Worthen, 25, **318**, 330–331, *417*

Sparrow, Yellow-browed, 26

Sparrow, Yolla Bolly Fox. *See* Thick-billed Fox Sparrow

Spizella, 25

Spizella atrogularis, 25, **303**

Spizella breweri, 25, **324**

Spizella pallida , 25, **308**

Spizella passerina , 25, **303**

Spizella pusilla , 25, **329**

Spizella taverneri , 25, **321**

Spizella wortheni, 25, **318**

Spizelloides, 20

Spizelloides arborea, 20, **188**

Spotted Towhee (*see also* Guadalupe Towhee *and* Socorro Towhee), 20, 138, 163–165, **167**, 181, 182, 185, *393*

Sprague Pipit, 298

Stripe-capped Sparrow, 27

Striped Sparrow, 15, **30**, 149, *376*

Stripe-headed Sparrow, 26–27, **346**, *422*

Sumichrast Sparrow. *See* Cinnamon-tailed Sparrow

Swamp Sparrow, 15, 18, 39, **46**, 53–54, 60, 154, 185, 190, *378*

Tanagers, Bush, 3

Tehuantepec Ground Sparrow. *See* Stripe-headed Sparrow

Thick-billed Fox Sparrow, 21, 200, 205, 207, **208**, *399*

Thrasher, Brown, 195

Timberline Sparrow (*see also* Brewer Sparrow), 25, 304, 310–311, **321**, 325, 326, 328, *418*

Towhee, Abert, 19, 130, 132, 137, **140**, *389*

Towee, Brown. *See* Canyon Towhee *and* California Towhee

Towhee, California, 19, 131–132, **135**, 141, *388*

Towhee, Canyon, 19, **130**, 137–138, 140, 141, 143, 144, 147, 155, 368, *388*

Towhee, Collared, 20, 170–171, **180**, 185, 283–284, *395*

Towhee, Eastern, 20, **161**, 170, 172, 352, *392*

Towhee, Green-tailed, 20, 150, 157–158, **184**, 259, *396*

Towhee, Guadalupe, 169, **176**, *395*

Towhee, Rufous-sided. *See* Eastern Towhee *and* Spotted Towhee

Towhee, Socorro, 168, **178**, *395*

Towhee, Spotted (*see also* Guadalupe Towhee *and* Socorro Towhee), 20, 138, 163–165, **167**, 181, 182, 185, *393*

Towhee, White-throated, 19, 132, **143**, *390*

Townsend Bunting , 269

Townsend Finch. *See* Townsend Bunting

Townsend Junco. *See* Oregon Junco

Tree Sparrow. *See* Sparrow, Eurasian Tree, *and* Sparrow, American Tree

Tricolored Blackbird, 12

Trinity Fox Sparrow. *See* Thick-billed Fox Sparrow

Tumbes Sparrow, 27

Varied Bunting, 10

Vesper Sparrow, 17, 38–39, 81–82, 90, **91**, *383*

Warbler, Black-throated Blue, 264

Warbler, Blue-winged, 311

Warbler, Orange-crowned, 233, 238

Warbler, Palm, 98

Warbler, Pine, 246, 306, 368

Warbler, Worm-eating, 306, 368

Warbling Finch, Rufous-sided, 119

White-crowned Sparrow, 5, 6, 22, 185, 189, 256, 258, **261**, 273, 276, *407*

White-eared Ground Sparrow, 180

White-naped Brush Finch, 20, **159**, *392*

White-throated Sparrow, 22, 185, 256, 263, 273, **275**, 314, *410*

White-throated Towhee, 19, 132, **143**, *390*

White-winged Junco, 21–22, 232, **249**, *405*

Whydah, Pin-tailed, 299

Worm-eating Warbler, 306, 368

Worthen Sparrow, 25, **318**, 330–331, *417*

Wren, Bewick, 219

Wren, House, 155, 216, 218

Wren, Rock, 212

Yellow-breasted Chat, 301

Yellow-browed Sparrow, 26

Yellow-eyed Junco (*see also* Guadalupe Junco, Baird Junco, and Red-backed Junco), 21–22, 216, **217**, 222, 225–226, *400*

Yellowhammer, 90

Yellowthroat, Common, 98

Yellow-throated Brush Finch, 159, 160

Yolla Bolly Fox Sparrow. *See* Thick-billed Fox Sparrow

Zonotrichia, 22

Zonotrichia albicollis, 22, **275**

Zonotrichia atricapilla, 22, **254**

Zonotrichia capensis, 22, **280**

Zonotrichia leucophrys, 22, **261**

Zonotrichia querula, 22, **269**

PETERSON FIELD GUIDES®

Roger Tory Peterson's innovative format uses accurate, detailed drawings to pinpoint key field marks for quick recognition of species and easy comparison of confusing look-alikes.

BIRDS

Birds of Northern Central America

Bird Sounds of Eastern North America

Bird Sounds of Western North America

Birds of North America

Birds of Eastern and Central North America

Western Birds

Feeder Birds of Eastern North America

Hawks of North America

Hummingbirds of North America

Warblers

Eastern Birds' Nests

PLANTS AND ECOLOGY

Eastern and Central Edible Wild Plants

Eastern and Central Medicinal Plants and Herbs

Western Medicinal Plants and Herbs

Eastern Forests

Eastern Trees

Western Trees

Eastern Trees and Shrubs

Ferns of Northeastern and Central North America

Mushrooms

Venomous Animals and Poisonous Plants

Wildflowers of Northeastern and North-Central North America

MAMMALS

Animal Tracks

Mammals

INSECTS

Insects

Eastern Butterflies

Moths of Northeastern North America

Moths of Southeastern North America

REPTILES AND AMPHIBIANS

Eastern Reptiles and Amphibians

Western Reptiles and Amphibians

FISHES

Freshwater Fishes

SPACE

Stars and Planets

GEOLOGY

Rocks and Minerals

PETERSON FIRST GUIDES®

The first books the beginning naturalist needs, whether young or old. Simplified versions of the full-size guides, they make it easy to get started in the field, and feature the most commonly seen natural life.

Astronomy

Birds

Butterflies and Moths

Caterpillars

Clouds and Weather

Fishes

Insects

Mammals

Reptiles and Amphibians

Rocks and Minerals

Seashores

Shells

Trees

Urban Wildlife

Wildflowers

PETERSON FIELD GUIDES FOR YOUNG NATURALISTS

This series is designed with young readers ages eight to twelve in mind, featuring the original artwork of the celebrated naturalist Roger Tory Peterson.

Backyard Birds

Birds of Prey

Songbirds

Butterflies

Caterpillars

PETERSON FIELD GUIDES® COLORING BOOKS®

Fun for kids ages eight to twelve, these color-your-own field guides include color stickers and are suitable for use with pencils or paint.

Birds

Butterflies

Reptiles and Amphibians

Wildflowers

Shells

Mammals

PETERSON REFERENCE GUIDES®

Reference Guides provide in-depth information on groups of birds and topics beyond identification.

Behavior of North American Mammals

Birding by Impression

Woodpeckers of North America

Sparrows of North America

PETERSON AUDIO GUIDES

Birding by Ear: Eastern/Central

Bird Songs: Eastern/Central

PETERSON FIELD GUIDE / *BIRD WATCHER'S DIGEST* BACKYARD BIRD GUIDES

Identifying and Feeding Birds

Bird Homes and Habitats

The Young Birder's Guide to Birds of North America

The New Birder's Guide to Birds of North America

DIGITAL

App available for Apple and Android.

Peterson Birds of North America

Peterson Mammals of North America

E-books

Birds of Arizona

Birds of California

Birds of Florida

Birds of Massachusetts

Birds of Minnesota

Birds of New Jersey

Birds of New York

Birds of Ohio

Birds of Pennsylvania

Birds of Texas